Study Guide and Additional Drill Problems *for* ORGANIC CHEMISTRY

6TH EDITION

Study Guide and Additional Drill Problems *for* ORGANIC CHEMISTRY

6TH EDITION

RALPH J. FESSENDEN
University of Montana
JOAN S. FESSENDEN

Brooks/Cole Publishing Company
I(T)P® *An International Thomson Publishing Company*

Pacific Grove • Albany • Belmont • Bonn • Boston • Cincinnati • Detroit
Johannesburg • London • Madrid • Melbourne • Mexico City • New York
Paris • Singapore • Tokyo • Toronto • Washington

Sponsoring Editor: *Beth Wilbur*
Editorial Associate: *Nancy Conti*
Production: *Dorothy Bell*
Cover Design: *Jamie Dagdigian*
Cover Photo: *Dennis Kunkel/Phototake*
Printing and Binding: *Malloy Lithographing*

For more information, contact:

BROOKS/COLE PUBLISHING COMPANY
511 Forest Lodge Road
Pacific Grove, CA 93950
USA

International Thomson Editores
Seneca 53
Col. Polanco
11560 México, D. F., México

International Thomson Publishing Europe
Berkshire House 168-173
High Holborn
London WC1V 7AA
England

International Thomson Publishing GmbH
Königswinterer Strasse 418
53227 Bonn
Germany

Thomas Nelson Australia
102 Dodds Street
South Melbourne, 3205
Victoria, Australia

International Thomson Publishing Asia
221 Henderson Road
#05-10 Henderson Building
Singapore 0315

Nelson Canada
1120 Birchmount Road
Scarborough, Ontario
Canada M1K 5G4

International Thomson Publishing Japan
Hirakawacho Kyowa Building, 3F
2-2-1 Hirakawacho
Chiyoda-ku, Tokyo 102
Japan

Printed in the United States of America

10 9 8 7 6 5 4 3

ISBN 0-534-35201-4

To the Student

This study guide is intended to accompany the text *Organic Chemistry*, 6th Edition. The introductory material in this guide contains hints to help you study organic chemistry. As you progress in your study of organic chemistry, we suggest that you reread these suggestions.

The study guide is divided into two parts. Part I is a chapter-by-chapter discussion describing the most important features of each chapter along with supplemental instructional material. Part I also contains additional drill problems that you can use to test your knowledge of the material in the text. Part II contains the solutions to these additional drill problems.

Studying Organic Chemistry

Sometimes organic nomenclature, bonding, reactions, etc. seem perfectly clear to a student, yet the student does poorly on an examination. Additional study time may not be the answer to this problem. Instead, the manner in which this student uses study time may need revamping.

What is the correct way to study organic chemistry? There actually is no single correct way to study organic chemistry--different students may study in different ways, any of which might be correct. However, all correct study techniques incorporate two important features: *understanding* followed by *memorization*. Rote memorizing of organic chemistry without first understanding the material is a waste of time. On the other hand, studying only to the point of understanding does not result in good examination grades. Only by coupling understanding and memorizing can you progress to thinking in the language of organic chemistry.

Two common study aids are flash cards and outlines. Flash cards are useful for a random memory drill. Outlines force you to select the salient features and list them in a logical organizational framework. Although you can use flash cards to help you memorize organic chemistry, you will probably do better by outlining the material. Students who use flash cards often have no trouble at the start of the course but find themselves in serious trouble as the sheer volume of the material accumulates. There seems to be a point at which the number of flash cards becomes excessive and the student does not have enough time to thumb through them.

The outline method of study is an associative technique. Each major heading will help you recall the subheadings. When subject material is organized in this fashion, your chances of recalling a fact correctly are enhanced. Also, when you find yourself unable to remember a fact in an examination, you can mentally scan your outline and, surprisingly, often find the information you need. (If you have used flash cards, you must sit there waiting for the fact to "pop" into your thoughts.)

Working the problems found in the text and in this study guide provides a straightforward self-testing procedure for studying organic chemistry.

Be sure to answer the problems yourself, and *then* check the solutions and the text. (You will gain nothing by checking the solutions before you have tried to answer the problems.) If your answers are correct, you have a reasonable mastery of the material. If you have several incorrect answers, then you need additional study.

When working the problems, studying the text, or studying your lecture material, write and rewrite the formulas. Whenever possible, verbalize new words and explain the concepts to yourself aloud. By doing so, you reinforce the material you are learning and you enhance your chances of recalling the material.

Be sure you understand each definition and each reaction in the chapter summary and summary tables. Be sure you can extrapolate from the generalities of the summaries to the specific reactions mentioned in the chapter. Then be sure you can extrapolate from the specific reactions in the chapter to reactions of new and different compounds.

General: $RCO_2 + OH^- \longrightarrow RCO_2^- + H_2O$

Specific: $CH_3CO_2H + OH^- \longrightarrow CH_3CO_2^- + H_2O$

Extrapolated: $CH_3CH(CH_3)CH(CH_3)CH(C_6H_5)CO_2H + OH^- \longrightarrow CH_3CH(CH_3)CH(CH_3)CH(C_6H_5)CO_2^- + H_2O$

We would wish you luck in your venture into organic chemistry, but we believe that perseverance and a desire to learn are truly more important than "luck". We do hope you find organic chemistry as fascinating a language as we do.

Contents

Part I Chapter-by-Chapter Discussion and Additional Drill Problems

PART I

Chapter-by-Chapter Discussion and Additional Drill Problems

1

Atoms and Molecules — A Review

Some Important Features

Electrons are found in shells surrounding the nucleus of an atom. Each shell is composed of one or more atomic orbitals. The first shell contains a 1*s* orbital (spherical); the second shell contains one 2*s* orbital (spherical) and three 2*p* orbitals (dumbbell-shaped and mutually perpendicular). Each orbital can hold zero, one, or two electrons. Electrons are usually contained in the lowest-energy orbitals possible (1*s*, then 2*s*, then 2*p*).

The halogens, oxygen, and nitrogen have fairly high electronegativities (attraction for outer electrons). The metals have low electronegativities, while carbon and hydrogen have intermediate electronegativities.

Chemical bonds are formed by electrons in the outer shell of an atom. Whether ionic or covalent bonds are formed depends on the electronegativity difference between two atoms. Carbon forms covalent bonds with other elements. These covalent bonds may be nonpolar (C-C or C-H) or polar (C-O, C-N, or C-Cl), depending on the electronegativity differences between C and the other element.

Because organic compounds are covalent compounds, the numbers and types of covalent bonds that different elements form are very important. Memorize the number of covalent bonds the following elements form:

carbon (C):	4	nitrogen (N):	3 (or 4 in a cation)
oxygen (O):	2	hydrogen and halogens (H, F, Cl, Br, I):	1

The formal charge of O and N in compounds is often easily determined simply by counting the bonds. The commonly encountered bondings for N and O follow:

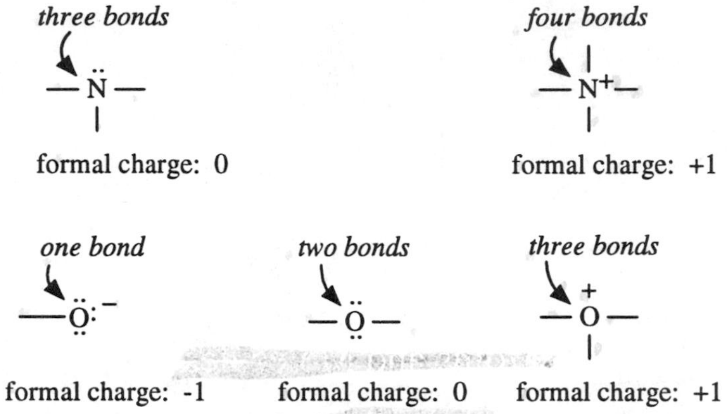

three bonds

$-\ddot{N}-$ (with a bond below)

formal charge: 0

four bonds

$-N^{+}-$ (with bonds above and below)

formal charge: +1

one bond

$-\ddot{\underset{..}{O}}:^{-}$

formal charge: -1

two bonds

$-\ddot{\underset{..}{O}}-$

formal charge: 0

three bonds

$-\overset{+}{O}-$ (with a bond below)

formal charge: +1

Molecules with NH, OH, or HF bonds can form hydrogen bonds with each other or with other molecules containing N, O, or F atoms with unshared electrons.

An acid is a compound that can donate H^+ or accept electrons, while a base is a compound that has unshared electrons that can be donated to form a covalent bond. Common organic acids are the carboxylic acids, compounds with a $-CO_2H$ group. Common organic bases are amines, compounds with a nitrogen atom bonded to three other atoms.

$$CH_3CO_2H \quad + \quad (CH_3)_3N: \rightleftharpoons CH_3CO_2^- \quad + \quad (CH_3)_3\overset{+}{N}H$$

a carboxylic acid *an amine*

The strengths of acids and bases are indicated by their pK_a values or pK_b values. A smaller numerical value for pK_a means a stronger acid.

If your instructor uses pK_a values of conjugate acids in discussing base strength, study the examples of conjugate acids carefully.

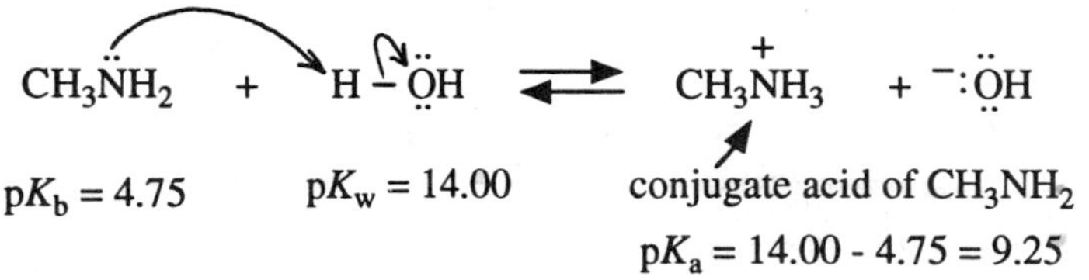

$$CH_3\ddot{N}H_2 \quad + \quad H-\ddot{\underset{..}{O}}H \rightleftharpoons CH_3\overset{+}{N}H_3 \quad + \quad ^{-}:\ddot{\underset{..}{O}}H$$

$pK_b = 4.75$ $pK_w = 14.00$

conjugate acid of CH_3NH_2

$pK_a = 14.00 - 4.75 = 9.25$

Other important topics covered in this chapter are ionic and covalent bonds, calculation of formal charge, types of chemical formulas, bond lengths and angles, and bond dissociation energies.

Reminders

- In stable compounds, carbon forms four bonds, hydrogen forms one bond, and oxygen forms two bonds.

```
                H                    O
                |                    ||
Correct:    H - C - O - H        H - C - H
                |
                H

                H                    H
                |                    |
Incorrect:  H - C - H - O        H - C = O
                |                    |
                H                    H
```

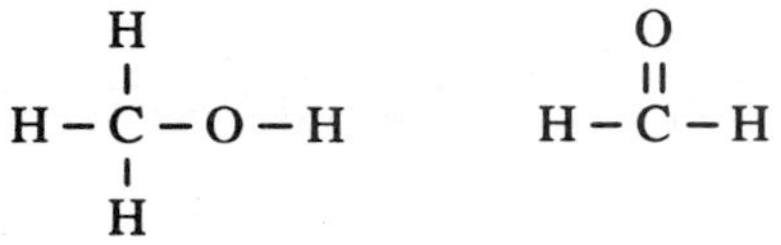

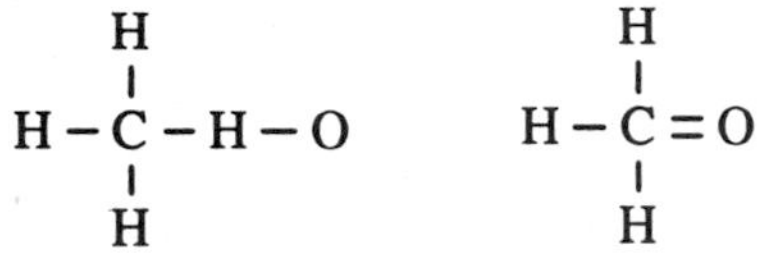

Hydrogen bonds are formed between unshared electrons on O, N, or F and a hydrogen attached to O, N, or F.

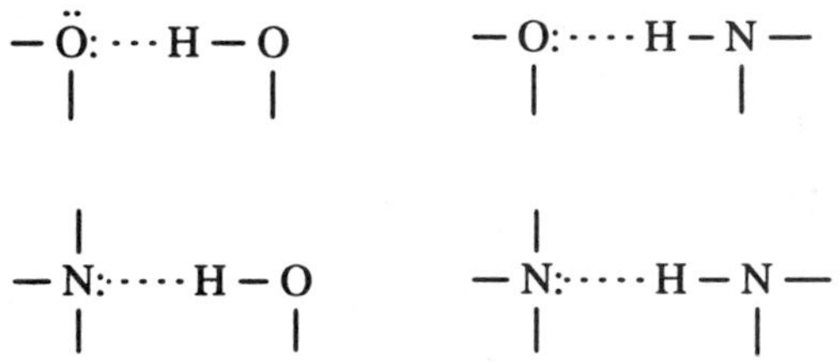

```
  ..
- O: ···H - O           - O:····H - N -
  |         |             |         |

  |                       |
- N:·····H - O          - N:·····H - N -
  |          |            |          |
```

Additional Drill Problems

1.1 Which element corresponds with each of the following configurations?

(a) $1s^2\,2s^2$ (b) $1s^2\,2s^2\,2p^6\,3s^2\,3p^4$

(c) $1s^2\,2s^2\,2p^6\,3s^2$ (d) $1s^2\,2s^2\,2p^6\,3s^2\,3p^6\,4s^1$

1.2 Write the electron configurations for the following pairs or groups:

(a) Na, Na^+ (b) O, O^{2-} (c) S, S^{2-} (d) H, H^+, H^-

1.3 Write the electron configuration for each of the following elements:

(a) Al (b) Ca (c) P (d) Ge

1.4 Draw the Lewis (dot) formula for each of the following molecules or ions:

(a) $MgCl_2$ (b) H_2SO_4 (c) $CH_3\overset{+}{O}H_2$

(d) HCO_3^- (e) NO_3^- (f) NO_2^-

1.5 Draw the Lewis (dot) formula for each of the following organic compounds:

(a) $CH_3C{\equiv}CCH_3$ (b) $CH_2{=}CH{-}CH{=}CH_2$ (c) CH_3NH_2

(d) $H-\overset{\overset{\large O}{\|}}{C}OH$ (e) $CH_3\overset{\overset{\large O}{\|}}{C}OH$ (f) $CH_3\underset{\underset{\large OH}{|}}{C}HCH_3$

(g) $CH_2{=}CHCl$ (h) $CHCl_3$

1.6 Redraw each of the following structures to show each bond as a line, and indicate the number of covalent bonds of each atom except hydrogen:

(a) CH_3NH_2 (b) $CH_2{=}C{=}CH_2$ (c) CH_3CO_2H

(d) $CH_3\overset{\overset{\large O}{\|}}{C}CH_3$ (e) $CH_2{=}CHCl$ (f) CH_3OCH_3

1.7 Write expanded structural formulas for the following compounds, showing each bond as a line:

(a) $(CH_3)_2CHOH$ (b) $(CH_3CH_2)_2O$

(c) $(CH_3)_2CH\overset{\overset{\large O}{\|}}{C}H$ (d) $CH_3CH_2CH{=}CH_2$

1.8 Write a condensed structural formula for each of the following formulas:

```
(a)     H       H              (b)     H   H              (c)    Cl       Cl
        |       |                      |   |                       \     /
    H - C - O - C - H              H - C - C - O⁻                   C = C
        |       |                      |   |                       /     \
        H       H                      H   H                     Cl       H
```

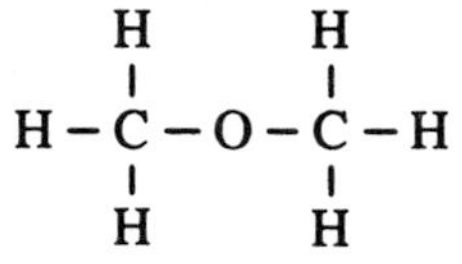

```
(d)  H      H H      H         (e)      O              (f)          O
      \    /   \    /                   ||                       +//
       C = C    C = C              H - O - S - O - H        H - O - N
      /     \  /     \                  ||                            \
     H       N        H                 O                              O⁻
             |
             H
```

```
(g)     H   H   O       H              (h)      H   O     H
        |   |   ||      |                       |   ||   /
    H - C - C - C - O - C - H               H - C - C - N
        |   |           |                       |        \
        H   C           H                       H         H
          / | \
         H  |  H
            H
```

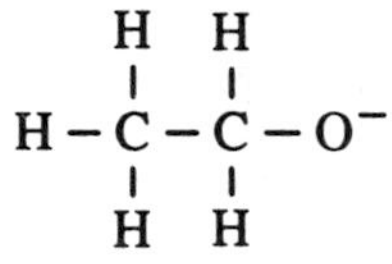

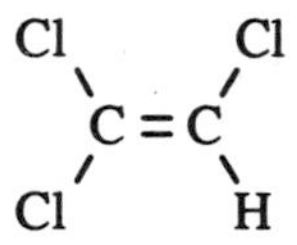

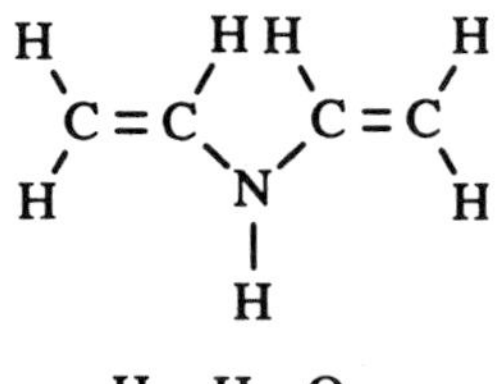

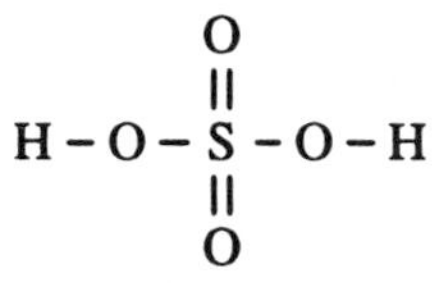

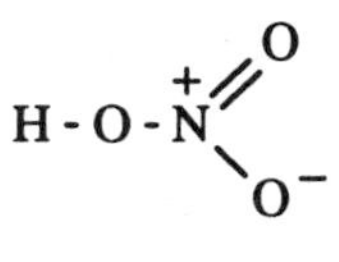

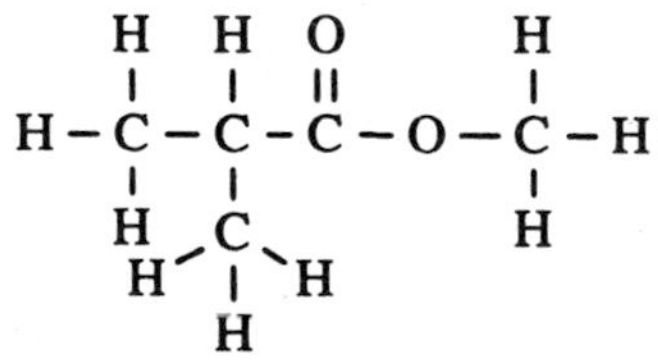

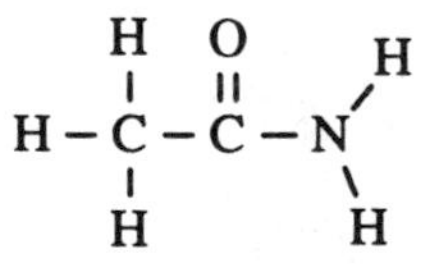

(i)

```
 H  H   HH   H  H
  \ |  /  \  | /
    C    O   C
    |    ||  |
 H- C  - C - C -H
    |        |
    C        C
  / | \   /  | \
 H  H  HH    H  H
```

1.9 Redraw each of the following structures inserting all pairs of unshared valence electrons as dots:

(a) $CH_3C(=O)CH_3$ (b) $ClCH_2C(=O)NHCH_3$ (c) $CH_3C(=O)OH$

1.10 Draw each of the following polygon formulas as an expanded formula, showing each atom and each bond:

(a) 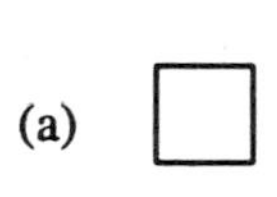

(b) 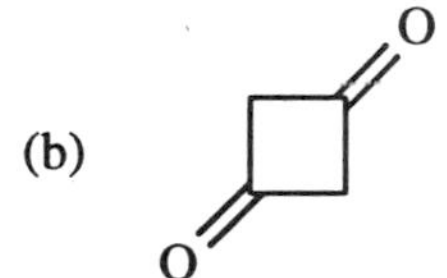

(c) 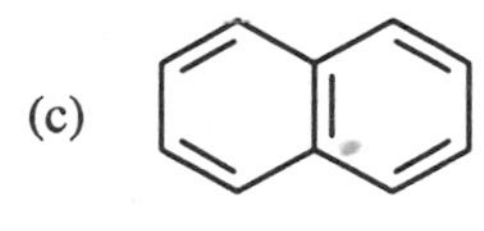

1.11 Draw polygon formulas for each of the following structures:

(a)

```
   H     CH3
     \  /
      C
     / \
 H - C - C - H
     |   |
     H   H
```

(b)

```
     H   CH3
     |   |
 H - C - C - H
      \ /
       O
```

(c)

```
 H             H
   \         /
    C ——— C
    ||    ||
    C     C
  /  \   /  \
 H     S      H
```

1.12 Which element in each of the following pairs has the greater electronegativity? (Refer to Figure 1.5 in the text.)

(a) C or Mg (b) C or Li (c) C or I (d) I or Cl

1.13 For each, identify the more electronegative element:

(a) C-O (b) C-S (c) C-Mg (d) C-Cl

(e) C-Li (f) N-O (g) O-H (h) N-H

1.14 Arrange the following bonds in order of increasing ionic character (least ionic first):

(a) N-O (b) C-O (c) C-Mg (d) C-H

1.15 Assign δ+ and δ- to each atom joined by the covalent bond shown in the following structures:

(a) CH_3-Br (b) CH_3-OCH_3 (c) CH_3-NH_2

1.16 Draw formulas that show the hydrogen bonding between the following pairs of compounds. Show the nonbonding electrons in your formulas.

(a) $(CH_3)_2NH$ and CH_3OH (b) CH_3CH_2OH and CH_3OH

1.17 Which of the following substances can undergo hydrogen bonding with water?

(a) $CH_3\overset{\overset{O}{||}}{C}OH$ (b) $CH_3\overset{\overset{O}{||}}{C}H$ (c) $CH_3CH_2CH_2CH_3$

1.18 Write an equation for the homolytic cleavage of the indicated bond in the following structures. Use fishhook arrows in you equations.

(a) CH_3-H (b) CH_3O-OCH_3 (c) H-I

1.19 Write an equation showing the heterolytic dissociation of the covalent bond shown for each of the following structures. Use electron-shift arrows in your equations.

(a) $(CH_3)_2CH$-$\overset{+}{O}H_2$ (b) $(CH_3)_3C$-Br

(c) H-OSO_3H (d) CH_3CH_2-Li

1.20 Which member of each of the following pairs would you expect to have the higher boiling point? Why?

(a) $CH_3OCH_2CH_2OCH_3$ or $HOCH_2CH_2CH_2CH_2OH$

(b) $CH_3\overset{\overset{O}{||}}{C}OH$ or $H\overset{\overset{O}{||}}{C}OCH_3$

(c) $ClCH_2CH_2OH$ or $HOCH_2CH_2OH$

(d) $CH_3CH_2CH_3$ or $CH_3CH_2NH_2$

1.21 Which member of each of the following pairs would you expect to be the more water soluble? Why?

(a) CH_3CH_2OH or $CH_3CH_2CH_3$ (b) $CH_3\overset{O}{\overset{\|}{C}}OH$ or $CH_3\overset{O}{\overset{\|}{C}}OCH_3$

1.22 Insert the missing entries in the following table:

acid	+	base	→	conjugate base of the acid	+	conjugate acid of the base
H_2O	+	H_2O	→	(a)	+	H_3O^+
H_2O	+	$H:^-$	→	(b)	+	H-H
H_2SO_4	+	$CH_3\overset{O}{\overset{\|}{C}}CH_3$	→	(c)	+	(d)
H_2SO_4	+	CH_3OH	→	(e)	+	(f)

1.23 Compounds (a), (c), and (d), because these compounds contain empty orbitals that can accept electrons. Compound (b) is not usually classified as a Lewis acid; however, under the proper reaction conditions, it can act as an electron acceptor in a Lewis acid-base reaction.

insert image

1.24 Convert each of the following K_a values to a pK_a:

(a) 5.9×10^{-2} (b) 1.5×10^{-14} (c) 8.5×10^{-6}

1.25 Assign an ionic charge to the indicated atoms in the following structures. (All unshared valence electrons on these atoms are shown.)

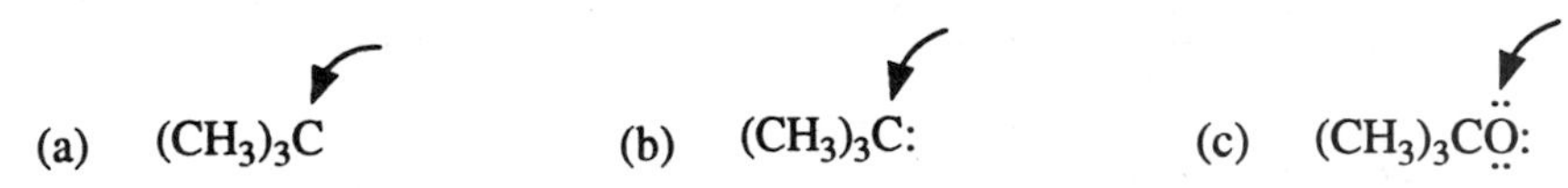

1.26 Each of the following structures contains a double or triple bond. Draw the Lewis (dot) formula and the line-bond formula for each:

(a) $CH_3CH_2CO_2H$ (b) H_2CO (c) HCO_2CH_3

(d) C_3H_6 (e) C_3H_4 (f) C_2H_2

1.27 Determine the order of increasing basicity (least basic first) of the following compounds from the pK_a values for their conjugate acids:

(a) $(CH_3)_3N$: pK_a: 9.8

(b) $C_6H_5\ddot{N}H_2$ (aniline) 4.6

(c) piperidine (N–H) 11.1

1.28 Complete the following table:

Structure	K_a	pK_a
ICH_2CO_2H	7.5×10^{-4}	(a)
$ClCH_2CO_2H$	(b)	2.8
$CH_3CH_2CH_2CO_2H$	1.5×10^{-5}	(c)

1.29 Write an equation for the acid-base reaction that would take place when each of the following Brønsted acids is treated with an aqueous solution of NaOH.

(a) cyclohexyl–COOH (C=O)

(b) $CH_3S(=O)_2-OH$

(c) cyclohexyl–$\overset{+}{O}H_2$

(d) $CH_3CH_2CH_2\overset{+}{N}H_3$

(e) $CH_3CH(\overset{+}{N}H_3)COOH$ (use excess NaOH)

1.30 Write an equation for the acid-base reaction when each of the following Brønsted bases is treated with concentrated HCl.

(a) $C_6H_5NH_2$

(b) $(CH_3CH_2)_3COH$

(c) piperidine (N–H)

(d) $C_6H_5CO_2^-$ (C=O)

Orbitals and Their Role in Covalent Bonding

Some Important Features

A bonding molecular orbital is formed by the overlap of two atomic or hybrid orbitals of the same phase. A bonding molecular orbital is of lower energy (more stable) than the two original orbitals. An antibonding molecular orbital is formed by interference of two atomic or hybrid orbitals of opposite phase and is of higher energy (less stable) than the two original orbitals.

A sigma (σ) molecular orbital results from the end-to-end overlap of atomic or hybrid orbitals. A pi (π) molecular orbital results from side-to-side overlap of *p* orbitals. A pi orbital is more exposed and is usually of higher energy (more reactive) than a sigma orbital.

A carbon atom can hybridize its atomic orbitals in one of three ways:

sp^3:	four tetrahedral single bonds
sp^2:	two single bonds and one double bond
sp:	one single bond and one triple bond

The bond length from a carbon atom decreases with increasing *s* character - that is, *sp* bonds are the shortest and sp^3 bonds are the longest.

Some important functional groups are listed in Tables 2.2 and 2.3 in the text. Pay particular attention to the different ways formulas for functional groups can be written. These different partial formulas are interchangeable and often, at the start, confuse students.

A nitrogen atom with three bonds to other atoms (as in NH_3 or RNH_2) has a pair of unshared electrons that can be donated to an electron-deficient species: $\ddot{N}H_3 + H^+ \longrightarrow NH_4^+$. An oxygen atom in compounds has two bonds to other atoms (as in H_2O, ROH, or $R_2C{=}O$) and has two pairs of unshared electrons. A carbon-oxygen or a carbon-nitrogen bond is polar because oxygen and nitrogen are more electronegative than carbon.

Conjugated double bonds can occur only if two *p* orbitals involved in two different pi bonds can partially overlap their sides. For this reason, the pi bonds must join adjacent atoms for conjugation to occur. Benzene (C_6H_6) is a symmetrical cyclic molecule in which six *p* electrons are completely delocalized (see Figure 2.23, Section 2.8).

Resonance structures are used to show complete or partial delocalization of electronic charge, and differ only in the positions of electrons. The major contributors are the resonance structures of lowest energy (greatest stabilization). Reread the "rules" in Section 2.9C.

Reminders

- Remember the numbers of covalent bonds for elements in stable compounds: C = 4, O = 2, H = 1.

- To determine the approximate bond angles around a carbon atom, ask yourself "What is the hybridization of that carbon?" If the answer is sp^3, then the bond angles are approximately 109°; if the answer is sp^2, 120°; if sp, 180°.

- A doubly bonded carbon atom is sp^2 hybridized and has three planar, trigonal sigma bonds plus a pi bond.

- Study the different ways of drawing electron-shift arrows in resonance structures presented in Section 2.9A:

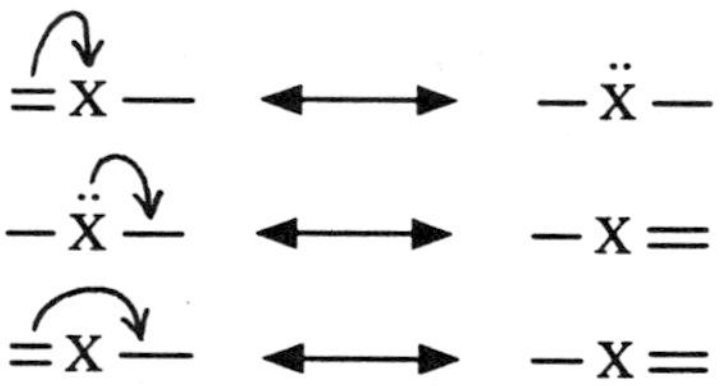

$$=X- \longleftrightarrow -\ddot{X}-$$

$$-\ddot{X}- \longleftrightarrow -X=$$

$$=X- \longleftrightarrow -X=$$

- After you draw each potential resonance structure, check it for reasonability. Are the electron-shift arrows used correctly? Does each atom possess an appropriate number of electrons? Are any formal charges included? Practice writing resonance structures with simple structures such as benzene and pyridine. Always include any unshared valence electrons in the formulas for resonance structures.

Additional Drill Problems

2.1 Draw a potential energy-orbital diagram using boxes for orbitals and arrows (↑↓) for electrons, for:

(a) an *sp* nitrogen (b) an sp^3 nitrogen (c) an sp^2 nitrogen

2.2 Identify the atom(s) that has undergone a change in hybridization in each of the following reactions. State the change in hybridization.

(a) cyclohexanone (=O) + HC≡N $\xrightarrow{^{-}C\equiv N}$ cyclohexane bearing OH and C≡N

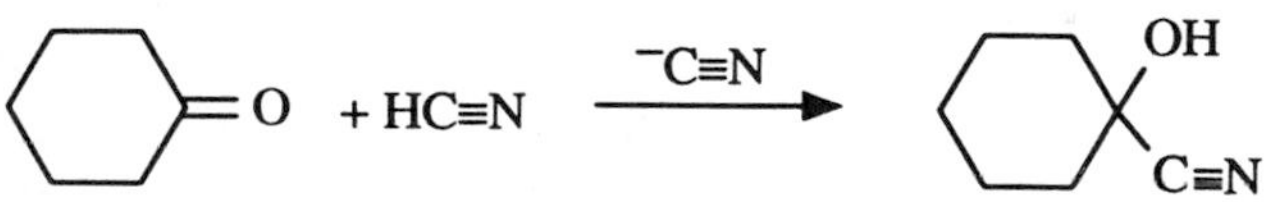

(b) $(CH_3)_3CCl + CH_3O^- \longrightarrow (CH_3)_2C{=}CH_2 + Cl^- + CH_3OH$

(c) $HC{\equiv}CH + H_2O \xrightarrow{H^+,\ Hg^{2+}} CH_3\overset{O}{\overset{\|}{C}}H$

(d) $CH_3C{\equiv}N + 2\ H_2O + H^+ \longrightarrow CH_3\overset{O}{\overset{\|}{C}}OH + NH_4^+$

2.3 Label each covalent bond in the following structures as sigma or pi:

(a) $O{=}C{=}O$ (b) $N{\equiv}C\text{-}CH_2\text{-}C{\equiv}N$ (c) $CH_3\overset{CH_3}{\overset{|}{C}}{=}N-OH$

2.4 Predict all the sigma-bond angles in the following structures:

(a) $H\text{-}C{\equiv}C\text{-}CH{=}CH\text{-}CH_2\text{-}\overset{O}{\overset{\|}{C}}\text{-}CH_3$ (b)

2.5 State the percent *s* character for each hybrid orbital originating at the indicated atom.

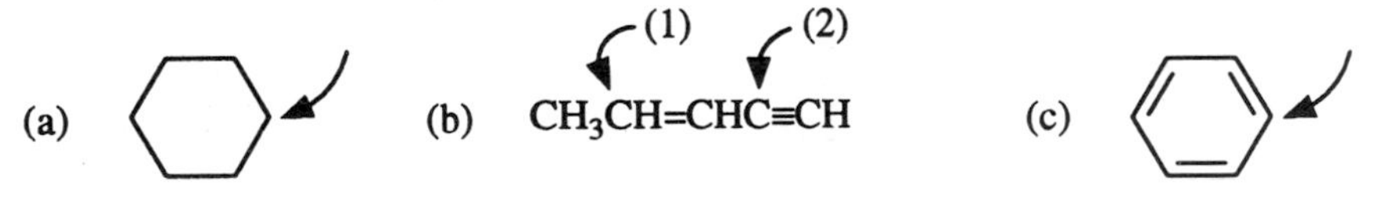

2.6 For each carbon atom in the following structures, state if its single-bond geometry is tetrahedral, trigonal, or linear:

(a) $CH_3CH_2CH_3$ (b) C≡C

(c) $(H_3C)_2CH\text{-}C{\equiv}H$

2.7 In each of the following structures, which of the two indicated covalent bonds has the greater bond length?

(a) 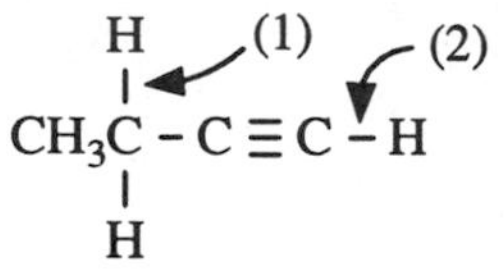 $CH_3CH_2-C\equiv C-H$ (the two H atoms drawn above and below the CH_2 carbon); arrow (1) points to the C–H bond of the CH_2 group; arrow (2) points to the terminal C–H bond

(b) 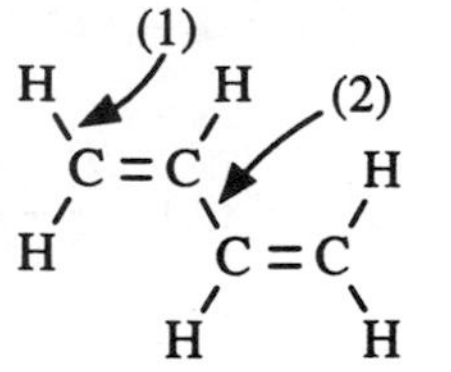 $H_2C=CH-CH=CH_2$ (drawn with all H atoms shown); arrow (1) points to a C–H bond of the first C=C; arrow (2) points to the C–C single bond between the two C=C groups

2.8 Draw the complete structural formula for each of the following compounds. Indicate the hybridization of each atom.

(a) 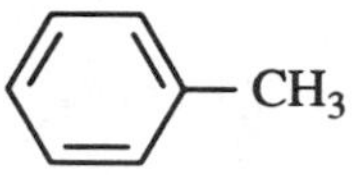 $C_6H_5-CH_3$ (benzene ring–CH_3) (b) $CH_3C{\equiv}CCH_2CN$ (c) 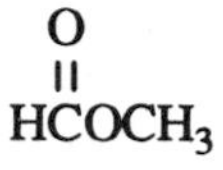 $HC(=O)OCH_3$ (O drawn above the C: $\overset{O}{\overset{\|}{HCOCH_3}}$)

2.9 What is the hybridization of each nitrogen atom in the following structures?

(a) $CH_3N{=}C{=}NCH_3$ (b) CH_3OCH_2CN (c)  pyrimidine (six-membered ring with two N atoms)

2.10 What hybrid or atomic orbitals are used to form the indicated covalent bonds?

(a) 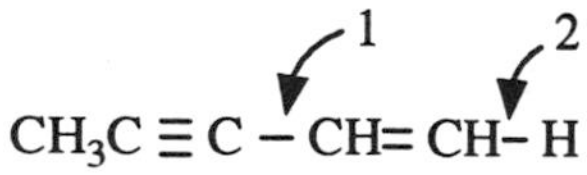 $CH_3C\equiv C-CH=CH-H$; arrow 1 points to the $C-CH$ bond; arrow 2 points to the $CH-H$ bond

(b) $CH_3-C(=O)-H$; arrow 1 points to the CH_3-C bond; arrow 2 points to the $C-H$ bond

(c) $(CH_3)_2N-CH_2-CH=N-CH_3$; arrow 1 points to the $N-CH_2$ bond; arrow 2 points to the $N-CH_3$ bond

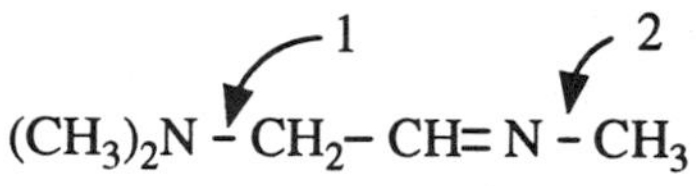

(d) 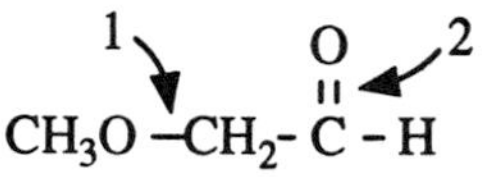 $CH_3O-CH_2-C(=O)-H$; arrow 1 points to the $O-CH_2$ bond; arrow 2 points to the C=O bond

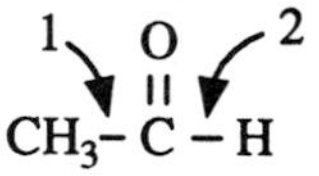

2.11 Circle and name the functional groups in the following structures:

(a) $HC{\equiv}CCH{=}CH_2$ (b) $CH_3CH_2CH_2OH$

(c) 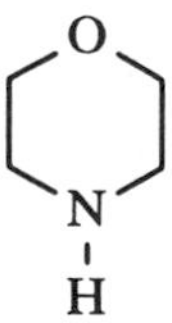 morpholine (six-membered ring containing O and N–H)

(d) 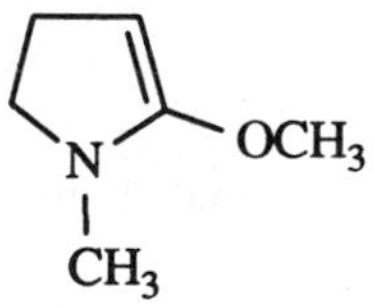 five-membered ring with N–CH_3 and a C=C double bond, with OCH_3 on the ring carbon next to N

2.12 For each of the general formulas, draw a specific compound where R = CH_3CH_2- and R′ = (cyclohexyl):

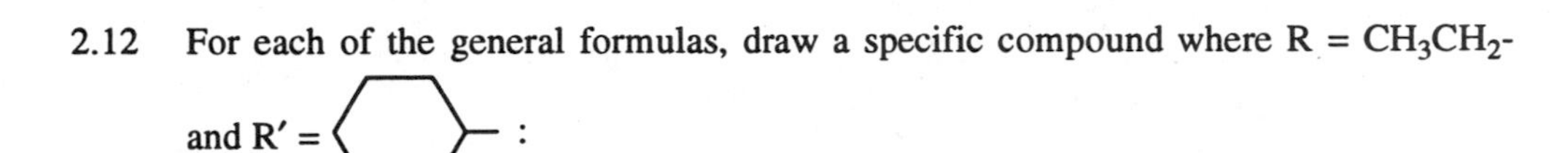

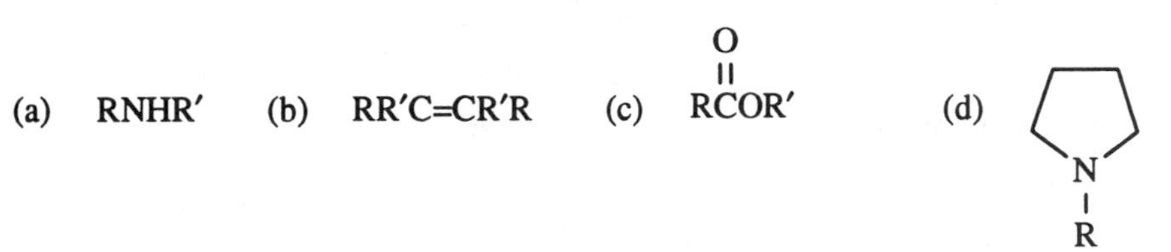

2.13 For each of the following compounds, write the general formula and the name of the class of compounds to which it belongs:

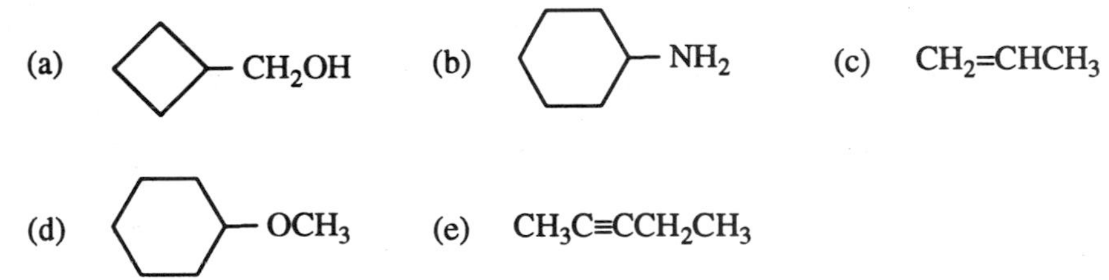

2.14 From the following orbital diagrams for the second energy level of neutral atoms, state the type of atom (carbon, oxygen, or nitrogen) and its hybrid state:

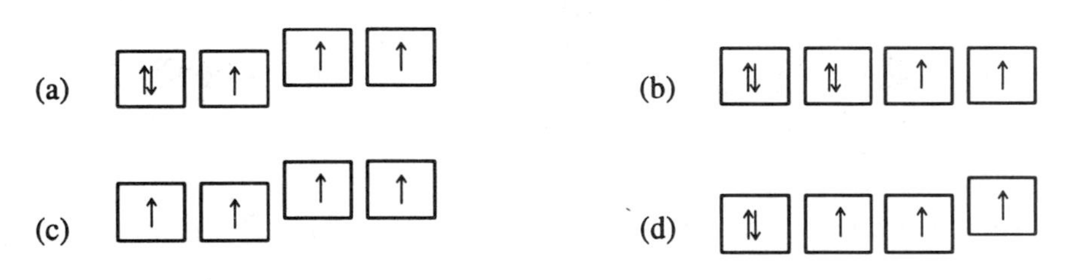

2.15 For each of the following nonconjugated dienes, draw a conjugated diene with the same carbon skeleton. (More than one correct answer may be possible.)

(a) CH_3CH=$CHCH_2CH$=CH_2 (b) CH_2=$C(CH_3)CH_2C(CH_3)$=CH_2 (c) (1,4-cyclohexadiene ring)

2.16 Which of the following pairs of structures are resonance structures? Explain your answer.

(a) $CH_3\overset{+}{N}$≡N ; CH_2=N=NH

(b)

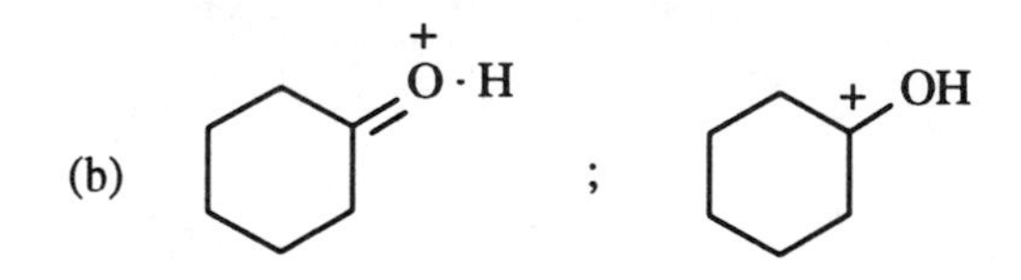

(c) $CH_2{=}CH{-}N(CH_3)_2$; $\overset{-}{C}H_2{-}CH{=}\overset{+}{N}(CH_3)_2$

(d) ;

2.17 Draw resonance structures for the following aromatic ring systems:

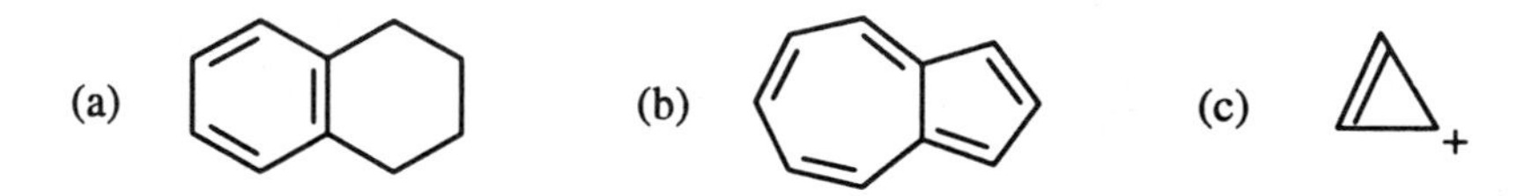

2.18 Draw resonance structures for the following compounds or ions:

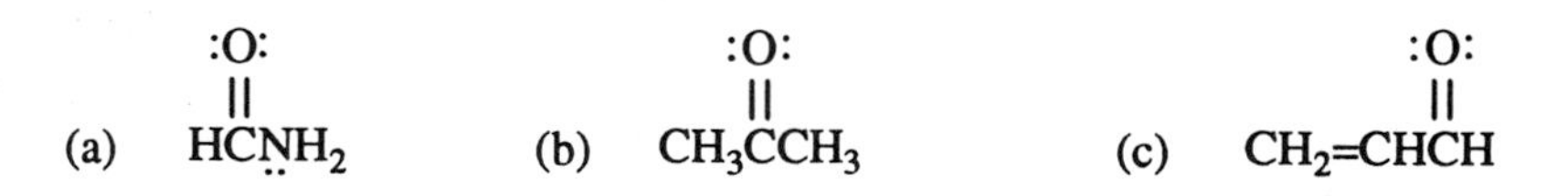

(d) CH_3COCH_3 (e) $CH_3{-}\ddot{N}{=}C{=}\ddot{O}{:}$ (f) $^{-}{:}CH_2{-}\overset{+}{N}{\equiv}N{:}$

(g)

2.19 Each of the following structures can be a reactive intermediate in an organic reaction. Draw resonance structures with electron-shift arrows for each.

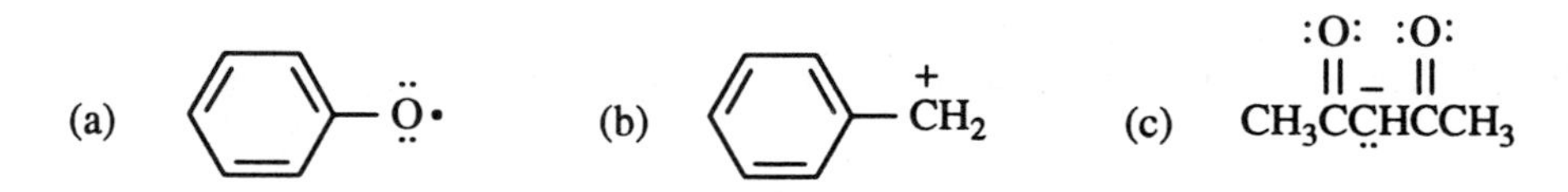

(d) [cyclohexadienyl cation with H and Br] (e) $\overset{+}{C}H_2CH{=}CH_2$ (f) $:N{\equiv}C\overset{-}{\underset{..}{C}}HC{\equiv}N:$

(g) $^{-}:CH_2\overset{+}{N}(=\ddot{O}:)(-\ddot{O}:^{-})$ (h) $C_6H_5C(=O)\ddot{O}\cdot$

2.20 Which of the following compounds will require resonance structures to adequately describe its structure? Draw these resonance structures, and lable the major contributors.

(a) $HCOH$ (with C=O) (b) $CH_3C{\equiv}CCH_2CH{=}CH_2$ (c) 1-methoxycyclopentene (OCH_3)

2.21 Write the acid-dissociation reaction for each of the following.

(a) $C_6H_5CH_2OH$ (b) cyclohexyl–COH (with C=O) (c) naphthalene–OH

2.22 Which member of each pair would have the more resonance stabilized conjugate base?

(a) p-nitrophenol (NO_2, OH) or m-nitrophenol (NO_2, OH)

(b) $HO{-}C_6H_4{-}C{=}O$ with CH_3 or $HO{-}C_6H_4{-}CH_2{-}C{=}O$ with CH_3

(c) 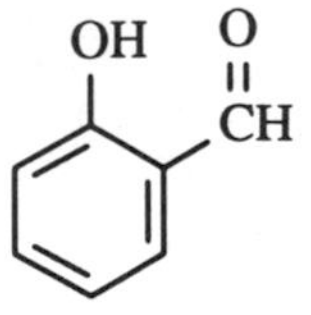 or 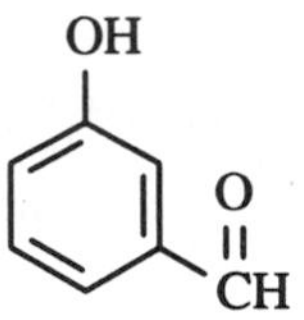

2.23 Which of the following has a conjugate base that is resonance stabilized?

(a) $O_2NCH_2CH_2OH$

(b) 

(c) $O_2NCH_2\overset{O}{\overset{\|}{C}}OH$

2.24 Rank the following in order of increasing acid strength (most acidic first). Give reasons for your ranking.

(a) $CH_3\underset{Cl}{\underset{|}{C}}H\overset{O}{\overset{\|}{C}}OH$

(b) 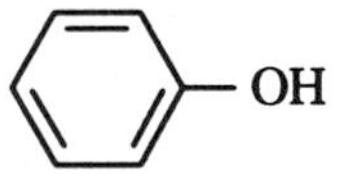

(c) 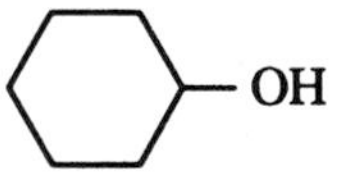

2.25 For each of the following, change the position of the circled group to obtain another structure that would be more acidic. For example,

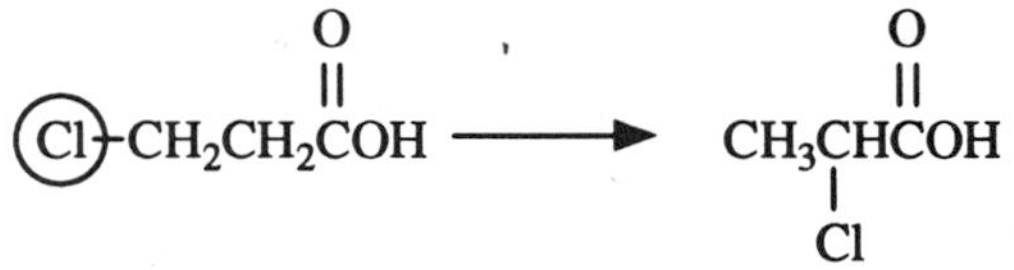

Would be more acidic because the electronegative group is closer to the carboxyl group.

(a) 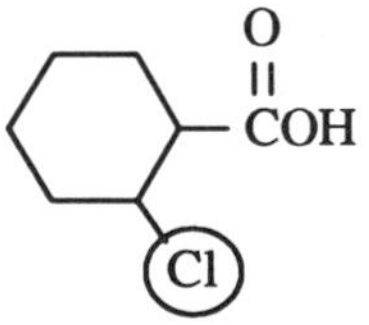

(b) 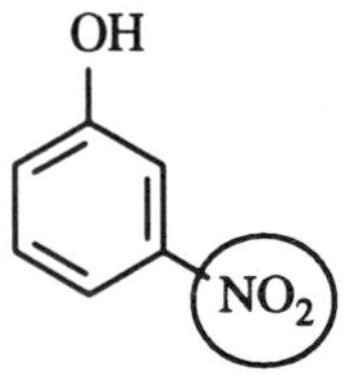

(c) 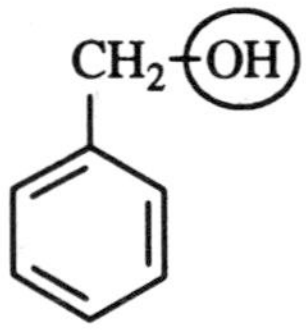

3

Structural Isomerism, Nomenclature, and Alkanes

Some Important Features

Structural isomers are compounds with the same molecular formula but different structures (sequence of atoms).

A saturated open-chain alkane (branched or not) has the general formula C_nH_{2n+2}. We subtract two hydrogen atoms from the general formula for each double bond or for each ring.

We will not reiterate the nomenclature of organic compounds here, but refer you to Chapter 3 in the text. (For quick reference, see the appendix in the Solutions Manual.) Remember to: (1) number a chain or ring to give the lowest number to the principal functional group, and (2) group like substituents together.

Alkanes are generally nonreactive, but they do undergo combustion and reaction with halogens. The heat of combustion for alkanes increases with increasing molecular weight and also with increasing energy contained in the bonds.

Reminders

The same molecule may be drawn in a number of ways:

Not isomers: (cyclohexane)–OH ≡ OH ≡ HO

To be structural isomers, two molecules must have the atoms attached in different sequences.

Isomers: OH CH_3 OH

- Some instructors suggest naming potential isomers; if the names are the same, the structures represent the same compound. (This is a valid approach only if you name both structures correctly.)

- When drawing isomers for a particular molecular formula, use the general formula. If a compound contains a ring, a structural isomer can contain a double bond instead.

and $CH_2{=}CHCH_3$: both C_nH_{2n}

- General formulas may be extrapolated to compounds other than hydrocarbons:

$$CH_3CH_2CH_2OH = C_nH_{2n+2}O$$

no double bond or ring

$$CH_3CH_2\overset{\overset{\displaystyle O}{\|}}{C}H = C_nH_{2n}O$$

one double bond or ring

- Because nitrogen has a valence of 3, take special care in using general formulas with nitrogen compounds.

$CH_3CH_2NH_2$	$CH_2{=}CHNH_2$ or $CH_3CH{=}NH$	$HC{\equiv}CNH_2$ or $CH_3C{\equiv}N$
C_2H_7N	C_2H_5N	C_2H_3N
$C_nH_{2n+3}N$	$C_nH_{2n+1}N$	$C_nH_{2n-1}N$

Additional Drill Problems

3.1 Label each of the following compounds as saturated or unsaturated:

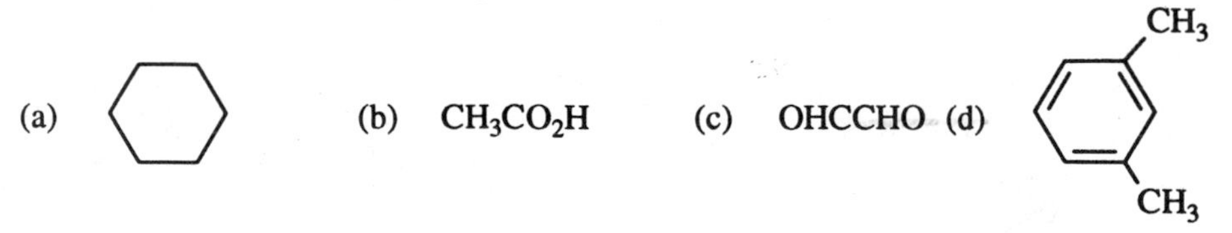

3.2 Which of the following are aliphatic compounds?

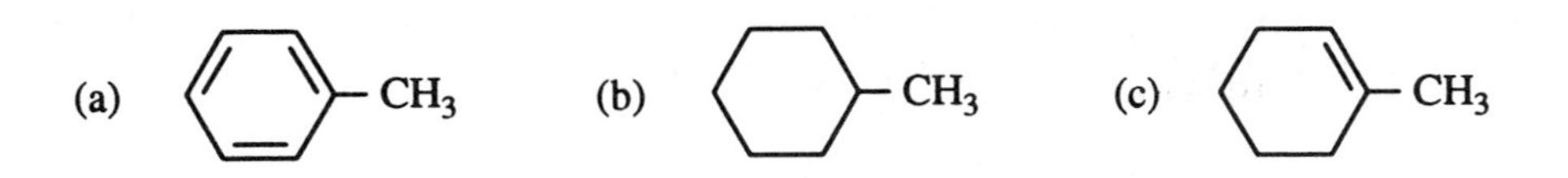

3.3 All the following formulas represent only two different compounds. Group these formulas into two sets, each set representing one compound.

(a) $CH_3CHCH_2CH_2CHCH_3$
    CH_3 CH_3

(b) $CH_3CH_2CHCH_2CHCH_3$
    CH_3 CH_3

(c) $CH_3CHCH_2CHCH_3$
    CH_3 CH_2CH_3

(d) CH_3
    $CH_3CHCH_2CHCH_2CH_3$
    CH_3

(e) $(CH_3CHCH_2)_2$
    CH_3

(f) $(CH_3)_2CHCH_2CHCH_2CH_3$
    CH_3

(g) H_3C CH_3
    CH
    CH_2CH_2
    CH
    H_3C CH_3

(h) $CH_3CHCH_2CH(CH_3)_2$
    CH_2CH_3

3.4 For each of the following structures, draw a structural isomer that has a different functional group. (There may be more than one correct answer.)

(a) $CH_3C{\equiv}CH$ (b) $CH_3CH_2CH_2OH$ (c) $H\overset{O}{\overset{\|}{C}}OCH_2CH_3$

(d) $CH_3OCH_2\overset{O}{\overset{\|}{C}}H$ (e) O (f) $CH_2\overset{O}{\overset{\|}{C}}OH$

3.5 Calculate the number of rings and/or multiple bonds (double bond, triple bond, *etc.*) for each of the following molecular formulas:

(a) $C_{16}H_{10}$ (b) C_5H_8 (c) $C_{13}H_{20}$ (d) C_9H_{14}

3.6 A compound of unknown structure has the molecular formula C_4H_8. What are the five possible structures of this compound?

3.7 What are the possible structures represented by C_4H_7Cl?

3.8 Match the compound on the left with its structural isomer on the right.

(1) 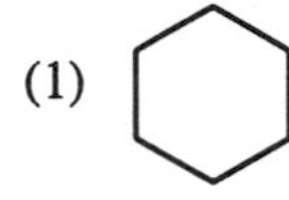

(a) $CH_3CH_2CH{=}CHC{\equiv}CH$

(2) 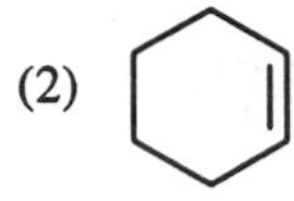

(b) $CH_2{=}CHCH{=}CHCH_2CH_3$

(3) 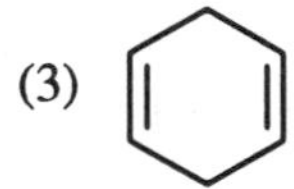

(c) $CH_2{=}CHCH_2CH_2CH_2CH_3$

(d) $CH_3CH_2CH_2CH_2CH_2CH_3$

3.9 Suggest two different ways of arranging the atoms in:

(a) CH_3NO_2 (b) C_3H_7Cl

3.10 Draw the four isomers for C_8H_{18} that have only a single branch on a continuous carbon chain.

3.11 Circle and name the functional groups in the following compounds:

(a) O O O

(b) $CH_3OCH_2CH_2\underset{NH_2}{\underset{|}{C}}H\overset{O}{\overset{||}{C}}OH$

(c) $-O\overset{O}{\overset{||}{C}}-$ $-\overset{O}{\overset{||}{C}}CH_2\overset{O}{\overset{||}{C}}H$

3.12 Draw each of the following as indicated:

(a) a four-carbon cycloalkane with a carboxylic acid group

(b) an isobutane with a double bond (isobutene)

(c) a continuous-chain, five-carbon hydrocarbon that has a hydroxyl group at one end and a chloro group at the other end

(d) a compound with a ketone group and an aldehyde group separated by three carbon atoms

3.13 Draw all possible isomers that contain only sp^3-hybridized carbons for the following molecular formulas:

(a) C_5H_{12} (b) C_5H_{10} (c) C_5H_8

3.14 Which of the following molecular formulas can represent only one structure?

(a) C_2H_5NO (b) C_2H_7NO (c) C_3H_8

3.15 Draw formulas for each of the following sets of isomers:

(a) six isomers of propenamine (C_3H_7N)

(b) three isomers of pentane (C_5H_{12})

3.16 Draw a cycloheptane structure that has each of the following substituents. Name the substituted cycloheptanes.

(a) $(CH_3)_2CH$- (b) C_6H_5- (phenyl ring drawn) (c) $CH_3CH_2CH_2$-

(d) $CH_3CH_2CH(CH_3)$- (e) $(CH_3)_3C$- (f) $(CH_3)_2CHCH_2$-

(g) $-NO_2$ (h) -F

3.17 Write names and structural formulas for all the monochloro isomers with the same carbon skeleton of the following hydrocarbons:

(a) $(CH_3)_2CHCH_2CH_3$ (b) $(CH_3)_3CCH(CH_3)_2$

3.18 Draw a condensed structural formula for each of the following:

(a) 3-ethylpentane

(b) 2,3-dimethylbutane

(c) 5-ethyl-1,2,3-trimethylcyclohexane

(d) *tert*-butylcyclopentane

(e) 2,2,4,4-tetramethylheptane

(f) 2,2,3,4,4-pentamethylpentane

3.19 Name the following hydrocarbons by the IUPAC system:

(a) Cyclohexane ring bearing CH_3, $-CHCH_2CH_3$ (with CH_3 on the CH), and CH_3 on adjacent carbons

(b)
$$\underset{\displaystyle CH_3}{|}$$
$CH_3CH\text{-}CH{=}CH\text{-}(CH_2)_2CH{=}CHCH{=}CH_2$ (with CH_3 on the second carbon)

(c) $(CH_3)_3C-$ on a cyclohexa-1,4-diene ring bearing two CH_3 groups on the same carbon

(d) Three cyclopentane rings joined in a chain

3.20 Write an IUPAC name for each of the following compounds:

(a) $CH_3CHICHCH_2CH_3$ (with OH on the third carbon)

(b) $HC{\equiv}CCHCH_2CHCH{=}CH_2$ (with Cl on the third carbon and CH_3 on the fifth carbon)

(c) $(CH_3)_2CHCCH(CH_3)_2$ (with $CH(CH_3)_2$ above and $CH(CH_3)_2$ below the central C)

(d) $CH_3CH[C(CH_3)_3]_2$

(e) $(CH_3)_3CCOH$ (with $=O$ on the carbonyl C)

(f) CH_3CHCOH (with $=O$ on the carbonyl C and a cyclopentyl group on the CH)

(g) $CH_3CH_2CH_2CHCH_3$ with $CH_3CH_2CHCH_3$ attached at the CH (structure)

$$\begin{array}{l} CH_3CH_2CH_2\underset{|}{C}HCH_3 \\ \quad\quad\quad CH_3CH_2\overset{}{C}HCH_3 \end{array}$$

3.21 Write condensed structural formulas for the following compounds:

(a) 2,4-dimethyl-1-hexanol

(b) 1,3,5-triethylbenzene

(c) 2-phenylethanol

(d) 2,2-dichloropentanal

(e) 3-isopropylcyclopentanone

(f) 3-*sec*-butylcyclooctene

(g) 1,4-di-*tert*-butylbenzene

3.22 Which group in each of the following pairs has the higher nomenclature priority?

(a) $-NO_2$ or $>C=C<$

(b) $>C=C<$ or $-C\equiv C-$

(c) $-OH$ or $-\overset{O}{\overset{\|}{C}}OH$

(d) C_6H_5- (phenyl) or $-\overset{O}{\overset{\|}{C}}H$

(e) $-\overset{O}{\overset{\|}{C}}-$ or $-\overset{O}{\overset{\|}{C}}H$

(f) $-CH_3$ or $-\overset{O}{\overset{\|}{C}}H$

3.23 Draw and name by the IUPAC system all possible isomers for each of the following molecular formulas:

(a) $C_2H_4Cl_2$

(b) dienes with the formula C_4H_6

(c) $C_5H_{11}Cl$

(d) aldehydes with the formula C_4H_6O

3.24 Give an IUPAC name for each of the following:

(a) $(CH_3)_3C\overset{O}{\overset{\|}{C}}H$

(b) $F_3C\overset{O}{\overset{\|}{C}}CF_3$

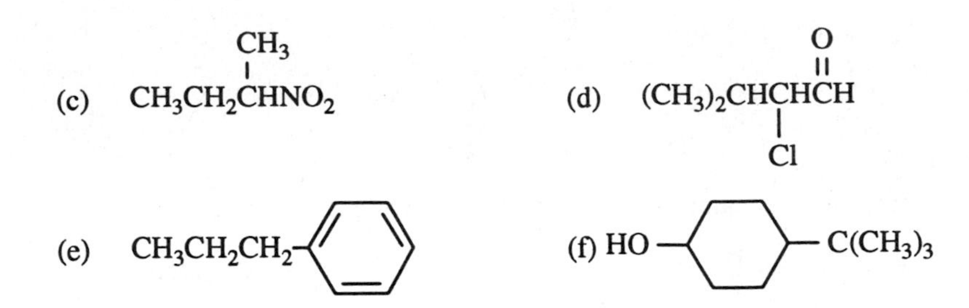

3.25 What is the difference between catalytic cracking and catalytic reforming?

3.26 (a) What is the octane rating of a mixture of heptane and isooctane that contains 10% heptane

(b) Would you expect to find a gasoline containing such a mixture at your local gas station?

Stereochemistry

Some Important Features

Geometric or (*cis-trans*) isomerism arises from substituents being on the same side or on opposite sides of a double bond or ring.

cis:

```
CH3    CH3
   \  /
   C=C
   /  \
  H    H
```

trans:

```
CH3    H
   \  /
   C=C
   /  \
  H    CH3
```

```
 cis ring: H (up), CH3 (down) on one carbon; H (up), CH3 (down) on the other
        H
   H   |
       CH3
  CH3
```

```
 trans ring:
          CH3
   H
          H
  CH3
```

Geometric isomerism is not possible around a double bond if two groups on one carbon atom are the same (two CH_3 groups in the following example).

```
CH3CH2    CH3                 H        CH3
      \  /                     \      /
      C=C          ≡           C=C
      /  \                     /      \
     H    CH3            CH3CH2        CH3
```

The above two formulas do not represent isomers because they are superposable. Any superposable molecules represent the same compound, not isomers. If this feature is not clear to you make molecular models of the above two formulas and prove to yourself that they are the same.

The priority rules for (*E*) and (*Z*) system are listed in Section 4.1B. The most important feature is that *higher atomic number means higher priority.* (In the case of isotopes, *higher atomic mass* means higher priority.) If the atoms attached to the sp^2 carbons are the same, proceed along the chain to the first point of difference.

higher atomic number,

higher priority than F

greater mass,

higher priority than H

Cl D
C=C
F H

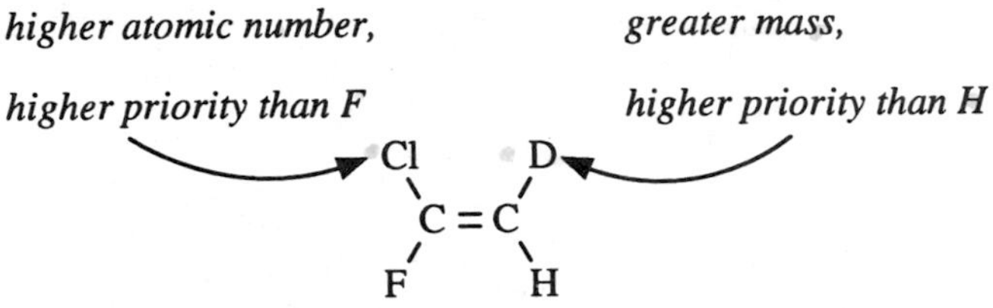

(*Z*)-1-chloro-2-deuterio-1-fluoroethene

higher priority, same side

H CH_2CH_2(H)
C=C
CH_3 CH_2CH_2(Cl)

first point of difference

higher priority than H

higher priority than CH_2CH_3

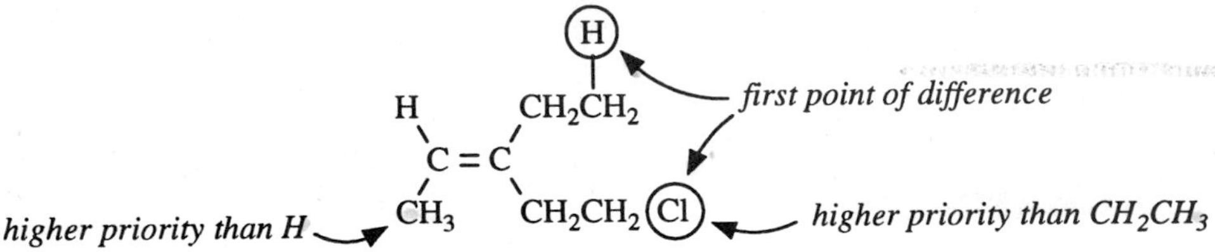

(*Z*)-5-chloro-3-ethyl-2-pentene

higher priority, same side

A common student error is to sum atomic numbers to arrive at priorities of groups. This is incorrect. Look at individual atoms at the first point of difference, not at groups of atoms.

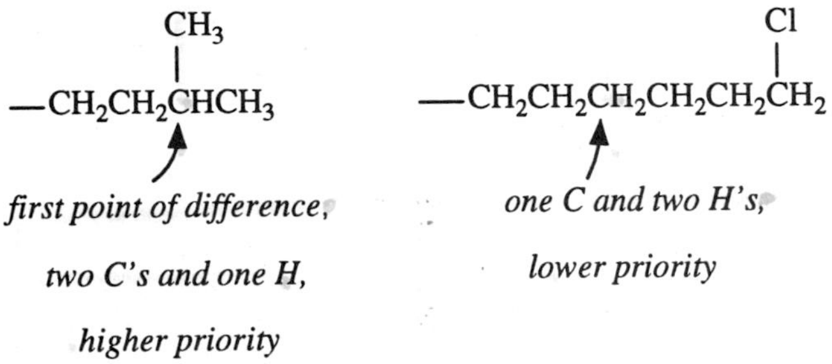

$—CH_2CH_2CH(CH_3)CH_3$

first point of difference,

two C's and one H,

higher priority

$—CH_2CH_2CH_2CH_2CH_2CH_2Cl$

one C and two H's,

lower priority

Conformations of molecules are the different shapes they can assume. *Anti* conformations are generally more stable than eclipsed conformations. (Compare the following Newman projections with molecular models.)

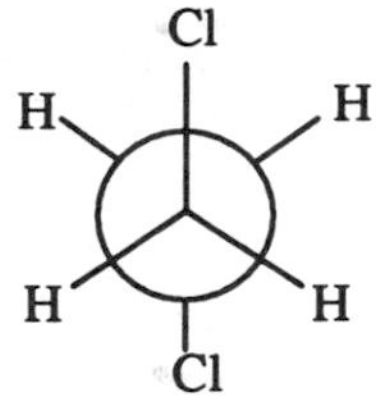

Cl
H H
H H
Cl
anti

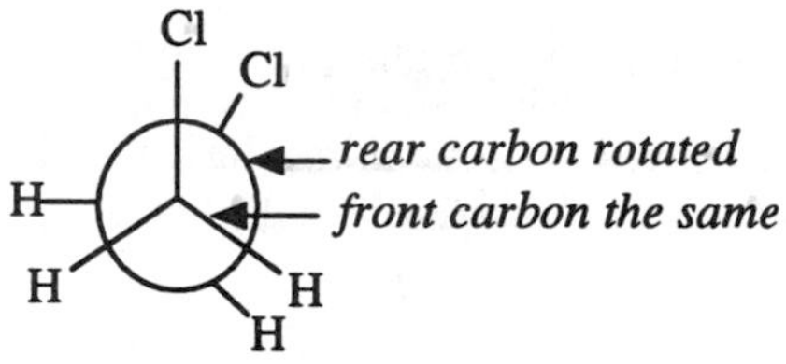

Cl
Cl
H
H H
H
rear carbon rotated
front carbon the same
eclipsed

Conformations of cyclic compounds (six-membered rings particularly) should be studied carefully. A molecular model of cyclohexane shows ring conformation much better than a "paper formula" does. (Boat, chair, etc.)

Substituents on the cyclohexane ring in the chair form can be equatorial or axial. (The axial substituents are especially easy to see with models.) Although the following conformers are interconvertible, the conformer with the bulkier substituent *equatorial* is more stable (that is, of lower energy)

Different conformers of the same molecule:

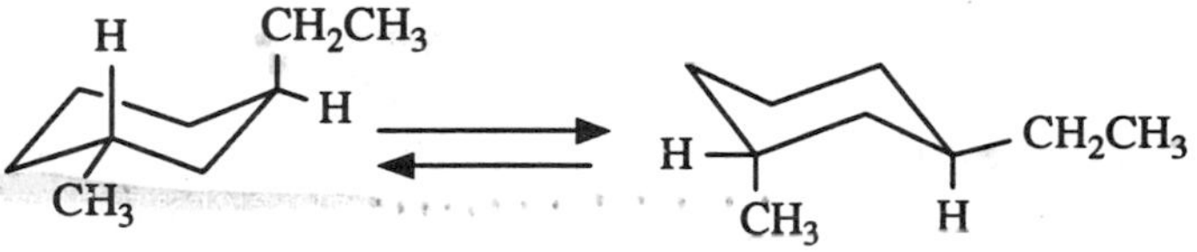

H CH_2CH_3 H CH_3

H CH_2CH_3 CH_3 H

axial CH_3CH_2 *equatorial* CH_3CH_2 *(favored)*

Molecular models are also indispensable in a study of chirality of molecules. A chiral carbon atom is one with four different substituents. The presence of one chiral carbon means that a molecule is chiral, or *not superposable on its mirror image,* and is capable of rotating the plane of polarization of plane-polarized light.

A pair of nonsuperposable mirror images are enantiomers of each other. The following formulas show two different ways of representing the enantiomers of 2-chlorobutane.

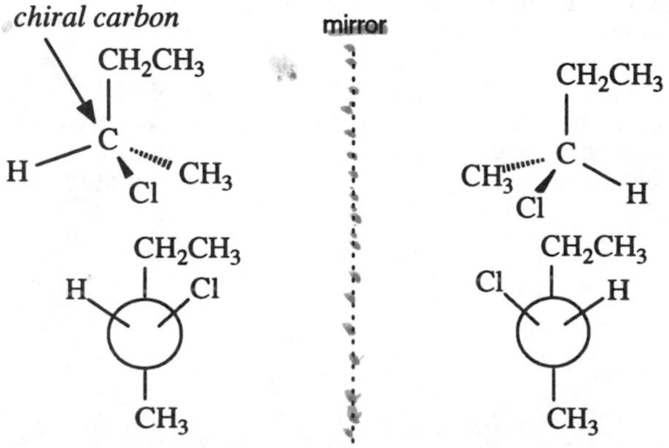

chiral carbon

mirror

CH_2CH_3
C
H CH_3
Cl

CH_2CH_3
C
CH_3 H
Cl

CH_2CH_3
H Cl
CH_3

CH_2CH_3
Cl H
CH_3

If a carbon atom is sp^2-hybridized, or if it has two identical substituents, the carbon atom is *achiral* (not chiral). (The chirality of a molecule as a whole, however, is determined by the lack of superposability of the mirror image.) The following two examples are superposable on their mirror images. (Try it with models.)

CH_2CH_3
H — C — CH_3
CH_2CH_3

achiral
(two CH_2CH_3 groups)

CH_3 H
C=C
Br Cl

achiral
(sp^2 carbons)

If a molecule contains more than one chiral carbon, there is a possibility of an internal plane of symmetry. A molecule with chiral carbons but with an internal plane of symmetry is a *meso* form of the compound and cannot rotate the plane of polarization of plane-polarized light.

CH_3
Cl — C — H
H — C — Cl
CH_3

a chiral molecule

CH_3
H — C — Cl
plane of symmetry
H — C — Cl
CH_3

an achiral, or meso, molecule

A *meso* molecule has a "top half" that is the mirror-reflective image of the "bottom half." Because groups can rotate around their bonds, it is not always easy to see at a glance if a structure is a *meso* form. Molecular models are useful here. Construct the models of the preceding two compounds and their mirror-reflective images, and verify the chirality or achirality of the two. Then, rotate groups around the bonds to see the different conformations each molecule can assume.

The use of Fischer projections is discussed in Section 4.6C. Remember that these projections are simply a shorthand way of representing ball-and-stick or three-dimensional formulas. For this reason, you can check any formula in a Fischer projection simply by converting it to a three-dimensional formula. If you do this operation, you are less likely to make errors.

The *(R)* and *(S)* system of assigning the absolute configuration around chiral carbon atoms is discussed in Section 4.8. You will probably find assignment of configuration easiest to do with models.

For chiral cyclic compounds, it is easiest to work with polygon formulas. Whether a carbon is bonded to four different atoms or groups is determined in the same way as it is in open-chain compounds -- by listing the four atoms or groups as is shown at the start of Section 4.9D. To see whether a cyclic compound with chiral carbons contains a plane of symmetry, try drawing the plane.

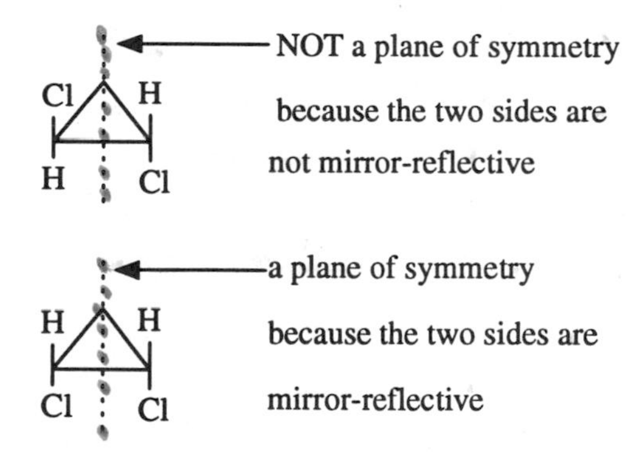

Reminders

- To draw a chair form of cyclohexane quickly, follow the steps below (or purchase a template).

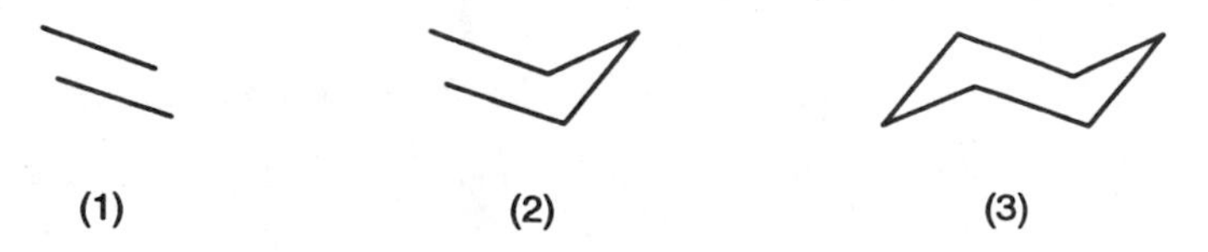

- In drawing equatorial and axial substituents, first draw the axial bonds, which are vertical.

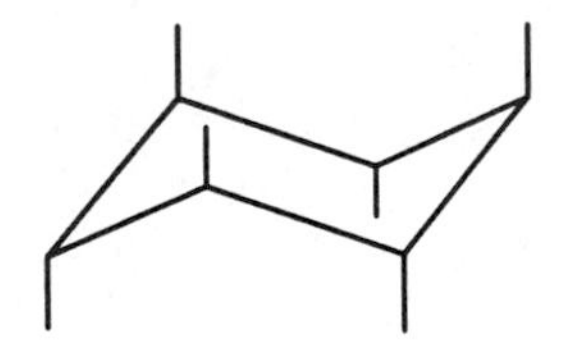

Then draw the equatorial bonds, keeping the bond angles at approximately 109°.

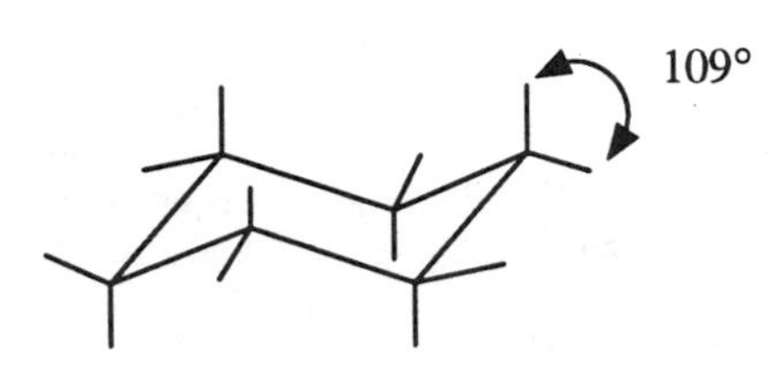

To determine *cis* and *trans* on a ring, decide if the groups are on the same side or on opposite sides.

X
X
H
H
≡
X
H
X
H
cis (both on "top")

To draw the enantiomer of a three-dimensional formula, simply reverse two groups, usually those attached to the "front" bonds (B, following).

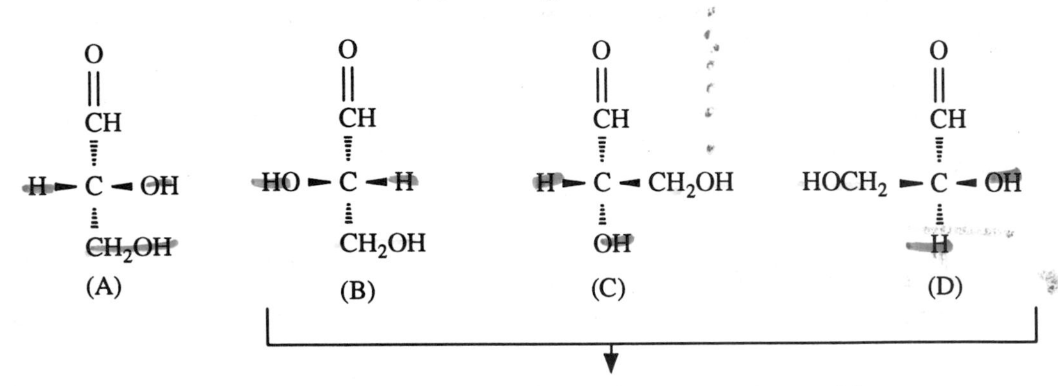

Different formulas for the one enantiomer of A

(Models will show that B, C, and D are identical.)

Additional Drill Problems

4.1 Define the following terms, using examples in your answers:

(a) *cis-trans* isomers
(b) geometric isomers
(c) structural isomers
(d) stereoisomers
(e) enantiomers
(f) diastereomers

4.2 Which of the following compounds can exist as a pair of geometric isomers? Draw their structures and label them as *cis* or *trans*.

(a) $ClCH_2C{\equiv}CCH_3$

(b) $ClCH_2CH{=}CHCH_2Cl$

(c) 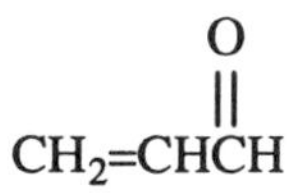 $CH_2{=}CHCH{=}O$ (O double-bonded to the terminal C)

(d) $ClCH{=}CHCH_2CH_2CH_2Cl$

(c) $HO_2CC{\equiv}CCO_2H$

(f) 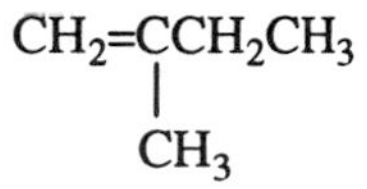 $CH_2{=}C(CH_3)CH_2CH_3$ (CH_3 on the second C)

4.3 Identify each of the following alkenes as being either *cis* or *trans*. For each, draw the other geometric isomer.

(a) 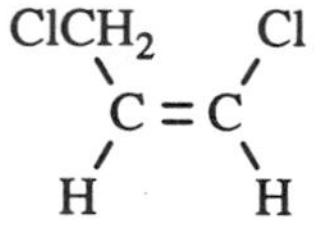 $ClCH_2(H)C{=}C(Cl)(H)$ — $ClCH_2$ top left, Cl top right, H bottom left, H bottom right

(b) $Cl(H)C{=}C(Br)(H)$ — Cl top left, Br top right, H bottom left, H bottom right

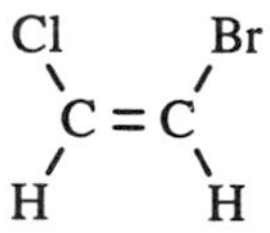

(c) $OHC(H)C{=}C(H)(CHO)$ — OHC top left, H top right, H bottom left, CHO bottom right

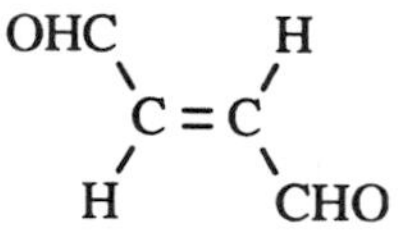

4.4 For each, select the group from the pair in parenthesis that has the higher priority. For example, for -Cl (-I, -F), -I is the correct selection.

(a) $-OH$ ($-OCH_3$, $-SH$)

(b) $-CH_2CH_3$ ($-CH_3$, $-CH_2CH_2CH_3$)

(c) 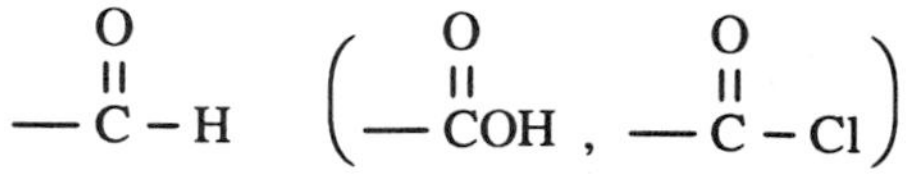 $-C(=O)-H$ ($-C(=O)OH$, $-C(=O)-Cl$)

(d) $-CH_2CH(CH_3)_2$ ($-CH(CH_3)_2$, $-CH_2\text{-}CH{=}CH_2$)

4.5 For each of the following structures, draw all geometric isomers; label each according to the *(E) (Z)* system of nomenclature; and label each as *cis* or *trans* wherever appropriate.

(a) $HOCH_2CH{=}C(CH_3)CH_2CH_2OH$

(b) $CH_3CH{=}CHCO_2H$

(c) $CH_3CH{=}C(CHO)CH_2Cl$

4.6 Draw and label the *cis* and *trans* isomers of the following structures. (Ignore the possibility of optical isomers.)

(a) OH Br

(b) $HOCH_2$ O OH

(c) CH_3 CH_3 CH_3 OH

(d) O CH_3 CH_3

4.7 Draw a formula for each of the following:

(a) *cis*-1,2-dichlorocyclohexane

(b) *trans*-2-aminocycloheptanol

(c) (2*E*, 4*Z*)-hexadienoic acid

(d) (3*E*)-2-chloro-3-methyl-1,3-pentadiene

(e) (*Z*)-3-chloro-2-methyl-2-buten-1-ol

(f) (2*E*, 4*Z*, 6*E*)-octatriene

4.8 Draw the formulas for each of the following conformations:

(a) the staggered conformation of ethane

(b) the eclipsed conformation of ethane

(c) the anti conformation of butane

(d) the gauche conformation of butane

(e) an eclipsed conformation of $HC{\equiv}CCH_2CH_3$

4.9 Using Newman projections, draw the important conformations of:

(a) 2-methylbutane (Use the C2-C3 bond for the Newman projection.)

(b) 2,3-dimethylbutane (Use the C2-C3 bond.)

4.10 Draw Newman projections of the most stable conformation and the least stable conformation of each of the following compounds. Use the circled two carbons for the projections. Place the lower nomenclature priority carbon as the "front" carbon in the formula.

(a) $ClCH_2\underset{}{\overset{H_3C\ \ CH_3}{(CHCH)}}CH_2Cl$

(b) $\overset{OH}{CH_2}\overset{OH\ \ OH}{(CH\text{-}CH)}CH_2OH$

(c) $CH_3CH_2\,(CH(CH_3)_2 / CH)\,CH_3$ — circled: the CH bearing $(CH_3)_2$ bonded to the CH bearing CH_3

(d) $CH_3\overset{^+NH_3}{(CHCH_2)}\overset{O}{\overset{\|}{C}}O^-$

4.11 Which of the following compounds would you expect to contain ring strain? Indicate the strained ring(s).

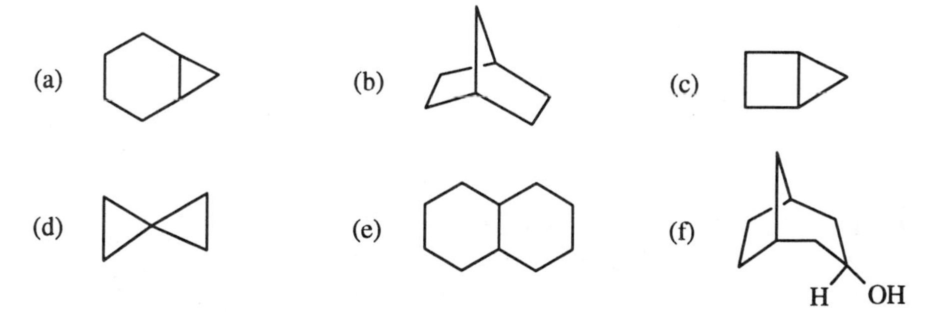

4.12 Redraw each of the following structures as a flat-ring projection formula, with the isopropyl group and the methyl group *cis* to each other and the hydroxyl group *trans* to the alkyl groups. (Ignore the possibility of optical isomers.)

(a) CH_3, OH, $CH(CH_3)_2$

(b) H_3C, $CH(CH_3)_2$, OH

4.13 Draw the *cis* and *trans* isomers for 1,3- and 1,4-dichlorocyclohexane and label each hydrogen in the structure as either equatorial (*e*) or axial (*a*).

4.14 In the following structure, label each indicated position as equatorial (*e*) or axial (*a*).

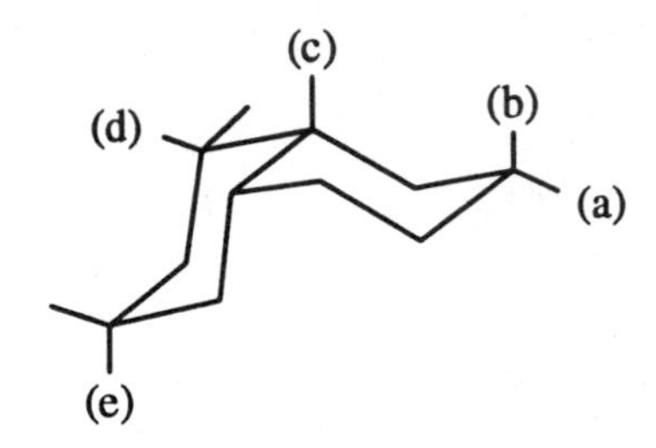

4.15 Draw a formula for the most stable chair conformation of each of the following substituted cyclohexanes:

(a) H, $CH(CH_3)_2$, CH_3, H

(b) H, CH_3, H_3C, $CH(CH_3)_2$

4.16 Draw equations illustrating the chair ⇄ boat ⇄ chair equilibria for each of the following substituted cyclohexanes.

(a) Cl, CH_2CH_3

(b) CO_2H, OH

(c) 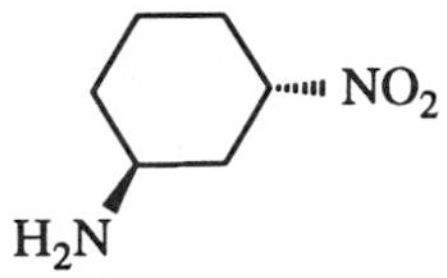 H_2N (wedge) / NO_2 (hashed) on cyclohexane

(d) 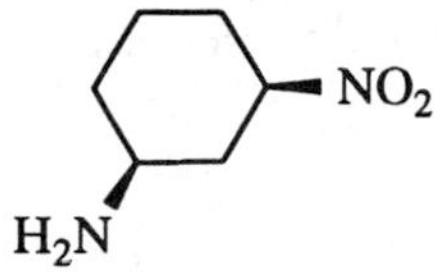 H_2N (wedge) / NO_2 (wedge) on cyclohexane

4.17 Draw formulas for all the compounds with the following molecular formulas that have at least one chiral carbon.

(a) C_7H_{16} (b) $C_4H_8Br_2$ (c) $C_3H_6O_3$

4.18 Star the chiral carbons, if any, in the following structures.

(a) 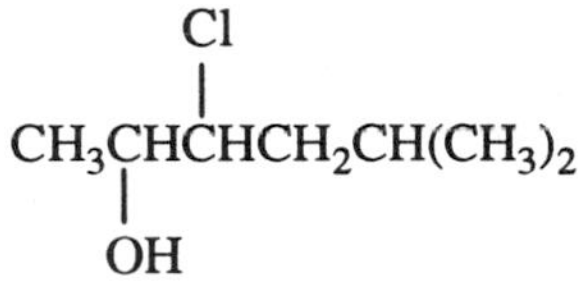 $CH_3CH(OH)CH(Cl)CH_2CH(CH_3)_2$

(b) 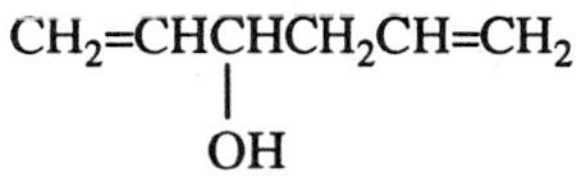 $CH_2{=}CHCH(OH)CH_2CH{=}CH_2$

(c) 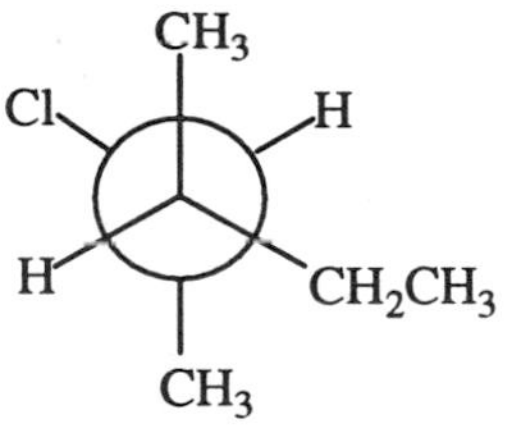 Newman projection: front carbon: CH_3, H, CH_2CH_3; back carbon: Cl, H, CH_3

(d) 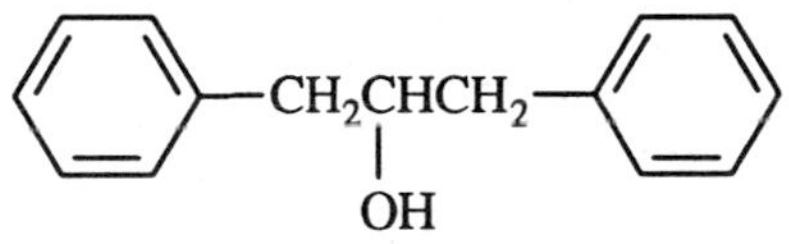 C_6H_5–$CH_2CH(OH)CH_2$–C_6H_5

4.19 Which of the following pairs of structures represent a pair of enantiomers and which represent the same compound?

(a) 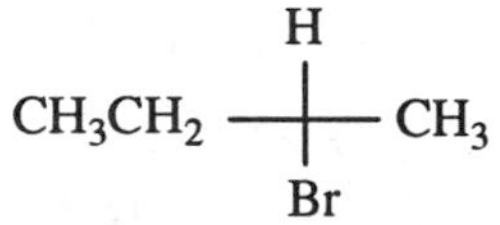 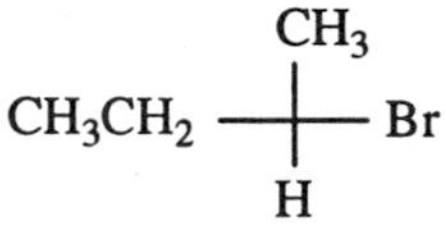 Fischer projections: top H, left CH_3CH_2, right CH_3, bottom Br; and top CH_3, left CH_3CH_2, right Br, bottom H

(b) 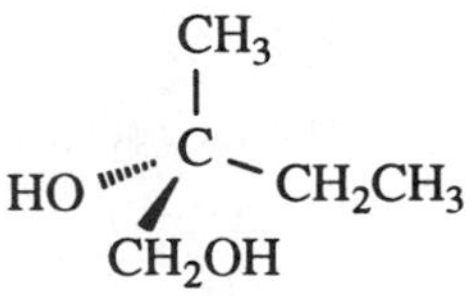 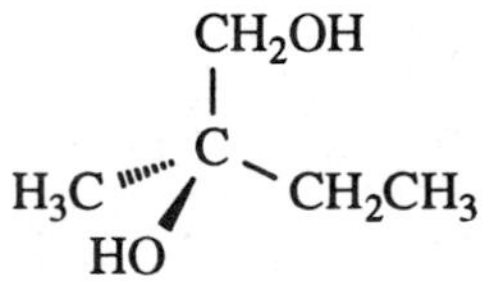 C with CH_3 (top), HO (hashed), CH_2OH (wedge), CH_2CH_3; and C with CH_2OH (top), H_3C (hashed), HO (wedge), CH_2CH_3

(c) 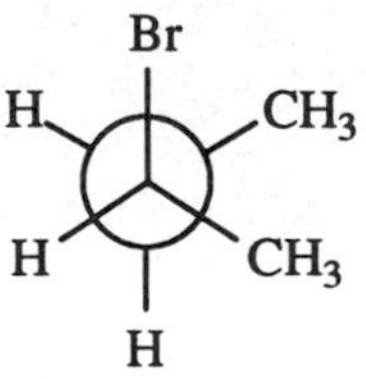 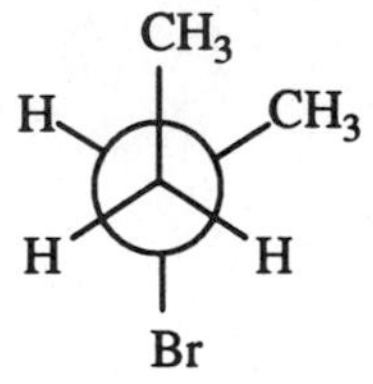 Newman projections: Br, H, CH_3, H, CH_3, H; CH_3, H, CH_3, H, H, Br

(d) 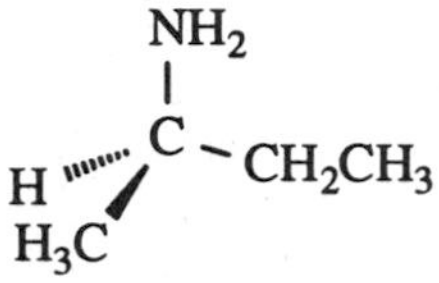 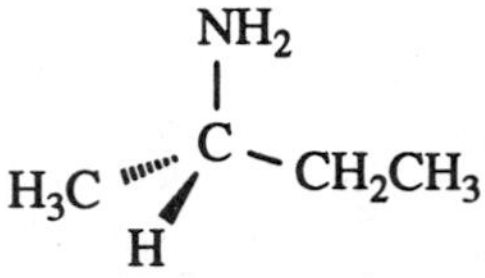 NH_2, H, C, CH_2CH_3, H_3C; NH_2, H_3C, C, CH_2CH_3, H

(e) 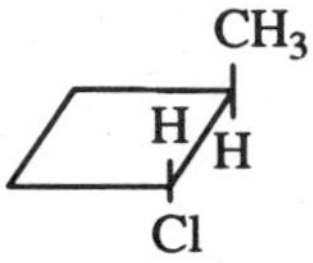 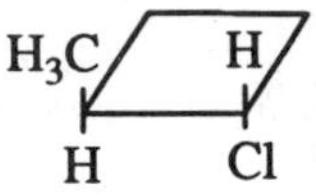 CH_3, H, H, Cl; H_3C, H, H, Cl

(f) 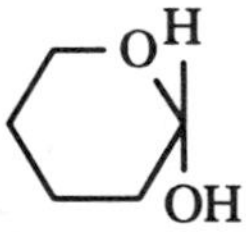 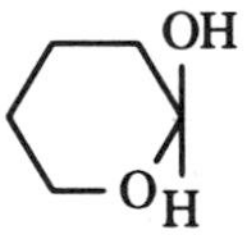 OH, OH; OH, OH

4.20 Draw Fischer projections for the enantiomers of the following structures:

(a) $CH_3CH_2CH(CH_3)CO_2H$

(b) $CH_3CH(NH_2)CO_2H$

(c) CH_2OCH_3–CHOH–CH_2OH

(d) $HOC(=O)-C(CH_3)(CH_2Cl)-C(=O)NH_2$

4.21 What is the maximum number of stereoisomers for each of the following structures:

(a) 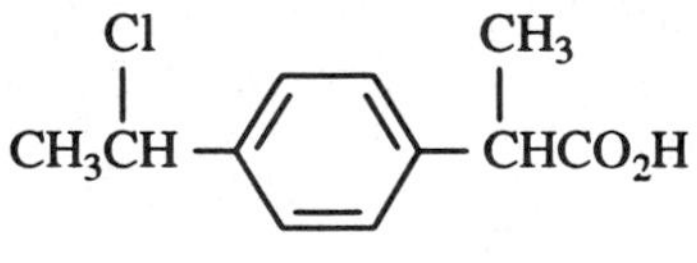 $CH_3CH(Cl)-C_6H_4-CH(CH_3)CO_2H$

(b) $CH_3CH(OH)CHClCH_3$

(c) 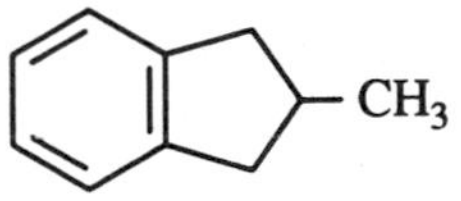 [structure: 2-methylindane, with CH_3 substituent]

(d) 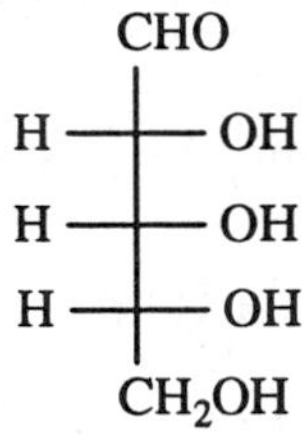 Fischer projection:

$$\begin{array}{c} CHO \\ H - \!\!\!\!\!\!\! + \!\!\!\!\!\!\! - OH \\ H - \!\!\!\!\!\!\! + \!\!\!\!\!\!\! - OH \\ H - \!\!\!\!\!\!\! + \!\!\!\!\!\!\! - OH \\ CH_2OH \end{array}$$

4.22 Calculate the specific rotations:

(a) A solution containing 1.3 g of a compound in 25 mL of ethanol shows a rotation of +2.5° in a 2.0-dm tube at 25°.

(b) A solution containing 0.80 g of a compound in 10.0 mL of solvent gives an observed rotation of -1.60° in a 5.0-cm cell at 20°.

4.23 Draw formulas for the (*R*) enantiomer of all optically active isomers of $C_5H_{11}Br$ containing only one chiral carbon.

4.24 Assign each chiral carbon in the following structures as (*R*) or (*S*).

(a) 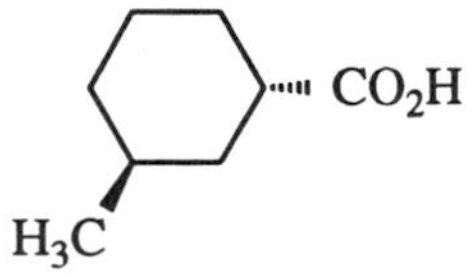 [structure: cyclohexane ring with CO_2H (hashed bond) and H_3C (wedged bond)]

(b) 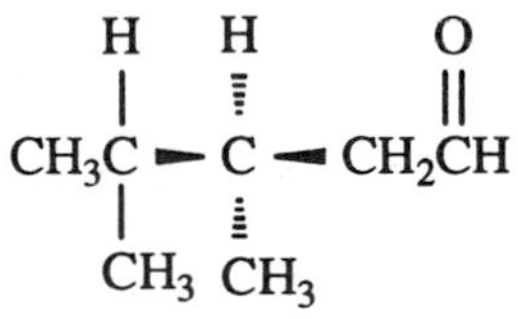 [structure: $CH_3C(H)(CH_3)$—$C(H)(CH_3)$—CH_2CH(=O), with H above and CH_3 below each carbon; labels: H, H, O, CH_3C, C, CH_2CH, CH_3, CH_3]

(c) [structure: H ····· C ····· $OCCH_2CH_3$ (with C=O), C bonded above to CH_2OCCH_3 (with C=O) and below to CH_2-$OCCH_2CH_2CH_3$ (with C=O)]

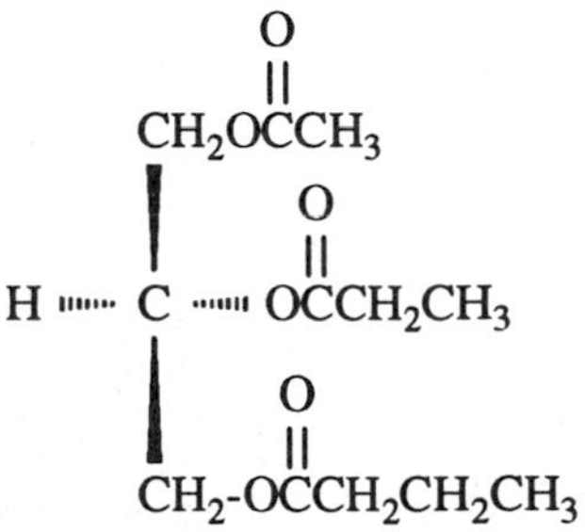

(d) 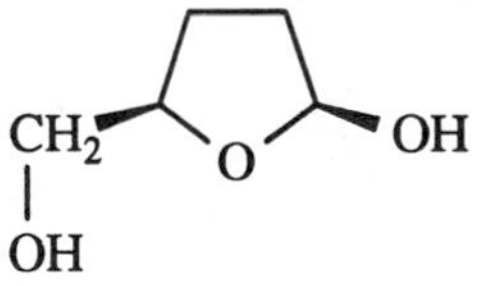 [structure: tetrahydrofuran ring (O) bearing CH_2—OH (wedged) and OH (wedged)]

4.25 Draw Fischer projections for the following names:

(a) (2*R*,3*S*,4*R*,5*R*)-2,3,4,5,6-pentahydroxyhexanal

(b) (2*S*,3*S*)-2,3-dichlorobutane

(c) (2*S*,3*S*)-2,3-pentanediol

(d) (*S*)-1-bromo-3-chloro-2-methylpropane

4.26 Redraw each of the following structures in a Fischer projection as a *meso* form:

(a) 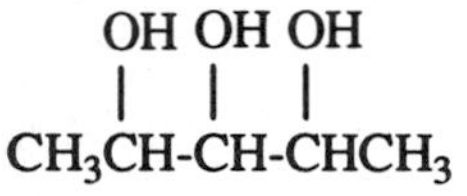 $CH_3CH(OH)-CH(OH)-CH(OH)CH_3$
(Draw two *meso* forms.)

(b) 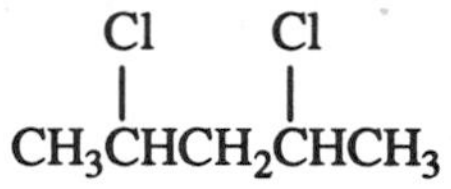 $CH_3CH(Cl)CH_2CH(Cl)CH_3$

(c) 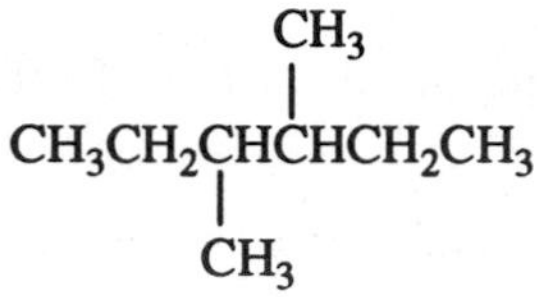 $CH_3CH_2CH(CH_3)CH(CH_3)CH_2CH_3$

(d) 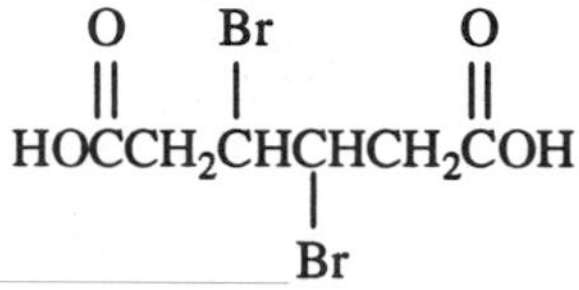 $HOC(=O)CH_2CH(Br)CH(Br)CH_2C(=O)OH$

4.27 Draw and enantiomer and a diastereomer for each of the following structures.

(a) 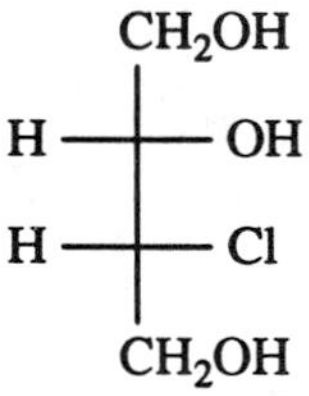 Fischer projection: CH_2OH (top); H—C—OH; H—C—Cl; CH_2OH (bottom)

(b) Ring structure with CH_2OH, H, O, H, OH

(c) Cyclohexane chair with Cl, H, OH, H

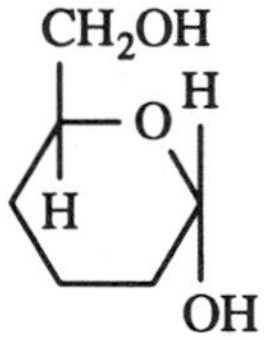

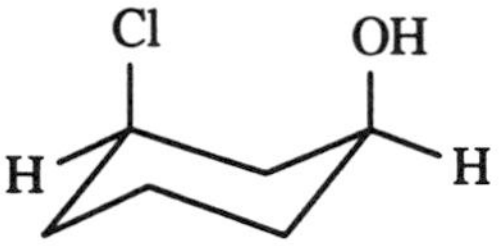

(d) 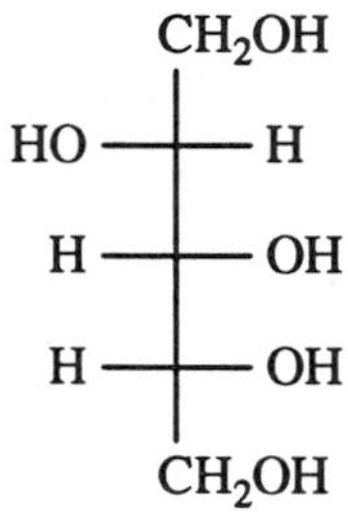 Fischer projection: CH_2OH (top); HO—C—H; H—C—OH; H—C—OH; CH_2OH (bottom)

4.28 Designate each of the chiral carbons in the following structures as (*R*) or (*S*):

(a) 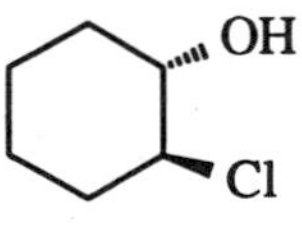 Cyclohexane with OH and Cl

(b) 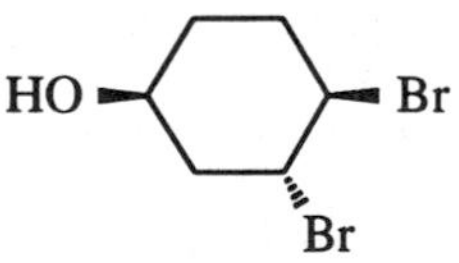 HO, Br, Br substituted cyclohexane

(c) 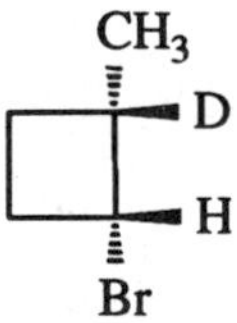 Cyclobutane with CH_3, D, H, Br

4.29 Draw flat-projection formulas for the two enantiomers of each of the following structures:

(a) 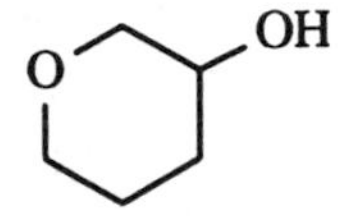 (b) 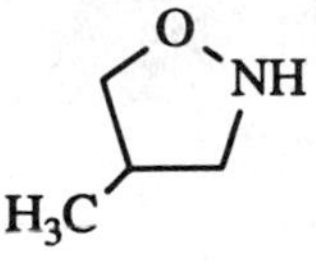 H_3C (c)

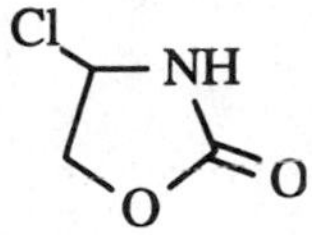

4.30 Star the chiral carbons, if any, in the following structures.

(a) 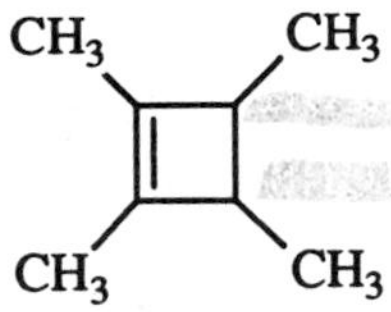 (b)

(c) 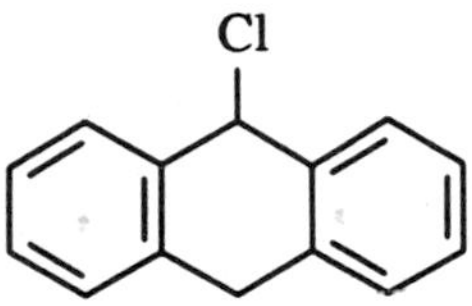 (d)

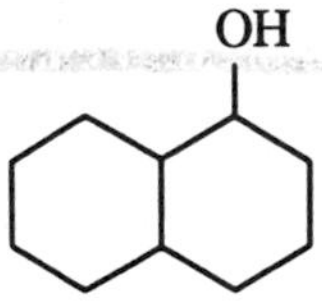

Alkyl Halides; Substitution and Elimination Reactions

Some Important Features

Alkyl halides, but not most aryl halides, can undergo substitution or elimination when treated with a nucleophile or base. Primary alkyl halides and, to an extent, secondary alkyl halides generally react by an S_N2 path. (Tertiary alkyl halides do not react by an S_N2 path.) Inversion of configuration at the reacting carbon is observed in a typical S_N2 reaction if that carbon is chiral.

$$S_N2\text{: } Nu{:}^- + \underset{\substack{1^\circ \\ (\text{and } 2^\circ)}}{R{-}\ddot{\underset{..}{X}}{:}} \longrightarrow Nu{-}R + {:}\ddot{\underset{..}{X}}{:}^-$$

When treated with a very weak nucleophile, such as H_2O or ROH, 2° or 3° alkyl halides can react by an S_N1 path. Primary alkyl halides (unless allylic or benzylic) do not react by this pathway.

$$S_N1\text{: } \underset{\substack{3^\circ \\ (\text{and } 2^\circ)}}{R{-}\ddot{\underset{..}{X}}{:}} \xrightarrow{-{:}\ddot{\underset{..}{X}}{:}^-} [R^+] \xrightarrow{H_2\ddot{O}{:}} [R\overset{+}{\underset{..}{O}}H_2] \xrightarrow{-H^+} R\ddot{\underset{..}{O}}H$$

Carbocation reactions have some disadvantages:

1. Racemization usually occurs because a carbocation is planar and therefore achiral.

2. Rearrangement can occur if a more stable carbocation can be formed by a 1,2-shift.

$$(CH_3)_2\overset{H}{\overset{|}{C}}-\overset{+}{C}HCH_3 \longrightarrow (CH_3)_2\overset{+}{C}-\overset{H}{\overset{|}{C}}HCH_3$$

2° → 3°

3. S_N1 reactions are accompanied by elimination (by an E1 path); therefore, mixtures of products are observed.

$$\text{E1: } (CH_3)_2\overset{+}{C}-\overset{H}{\overset{|}{C}}HCH_3 \xrightarrow{H_2\ddot{O}:} (CH_3)_2C{=}CHCH_3 + H_3\overset{+}{O}:$$

a carbocation

Tertiary (and 2°) alkyl halides undergo elimination by an E2 path when they are heated with a strong base. The principal alkene product of either an E1 or E2 reaction is usually the more substituted, *trans* alkene.

on adjacent carbons

$$CH_3\overset{H}{\overset{|}{C}}H\overset{Br}{\overset{|}{C}}HCH_3 \xrightarrow[-Br^-,\ -H_2O]{OH^-} \underset{trans}{(CH_3)HC{=}CH(CH_3)}$$

is favored over

$$\underset{cis}{(CH_3)HC{=}CH(CH_3)} \quad \text{or} \quad \underset{\text{less substituted}}{CH_3CH_2CH{=}CH_2}$$

An E2 reaction proceeds by *anti*-elimination of H^+ and X^- on adjacent carbons. If the halogen is attached to a ring carbon, the H and X must be *trans* and diaxial in order to be eliminated readily.

(axial :X: and axial H on adjacent ring carbons; $^-:\ddot{O}H$ removes H) → cyclohexene $+ H_2\ddot{O} + :\ddot{X}:^-$

A properly selected alkyl halide yields a single geometric isomer in an E2 reaction.

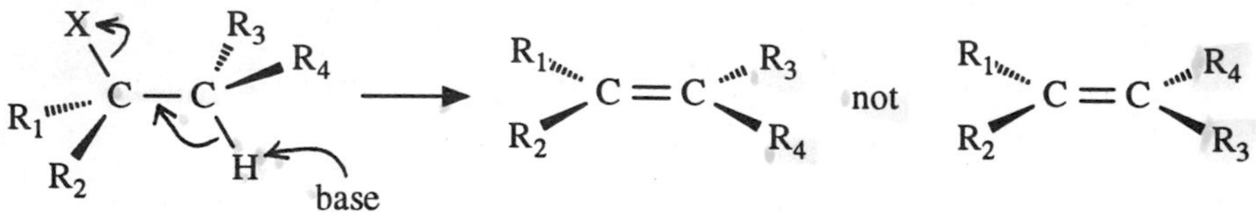

the only reactive conformation

Every reaction step proceeds through a transition state. A lower-energy (more stable) transition state means a lower E_{act} and therefore a faster reaction.

In a typical S_N1 or E1 reaction, the rate-determining (slower) step is the formation of the carbocation. Any feature that stabilizes the carbocation also stabilizes the transition state and leads to a faster reaction.

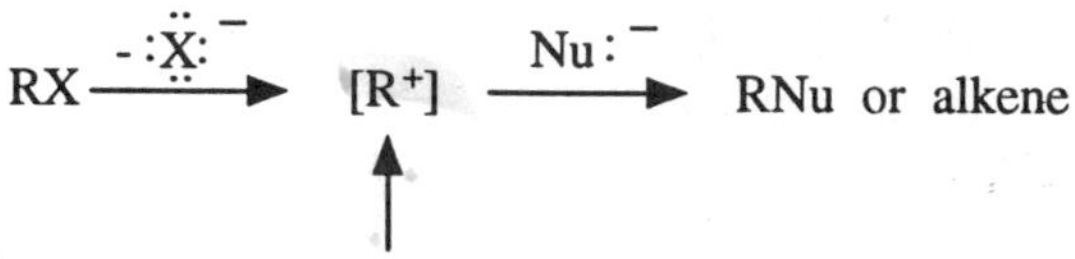

$$RX \xrightarrow{-\,:\ddot{X}:^-} [R^+] \xrightarrow{Nu:^-} \text{RNu or alkene}$$

The rate is increased if R^+ is stabilized by the inductive effect, resonance, or a highly polar solvent.

Other important topics discussed in this chapter are how a carbocation is stabilized; first- and second-order rates; the kinetic isotope effect; and the effects of steric hindrance on the reactivity of alkyl halides and on the products of E2 reactions (Saytzeff *versus* Hofmann products). You may also want to review the rules for writing resonance structures in Chapter 2.

Reminders

S_N2 reactions are the only group of reactions discussed in this chapter that can yield optically active products from optically active reactants; S_N1, E1, and E2 reactions all lead to racemization or loss of chirality at the reacting carbon.

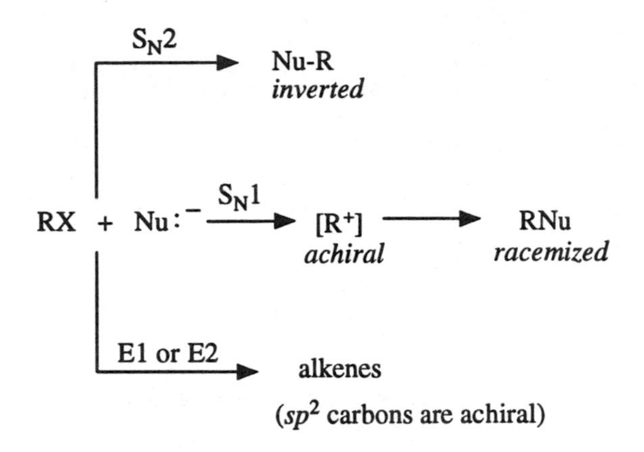

Remember the following general rules of reactivity:

S_N2:	CH_3X > 1° RX > 2° RX
S_N1 (or E1):	3° RX > 2° RX
E2:	3° RX > 2° RX

Strong Nu:⁻, high concentration: S_N2 for 1° and 2° RX
Weak Nu:⁻ (H_2O, *etc.*): S_N1 for 2° and 3° RX
Strong base: E2 for 2° and 3° RX

Additional Drill Problems

5.1 Draw structures for the following trivial names:

(a) allyl chloride (b) vinyl bromide
(c) benzyl bromide (d) butyl iodide
(e) isobutyl bromide (f) cyclohexyl bromide

5.2 Write a trivial name for each of the following organohalogen compounds:

(a) $(CH_3)_3CBr$ (b) $(CH_3)_2CHCl$ (c) CH_2Cl_2
(d) $CH_3CHICH_2CH_3$ (e) $C_6H_5CH_2Cl$ (f) $CH_2{=}CHCH_2Br$

5.3 Write IUPAC names for all isomers with the following molecular formulas, and star all chiral carbons.

(a) C_3H_7F (b) C_4H_9Br

5.4 Draw the structures of the following organohalogen compounds:

(a) (4*S*)-4-bromo-2-chloro-(*E*)-2-pentene

(b) (*R*)-2-bromobutanoic acid

(c) (1*E*,3*E*)-1-chloro-1,3-pentadiene

(d) (2*R*,3*R*,4*S*)-2,3-dibromo-4-chlorohexane

5.7 Identify each halogen in the following compounds as alkyl, aryl, or vinylic:

(a) 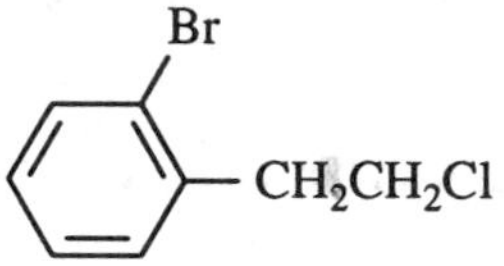 Br; $-CH_2CH_2Cl$

(b) 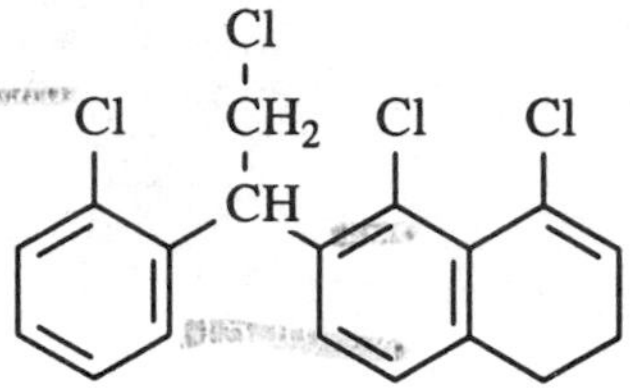 Cl; Cl; CH_2; CH; Cl; Cl

(c) $BrCH{=}CHCH{=}CHBr$

(d) 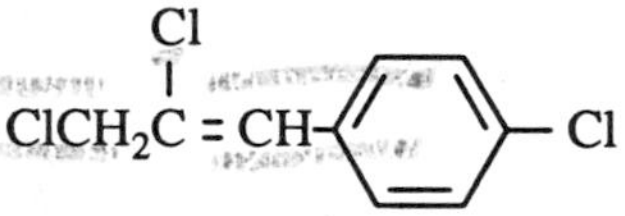 $ClCH_2C(Cl){=}CH-$ ... $-Cl$

5.6 Classify each of the following halides as primary (1°), secondary (2°), or tertiary (3°):

(a) 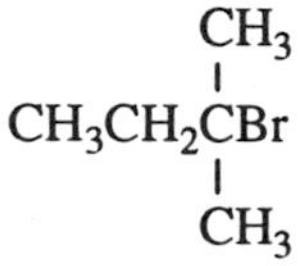 $CH_3CH_2C(CH_3)_2Br$

(b) 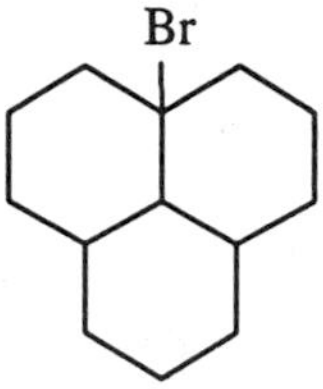 Br

(c) 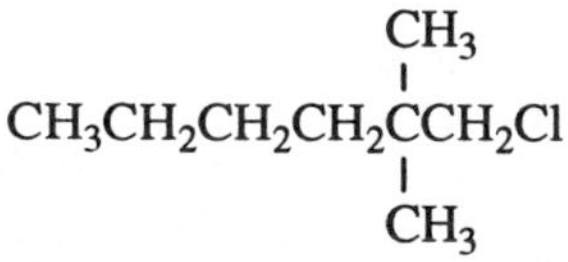 $CH_3CH_2CH_2CH_2C(CH_3)_2CH_2Cl$

5.7 Identify each halogen in the following compounds as primary (1°), secondary (2°), tertiary (3°), benzylic, allylic, or as a combination of these classifications:

(a) 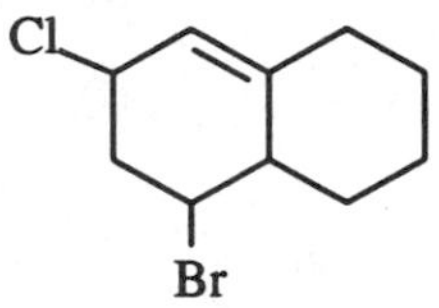 [structure: bicyclic ring with Cl and Br substituents]

Cl

Br

(b) $CH_3\overset{Br}{\overset{|}{C}}HCH_2CH{=}CH_2$

(c) 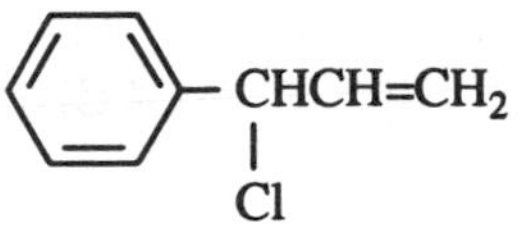 $C_6H_5\underset{Cl}{\underset{|}{C}}HCH{=}CH_2$

5.8 In each of the following reactions, designate whether the methoxide ion (CH_3O^-) is acting as a nucleophile or as a base. Explain.

(a) $CH_3O^- + H_2O \rightleftharpoons CH_3OH + {}^-OH$

(b) $CH_3O^- + CH_3I \longrightarrow CH_3OCH_3 + I^-$

(c) $CH_3O^- + CH_3\overset{Br}{\overset{|}{C}}HCH_3 \longrightarrow CH_3OH + CH_3CH{=}CH_2$

(d) $CH_3O^- + CH_3\overset{Br}{\overset{|}{C}}HCH_3 \longrightarrow CH_3\overset{OCH_3}{\overset{|}{C}}HCH_3 + Br^-$

5.9 Circle the leaving group in the starting alkyl halide in each of the following substitution or elimination reactions:

(a) $CH_3\underset{CH_3}{\underset{|}{\overset{Br}{\overset{|}{C}}}}CH_3 + CH_3O^- \longrightarrow CH_2{=}\underset{CH_3}{\underset{|}{C}}CH_3 + CH_3OH + Br^-$

(b) $C_6H_5CH_2Cl + CH_3S^- \longrightarrow C_6H_5CH_2SCH_3 + Cl^-$

5.10 Using 1-bromopentane, write an equation for a substitution reaction with each of the following reagents:

(a) $Na^+ \ ^-OH$ (b) $Na^+ \ ^-SCH_2CH_3$ (c) $Na^+ \ ^-CN$

(d) $Na^+ \ I^-$ (e) $Na^+ \ ^-SCN$ (f) $HC{\equiv}C^- \ Na^+$

5.11 Insert the correct nucleophile in each of the following equations.

(a) 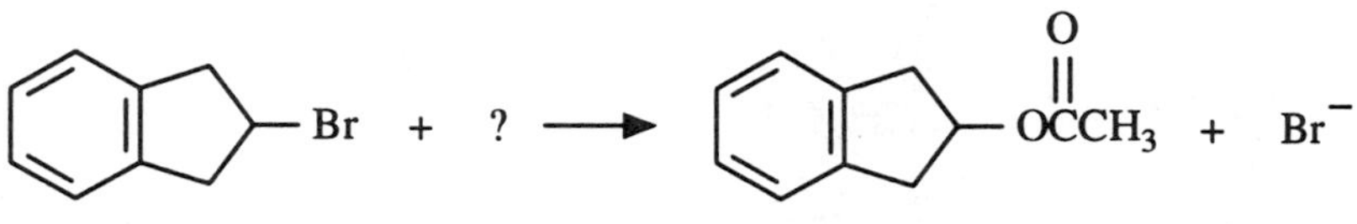 (2-bromoindane) + ? → (indan-2-yl $OCCH_3$, with C=O) + Br^-

(b) $C_6H_5CH_2CH_2Br$ + ? → $C_6H_5CH_2CH_2CN$ + Br^-

(c) $(CH_3)_3C-Cl$ + ? → $(CH_3)_3C-OH$ + Cl^-

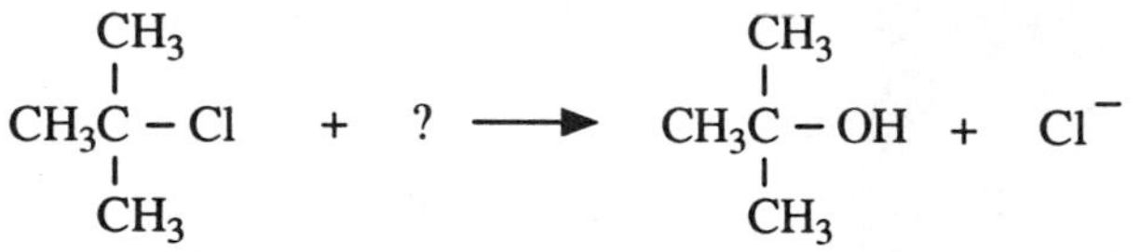

(d) 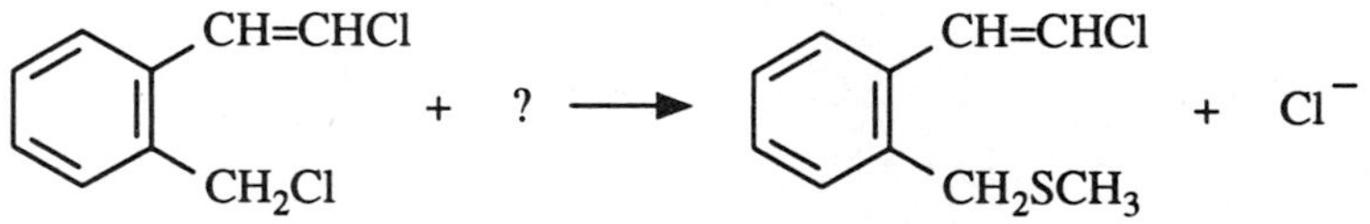 (benzene ring with CH=CHCl and CH_2Cl) + ? → (benzene ring with CH=CHCl and CH_2SCH_3) + Cl^-

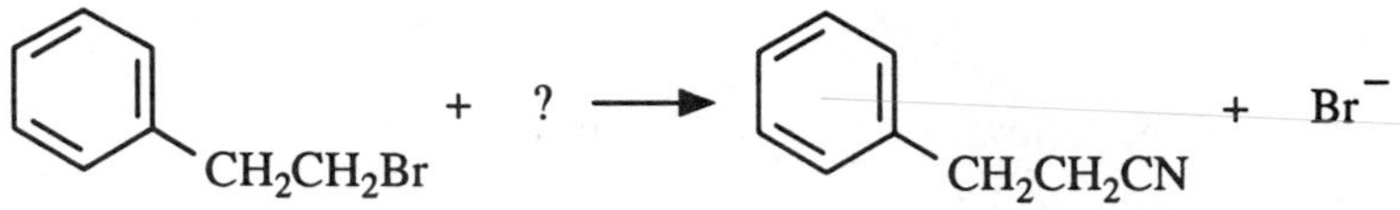

5.12 Write formulas of the S_N2 transition states and products for the following reactions. (Show unshared pairs of valence electrons as dots.)

(a) 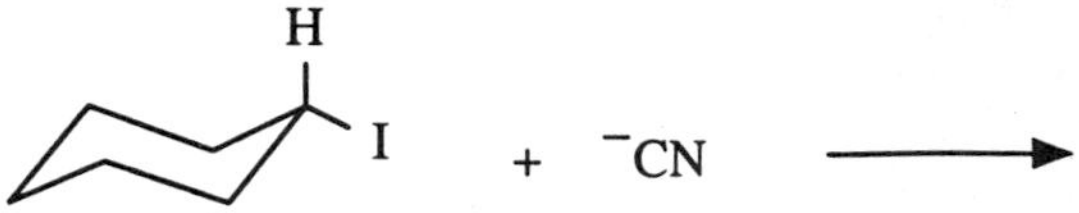 (cyclohexane chair with H and I) + ^-CN →

(b) 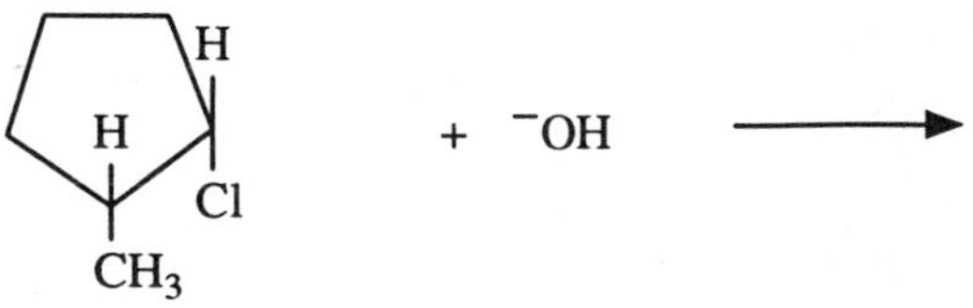 (cyclopentane with H, Cl and H, CH_3) + ^-OH →

(c) 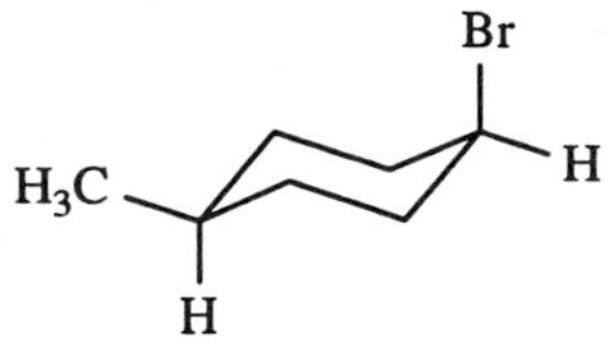 H_3C (cyclohexane chair, H; Br, H) + $^-SCH_3$ ⟶

(d) 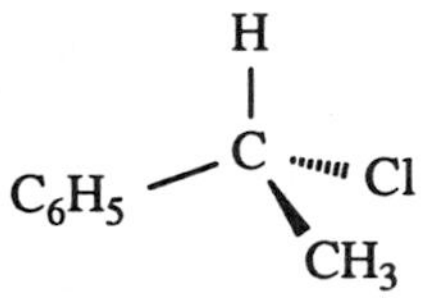 C_6H_5–C(H)(Cl)(CH_3) + $^-OCH_2CH_3$ ⟶

(e) 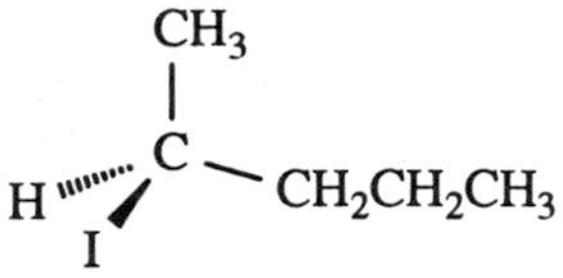 H, I, C(CH_3)–$CH_2CH_2CH_3$ + ^-CN ⟶

5.13 Each of the following reactions can be expected to yield an S_N2 product. Write the formula for the expected product along with its stereochemistry, where appropriate.

(a) 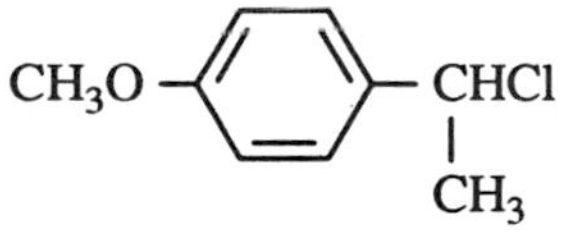 CH_3O–C_6H_4–$CHCl$(CH_3) + NaN_3 ⟶

(b) 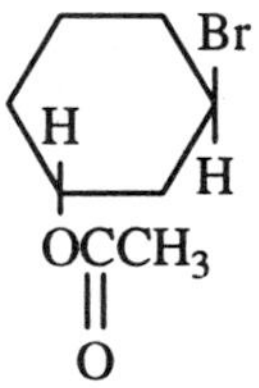 (cyclohexane with Br, H, H, $OCCH_3$ (C=O)) + $NaOCH_2CH_3$ ⟶

(c) 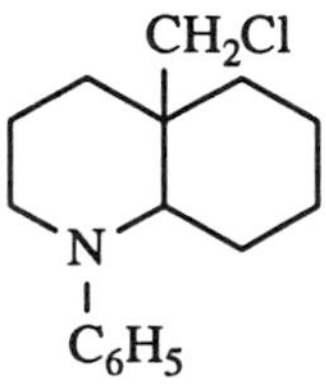 (CH_2Cl, N–C_6H_5 decahydroquinoline) + $CH_3S^-\,Na^+$ ⟶

(d) CH_3Br + $CH_3C{\equiv}C^-\,Na^+$ ⟶

(e) 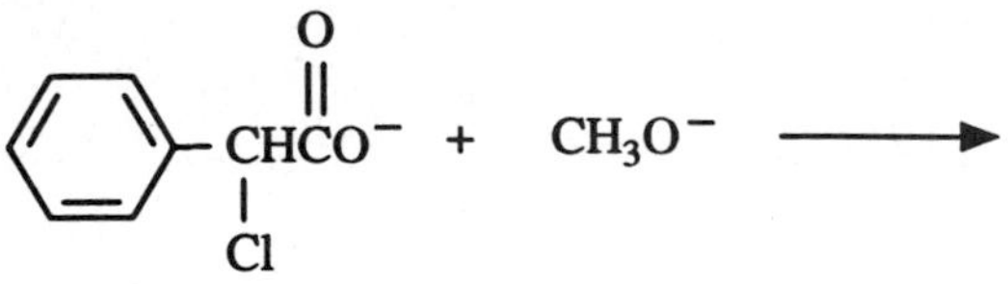

5.14 Which member of each of the following pairs would undergo the faster S_N2 reaction?

(a) 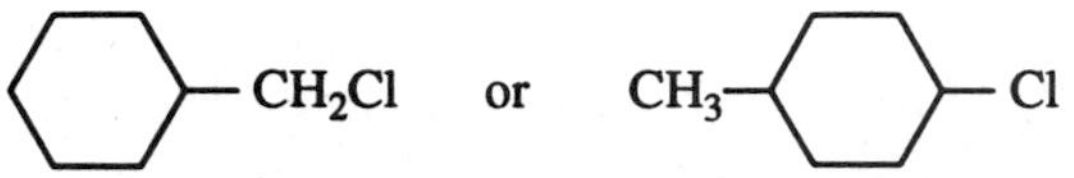

(b) 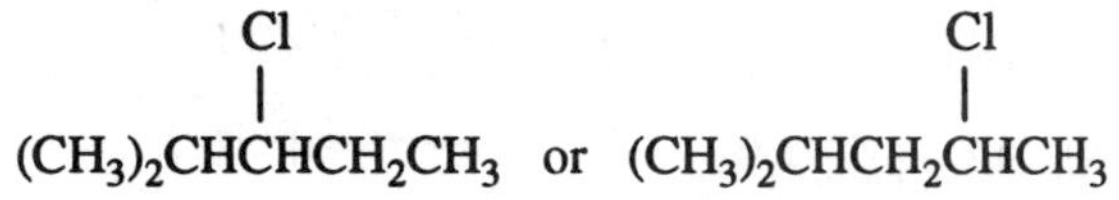

(c) 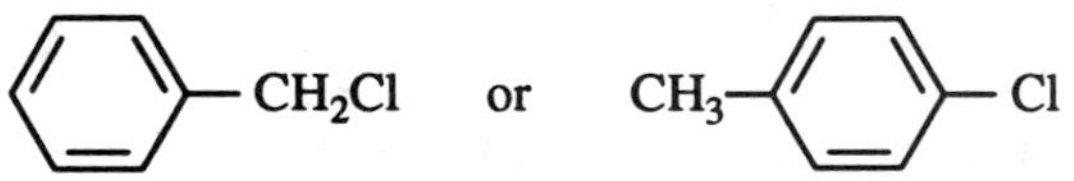

5.15 Predict the stereochemistry of the product of each of the following S_N2 reactions:

(a) 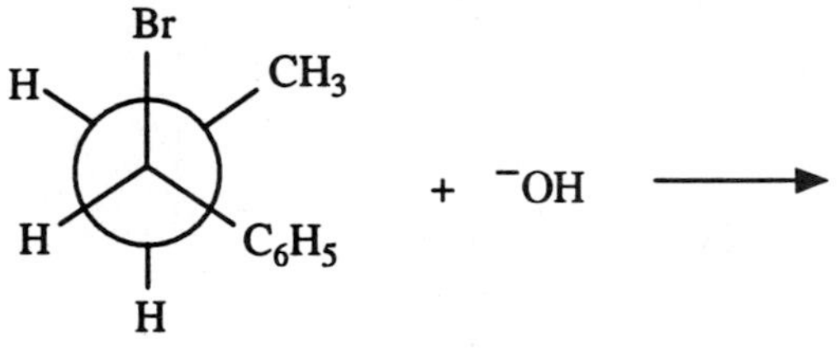

(b) 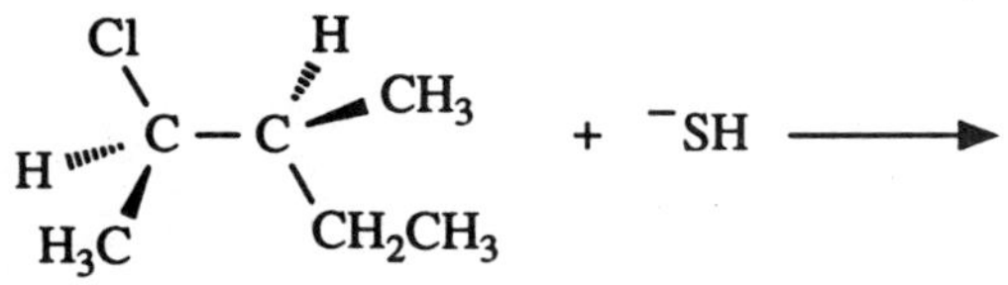

(c) 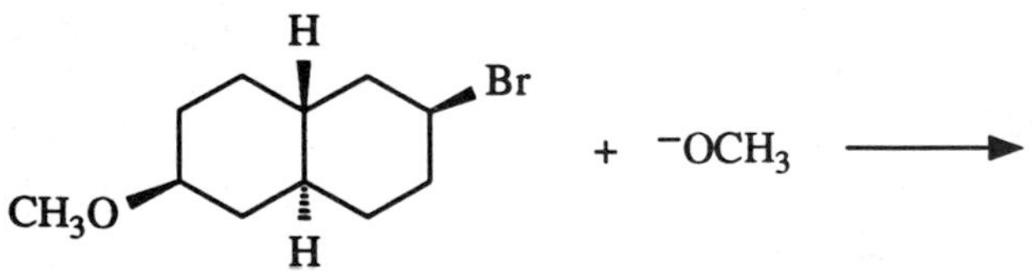

(d) 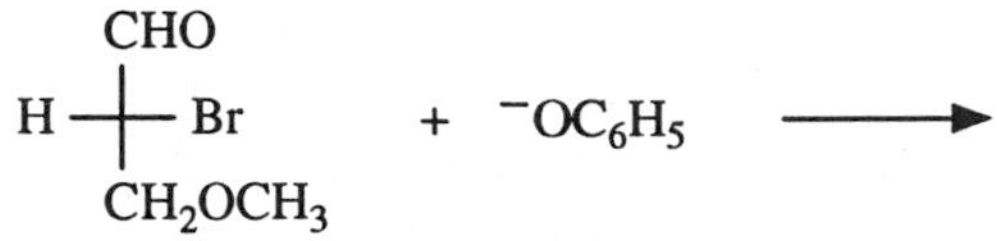 Fischer projection: CHO on top, H on left, Br on right, CH_2OCH_3 on bottom $+\ ^{-}OC_6H_5 \longrightarrow$

5.16 Write equations (including electron dots) to show each step in the solvolysis of 2-iodo-2-methylbutane with methanol. Label which step determines the overall rate of reaction.

5.17 Which member of the following pairs of structures would you expect to undergo an S_N1 reaction at the faster rate. Explain.

(a) $CH_3CH_2CH_2Cl$ or CH_3CHCH_3 (with Cl on the middle C)

(b) $(CH_3)_3CBr$ or $(CH_3CH_2)_2CHCH_2Br$

(c) 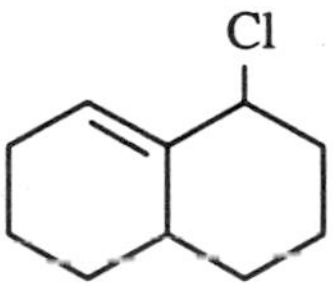 Cl-substituted bicyclic structure or Cl-substituted bicyclic structure

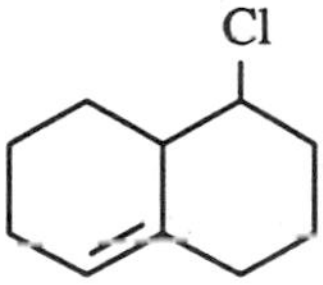

5.18 Draw the structures of the intermediate carbocation and the expected S_N1 product, with stereochemistry where appropriate, for each of the following reactions.

(a) 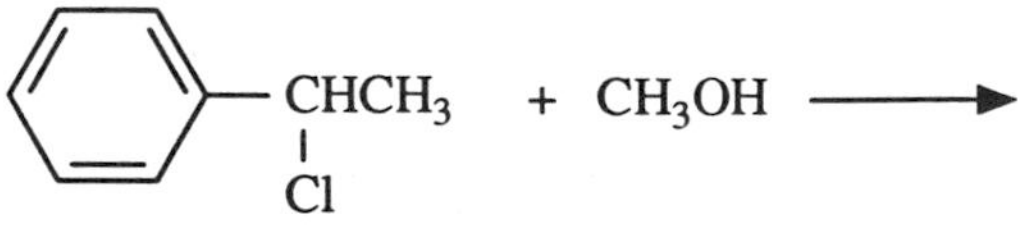 C_6H_5–$CHCH_3$ (with Cl on CH) $+ CH_3OH \longrightarrow$

(b) 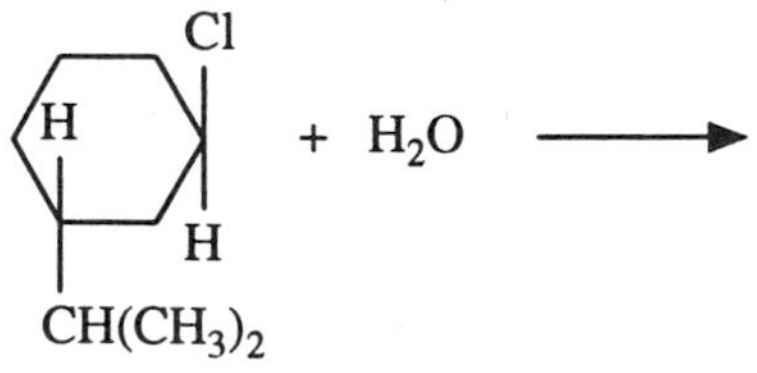 Cyclohexane chair with Cl, H, H and $CH(CH_3)_2$ substituents $+ H_2O \longrightarrow$

5.19 List the following structures in order of increasing S_N1 reactivity (least reactive first). Explain your answer.

(a) C_6H_5–CH(Cl)CH_2CH_3 (b) C_6H_5–CH_2CH(Cl)CH_3

(c) C_6H_5–$CH_2CH_2CH_2Cl$ (d) Cl–C_6H_4–$CH_2CH_2CH_3$

5.20 Write resonance structures for the following two carbocations. Which of the two would you expect to be the more stable? Why?

(a) CH_3O–C_6H_4–$\overset{+}{C}H_2$ (b) O_2N–C_6H_4–$\overset{+}{C}H_2$

5.21 Write equations for steps in the mechanism of the elimination (E1) reaction of 2-chloro-2-methylbutane in water. Show only the most stable alkene product in your answer.

5.22 Write flow equations showing the steps in the mechanisms, including structures of the transition states, for the following E2 elimination reactions.

(a) $CH_3CH_2CH_2CH(CH_3)Cl + {}^{-}OH \xrightarrow{\text{heat}}$

(b) $C_6H_5CH_2CH(Br)CH_3 + {}^{-}OH \xrightarrow{\text{heat}}$

5.23 Show all possible alkene products in the following elimination reactions and state which would be expected to be formed in the greatest yield.

(a) C_6H_5–$CH_2CH_2C(Br)(H)$-$CH_3 + {}^{-}OH \xrightarrow{\text{heat}}$

(b) $CH_3CH_2CH_2C(Br)(CH_3) - CH_3 + {}^{-}OH \xrightarrow{\text{heat}}$

5.24 Rank the following alkyl halides in order of increasing E2 rate (slowest first) when heated with KOH in ethanol. Give reasons for your ranking.

(a) $(CH_3CH_2)_2CHBr$ (b) $CH_3CH_2CH_2Br$ (c) $(CH_3CH_2)_3CBr$

5.25 Rank the following alkenes in order of stability (least stable first). Give reasons for your ranking.

(a) *cis*-$CH_3CH_2CH{=}CHCH_3$ (b) $CH_3CH_2CH_2CH{=}CH_2$

(c) *trans*-$CH_3CH_2CH{=}CHCH_3$ (d) *trans*-$C_6H_5CH{=}CHCH_3$

5.26 Predict the dehydrohalogenation product(s) when the following compounds are heated with base.

(a) 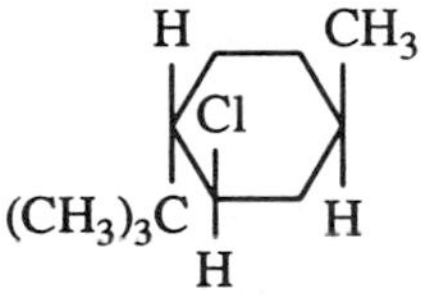 H CH$_3$ Cl (CH$_3$)$_3$C H H

(b) 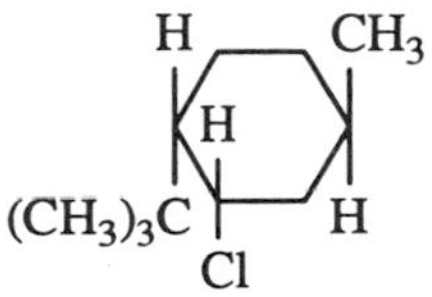 H CH$_3$ H (CH$_3$)$_3$C H Cl

5.27 For each of the following elimination reactions, write a formula for an alkyl halide that would yield the alkene given.

(a) ? + $(CH_3)_3CO^-$ ⟶ (cyclopentene)

(b) ? + $(CH_3)_3CO^-$ ⟶ C_6H_5–CH=CH–C_6H_5

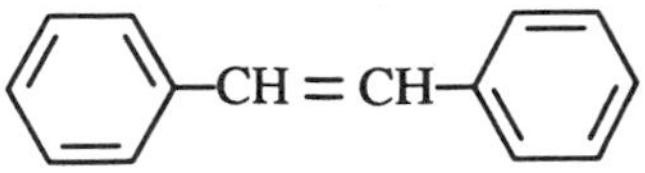

(c) ? + $(CH_3)_3CO^-$ ⟶ $CH_3CH{=}C(CH_2CH_3)_2$

(d) ? + $(CH_3)_3CO^-$ ⟶ (cyclohexene ring bearing H$_3$C, CH$_3$, H$_3$C, CH$_3$, H$_3$C, CH$_3$)

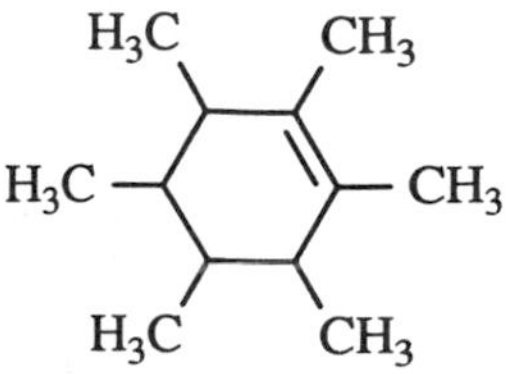

5.28 Write formulas for the substitution and elimination products for the following equations:

(a) 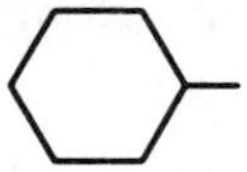 C_6H_{11}–CH_2Br + $^{-}OH \longrightarrow$ (cyclohexyl–CH_2Br)

(b) $CH_3CH_2CH_2CH_2I + (CH_3)_3CO^{-} \longrightarrow$

(c) $CH_3C(CH_3)_2Br + {}^{-}SCH_3 \longrightarrow$

(d) $C_6H_5CHClCH_3 + {}^{-}CN \longrightarrow$

5.29 Predict the Hofmann product of the following elimination and explain why this product would predominate.

$$CH_3CH_2CH_2CBr(CH_3)_2 + {}^{-}OC(CH_3)_3 \xrightarrow{\text{heat}}$$

5.30 Predict the order of increasing reactivity with a nucleophile in an S_N2 reaction (least reactive first):

(a) $CH_3CH_2CH_2CH_2Cl$

(b) $CH_3CH_2CH_2CH_2I$

(c) $(CH_3)_3CCH_2Cl$

(d) $CH_3CH_2C(CH_3)_2Cl$

5.31 Predict the principal product(s), if any, of the following reactions. State by which mechanistic path each product would be formed.

(a) $CH_3CH_2Br + Na^{+}\ {}^{-}OCH_3 \xrightarrow{CH_3OH}$

(b) $(CH_3)_3CBr + Na^{+}\ {}^{-}OCH_2CH_3 \xrightarrow[\text{warm}]{CH_3CH_2OH}$

(c) $(CH_3)_3CBr + K^{+}\ {}^{-}OH \xrightarrow[\text{warm}]{CH_3CH_2OH}$

(d) $C_6H_5Br + K^{+}\ {}^{-}OH \xrightarrow[\text{warm}]{CH_3CH_2OH}$

(e) $CH_3CH(Br)CH_3 + Na^+ \ ^-CN \longrightarrow$

(f) CH_3CH_2I + excess $NH_3 \longrightarrow$

(g) $CH_3CH(Br)CH_3 + Na^+ \ ^-OC(CH_3)_3 \xrightarrow[\text{warm}]{(CH_3)_3COH}$

(h)  1-bromo-1-methylcyclohexane (cyclohexane ring bearing CH_3 and Br on the same carbon) $+ \ H_2O \longrightarrow$

(i) $CH_2{=}CHCH_2Cl + K^+ \ ^-OH \xrightarrow{H_2O}$

(j) $(CH_3)_2C(Cl)CH_2CH_3 + Na^+ \ ^-OCH_2CH_3 \xrightarrow{CH_3CH_2OH}$

(k) $C_6H_5C(Cl)(CH_3)CH_2CH_3$ + dilute $Na^+ \ ^-CN + CH_3CH_2OH \longrightarrow$

(l) $(CH_3)_2CHCH(Br)CH_2CH_3 + K^+ \ ^-OH \xrightarrow[\text{heat}]{CH_3CH_2OH}$

5.32 Predict the principal product(s), if any, of the following reactions. State by which mechanistic path each product would be formed.

(a) $C_6H_5O^- \ Na^+$ + cyclopropyl$-CH_2CH_2Br \longrightarrow$

(b) CH_3I + piperidine (six-membered ring with NH) $\longrightarrow$

(c) 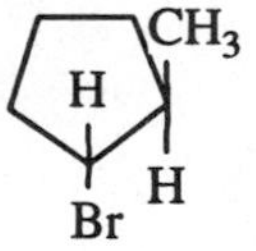 [cyclopentane ring with H, CH_3, H, Br substituents] + $C_6H_5S^- Na^+$ ⟶

(d) $(CH_3)_3CCHClCH_2Cl$ + $K^+ \ ^-OH$ $\xrightarrow[\text{heat}]{CH_3CH_2OH}$

(e) 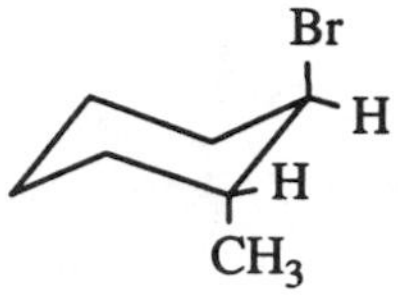 [cyclohexane chair with Br, H, H, CH_3 substituents] + $K^+ \ ^-OH$ $\xrightarrow[\text{heat}]{CH_3CH_2OH}$

(f) $(CH_3)_2CCl\overset{O}{\overset{\|}{C}}H$ + $^-CH(\overset{O}{\overset{\|}{C}}OCH_2CH_3)_2$ ⟶

5.33 Predict which reaction in each pair would give the larger percent of substitution as opposed to elimination. Give reasons for your answers.

(a) treatment of $CH_3CHBrCH_3$ with I^- or with Cl^-

(b) treatment of $CH_3CHBrCH_3$ with $^-SCH_3$ or with $^-OCH_3$

(c) heating $(CH_3)_2CHCBr(CH_3)_2$ or $CH_3CH_2CH_2CBr(CH_3)_2$ in aqueous ethanol

(d) heating $C_6H_5CH_2CH_2Br$ or $CH_3CH_2CH_2Br$ with $NaOCH_2CH_3$ in ethanol

(e) heating bromocyclohexane with $NaOCH_3$ in methanol or with $NaOC(CH_3)_3$ in *tert*-butyl alcohol

5.34 Write equations to show how you would make the following conversions:

(a) $C_6H_5OCH(CH_3)_2$ from 2-bromopropane

(b) $(CH_3)_2CHOCH_3$ from 2-bromopropane

(c) $(CH_3CH_2CH_2)_2O$ from 1-chloropropane

(d) CH_3OH from iodomethane

(e) CH_3CN from iodomethane

(f) *trans*-2-butene from 2-bromobutane

(g) $(CH_3CH_2CH_2CH_2)_2S$ from 1-bromobutane

(h) cyclohexene from bromocyclohexane

5.35 Suggest syntheses for the following compounds:

(a) $CH_3\overset{\overset{\large O}{\|}}{C}OCH_2C_6H_5$ from an organohalogen compound

(b) $C_6H_5CH_2I$ from a bromo compound

(c) $C_6H_5NHCH_3$ from iodomethane

(d) $C \equiv N$ / $CH_2N(CH_3)_2$ from a bromo compound

(e) S from 1,4-dibromobutane

6

Free-Radical Reactions

Some Important Features

A radical, or free radical, is an atom or group of atoms containing one or more unpaired electrons; most radicals are highly reactive intermediates. An alkyl radical is achiral around the reacting carbon.

planar around C•,

therefore achiral

Free-radical reactions occur stepwise: (1) initiation (initial generation of radicals); (2) propagation (reaction of radicals yielding new reactive radicals); (3) termination (destruction of radicals, often by the joining together of two radicals or by the formation of a relatively stable, nonreactive radical).

Different hydrogen atoms in a structure may be replaced by halogen atoms at different rates. If the intermediate radical is stabilized, then reaction rate at that position is enhanced.

CH_4 RCH_3 R_2CH_2 R_3CH allylic and benzylic

→

Increasing rate of free-radical reaction and increasing stability of radical when H• is removed

Because of the enhanced reactivity of some H atoms, we can often predict the principal free-radical product.

$$\overset{1^\circ\ 2^\circ}{CH_3CH_2CH_3} \xrightarrow[-H\bullet]{X_2,\ h\nu} [CH_3\dot{C}HCH_3 \text{ favored over } CH_3CH_2\dot{C}H_2] \xrightarrow[X_2]{-X\bullet} CH_3\underset{}{\overset{X}{C}}HCH_3$$

The reagent also affects the product ratio. For example, a less reactive radical reagent is more selective.

$$CH_3CH_2CH_3 \xrightarrow[\text{(more reactive)}]{Cl_2,\ h\nu} CH_3CHClCH_3 + CH_3CH_2CH_2Cl$$

$$CH_3CH_2CH_3 \xrightarrow[\text{(less reactive)}]{Br_2,\ h\nu} CH_3CHBrCH_3$$

The mechanisms of free-radical reactions are discussed in the text, including the reasons for the reactivity of different H atoms in a structure, the different reactivities of halogenating agents, and halogenation with NBS. Pyrolysis, autooxidation and polymerization are also covered in this chapter. Free-radical initiators and inhibitors are mentioned.

Reminders

Be sure you know how to write resonance structures for free-radical intermediates. In these cases, we shift only one electron (not two, as in carbocation intermediates).

allylic H

$$R_2C{=}CHCH_2R \xrightarrow[\text{-HX}]{X\cdot} [R_2C{=}CH{-}\dot{C}HR \longleftrightarrow R_2\dot{C}{-}CH{=}CHR]$$

an allylic radical

a resonance-stabilized intermediate

$$R\ddot{\underset{\cdot\cdot}{O}}CH_2R \xrightarrow[\text{-HX}]{X\cdot} [R\ddot{\underset{\cdot\cdot}{O}}{-}\dot{C}HR \longleftrightarrow R\dot{\underset{\cdot\cdot}{O}}{=}CHR]$$

an ether *a resonance-stabilized intermediate*

Additional Drill Problems

6.1 Write Lewis formulas for the following radicals:

(a) $\cdot CH_3$ (b) $CH_3\dot{C}HCH_3$ (c) $CH_3O\cdot$

(d) cyclohexenyl radical (ring structure) (e) $CH_3\overset{O}{\overset{\|}{C}}O\cdot$ (f) $C_6H_5{-}O\cdot$ (ring structure)

6.2 Using electron dots, write equations showing homolytic cleavage at the indicated bonds:

(a) $Br\text{-}Br$ (b) $HO\text{-}OH$ (c) $C_6H_5\overset{O}{\overset{\|}{C}}O\text{—}O\overset{O}{\overset{\|}{C}}C_6H_5$

(d) $CH_3CH_2\text{-}H$ (e) $C_6H_5CH_2\text{-}H$ (f) $CH_2{=}CHCH_2\text{—}H$

6.3 Write equations for each step in the monochlorination of toluene. (Use curved arrows to show electron movement.) Label each step as initiation, propagation, or termination.

6.4 Write formulas for all possible monochlorination products that could be obtained from the following hydrocarbons:

(a) $(CH_3)_3CH$ (b) $(CH_3)_3CCH_2CH(CH_3)_2$ (c) methylcyclohexane (C_6H_{11}—CH_3)

6.5 Circle and label each 1°, 2°, and 3° hydrogen in the following hydrocarbons:

(a) ethylcyclohexane (C_6H_{11}—CH_2CH_3) (b) decalin (two fused cyclohexane rings)

(c) $CH_3C(CH_3)_2CH_2CH(CH_3)CH_3$

6.6 Which of the following compounds would react more rapidly with Br_2 in sunlight? Explain.

(a) $CH_3CH_2C(CH_2CH_3)_2(CH_2)_3CH_3$ (b) $CH_3CH_2CH(CH_2CH_3)CH(CH_2CH_3)(CH_2)_2CH_3$

6.7 For each of the following hydrocarbons, give the ratio of the 1°: 2°: 3° hydrogens.

(a) 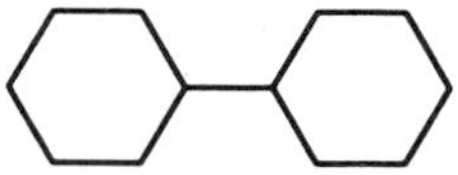 [bicyclohexyl structure]

(b) 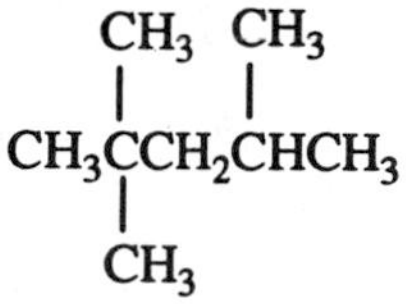 $CH_3C(CH_3)_2CH_2CH(CH_3)CH_3$

(c) 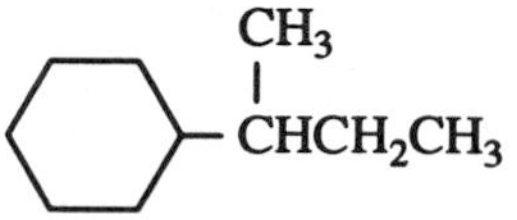 [cyclohexyl–$CH(CH_3)CH_2CH_3$ structure]

(d) $CH_3CH_2CH_2CH_3$

6.8 Which of the two indicated hydrogen atoms would be abstracted at a more rapid rate by Br_2 in the presence of light?

(a) 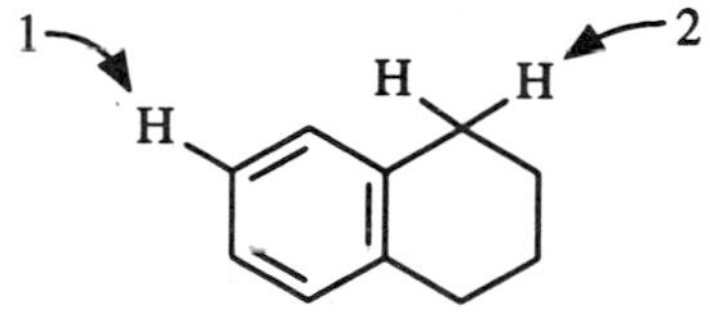 [tetralin structure with aromatic H labeled 1 and benzylic H labeled 2]

(b) 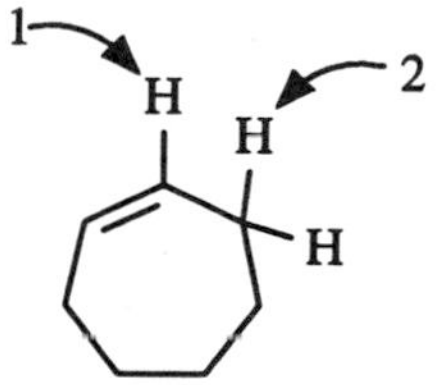 [cycloheptene structure with vinylic H labeled 1 and allylic H labeled 2]

6.9 Given the following bond dissociation energies, calculate the heats of reaction for the following reactions. Indicate whether each reaction is exothermic or endothermic.

$C_6H_5CH_2$-H, 66 kcal/mol RS-H, 89 kcal/mol CH_3-H, 104 kcal/mol

(a) $\bullet CH_3 + RSH \longrightarrow CH_4 + RS\bullet$

(b) $\bullet CH_3 + C_6H_5CH_3 \longrightarrow CH_4 + C_6H_5\dot{C}H_2$

6.10 In the monochlorination of methylbutane, the relative rates of hydrogen abstraction are as follows:

3° RH: 4.5 2° RH: 3.0 1° RH: 1.0

What percentages of four isomeric monochlorination products would be obtained?

6.11 Write resonance arrows (using curved arrows) for the following radicals

(a) $\dot{C}H_2$-CH=CH-CH_2-CH(=O)

(b)

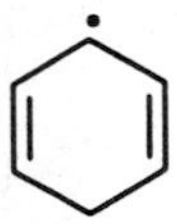

6.12 Predict the product(s) when each of the compounds is treated with *N*-bromosuccinimide and a peroxide catalyst:

(a) 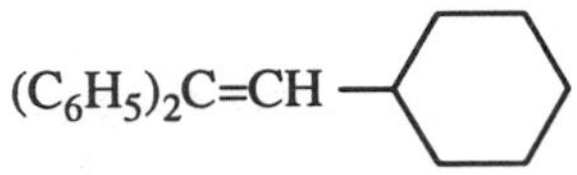 $(C_6H_5)_2C$=CH–

(b)

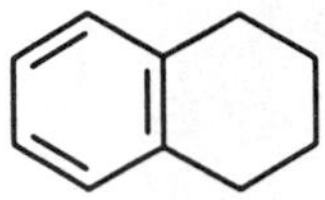

(c) *cis*-2-heptene

6.13 Predict the major products.

(a) CH_3-CH=CHC(=O)OCH_3 + NBS $\xrightarrow{h\nu}$

(b) 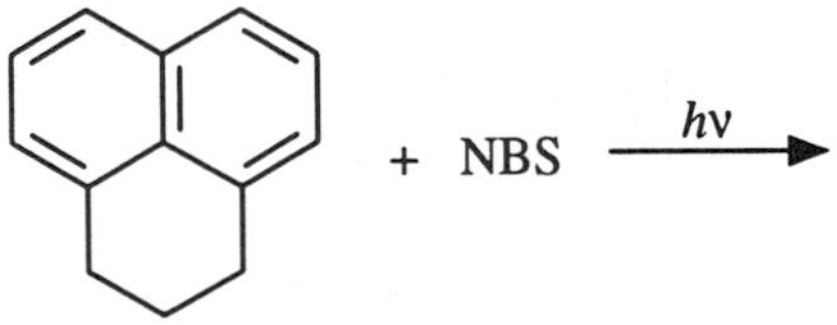 + NBS $\xrightarrow{h\nu}$

(c) 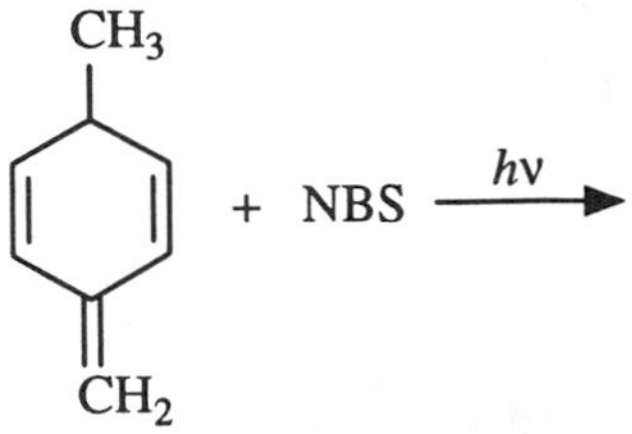 CH_3 ... CH_2 + NBS $\xrightarrow{h\nu}$

(d) 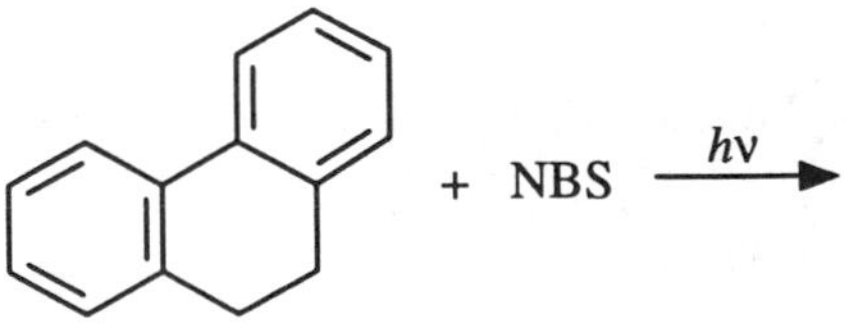 + NBS $\xrightarrow{h\nu}$

6.14 Write a free-radical mechanism that can be used to explain the following reaction.

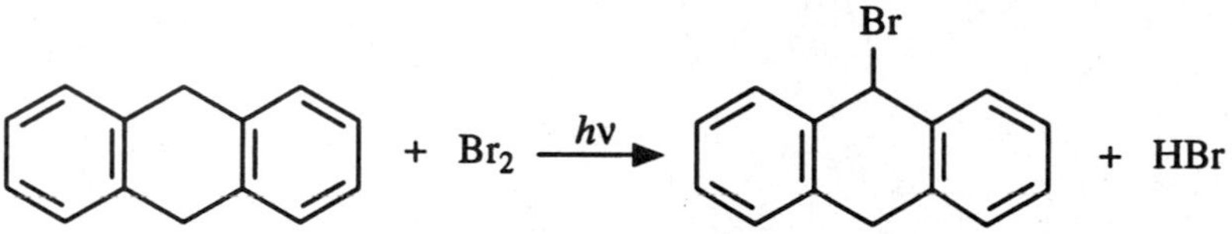

$$\text{(9,10-dihydroanthracene)} + Br_2 \xrightarrow{h\nu} \text{(9-bromo-9,10-dihydroanthracene)} + HBr$$

6.15 Write equations for auto-oxidation reactions of the following compounds:

(a) $CH_3OCH_2CH_3$

(b) $CH_3CH(CH_3)CH{=}O$ (CH_3CHCH with O double-bonded to the terminal C and CH_3 on the middle C)

(c) 1,4-dioxane (six-membered ring with two O atoms)

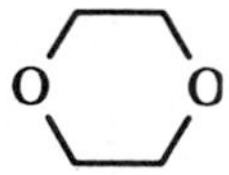

6.16 Write equations for the polymerization of the following alkenes in the presence of a peroxide catalyst.

(a) $CH_2{=}CHBr$

(b) $CH_2{=}CHOCH_3$

(c) 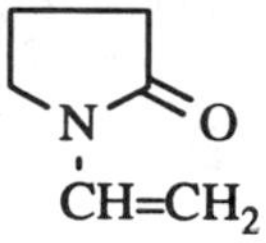 N-vinylpyrrolidone (five-membered lactam ring, N bearing $CH{=}CH_2$, C=O)

(d) $(CH_3)_3Si\text{-}CH{=}CH_2$

Alcohols

Some Important Features

One of the most important organometallic compounds is the *Grignard reagent*, RMgX, which can be used to prepare complex alcohols. The Grignard reagent contains a carbanion-like organic group that reacts with a compound containing a partially positive hydrogen (as in H_2O or ROH) or a partially positive carbon (as in a carbonyl compound).

$$\overset{\delta-}{R}-\overset{\delta+}{MgX} + \overset{\delta+}{H}-\overset{\delta-}{\ddot{O}H} \longrightarrow RH + HO-MgX$$

$$R-MgX + \overset{\delta-}{:\ddot{O}}{=}\overset{\delta+}{C}< \longrightarrow -\underset{R}{\overset{:\ddot{O}:^{-}\ {}^{+}MgX}{C}}- \xrightarrow{H_2O,\ H^+} -\underset{R}{\overset{:\ddot{O}H}{C}}- + Mg^{2+} + X^{-}$$

The reaction of a Grignard reagent with an aldehyde or ketone, followed by hydrolysis, yields an alcohol. The type of alcohol produced depends on what was originally bonded to the carbonyl group.

$$-\overset{O}{\overset{\|}{C}}- \xrightarrow[(2)\ H_2O,\ H^+]{(1)\ RMgX} -\underset{R}{\overset{OH}{C}}-$$

Because the O in $R\ddot{O}H$ is electronegative and has unshared electrons, alcohols form hydrogen bonds with compounds that contain partially positive hydrogens.

$$\mathrm{R\ddot{O}H\cdots\cdots:\underset{..}{O}(H)R}$$

Alcohols, which become protonated in acidic solution, undergo substitution reactions with HX (3° ROH > 2° ROH > 1° ROH). Secondary and tertiary alcohols react by an S_N1 path with HX; therefore, PBr_3 and $SOCl_2$ are often used as reagents for converting alcohols to alkyl halides. Both PBr_3 and $SOCl_2$ result in stereospecific reactions without rearrangement.

$$R_2CH-\ddot{O}H \underset{}{\overset{H^+}{\rightleftharpoons}} R_2CH-\overset{H}{\overset{|+}{O}}H \overset{-H_2\ddot{O}:}{\rightleftharpoons} [R_2CH^+] \xrightarrow{:\ddot{X}:^-} R_2CH-\ddot{X}:$$

(1° and 2° alcohols require $ZnCl_2$ with HCl.)

$$R_2CHOH \xrightarrow{SOCl_2} R_2CHCl$$

$$R_2CHOH \xrightarrow{PBr_3} R_2CHBr$$

Dehydration of alcohols yields Saytzeff products. (Again, the relative rates are 3° ROH > 2° ROH > 1° ROH.)

Alcohols act as acids when treated with extremely strong bases (such as RMgX), with alkali metals (Na or K), or with alkali metal hydrides (such as NaH).

Oxidation of alcohols can yield carboxylic acids, ketones, or aldehydes.

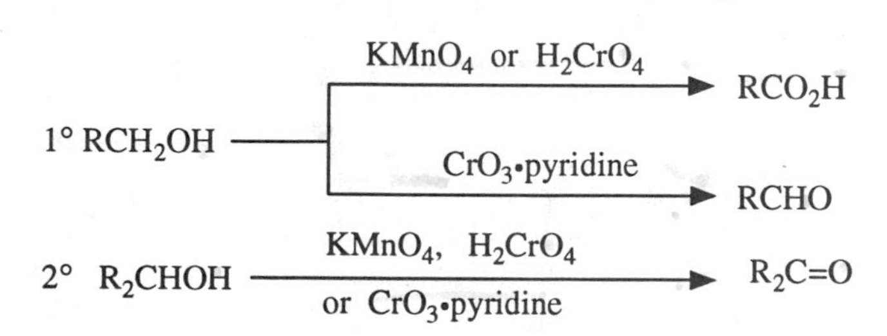

Reminders

A Grignard reagent is a very strong base that reacts with acidic hydrogens. Acid-base reactions, such as $RMgX + H_2O \rightarrow RH + HOMgX$, are fast compared to most organic reactions, such as that of RMgX with a carbonyl compound.

A common student error is to forget to look for acidic hydrogens in reactions involving Grignard reagents:

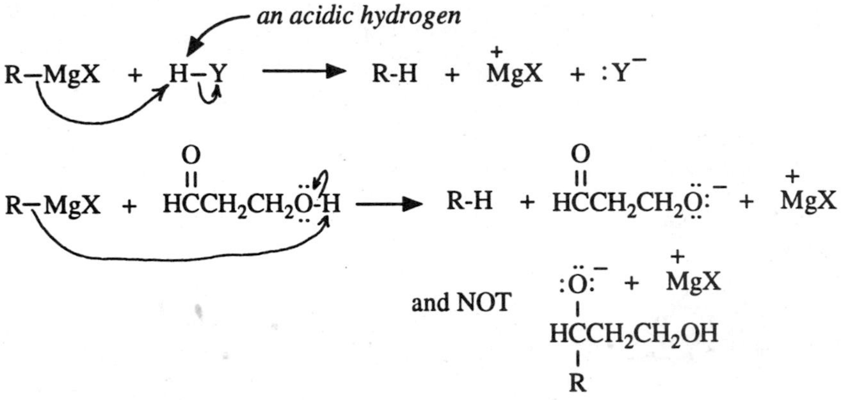

an acidic hydrogen

$$R{-}MgX + H{-}Y \longrightarrow R{-}H + \overset{+}{M}gX + :Y^{-}$$

$$R{-}MgX + HC(=O)CH_2CH_2\ddot{O}{-}H \longrightarrow R{-}H + HC(=O)CH_2CH_2\ddot{O}:^{-} + \overset{+}{M}gX$$

and NOT

$$HC(R)(:\ddot{O}:^{-})CH_2CH_2OH + \overset{+}{M}gX$$

- In acidic solution, an alcohol is protonated and can be attacked by a nucleophile such as Br^{-}.
- In dilute base, alcohols are not protonated and undergo no appreciable reaction.
- Reaction of a 3° or 2° alcohol in acidic solution yields *racemic* or *achiral* products.

Additional Drill Problems

7.1 Classify each hydroxyl group in the following structures as 1°, 2°, or 3°:

(a) 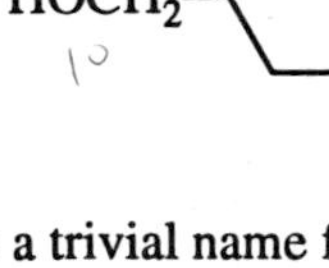 $CH_3C(CH_3)_2CH_2OH$ (1°)

(b) 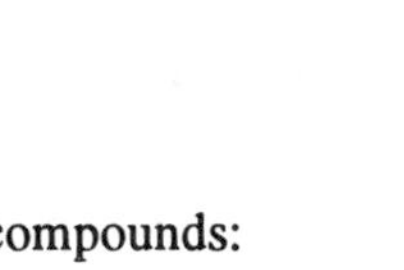 $CH_3CH(OH){-}C(CH_3)_2OH$ (2°, 3°)

(c) 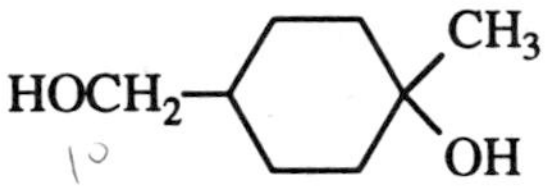 $HOCH_2$–cyclohexyl–$C(CH_3)OH$ (1°, 3°)

7.2 Write a trivial name for each of the following compounds:

(a) $CH_3CH_2CH_2OH$ — propyl alcohols

(b) $CH_2{=}CHCH_2OH$ — allyl alcohol

(c) 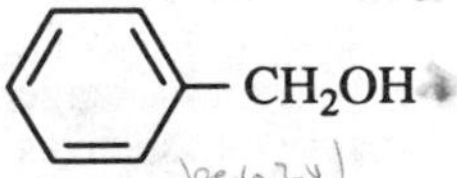 $C_6H_5CH_2OH$ — benzyl alcohol

(d) 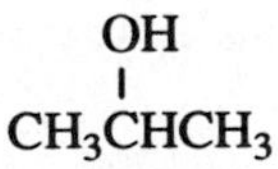 $CH_3CH(OH)CH_3$ — isopropyl alcohol

(e) 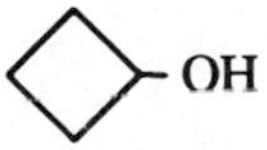 cyclobutyl–OH — cyclobutyl alcohol

(f) 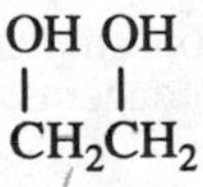 $CH_2(OH)CH_2(OH)$ — ethylene glycol

7.3 (1) For each of the following structures, which functional group has the highest nomenclature priority? (2) Write the IUPAC name of each structure.

(a) 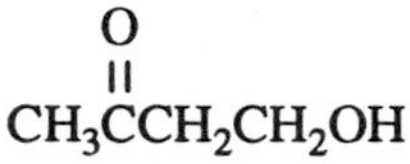 $CH_3C(=O)CH_2CH_2OH$

(b) 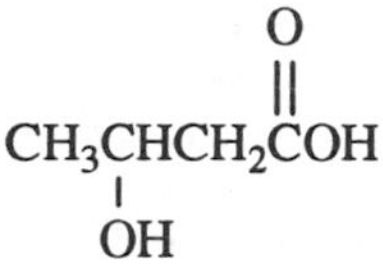 $CH_3CH(OH)CH_2C(=O)OH$

(c) 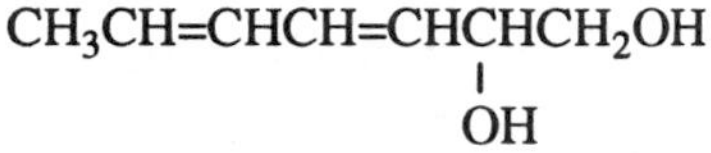 $CH_3CH{=}CHCH{=}CHCH(OH)CH_2OH$

(d) 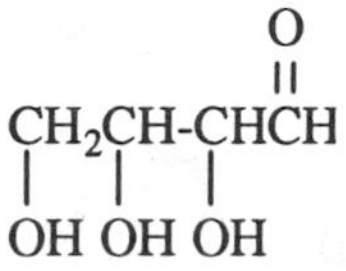 $CH_2(OH)CH(OH)\text{-}CH(OH)C(=O)H$

7.4 Classify the hydroxyl groups in the following structures as 1°, 2°, 3°, allylic, benzylic, or as a combination of these classifications:

(a) $HOCH_2CH{=}CHCH_2CH_2OH$

(b) 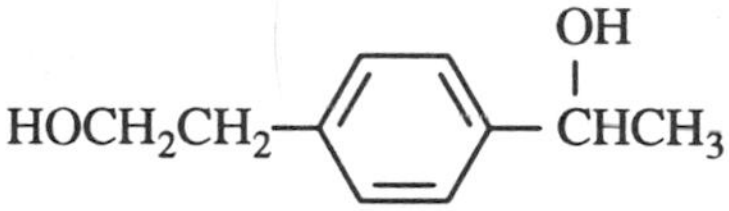 $HOCH_2CH_2\text{–}C_6H_4\text{–}CH(OH)CH_3$

(c) [structure: dihydronaphthalene ring with OH]

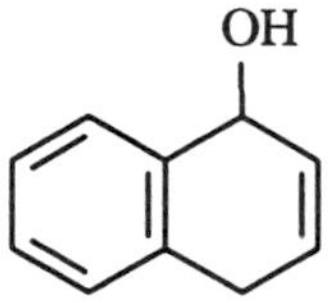

(d) 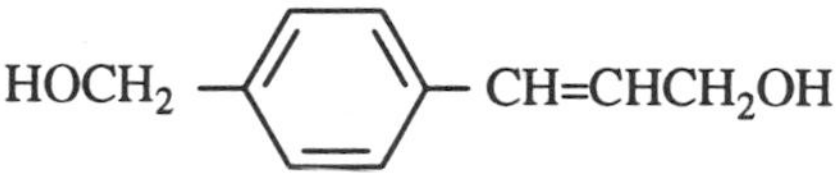 $HOCH_2\text{–}C_6H_4\text{–}CH{=}CHCH_2OH$

(e) 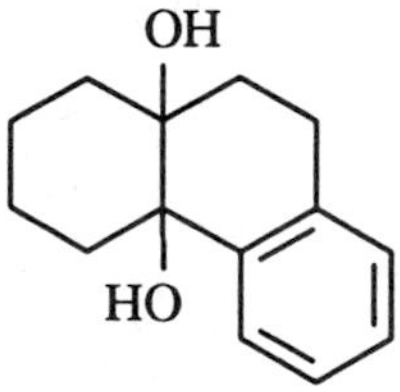 [structure: fused ring system with OH and HO]

(f) 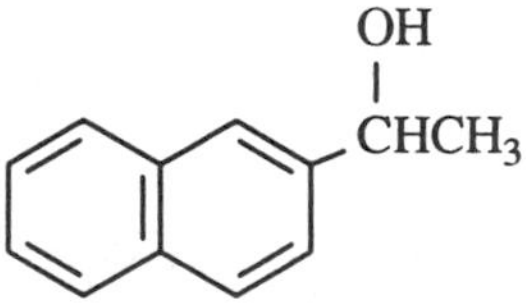 [structure: naphthalene with $CH(OH)CH_3$]

7.5 What alkyl halide would be required to prepare each of the following alcohols by an S_N2 reaction?

(a) 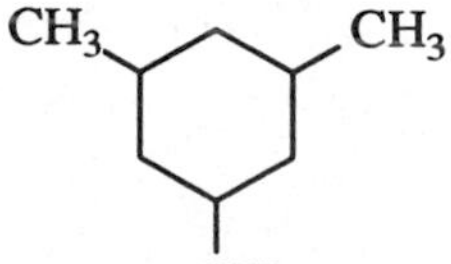 [structure: cyclohexane ring with two CH_3 groups and OH]

(b) $CH_3(CH_2)_{16}CH{=}CHCH_2CH_2OH$

(c) 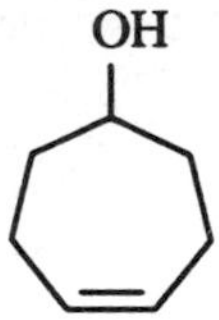

(d) 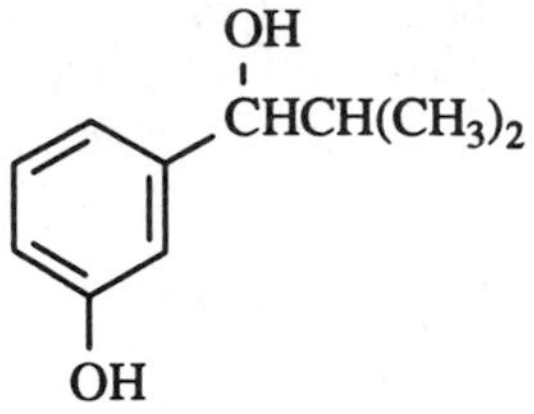

7.6 Suggest three different Grignard syntheses for each of the following alcohols.

(a) 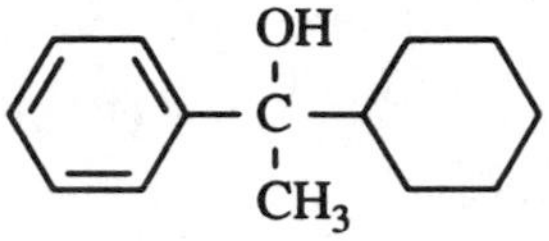

(b) 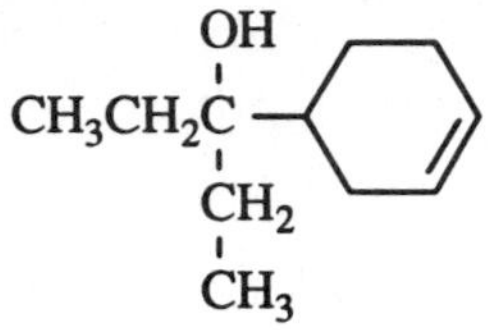

7.7 Complete the following equations:

(a) $CH_2{=}CHI + Mg \xrightarrow{\text{ether}}$ $CH_2{=}CHMgI$

(b) $CH_2{=}CHMgI + CH_3\overset{\overset{O}{\|}}{C}CH_3 \longrightarrow$ $CH_3CCH{=}CH_2$

(c) $CH_2{=}CHMgI + CH_3OH \longrightarrow$

(d) $CH_3CH_2CO_2^{-}\ {}^{+}MgBr + H^+ \longrightarrow$

7.8 Predict the organic product:

(a) $CH_3MgI \xrightarrow[\text{(2) } H_2O,\ H^+]{\text{(1) } H\overset{\overset{O}{\|}}{C}OCH_2CH_3}$ $CH_3\overset{\overset{OH}{|}}{C}H$

(b) $CH_3MgI \xrightarrow[\text{(2) } H_2O,\ H^+]{\text{(1) } CH_3\overset{\overset{O}{\|}}{C}OCH_2CH_3}$

(c) $$CH_3MgI \xrightarrow[(2)\ H_2O,\ H^+]{(1)\ CH_3\overset{O}{\overset{\|}{C}}CH_3}$$

(d) $$C_6H_5MgBr \xrightarrow[(2)\ H_2O,\ H^+]{(1)\ H\overset{O}{\overset{\|}{C}}H}$$

7.9 Complete the following equations:

(a) 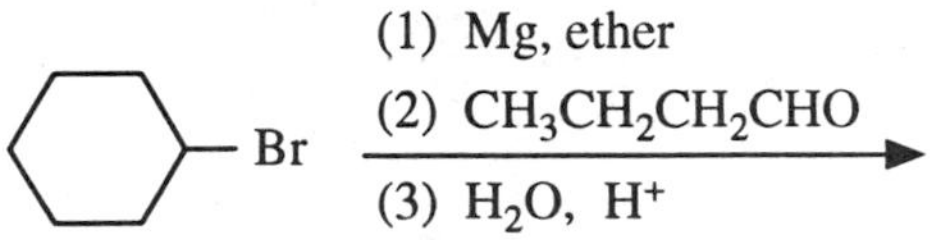
Br
(1) Mg, ether
(2) CH3CH2CH2CHO
(3) H2O, H+

(b) 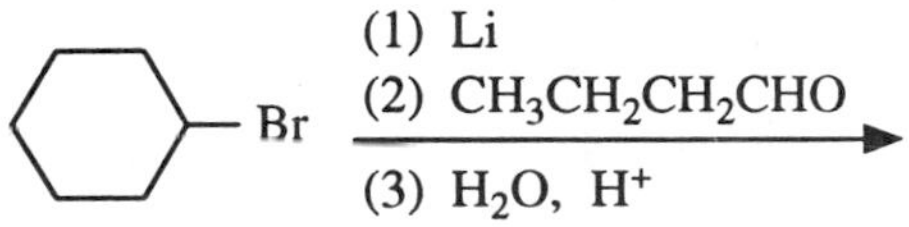
Br
(1) Li
(2) CH3CH2CH2CHO
(3) H2O, H+

7.10 Predict the major organic compounds:

(a) $$CH_3\overset{O}{\overset{\|}{C}}CH_2CH_3 \xrightarrow[(2)\ H_2O,\ H^+]{(1)\ CH_3CH_2MgBr}$$

(b) 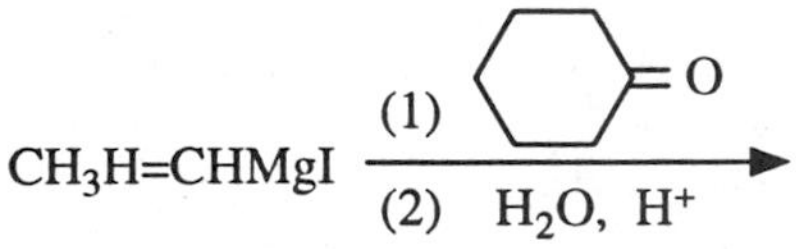
CH3H=CHMgI
(1)
O
(2) H2O, H+

(c) 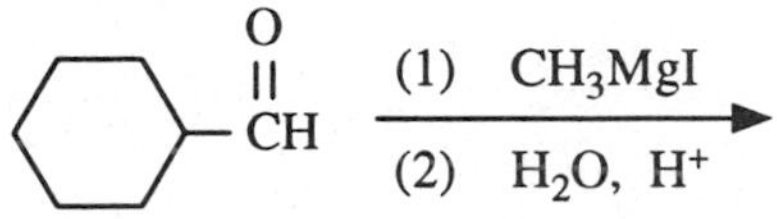
O
CH
(1) CH3MgI
(2) H2O, H+

(d) 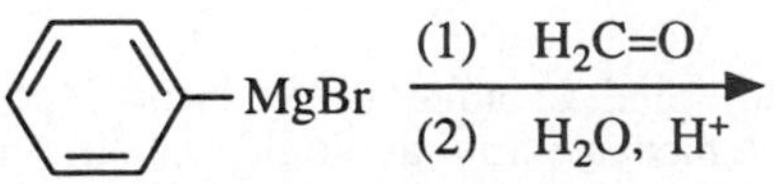
MgBr
(1) H2C=O
(2) H2O, H+

7.11 Write equations showing how each of the following alcohols could be prepared by a Grignard synthesis. (There may be more than one correct answer.)

(a) $(CH_3)_2CHCH_2OH$

(b)
$$\begin{array}{c} CH_2CH_3 \\ | \\ CH_3CCH_2CH_2CH_3 \\ | \\ OH \end{array}$$

(c)
$$\begin{array}{c} OH \\ | \\ (CH_3)_2CCH_2CH_3 \end{array}$$

7.12 Write equations to show how you would prepare the following compounds by Grignard reactions. (There may be more than one correct answer.)

(a) 2-hexanol

(b) 2,2-dimethyl-1-propanol

(c) diphenylmethanol

7.13 For each of the following alcohols, would you expect a substitution reaction with HBr to proceed by an S_N1 or by an S_N2 path?

(a) 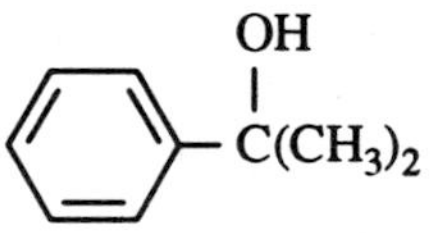 OH, $C(CH_3)_2$

(b) 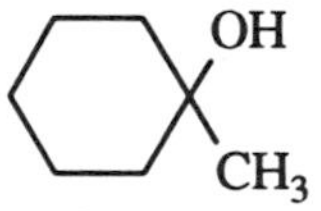 OH, CH_3

(c) CH_2OH

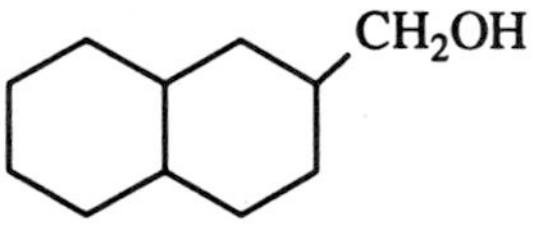

7.14 Write equations for the mechanisms of the reaction of HBr with (a) 1-butanol and (b) 2-methyl-2-butanol. Be sure to show all proton-transfer steps and to show the structures of the key transition states.

7.15 Predict the relative rates of reaction of the following alcohols with HBr:

(a) $(CH_3)_2CHCH_2CH_2OH$

(b)
$$\begin{array}{c} OH \\ | \\ CH_3CH_2C(CH_3)_2 \end{array}$$

(c) $(CH_3CH_2)_2CHOH$

7.16 When (*R*)-2-butanol is allowed to stand in aqueous acid for a period of time, a racemic mixture is formed. Write a mechanism that would explain this observation.

7.17 Each of the following reactions can be expected to yield both the expected substitution product and a rearranged product. Write formulas for the structures of these products.

(a) $(CH_3)_2CCH_2OH + HCl \xrightarrow{ZnCl_2}$

(b) $(C_6H_5)_3CCH_2OH + HCl \xrightarrow{H^+}$

(c) 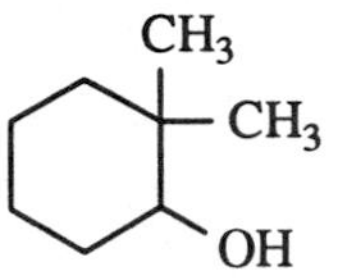 1-methyl(CH$_3$)-1-methyl(CH$_3$)-2-hydroxy(OH)cyclohexane + HBr $\longrightarrow$

7.18 Complete the following equations showing all elimination products. Indicate which would predominate in the reaction mixture.

(a) 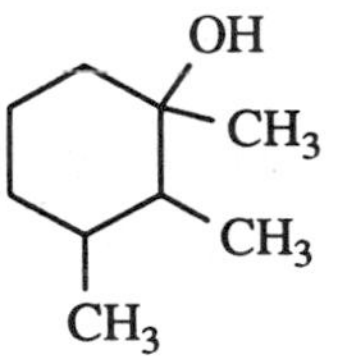 cyclohexane ring bearing OH and CH$_3$ on one carbon, CH$_3$ on the adjacent carbon, and CH$_3$ on the next carbon $\xrightarrow[\text{heat}]{H^+}$

(b) 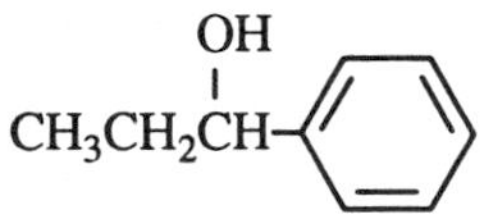 $CH_3CH_2CH(OH)-C_6H_5 \xrightarrow[\text{heat}]{H^+}$

7.19 Write a structure for each of the following names:

(a) 1-butyl tosylate (b) *sec*-butyl nitrate

(c) *tert*-butyl hydrogen sulfate

7.20 (1) Write an equation showing the formation of 2-propyl tosylate from an alcohol. (2) Write a flow equation to show the mechanism of the solvolysis of 2-propyl tosylate in water.

7.21 List the following compounds in order of increasing oxidation state (less oxidized first):

(a) 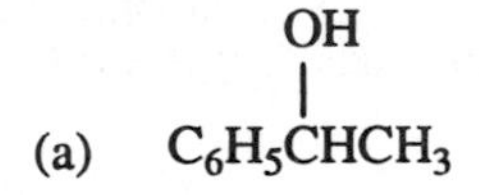

(b) $C_6H_5CCl_2CH_3$

(c) $C_6H_5\overset{O}{\overset{\|}{C}}OCH_3$

(d) $C_6H_5CH{=}CH_2$

(e) $C_6H_5CH_2CH_3$

(f) $C_6H_5\overset{O}{\overset{\|}{C}}CH_3$

(g) $C_6H_5C{\equiv}CH$

7.22 Write equations showing how you could prepare the following carbonyl compounds from alcohols in one step:

(a) $CH_3CH_2\overset{O}{\overset{\|}{C}}CH_3$

(b) $CH_3CH_2\overset{O}{\overset{\|}{C}}OH$

(c) $CH_3CH_2CH_2\overset{O}{\overset{\|}{C}}H$

7.23 For each of the following compounds, write the structure of a compound that would be at the next higher oxidation state and the structure of the compound that would be at the next lower oxidation state. (More than one answer may be correct.)

(a) (benzene ring)–$\overset{O}{\overset{\|}{C}}H$

(b) $CH_3CH{=}CHCH_3$

(c) $CH_3CH_2CH_2OH$

(d) (cyclohexane ring) Cl, Cl

7.24 Using a flow diagram, show how each of the following compounds could be prepared from alkyl halide or benzylic halide without using a Grignard reaction.

(a) (benzene ring)–$\overset{O}{\overset{\|}{C}}H$

(b) (cyclohexane ring)–OH

(c) $CH_3CH_2CH_2\overset{O}{\overset{\|}{C}}CH_3$

(d) $(CH_3)_2CH\overset{O}{\overset{\|}{C}}OH$

7.25 Predict the major organic products:

(a) (*R*)-$CH_3CH_2\underset{\displaystyle CH_3}{\underset{|}{C}}HOMgBr$ + H_2O ⟶

(b)  $-CH_2CH_2OH$ + $SOCl_2$ $\xrightarrow{\text{pyridine}}$

(c)  + $SOCl_2$ $\xrightarrow{\text{pyridine}}$

(d) $(C_6H_5)_2\overset{\displaystyle CH_2CH_3}{\overset{|}{C}}O^- K^+$ + H_2O ⟶

(e) 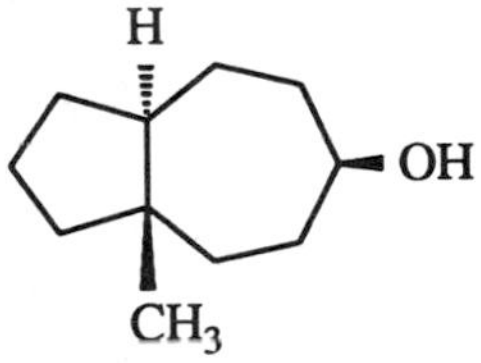 + 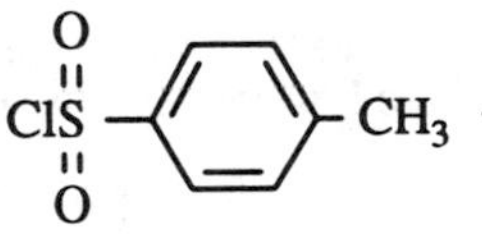 ⟶

(f) (*S*)-CH_3CHDOH + HCl $\xrightarrow{ZnCl_2}$

(g) $(CH_3)_2\overset{\displaystyle OH}{\overset{|}{C}}CH_2CH_2OH$ + $CrO_3 \cdot$ 2 pyridine ⟶

(h) (*R*)-2-butanol + Na ⟶

7.26 Write equations showing the probable mechanisms of the following reactions.

(a) $CH_3CH_2\overset{\displaystyle OH}{\overset{|}{C}}HCH_2CH_3$ + HBr ⟶ $CH_3CH_2\overset{\displaystyle Br}{\overset{|}{C}}HCH_2CH_3$

(b) (*R*)-2-butanol $\xrightarrow[\text{(2) aqueous NaOH}]{\text{(1) } CH_3SO_2Cl}$ (*S*)-2-butanol

7.27 What would be the best reagent(s) for each of the following conversions? Explain your choices.

(a)  cyclohexyl–CH_2OH ⟶ cyclohexyl–CH_2Cl

(b)  $CH_3CH_2CH(OH)CH_3$ ⟶ $CH_3CH_2CH(I)CH_3$

(c)  1-methylcyclohexanol (CH_3, OH) ⟶ 1-chloro-1-methylcyclohexane (CH_3, Cl)

(d) $CH_3CH_2CH_2OH$ ⟶ $CH_3CH_2CH_2Cl$

7.28 Write equations for the reactions of 2-butanol with the following reagents. If little or no reaction is expected, write "no reaction".

(a) Na (b) NaOH (c) HI

(d) H_2SO_4, heat (e) $KMnO_4$, ^-OH (f) $SOCl_2$

(g) NaH (h) NaCN (i) PCl_3

7.29 Suggest reagents for the conversion of 1-butanol to the following compounds:

(a) butanoic acid (b) 1-chlorobutane (c) 1-iodobutane

(d) 2-pentanone (e) butanal (f) 1-pentanol

(g) 2-hexanol (h) 5-nonanol (i) pentanenitrile ($CH_3CH_2CH_2CH_2CN$)

7.30 How would you synthesize the following compounds from acetone, using any other organic or inorganic compounds you wish?

(a) *tert*-butyl alcohol (b) 2-methyl-2-butanol

(c) 2-deuteriopropane

Ethers, Epoxides, and Sulfides

Some Important Features

Ethers can form hydrogen bonds with other compounds containing OH or NH, but not in the pure state.

Ethers and epoxides can be prepared by the Williamson synthesis:

$$RO^- + R_2CHX \longrightarrow RO\text{-}CHR_2 + X^-$$

Although most ethers are cleaved by strong acid, the more reactive epoxides can be ring-opened in dilute acid or base to yield a variety of 1,2-disubstituted products.

$$R_2C\overset{O}{\frown}CHR \xrightarrow{HY} R_2C(Y)-CH(OH)R$$

$$R_2C\overset{O}{\frown}CHR \xrightarrow[(2)\ H^+]{(1)\ Y^-} R_2C(OH)-CH(Y)R$$

Other topics covered in this chapter are crown ethers and some sulfur compounds (thiols, sulfides, sulfones, and sulfoxides).

Reminders

The reactions of a Grignard reagent with ethylene oxide is a standard reaction for extending a carbon chain by two.

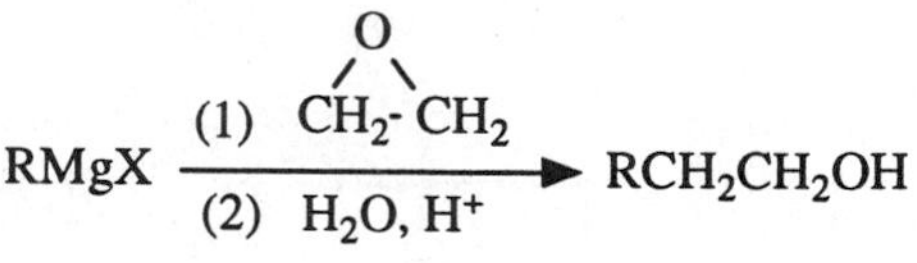

$$RMgX \xrightarrow[(2)\ H_2O,\ H^+]{(1)\ \overset{O}{\widehat{CH_2-CH_2}}} RCH_2CH_2OH$$

Additional Drill Problems

8.1 Write acceptable names for the following ethers:

(a) 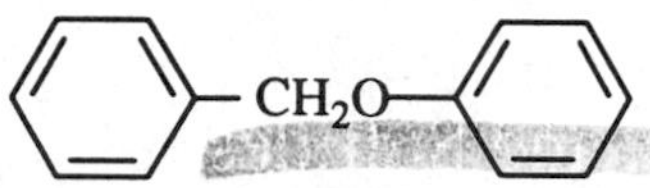 $C_6H_5CH_2O-C_6H_5$ (benzene ring–CH_2O–benzene ring)

(b) $CH_2{=}CHCH_2OCH_2CH{=}CH_2$

(c) 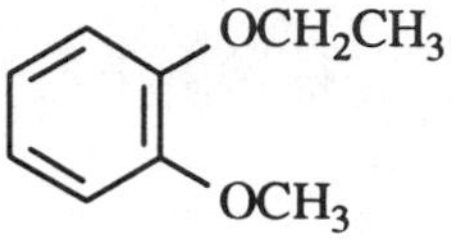 benzene ring with adjacent OCH_2CH_3 and OCH_3 substituents

(d) $(CH_3)_2CHOCH_2CH_2OH$

(e) $(CH_3)_2CHOCHCH_2CH_3$ with CH_3 on the CH

(f) H_3C (hashed bond), Cl (wedge), CH_2CH_3 on C; C–$CH_2CH{-}CH_2$ with O bridging CH and CH_2 (epoxide)

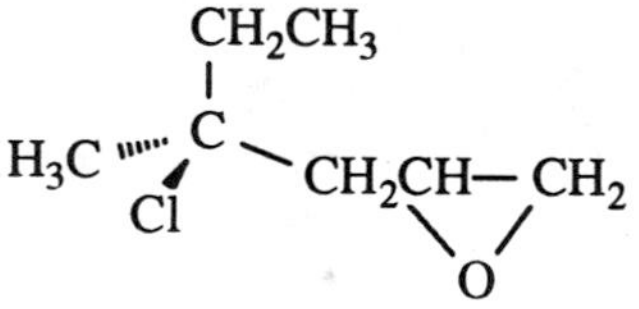

8.2 Draw formulas for the following ethers:

(a) dimethyl ether

(b) *trans*-2-ethyl-3-methyloxirane

(c) (*R*)-2-chloropropyl methyl ether

(d) cyclohexyl isopropyl ether

(e) 2-methoxyethanol

(f) 3-bromo-1-butoxy-3-ethylpentane

(g) cyclohexene oxide

(h) propyloxirane

(i) 15-crown-5

(j) 12-crown-4

8.3 Complete the following equations:

(a) $CH_3CH_2OSO_3H \xrightarrow{180°}$

(b) $CH_3CH_2OSO_3H + CH_3CH_2OH \xrightarrow{140°}$

(c) $CH_3CH_2OSO_3H + \text{excess } H_2O \longrightarrow$

8.4 Predict the principal organic product of the reaction, if any, of isobutyl propyl ether with the following reagents:

(a) dilute aqueous NaOH
(b) concentrated aqueous NaOH
(c) dilute aqueous H_2SO_4
(d) hot, concentrated HBr
(e) H_2CrO_4 at 25°
(f) NaCN in aqueous ethanol

8.5 Write equations for the Williamson synthesis of the following ethers:

(a) benzyl isopropyl ether
(b) *tert*-butyl butyl ether
(c) butyl phenyl ether

8.6 Write equations to show the mechanisms and products of the following reaction:

$$C_6H_5{-}OCH_2{-}C_6H_5 + HI \xrightarrow{\text{heat}}$$

8.7 Write equations for the reactions of ethylene oxide with the following reagents:

(a) HI
(b) $(CH_3)_2NH$
(c) C_6H_5MgBr
(d) CH_3CH_2OH, H^+
(e) aqueous NaOH
(f) Na^+ $^-OCH_2CH_3$ in CH_3CH_2OH

8.8 Write equations for the reactions of propylene oxide with each of the reagents in the preceding problem. (Problem 8.7)

8.9 Write equations showing the mechanism of the reaction of 1-methylcyclopentene oxide with ethanol and a trace of sulfuric acid.

8.10 Fill in the indicated blanks:

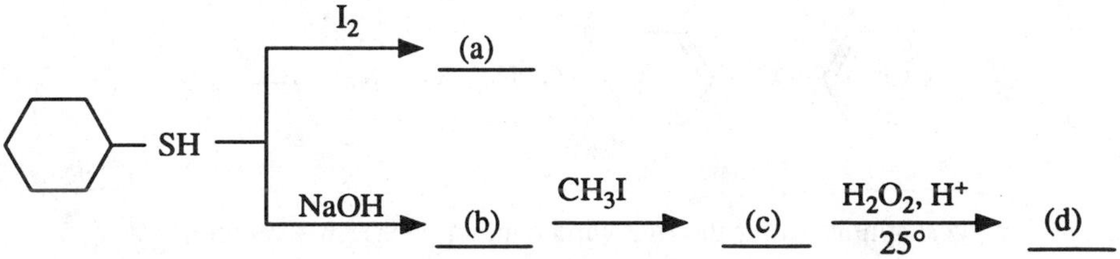

$$C_6H_{11}{-}SH \xrightarrow{I_2} \underline{\text{(a)}}$$

$$C_6H_{11}{-}SH \xrightarrow{NaOH} \underline{\text{(b)}} \xrightarrow{CH_3I} \underline{\text{(c)}} \xrightarrow[25°]{H_2O_2,\ H^+} \underline{\text{(d)}}$$

8.11 Beginning with Na_2S and other appropriate reagents, show how you could prepare diethyl sulfide.

8.12 Fill in the blanks:

$$CH_3CH_2CH_2CH_2OH \xrightarrow[-H_2]{K} \underline{(a)} \xrightarrow{(CH_3)_2CHBr}$$

$$\underline{(b)} \xrightarrow[-HBr]{Br_2,\ h\nu} \underline{(c)} \xrightarrow[heat]{(CH_3)_3CO^-K^+} \underline{(d)}$$

8.13 Predict the major organic products, if any:

(a) cyclohexene oxide $\xrightarrow[(2)\ H_2O,\ H^+]{(1)\ CH_3MgI}$

(b) benzene ring with OCH_3 and ortho CH_2OCH_3 + excess HI $\xrightarrow{heat}$

(c) $C_6H_5\overset{O}{CH-CH_2}$ (epoxide) $\xrightarrow[(2)\ H_2O,\ H^+]{(1)\ C_6H_5Li}$

(d) piperidine (ring NH) + $\overset{O}{CH_2-CH_2}$ (epoxide) $\longrightarrow$

(e) $\overset{O}{CH_2-CH_2}$ (epoxide) + HBr + H_2O $\longrightarrow$

(f) $C_6H_5\overset{O}{CH-CH}$–C_6H_4–OCH_3 (epoxide; para-substituted ring) + CH_3OH $\xrightarrow{H^+}$

(g) C_6H_5–O–C_6H_4–OCH_3 (para) + Na $\longrightarrow$

8.14 Suggest syntheses for the following compounds from 1-butanol:

(a) dibutyl ether (b) butanethiol (c) dibutyl sulfide

Spectroscopy I

Some Important Features

Electromagnetic radiation can be absorbed by organic compounds, and this absorption of energy results in increased energy of the molecules. Different compounds absorb electromagnetic radiation of different energy, and thus of different wavelength (λ) or frequency (ν). Radiation of shorter wavelength is of higher frequency and higher energy than radiation of longer wavelength.

$$\longrightarrow$$

increasing frequency
shorter wavelength
higher energy

Absorption of infrared radiation results in increased amplitudes of bond vibrations. Polar groups usually exhibit stronger peaks in the infrared spectrum than nonpolar groups. The usual positions of absorption are shown inside the cover of the text. The absorption positions (and peak appearances) of -OH, -NH, and $-NH_2$ (a double peak), $-CO_2H$, and C=O are particularly distinctive.

A proton NMR spectrum results from the change in spin states of hydrogen nuclei (protons). The position of absorption by a proton (the *chemical shift*) is affected by neighboring atoms and by other groups in the molecule. A nearby electronegative atom results in deshielding, and absorption is observed farther downfield.

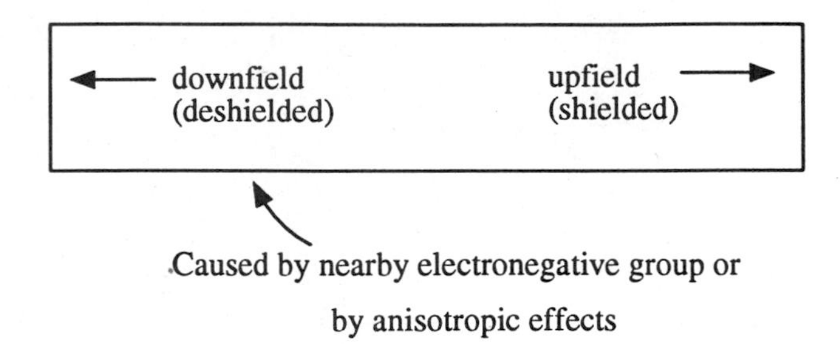

Other groups in the molecule affect the position of absorption by *anisotropic* effects. Absorption by protons bonded to aromatic rings, to aldehyde carbonyl groups, and to sp^2 carbons of alkenes is observed downfield. Figure 9.28 in the text shows some typical positions of proton absorption in NMR spectra.

Recognizing the equivalence or nonequivalence of protons in ^1H NMR spectroscopy is important in structure determination. We suggest that you review Section 9.8A.

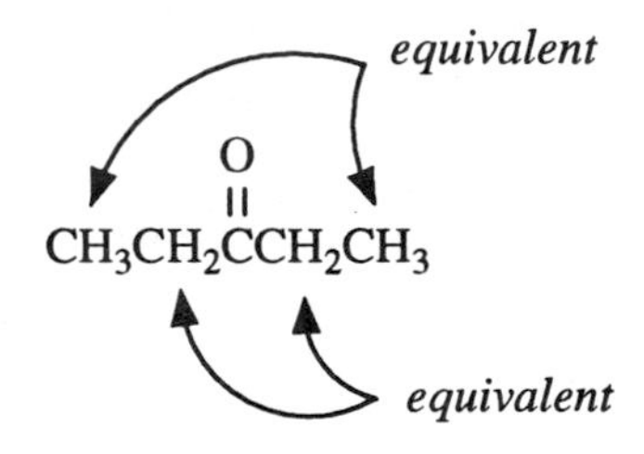

The area under the signal for a proton is proportional to the number of protons giving rise to that signal. In the above example, we observe two principal signals: the CH_3 signal and the CH_2 signal. The area ratio is 6 : 4, or 3 : 2.

Absorption peaks in the NMR spectrum may be split into multiple peaks. A group of magnetically equivalent protons absorbs radio waves at the same position of the NMR spectrum and do not split the signals of each other. Neighboring protons that are not magnetically equivalent to the proton in question do split the signal of the proton.

CH_3 signal split into three peaks by CH_2

$$CH_3CH_2\overset{\overset{\displaystyle O}{\|}}{C}CH_2CH_3$$

CH_2 signal split into four peaks by CH_3

If a signal is split by a group of neighboring protons equivalent to each other, the $n + 1$ rule is followed. In the above example, CH_3 is split by CH_2 (two protons equivalent to each other but not equivalent to the CH_3 protons). The splitting pattern for CH_3 is 2 + 1, or a triplet.

Another topic covered in the ^1H NMR discussion is *chemical exchange*, which may result in no splitting of (or by) and OH or NH proton.

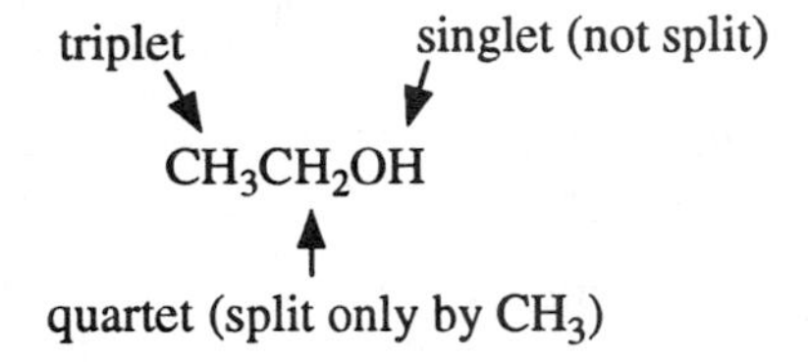

Coupling constants (J values) are covered only briefly in the text. Your instructor may wish to cover this topic in more detail.

Coupling constants allow us to determine which protons are coupled with one another in a complex spectrum. If the J values are the same, the protons may be coupled. If the J values are *different,* the protons in question are not splitting the signals of one another, but are being split by some other proton. For example, in the NMR spectrum of the partial formula $—CH_2CH<$, the CH_2 protons are split into a doublet by the CH protons and the CH proton, in turn, is split into a triplet by the CH_2 protons.

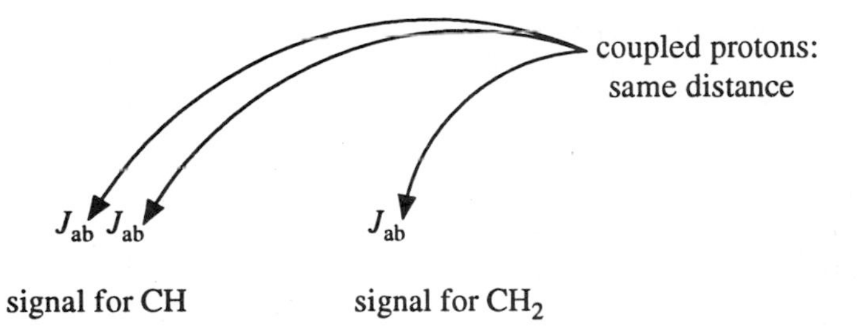

If the distances between peaks are *not* the same, then the protons in question are *not* splitting the signals of each other -- some other protons are causing the splitting.

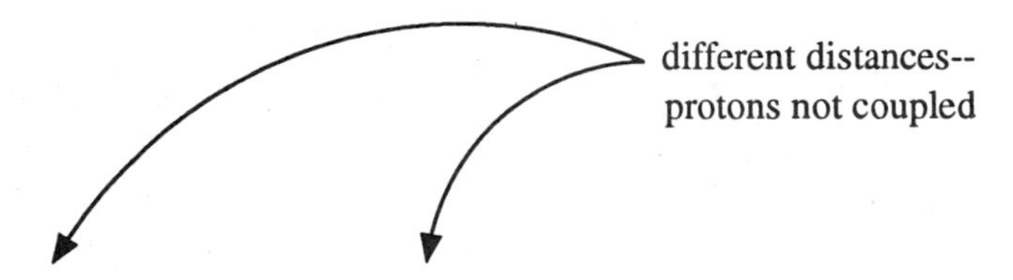

Section 9.13 in the text is an introduction to determination of structure from the molecular formula and spectral data. Study this section carefully, as well as the discussions in the answers to Problem 9.44 (after you have solved them or tried to solve them).

Before studying spectra in detail for clues about a structure, first examine the molecular formula and the spectra for obvious clues.

1. Does the structure contain an aromatic ring? Check the proton NMR spectrum for aryl protons.

2. Does the structure contain C=O, NH or OH, NH_2, or CO_2H? Check the infrared spectrum for these distinctive peaks.

3. How many rings or double bonds does the structure contain? Check the molecular formula. Remember that the C=O group is a site of unsaturation. Butanone has the general formula C_4H_8O, or $C_nH_{2n}O$. (Compare this formula with that of the butenes, C_4H_8, or C_nH_{2n}.) A phenyl group contains three sites of unsaturation *and* a ring. $C_6H_5CH_2CH_3$ has the general formula C_nH_{2n-6}.

4. Does the structure contain an ethyl group bonded to an electronegative atom? Check the ^1H NMR spectrum for an upfield triplet and a downfield quartet.

Carbon-13 NMR spectroscopy is becoming increasingly important in organic chemistry. If you understand the principles of ^1H NMR spectroscopy, which is more complex, you should have no problems with carbon-13 spectroscopy.

Additional Drill Problems

9.1 Convert each of the following values to μm:

(a) 3600 cm^{-1} (b) 1450 cm^{-1} (c) 1050 cm^{-1}

9.2 Convert each of the following values to cm^{-1}:

(a) 3.300 μm (b) 7.220 μm (c) 5.500 μm

9.3 Arrange the following types of electromagnetic radiation in terms of increasing energy (least energetic first):

(a) radio waves (b) visible light

(c) ultraviolet radiation (d) infrared radiation

9.4 Give the position of the characteristic infrared absorption that would allow you to distinguish the compounds in the following groups:

(a) cyclohexanone (ring=O) cyclopentyl–CH(=O) cyclopentyl–CH_2OH

(b) $CH_3C(=O)CH_3$ $CH_3C(=O)OH$ $CH_3C(=O)OCH_3$

(c) 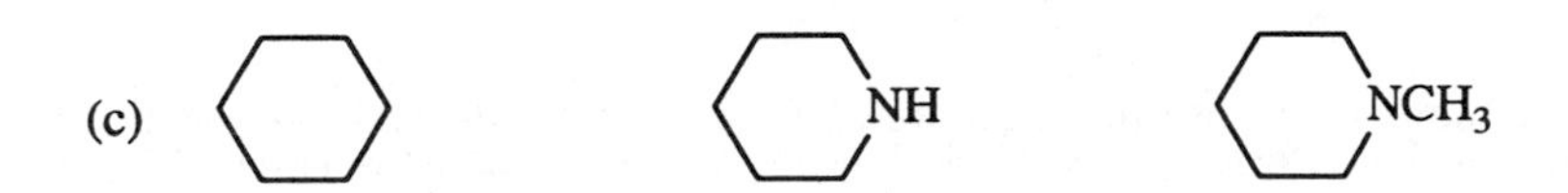

9.5 Which of the following pairs of compounds could *not* be distinguished by infrared spectroscopy? Explain your answers.

(a) $CH_3CH_2C(=O)NHCH_3$ and $CH_3CH_2C(=O)OCH_3$

(b) $C_6H_5CH(OH)CH_3$ and $C_6H_5CH_2CH_2OH$

(c) $CH_3CH{=}CHCH_3$ and $CH_3C{\equiv}CCH_3$

(d) piperidine (NH) and piperazine (HN, NH)

9.6 Indicate the position of the principal distinguishing infrared band for each of the following structures:

(a) $CH_3C{\equiv}CH$

(b) $CH_3C(=O)-C_6H_5$

(c) C_6H_5-OH

(d) $C_6H_5-NH_2$

9.7 What would you look for in the infrared spectrum in order to distinguish between each of the following compounds?

(a) 1-nitropropane and propylamine

(b) $CH_3CH{=}CHCH_2OH$ and $CH_3C{\equiv}CCHO$

(c) $C_6H_5-CH_2Cl$ and $C_6H_5-C(=O)Cl$

9.8 The C-O stretching absorption for an alcohol is at 1050-1150 cm^{-1} (8.6-9.5 μm). Would you expect this band to be shifted to higher or lower frequencies for phenol, C_6H_5OH? (*Hint*: Write all the resonance structures of phenol, including those that involve the unshared electrons of the oxygen.)

9.9 In the infrared spectra labeled 9.1 through 9.5, assign the bands arising from the following functional groups:

(a) $-\overset{|}{\underset{|}{C}}-O-H$ (b) $-\overset{O}{\overset{\|}{C}}-OCH_3$ (c) $-\overset{|}{\underset{|}{C}}-\overset{H}{\overset{|}{N}}-H$

(d) $-\overset{O}{\overset{\|}{C}}-OH$ (e) $-\overset{O}{\overset{\|}{C}}-H$

Spectrum 9.1

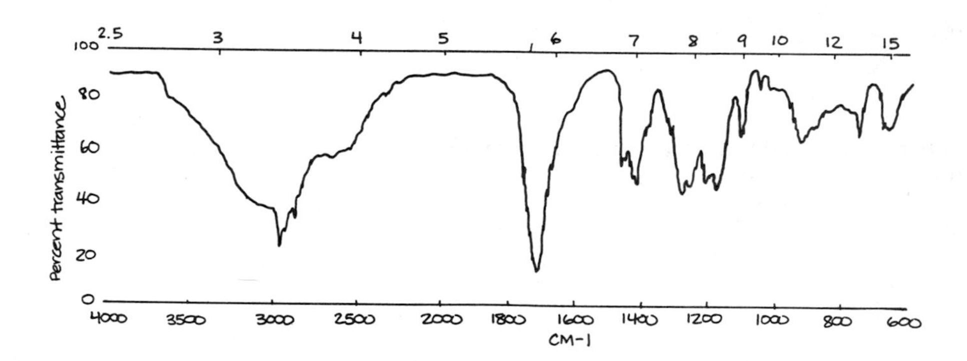

Spectrum 9.2

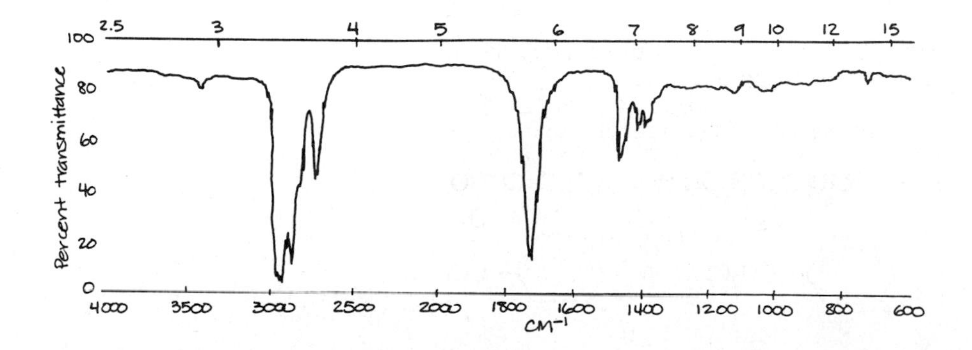

Spectrum 9.3

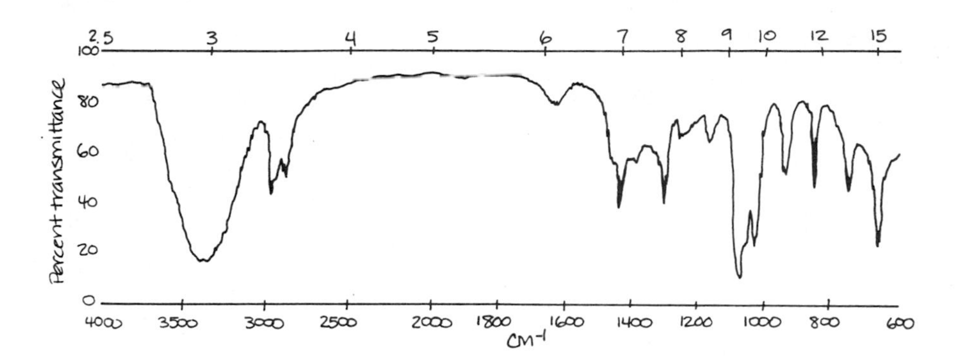

Spectrum 9.4

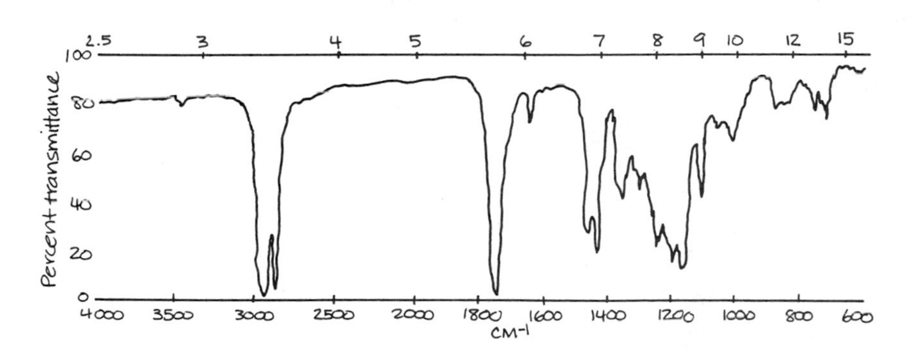

Spectrum 9.5

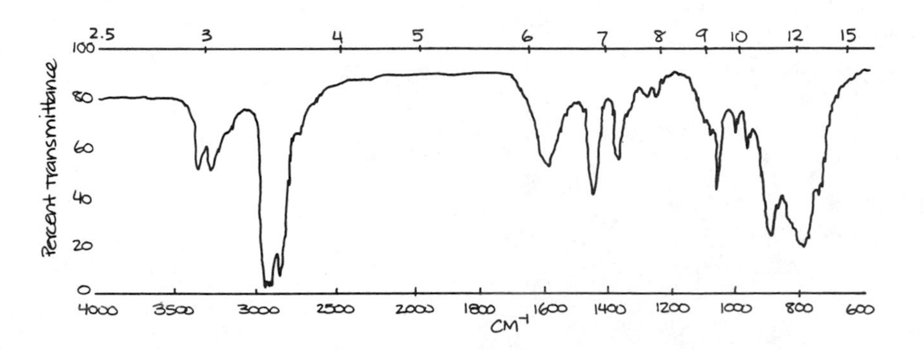

9.10 Match the infrared spectra labeled 9.6 through 9.10 with the following general formulas. (Not all of the general formulas will necessarily correspond to a spectrum.)

(a) RNH_2 (b) ArOH, where Ar = aryl (c) ROH

(d) RNHR′ (e) RCO_2R' (f) RCHO

Spectrum 9.6

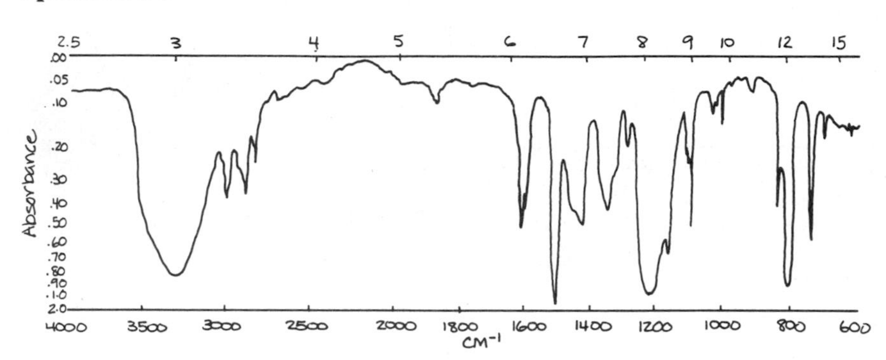

Spectrum 9.7

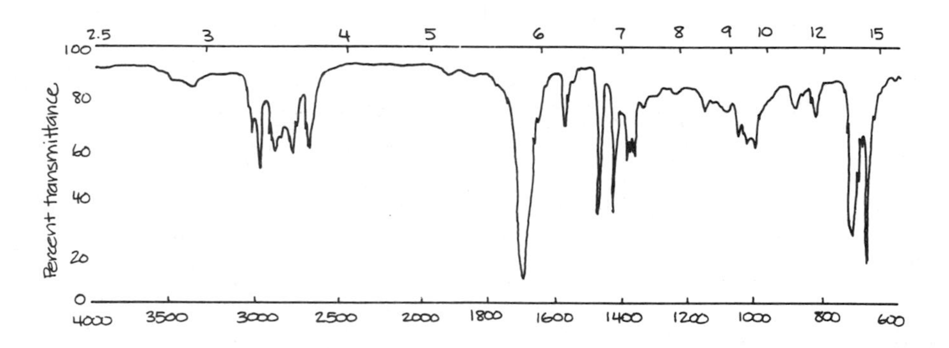

Spectrum 9.8

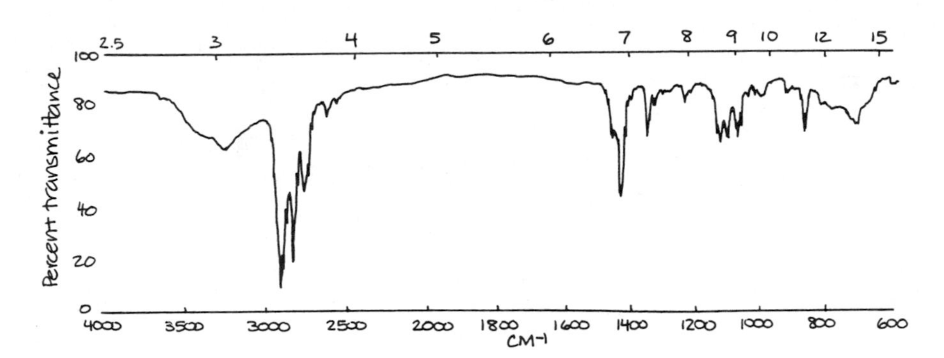

Spectrum 9.9

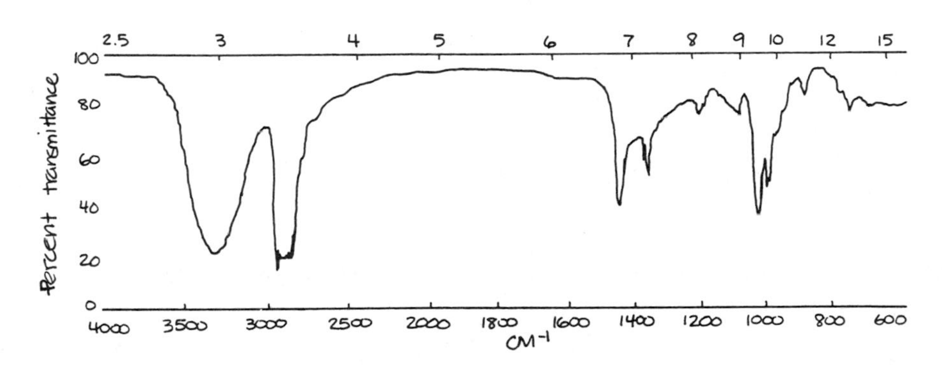

Spectrum 9.10

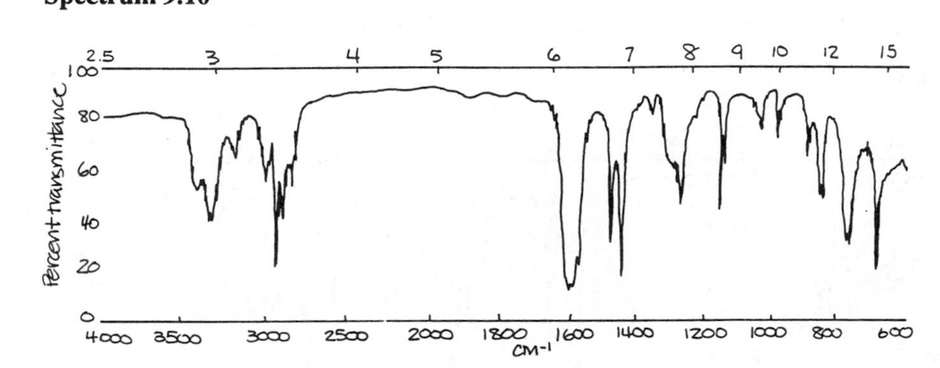

9.11 Match the infrared spectrum labeled 9.11 with one of the following structures:

(a) $C_6H_5CH_2COOH$ (benzene ring–$CH_2\overset{O}{\overset{\|}{C}}OH$)

(b) $HOCH_2CH_2CH_2OH$

(c) $CH_3\underset{CH_3}{\underset{|}{C}}HCH_2\overset{O}{\overset{\|}{C}}OH$

(d) $HOCH_2CH_2\overset{O}{\overset{\|}{C}}CH_3$

Spectrum 9.11

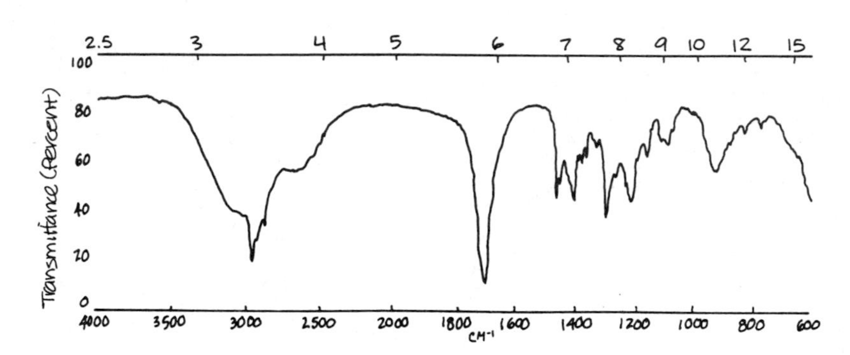

9.12 Match the infrared spectrum labeled 9.12 with one of the following structures:

(a) $CH_2{=}CHCH_2CH_3$

(b) $HC{\equiv}CCH_2CH_2\overset{O}{\overset{\|}{C}}CH_3$

(c) $CH_2{=}CHCH_2CH_2\overset{O}{\overset{\|}{C}}CH_3$

(d) $CH_3CH_2CH_2CH_2\overset{O}{\overset{\|}{C}}CH_3$

Spectrum 9.12

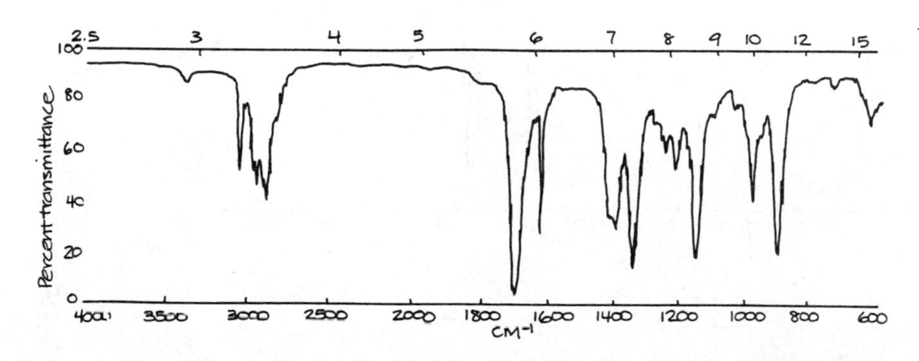

9.13 For each of the following structures, circle the proton(s), if any, that are chemically equivalent to the indicated proton(s).

(a) 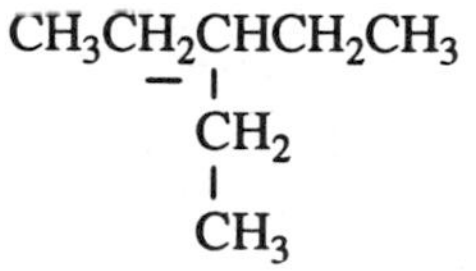 $CH_3CH_2CHCH_2CH_3$ with CH_2CH_3 substituent on the central carbon (indicated: a proton of the first CH_2)

(b) 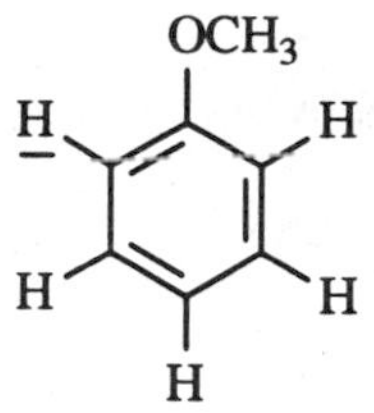 Methoxybenzene (OCH_3 on ring) with ring H atoms (indicated: one H ortho to OCH_3)

(c) $CH_2{=}C(H)-C(H){=}C(H)-C(H){=}CH_2$ (indicated: an H on the central C=C)

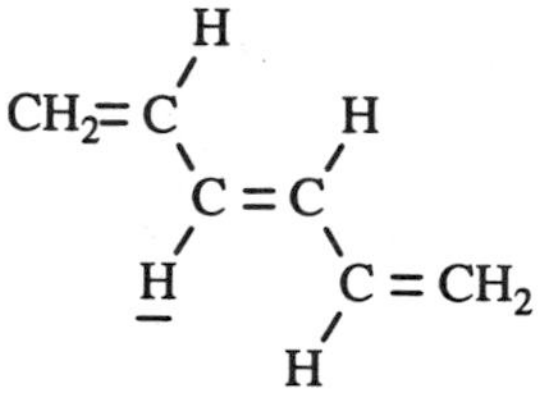

(d) 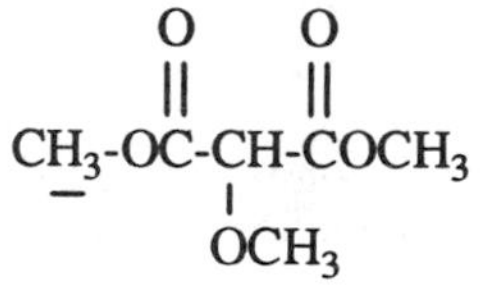 CH_3-OC(=O)-CH(OCH_3)-C(=O)OCH$_3$ (indicated: the first CH_3)

9.14 In each structure below, label each group of chemically nonequivalent protons.

Example: $CH_3CH_2OCH_2CH_3$ (a b b a)

(a) $CH_3CH_2CH(OH)CH_3$

(b) $CH_3CH(CH_3)CH_2CH_2OH$

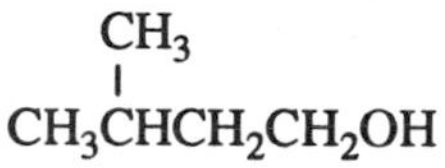

(c) 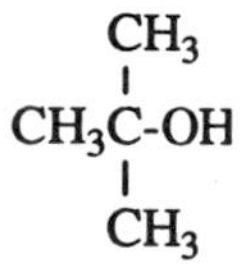 $CH_3C(CH_3)_2$-OH

(d) 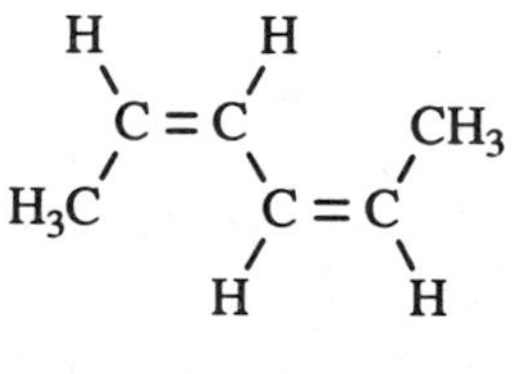 H_3C-CH=CH-CH=CH-CH_3

(e) 1,3-dimethylbenzene (CH_3, CH_3)

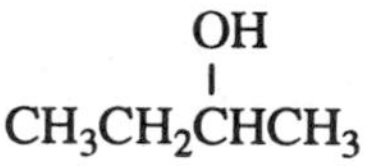

(f) 3-methylphenol (CH_3, OH)

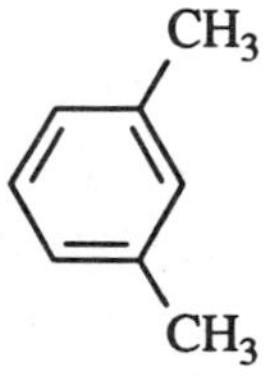

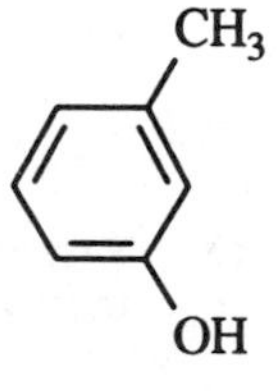

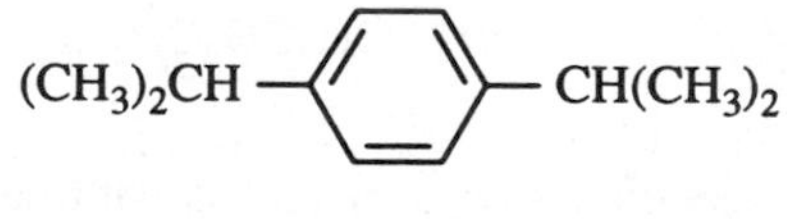

(g) $(CH_3)_2CH$–C_6H_4–$CH(CH_3)_2$

(h) CH_3O–C_6H_4–Br

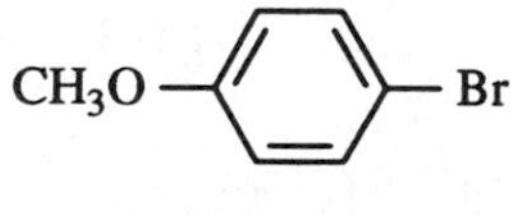

(i) $CH_3CH_2CH_2CH_2CH_2CH_2CH_3$

9.15 Why does the -OH proton of an alcohol become more deshielded as the concentration of the alcohol is increased?

9.16 For each of the following, which group of protons, 1 or 2, will have the greater chemical shift?

(a) $HO\overset{O}{\overset{\|}{C}}-CH_2\overset{H}{\underset{OH}{C}}-\overset{O}{\overset{\|}{C}}OH$ (1 → CH_2; 2 → H)

(b) $CH_2=\underset{H}{C}-CH_2Br$ (1 → H; 2 → CH_2Br)

9.17 Arrange the following compounds in order of increasing chemical shift of the indicated proton (smallest chemical shift first):

(a) C_6H_4Br (ring $\underline{H}$ ortho to Br)

(b) $C_6H_5\overset{O}{\overset{\|}{C}}\underline{H}$

(c) $C_6H_5\overset{O}{\overset{\|}{C}}O\underline{H}$

(d) $C_6H_5C\underline{H}_2\overset{O}{\overset{\|}{C}}OH$

(e) $C_6H_5CH{=}C\underline{H}\overset{O}{\overset{\|}{C}}OH$

(f) $C_6H_5CH_2C\underline{H}_3$

9.18 Arrange the following compounds in order of increasing chemical shift of the indicated proton:

(a) $CH_3CH_2C\underline{H}_2Cl$

(b) $CH_3C\underline{H}_2CH_2Cl$

(c) $CH_3CH_2C\underline{H}Cl_2$

(d) $C\underline{H}_3CH_2CH_2Cl$

9.19 The integration curve on an NMR spectrum shows two steps with heights of 62.5 mm and 20.8 mm. Which of the following compounds is compatible with this spectrum?

(a) $CH_3CH_2CH_2CH_3$ (b) $CH_3CH_2CH_3$ (c) CH_3CH_3

9.20 Predict the relative areas under the principal ^1H NMR signals for the following compounds:

(a) 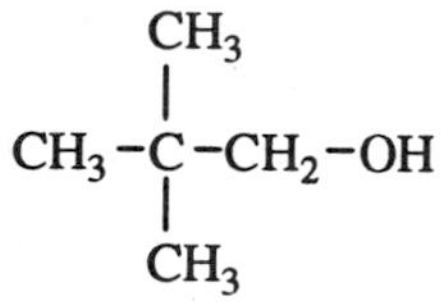 $CH_3-C(CH_3)_2-CH_2-OH$

(b) 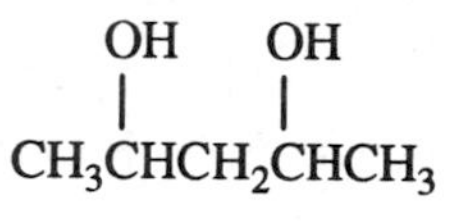 $CH_3CH(OH)CH_2CH(OH)CH_3$

9.21 Using the $n + 1$ rule, predict the splitting patterns for the indicated protons:

(a) $CH_3CH_2\underline{CH_2}Cl$

(b) $CH_3\underline{CH_2}CH_2Cl$

(c) $(CH_3)_2\underline{C}\underline{H}OH$

(d) $(CH_3)_2CHC\underline{H}BrCH_2Br$

(e) 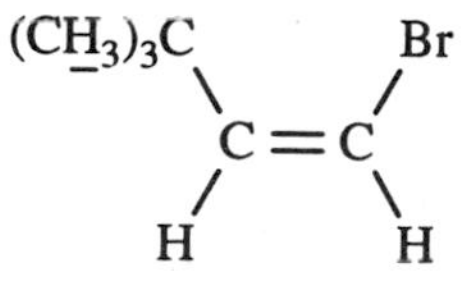 $(\underline{CH_3})_3C(H)C=C(H)Br$

(f) 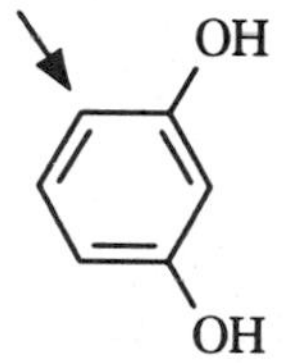 hydroquinone (1,4-dihydroxybenzene), arrow indicating a ring proton

9.22 Sketch the expected ^1H NMR spectrum for each of the following structures:

(a) $CH_3CH_2CCl_2CH_3$

(b) $(CH_3)_2NCH(CH_3)_2$

(c) 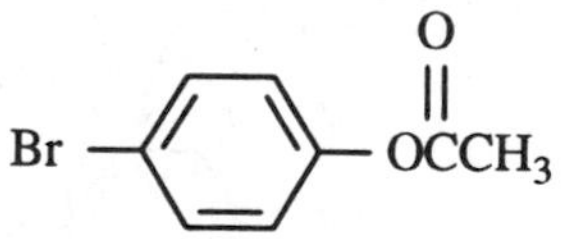 Br–C_6H_4–$OCCH_3$ (with C=O)

9.23 What would you look for in an NMR spectrum to differentiate between the following pairs?

(a) 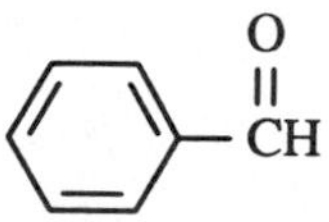 C_6H_5CHO and $C_6H_5CH_2CHO$

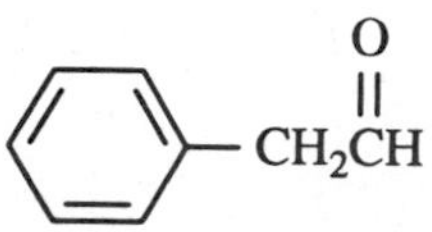

(b) 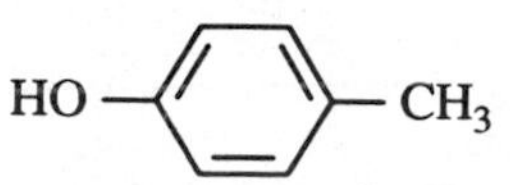 HO–C_6H_4–CH_3 (para) and HO–C_6H_4–CH_3 (meta)

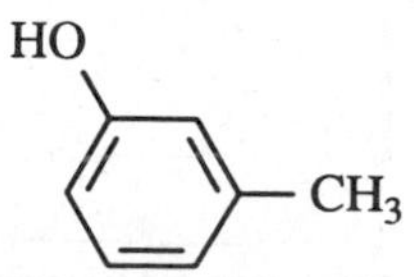

(c) $(CH_3)_2C{=}C(CH_3)_2$ and $(CH_3CH_2)_2C{=}CH_2$

(d) $CH_3CH_2CH_2Br$ and CH_3CHCH_3 (with Br on the middle carbon)

9.24 Match the ¹H NMR spectrum labeled 9.13 with one of the following compounds:

(a) $(CH_3)_3COH$

(b) $(CH_3)_3CCH_2Cl$

(c) $(CH_3)_3CNH_2$

(d) $(CH_3)_2CHCH(CH_3)_2$

Spectrum 9.13

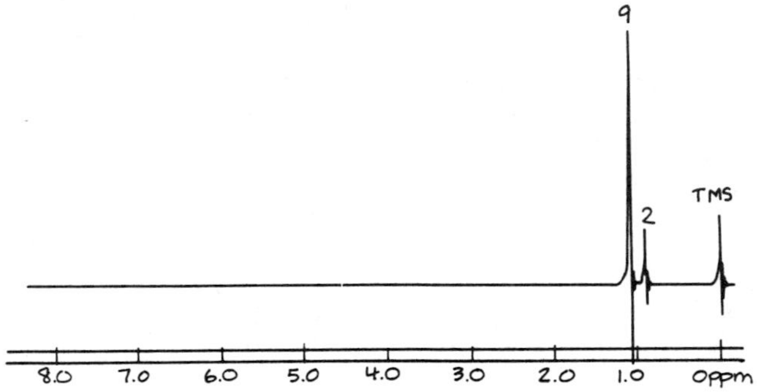

9.25 Match the ¹H NMR spectrum labeled 9.14 with one of the following compounds:

(a) 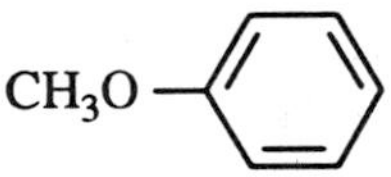

(b) 

(c) 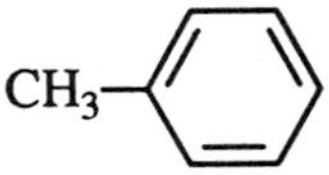

(d) 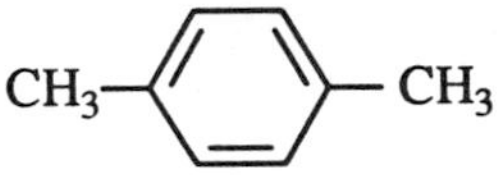

Spectrum 9.14

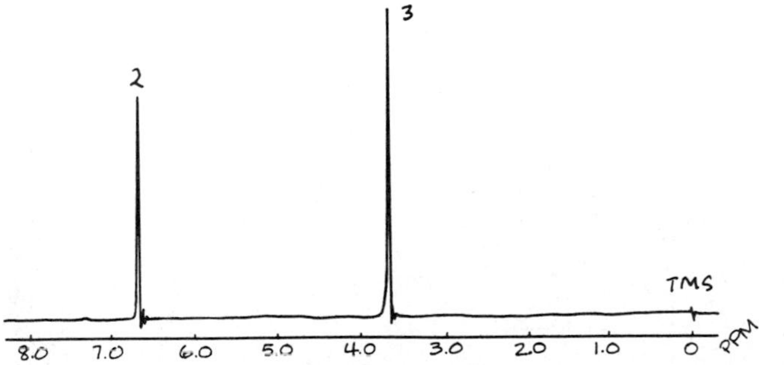

9.26 Using graph paper and a ruler, construct a spin-spin splitting diagram for the indicated proton in the following partial structure.

$$\begin{array}{ccc} R & & H_a \\ & C=C & \\ \underline{H}_c & & H_b \end{array}$$

H_c where $J_{ac} = 5$ Hz
and $J_{bc} = 10$ Hz

9.27 Match the ^{13}C NMR spectra labeled 9.15 with one of the following formulas:

(a) $CH_3CHClCHCl_2$ (b) $CH_3CCl_2CH_3$

(c) $ClCH_2CH_2CH_2Cl$ (d) $ClCH_2CCl_2CH_3$

Spectra 9.15

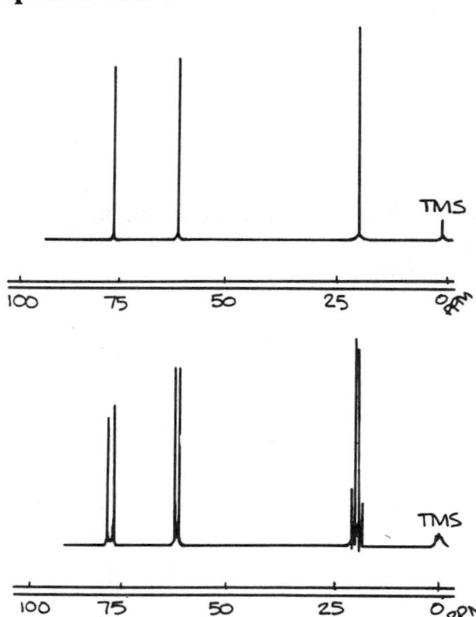

9.28 Match the ^{13}C NMR spectra labeled 9.16 with one of the following formulas:

(a) CH_3OH (b) $CH_3(CH_2)_3CHO$

(c) $(CH_3)_2C{=}O$ (d) CH_3CHO

Spectra 9.16

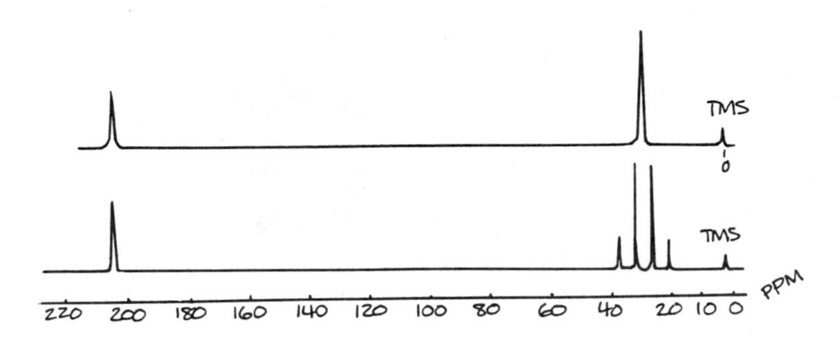

9.29 Match the ^{13}C NMR spectra labeled 9.17 with one of the following formulas:

(a) $CH_3CHClCH_3$ (b) $CH_3OCH_2CH_3$

(c) $(CH_3)_2CHOH$ (d) $(CH_3CH_2CH_2)_2O$

Spectra 9.17

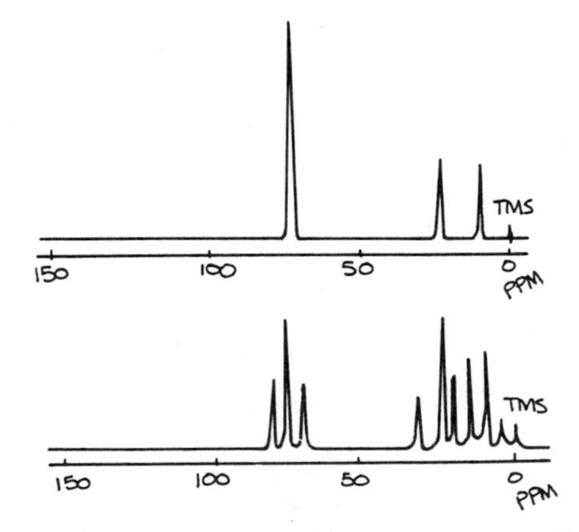

9.30 Match the ^{13}C NMR spectra labeled 9.18 with one of the following formulas:

(a) $CH_3\overset{O}{\overset{\|}{C}}CH_2CH_2CH_3$

(b) $(CH_3)_2CHCH_2CH{=}CH_2$

(c) $(CH_3)_2C{=}C(CH_3)_2$

(d) $(CH_3)_2CHCH{=}CHCH_3$

Spectra 9.18

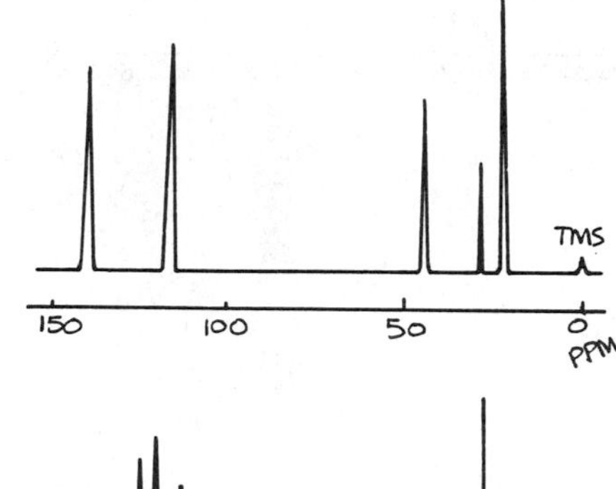

9.31 What is the structure of a compound ($C_{14}H_{14}O$) with the infrared and ^1H NMR spectra labeled 9.19?

Spectra 9.19

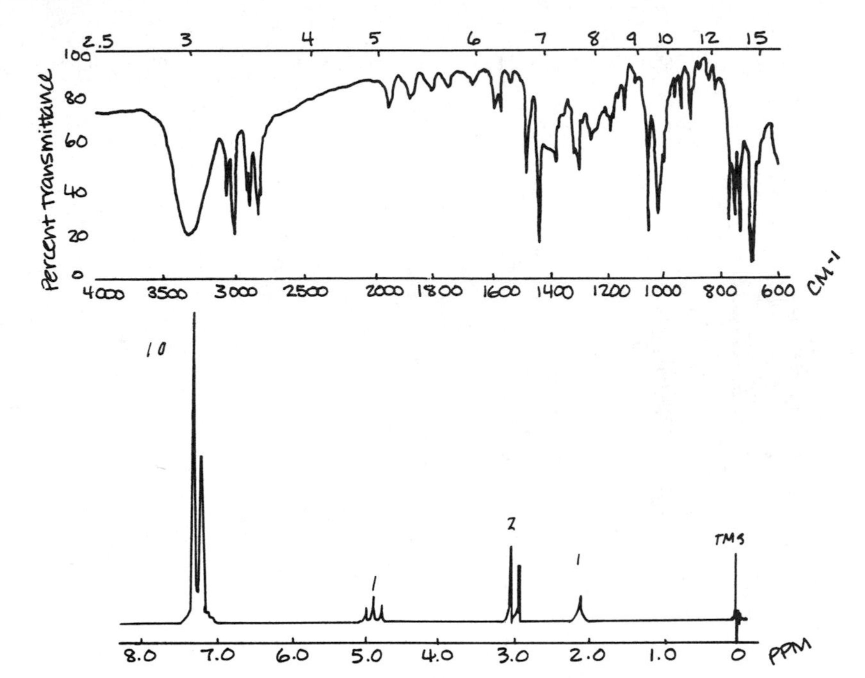

9.32 What is the structure of a compound (C_4H_6O) with the spectra labeled 9.20?

Spectra 9.20

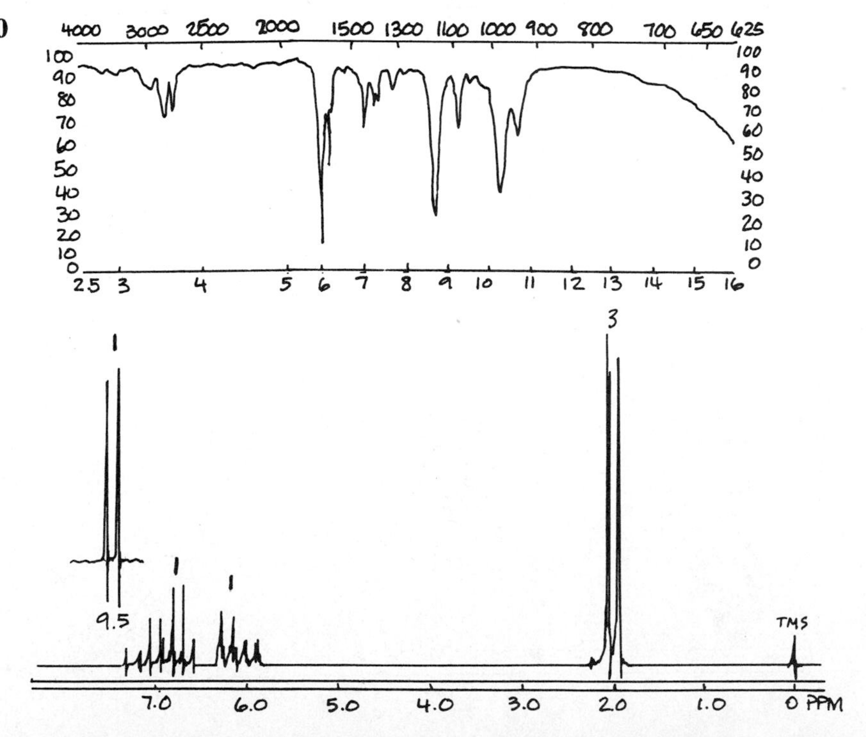

9.33 What is the structure of a compound (C_3H_7ClO) with the spectra labeled 9.21?

Spectra 9.21

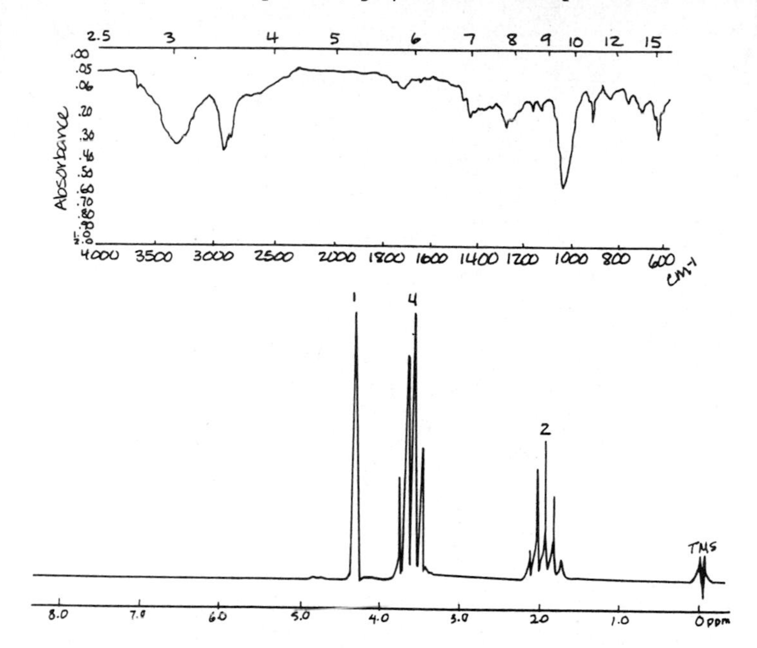

9.34 What is the structure of a compound ($C_8H_{10}O$) with the spectra labeled 9.22?

Spectra 9.22

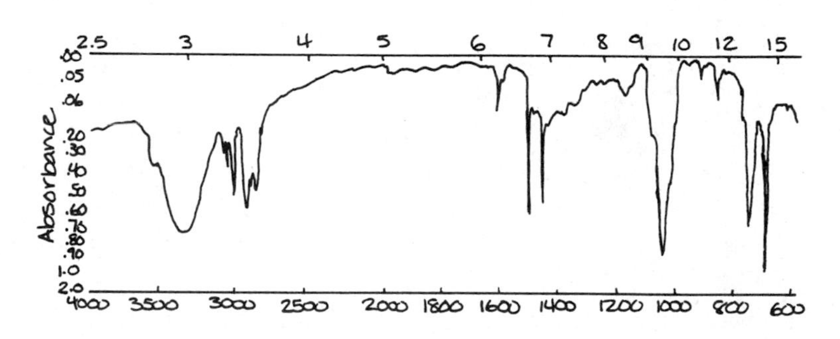

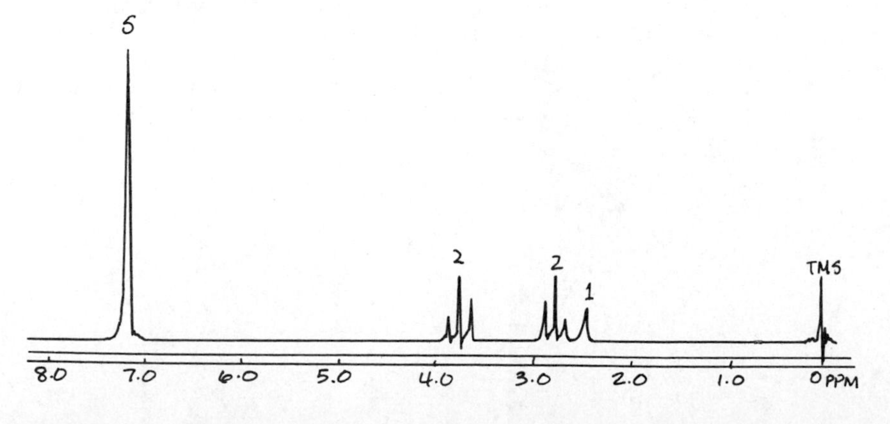

Alkenes and Alkynes

Some Important Features

The principal reactions discussed in Chapter 10 are additions of reagents to carbon-carbon double bonds. These addition reactions can be grouped together according to the types of intermediates formed. Addition of HX or H_2O usually proceeds by way of the more stable carbocation to yield Markovnikov products.

$$R_2C{=}CHR \xrightarrow{H^+} R_2\overset{+}{C}CH_2R$$

a 3° carbocation

$$R_2\overset{+}{C}CH_2R \xrightarrow{X^-} R_2C(X)CH_2R$$

$$R_2\overset{+}{C}CH_2R \xrightarrow[-H^+]{H_2O} R_2C(OH)CH_2R$$

A carbocation can undergo rearrangement if a more stable carbocation will result. In predicting products from carbocation reactions, always inspect the structure of the intermediate carbocation to see if it can rearrange to a more stable intermediate.

Addition reactions of Br_2 and $Hg(O_2CCH_3)_2$ proceed through bridged intermediates, which result in anti-additions. No rearrangements occur, but Markovnikov's rule is still followed.

$$R_2C{=}CHR \xrightarrow[-X^-]{X_2} R_2\overset{\delta+}{C}\cdots\overset{\delta+\ X}{|}\!-CHR \xrightarrow[\text{(such as } X^- \text{ or } H_2O)]{Nu:^-} R_2C(Nu)-CHR(X)$$

$$R_2C{=}CHR \xrightarrow{^+HgO_2CCH_3} R_2\overset{\delta+}{C}\cdots\overset{\delta+\ HgO_2CCH_3}{|}\!-CHR \xrightarrow[-H^+]{H_2O} R_2C(OH)-CHR(HgO_2CCH_3)$$

Other types of addition reactions are those of HBr with peroxides (a free-radical, anti-Markovnikov reaction); BH_3 (which yields what appear to be anti-Markovnikov products); singlet methylene (syn-addition); and H_2 (syn-addition).

Oxidation reactions of alkenes can lead to a variety of products, depending on the reagent and on the degree of alkene substitution.

$$R_2C{=}CHR \xrightarrow[\text{syn-addition}]{\text{cold } KMnO_4 \text{ solution or (1) } OsO_4,\ (2)\ Na_2SO_3} R_2C(OH)-CHR(OH)$$

$$R_2C{=}CHR \xrightarrow[\text{or } O_3\text{, oxidative workup}]{\text{hot } KMnO_4 \text{ solution}} R_2C{=}O + HO\overset{O}{\overset{\|}{C}}R$$

Reminders

Compounds of the type RC≡CH are acidic with respect to very strong bases ($NaNH_2$, RMgX, *etc.*). Metal acetylides such as RC≡CNa or RC≡CMgX are strong nucleophiles and strong bases.

In an addition reaction that proceeds by way of a carbocation (or a bridged intermediate with carbocation character), check the relative stabilities of all possible intermediates. The nucleophile will attack predominantly the most positive carbon.

$$\xrightarrow[\text{increasing carbocation stability}]{2^\circ \quad 3^\circ \quad \text{allylic and benzylic}}$$

Weak nucleophiles (like water, an alcohol, or an anion) can attack a carbocation.

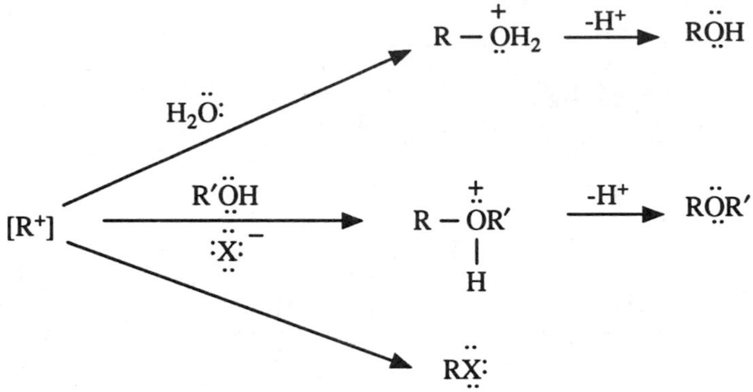

$$[R^+] \xrightarrow{H_2\ddot{O}:} R-\overset{+}{O}H_2 \xrightarrow{-H^+} R\ddot{O}H$$

$$[R^+] \xrightarrow{R'\ddot{O}H} R-\overset{+}{O}R'(H) \xrightarrow{-H^+} R\ddot{O}R'$$

$$[R^+] \xrightarrow{:\ddot{X}:^-} R\ddot{X}:$$

Reactions that proceed through true carbocations are not stereospecific. However, anti-addition reactions, which proceed by way of bridged ions, and syn-addition reactions are stereospecific.

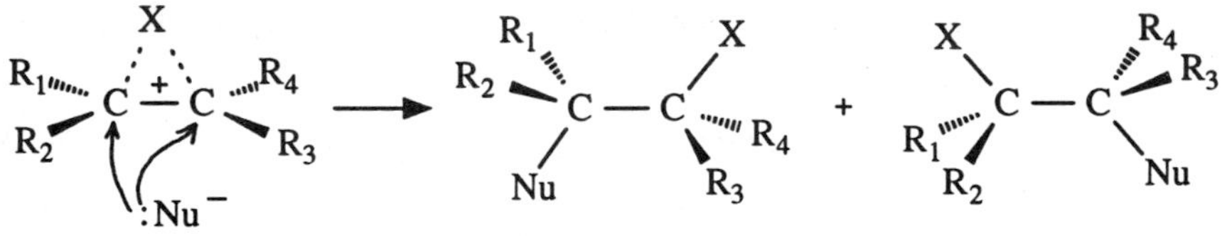

X
R_1 C + C R_4
R_2 R_3
:Nu$^-$
→
R_1, R_2, Nu–C–C–X, R_4, R_3 + X, R_1, R_2–C–C–Nu, R_4, R_3

Additional Drill Problems

10.1 Write the IUPAC names of the following structures:

(a) CH_2=CHI

(b) 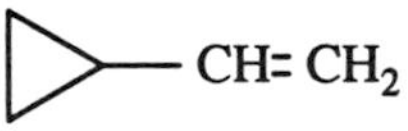 (cyclopropyl)–CH= CH_2

(c) 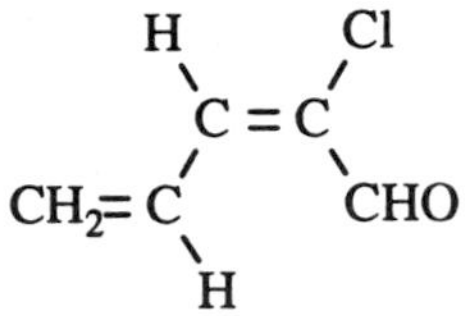 H, Cl; C=C; CH_2= C(H), CHO

(d) 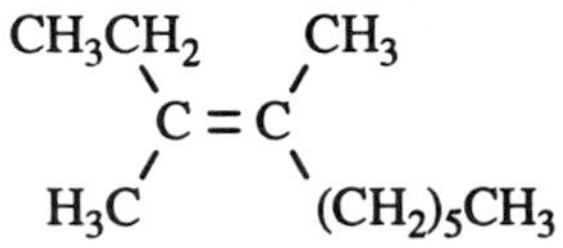 CH_3CH_2, CH_3; C=C; H_3C, $(CH_2)_5CH_3$

(e) CH_2=C=$CHCH_2CH_3$

(f) 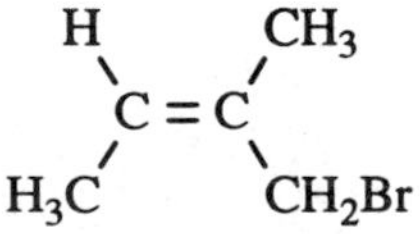 H, CH_3; C=C; H_3C, CH_2Br

(g) H, CH_2Br; C=C; H_3C, CH_3

(h) H, H; C=C; HO_2C, CO_2H

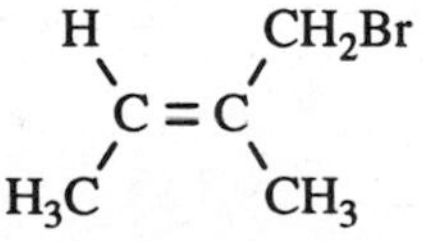

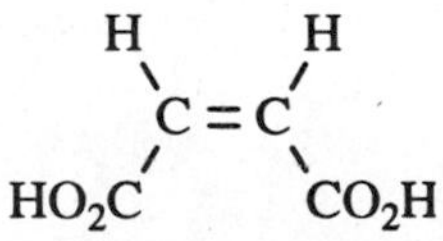

10.2 Draw the structure for each of the following names:

(a) (*E*)-4-methyl-2-hexene

(b) 3,4-diiodo-1,4-heptadiene

(c) *trans*-2-chloro-3-hexene

(d) (*Z*)-3,4-dimethyl-2-pentene

(e) (1*R*,2*S*)-2-methyl-3-cyclopentenol

(f) 3-ethyl-2-hexen-1-ol

10.3 Draw the structure for each of the following names:

(a) 3-methoxy-1-pentyne

(b) 1-bromo-2-butyne

(c) (*R*)-3-methyl-1-octyne

(d) (*Z*)-3,4-dibromo-3-penten-1-yne

10.4 Assign each carbon-carbon double bond as (*E*) or (*Z*):

(a) $CH_3CH_2CH_2$(top left), Br (top right), C=C, CH_3CH_2 (bottom left), H (bottom right)

(b) $BrCH_2$ (top left), CH_3 (top right), C=C, Cl (bottom left), H (bottom right)

(c) H (top left), $CH_2C(CH_3)_3$ (top right), C=C, F (bottom left), $CH_2CH(CH_3)_2$ (bottom right)

(d) HC(=O) (top left), C(=O)OCH_3 (top right), C=C, H_3CC(=O) (bottom left), CN (bottom right)

10.5 Assign each carbon double bond as *cis* or *trans*:

(a) $(CH_3)_2CH$ (top left), Br (top right), C=C, H (bottom left), H (bottom right)

(b) H (top left), C≡CH (top right), C=C, HC≡C (bottom left), H (bottom right)

(c) cyclopentyl–CH=CH–CH=CH_2 (each C bearing H)

10.6 The figure labeled Spectra 10.1 is the infrared and ^{1}H NMR spectra of a compound with the molecular formula $C_5H_8O_2$.

(a) What is the oxygen functional group?

(b) Can absorption by an alkenyl C-H be readily discerned in the infrared spectrum?

(c) Can absorption by an alkenyl C-H be readily discerned in the NMR spectrum?

(d) What is the structure of the compound?

Spectra 10.1

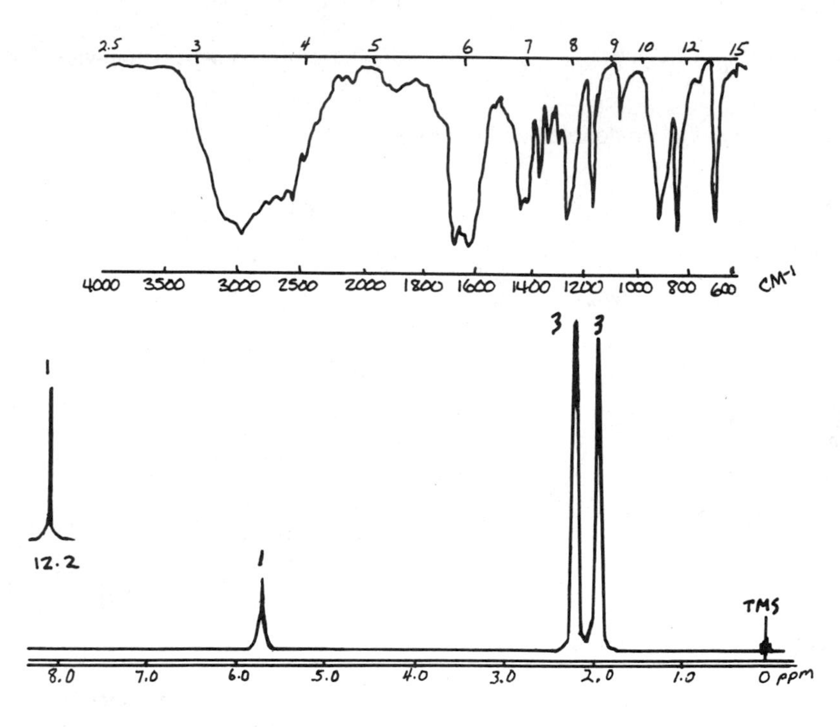

10.7 Match Spectrum 10.2 with one of the following structures:

(a) $CH_2{=}CHOC(CH_3)_3$

(b) $CH_3CH{=}CHOCH_2CH_2CH_2CH_3$

(c) $CH_2{=}\underset{\displaystyle CH_3}{\underset{|}{C}}OCH_2\underset{\displaystyle CH_3}{\underset{|}{C}}HCH_3$

(d) $CH_2{=}CHOCH_2CH_2CH_2CH_3$

Spectrum 10.2

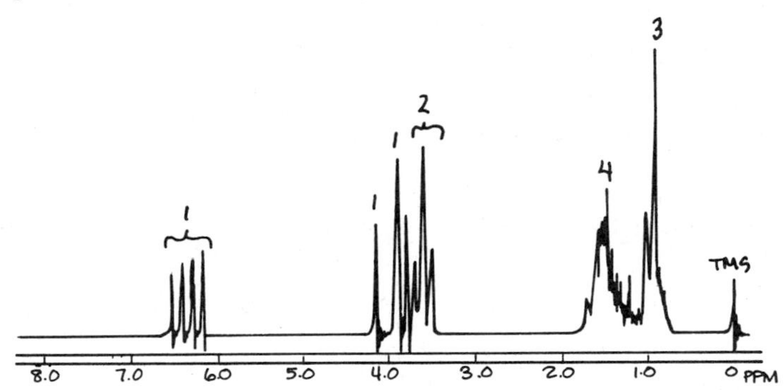

10.8 What organohalogen compound would be needed for each of the following reactions?

(a) $? + CH_3C{\equiv}C^- \ Na^+ \longrightarrow CH_3C{\equiv}CCH_2{-}C_6H_5$

(b) $? + HO^- \longrightarrow CH_2{=}\underset{\displaystyle CH_3}{\underset{|}{C}}{-}C_6H_5$

(c) $? \xrightarrow[(2)\ H^+]{(1)\ NaNH_2} CH_3CH_2CH_2CH_2CH_2C{\equiv}CH$

10.9 Show how the following conversions could be carried out·

(a) 1-bromopropane $\longrightarrow$ 1-pentyne

(b) 1-hexene $\longrightarrow$ 1-hexyne

10.10 Predict the products from the reaction of HCl with the following compounds:

(a) *cis*-2-butene (b) *trans*-2-butene (c) acetylene

(d) propene (e) 1,2-dimethylcyclopentene

10.11 Complete the following equations, showing any reaction intermediates in your answer:

(a) $CH_3CH_2CH_2CH{=}CH_2 + HBr \xrightarrow{\text{peroxide}}$

(b) $C_6H_5C{\equiv}CC_6H_5 + 2\ HCl \longrightarrow$

(c) *cis*-$CH_3CH{=}CHCH_2CH_2CH_3 + HBr \longrightarrow$

10.12 Predict the major products:

(a) $CH_3CH_2CH{=}CH_2 + H_2O \xrightarrow{H^+}$

(b) $(CH_3)_3CCH{=}CH_2 + H_2O \xrightarrow{H^+}$

(c) 1 [cyclopentene] + 1 $H_2O \xrightarrow{H^+}$

10.13 Arrange the following alkenes in order of increasing reactivity towards concentrated sulfuric acid (least reactive first):

(a) $CH_3CH_2CH{=}CH_2$ (b) $CH_3CH{=}CHCH_3$

(c) $CH_2{=}CH_2$ (d) $CH_3C(CH_3){=}CH_2$

10.14 Suggest syntheses for the following compound from alkenes or dienes containing six or fewer carbons and other appropriate reagents:

(a) 1-cyclohexyl-1-propanone

(b) cyclohexylmethanol

(c) *trans*-2-methyl-1-cyclohexanol

(d) *cis*-1,2-dibromocyclohexane

10.15 Predict the major organic product of the reaction of 1-methylcyclooctene with:

(a) $H_2O + H_2SO_4$

(b) (1) BH_3; (2) H_2O_2, ^{-}OH

(c) (1) $H_2O + Hg(O_2CCH_3)_2$; (2) $NaBH_4$

(d) BH_3; (2) Br_2; NaOH

(e) $CHCl_3 + K^+ \ {}^{-}OC(CH_3)_3$

(f) (1) OsO_4; (2) Na_2SO_3

(g) $C_6H_5CO_3H$

(h) CH_2I_2 + Zn(Cu) + $(CH_3CH_2)_2O$

10.16 Of each of the following pairs of compounds, tell which one would be more stable. Explain.

(a) 2-methylcyclohexane with $=CH_2$ (CH_3) or cyclohexane with $=CHCH_3$

(b) *cis*- or *trans*-$CH_3CH_2CH{=}CHCH_3$

(c) (*E*)- or (*Z*)-$CH_3\overset{O}{\overset{\|}{C}}OCH{=}CH\overset{O}{\overset{\|}{C}}OCH_3$

(d) $CH_2{=}CHCH_2\overset{O}{\overset{\|}{C}}OCH_3$ or $CH_2{=}CH\overset{O}{\overset{\|}{C}}OCH_2CH_3$

10.17 Write the formula for the hydrogenation product of (1.0 mol H_2) for each of the following alkenes. Show stereochemistry where appropriate.

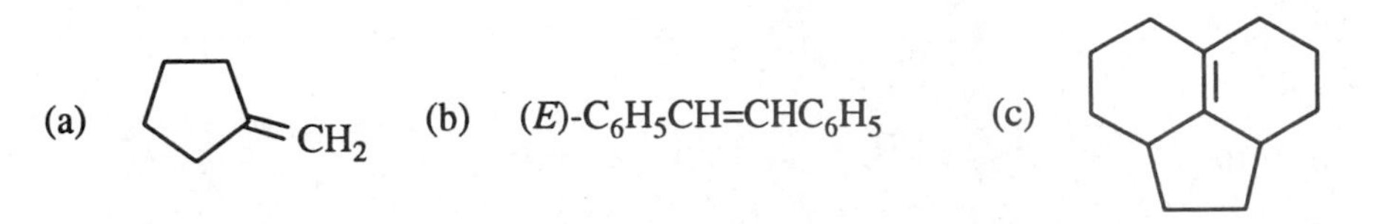

10.18 What would be the products if each of the following compounds were treated with ozone followed by oxidative work-up?

(a) styrene

(b) cycloheptene

(c) (*E*)-3-methyl-2-hexene

(d) *cis*-3,6-diphenylcyclohexene

10.19 Complete the following equations, showing only the major organic products:

(a) 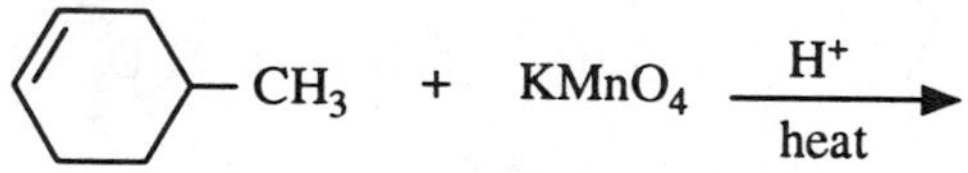 —CH_3 + $KMnO_4$ $\xrightarrow[\text{heat}]{H^+}$

(b) 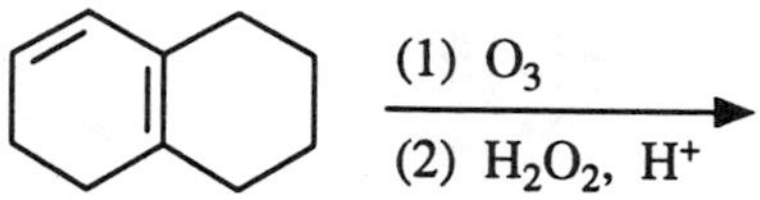 $\xrightarrow[(2)\ H_2O_2,\ H^+]{(1)\ O_3}$

(c) 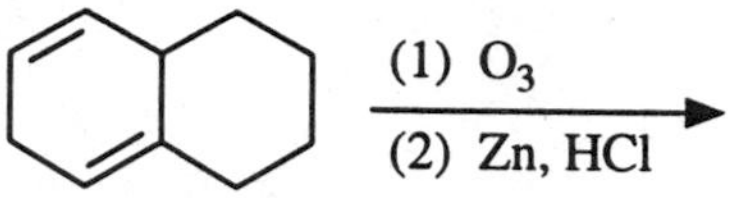 $\xrightarrow[(2)\ Zn,\ HCl]{(1)\ O_3}$

10.20 Write flow equations for the following reactions:

(a) 3,4-diethyl-3-hexene and ozone, followed by H_2O_2, H^+

(b) 2 butyne-1,4-diol and 1.0 mol of H_2 with Pd

(c) 2-butyne and H_2O, Hg^{2+}, H^+

10.21 Complete the following equations. Show stereochemistry where appropriate.

(a) 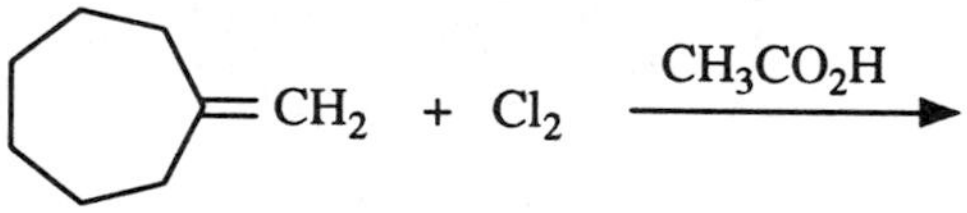 $=CH_2$ + Cl_2 $\xrightarrow{CH_3CO_2H}$

(b) $CH_3(CH_2)_4C(CH_3)=CH_2$ + $Hg(OCCH_3)_2$ (C=O) $\xrightarrow{H_2O}$

(c) product from (b) + $NaBH_4$ $\longrightarrow$

(d) 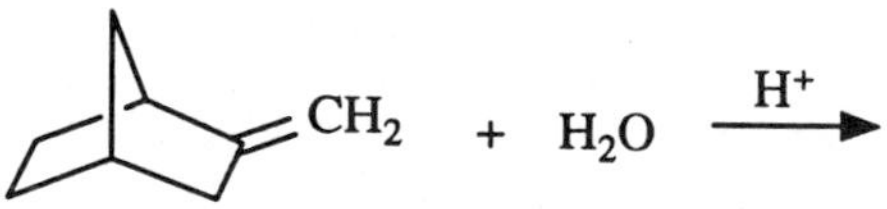 CH_2 + H_2O $\xrightarrow{H^+}$

10.22 Predict the major product(s) of the reaction of (1) *trans*-2-pentene and (2) *cis*-2-pentene with each of the following reagents:

(a) (1) O_3; (2) H_2O_2, H^+ (b) cold $KMnO_4$, OH^-

(c) $C_6H_5CO_3H$ (d) H_2/Pt

(e) Br_2 (f) hot $KMnO_4$, H^+

10.23 Predict the ozonolysis products when each of the following alkenes is subjected to (1) reaction with O_3, followed by (2) reaction with H_2O_2 and H^+:

(a) 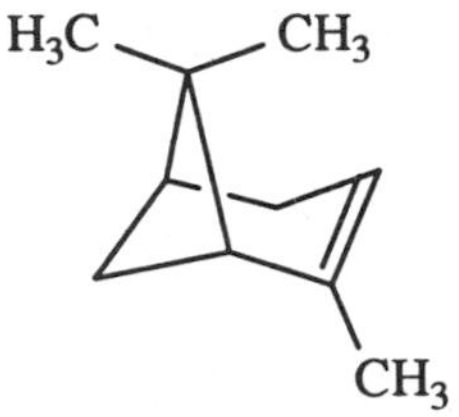

α-pinene

(b) 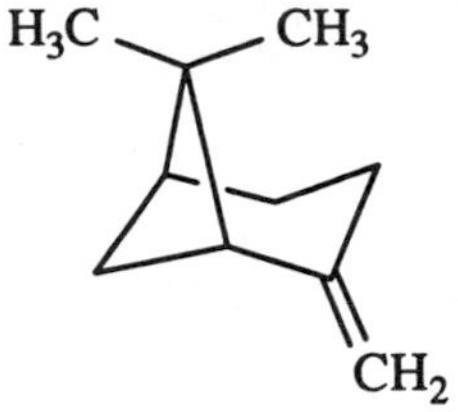

β-pinene

10.24 In each of the following equations, fill in the missing reagents:

(a) CH_2N_2 + ? ⟶ 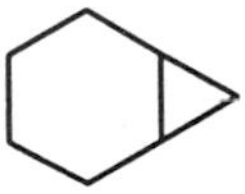

(b) ? + Br_2 $\xrightarrow{CCl_4}$ $CH_3CH_2CHBrCH_2Br$

(c) $CH_3CH_2CH{=}CH_2$ + ? ⟶ $CH_3CH_2CH_2CH_2Br$

(d) $CH_3CH_2CH{=}CH_2$ + ? ⟶ $CH_3CH_2CH(OH)CH_3$

(e) $CH_3CH_2CH{=}CH_2$ + ? ⟶ $CH_3CH_2CH_2CH_2OH$

(f)  + ? ⟶ 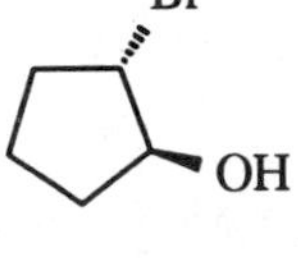

(g) ? + H_2O, H_2SO_4 $\xrightarrow{Hg^{2+}}$ $CH_3C(=O)CH_2CH_3$

10.25 Give reagents needed to carry out each of the following conversions:

(a) (cyclohexylidene)$=CH_2$ ⟶ (cyclohexyl)$-CH_2Br$

(b) $CH_3CH{=}CHCH_3$ ⟶ $CH_3CH(Cl)-CH(OCH_3)CH_3$

(c) (octahydronaphthalene, C=C at ring fusion) ⟶ (decalin with Br at each ring-fusion carbon, trans: Br wedge up, Br hashed down)

(d) $CH_3CH_2CH(CH_3)CH{=}CH_2$ ⟶ $CH_3CH_2CH(CH_3)CH_2CH_2OH$

(e) $CH_3CH_2CH{=}CH_2$ ⟶ $CH_3CH_2CBr_2CH_3$

10.26 Show how the following conversions could be carried out:

(a) an alkyne ⟶ *trans*-2-hexene

(b) the same alkyne as in (a) ⟶ *cis*-2-hexene

(c) $(CH_3)_3COH$ ⟶ $(CH_3)_3CH$

11

Aromaticity and Benzene; Electrophilic Aromatic Substitution

Some Important Features

An aromatic compound is one that is substantially stabilized by pi-electron delocalization. To be aromatic, a compound must be cyclic and planar; each ring atom must have a *p* orbital perpendicular to the plane of the ring; and the number of pi electrons in the ring must fit the formula $4n + 2$, where n is an integer.

Benzene, a typical aromatic compound, undergoes a variety of electrophilic aromatic substitution reactions.

E^+ $-H^+$ H   E   +   E

If the first substituent can donate electrons to the ring by resonance or by the inductive effect, a second substitution occurs *ortho* and *para* because of stabilization of the intermediate (see Section 11.9). Be sure you can draw resonance structures for the intermediates.

unshared e^-

$\ddot{E}$ E^+ / $-H^+$ E, $\ddot{E}$ + E, $\ddot{E}$

o   *p*

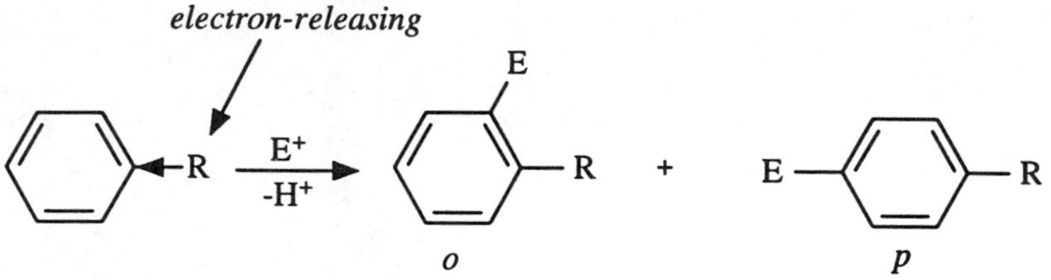

electron-releasing

E

R $\xrightarrow[-H^+]{E^+}$ R + E — R

o *p*

Except for the halogens, the *o,p*-directors activate the ring toward further electrophilic substitution.

If a first substituent cannot donate electronic charge to the ring, it deactivates the ring toward further electrophilic substitution. In this case, a second substitution occurs *meta* to the first substituent.

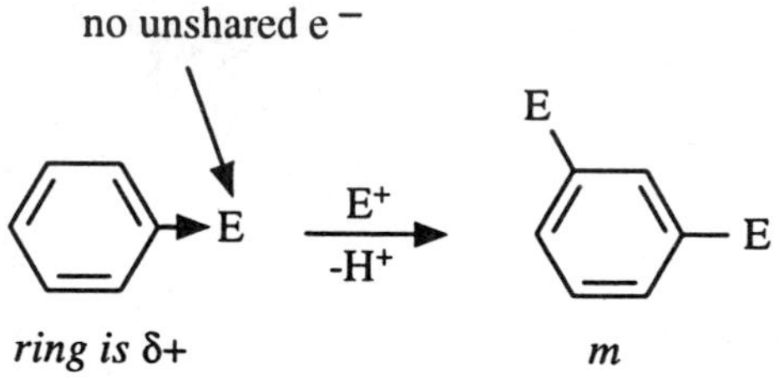

no unshared e$^-$

E

E $\xrightarrow[-H^+]{E^+}$ E

ring is δ+ *m*

Reminders

Except for alkyl and aryl groups, an *o,p*-director has unshared electrons on the atom adjacent to the ring.

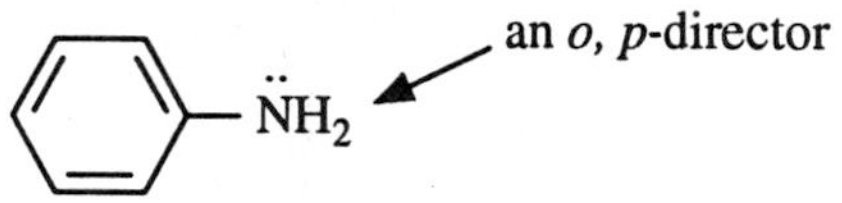

$\ddot{N}H_2$ ← an *o, p*-director

Release of electron density by a substituent activates a ring toward electrophilic substitution, while withdrawal of electrons deactivates a ring toward this type of substitution.

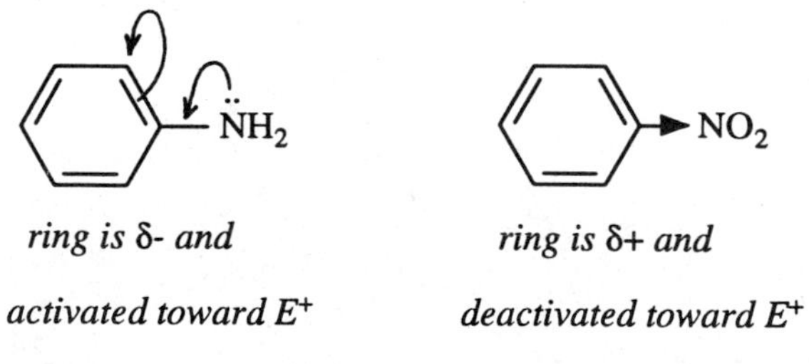

$\ddot{N}H_2$ NO_2

ring is δ- *and* *ring is* δ+ *and*

activated toward E^+ *deactivated toward* E^+

Additional Drill Problems

11.1 Draw structures for the following names:

(a) toluene

(b) *m*-bromostyrene

(c) *o*-methoxyphenol

(d) methyl phenyl ether

(e) benzyl methyl ether

(f) *p*-nitroaniline

11.2 Name the following structures:

(a) 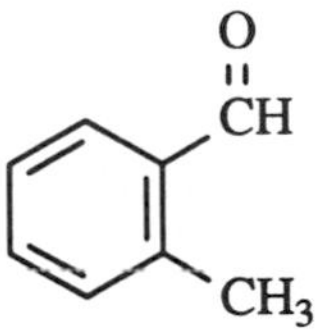 [benzene ring with CHO (O=CH) and ortho CH_3]

(b) 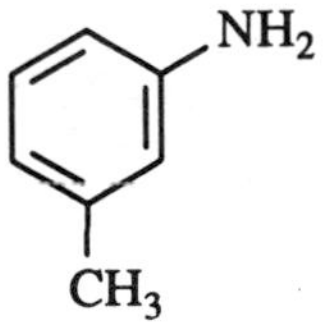 [benzene ring with NH_2 and meta CH_3]

(c) 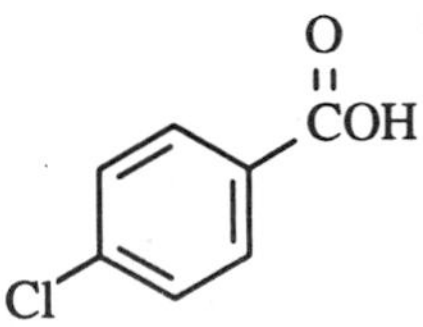 [benzene ring with COOH (O=COH) and para Cl]

(d) 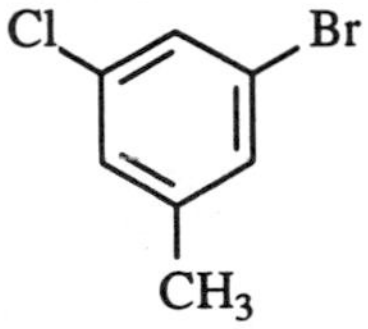 [benzene ring with Cl, Br, and CH_3 at 1,3,5 positions]

11.3 Match the infrared and ^{1}H NMR spectra labeled 11.1 with one of the following compounds:

(a) $C_6H_5CH_2CH(OH)C_6H_5$

(b) $C_6H_5CH_2CH(OH)CH_3$

(c) $(C_6H_5)_2CHC(=O)CH_3$

(d) $C_6H_5CH_2CH(C_6H_5)OC_6H_5$

Spectra 11.1

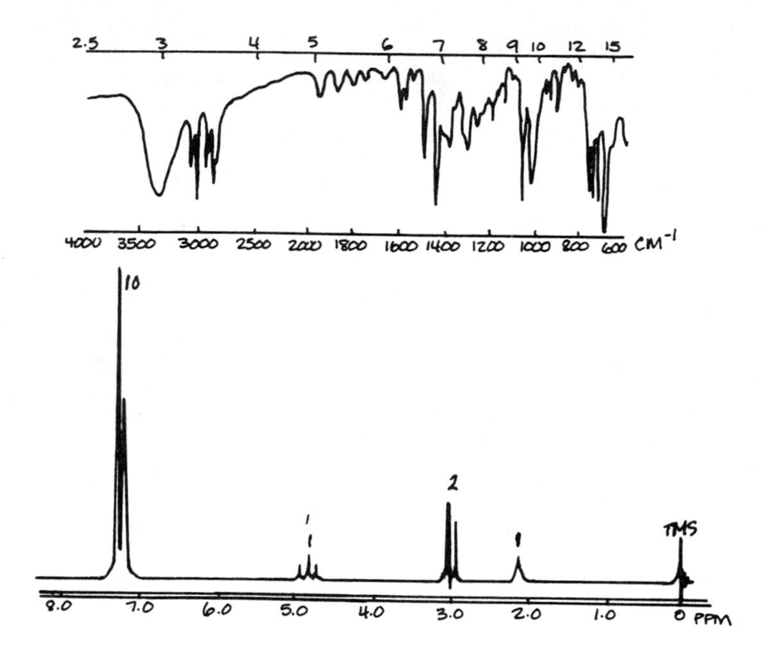

11.4 Match the ^{1}H NMR spectrum labeled 11.2 with one of the following compounds:

(a) $H_2N-C_6H_4-OH$ (para)

(b) $HOCH_2-C_6H_4-CH_2NH_2$ (para)

(c) $HOCH_2CH_2-C_6H_4-NH_2$ (para)

(d) C_6H_4 with NH_2 and CH_2CH_2OH (ortho)

Spectrum 11.2

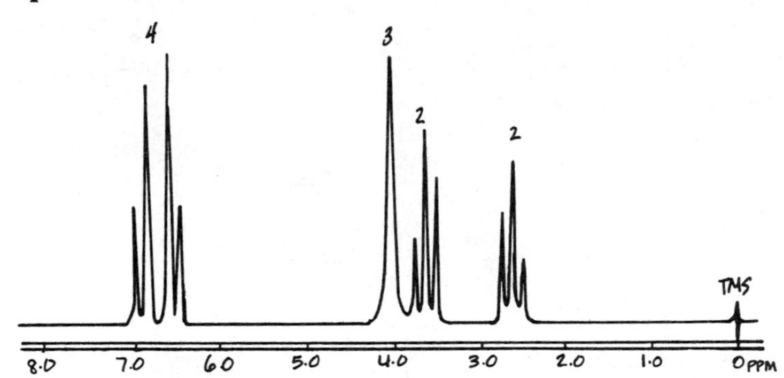

11.5 What is the structure of the compound $C_8H_{10}O$ with the ^1H NMR spectrum labeled 11.3?

Spectrum 11.3

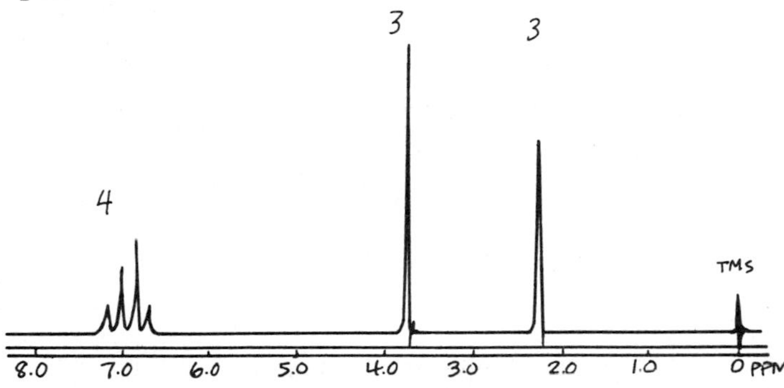

11.6 Which of the following structures would you expect to be aromatic?

(a)

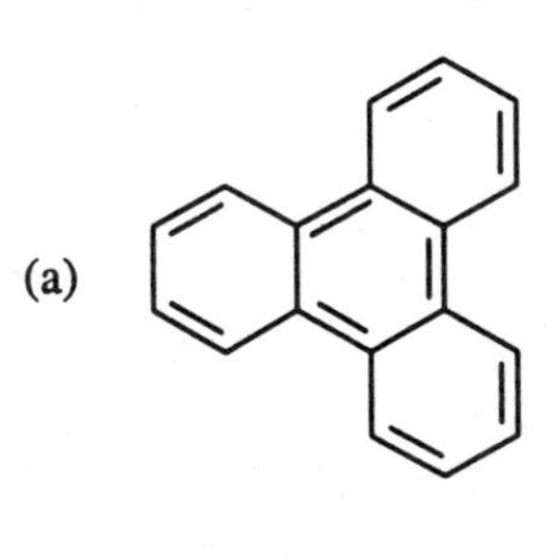

(b)

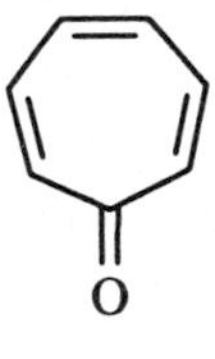

(c)

(d)

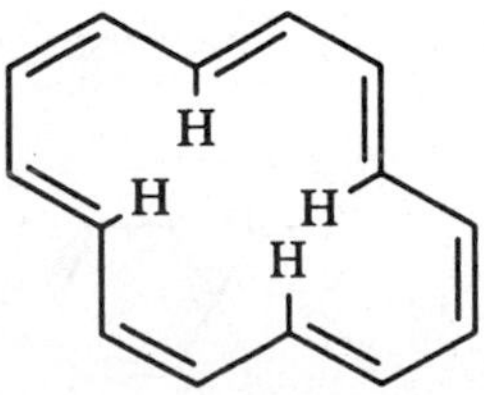

11.7 Write flow equations for the mechanism of the chlorination of benzene.

11.8 Identify each of the following compounds as containing an *o,p*-directing or *m*-directing substituent:

(a) 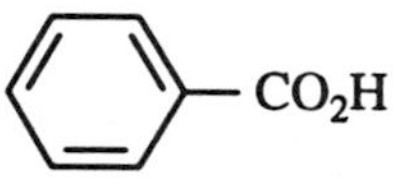

(b) 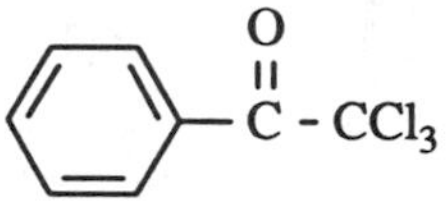

(c) 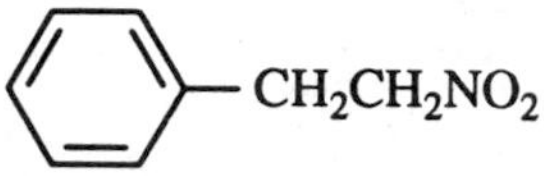

(d) 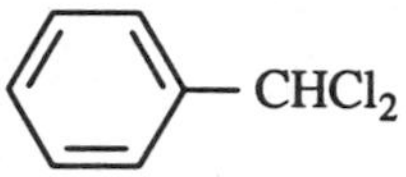

11.9 List the following compounds in order of increasing reactivity towards Br_2 + $FeBr_3$ (least reactive first):

(a) C_6H_5OH (b) $C_6H_5O\overset{O}{\overset{\|}{C}}CH_3$ (c) C_6H_5Br

(d) $C_6H_5\overset{O}{\overset{\|}{C}}OCH_3$ (e) $C_6H_5SO_3H$ (f) $C_6H_5CH_2CH_3$

11.10 Write one-step equations for preparing the following compounds:

(a) 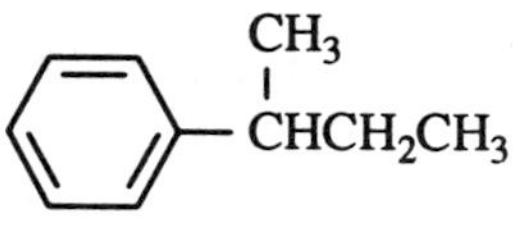

(b) 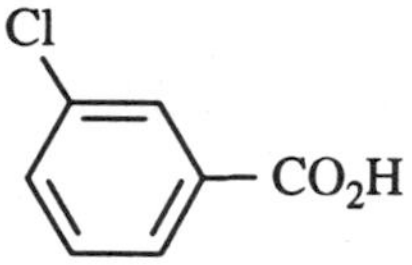

(c) 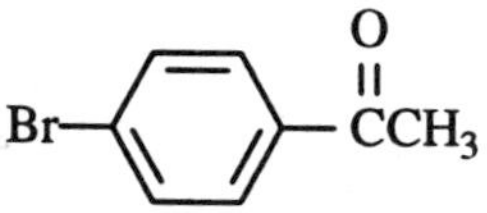

11.11 Draw resonance structures for the following intermediates:

(a) 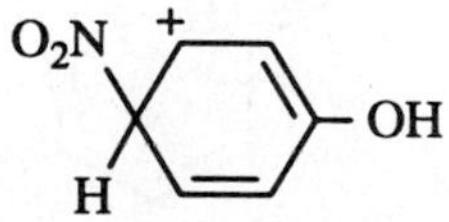

(b) 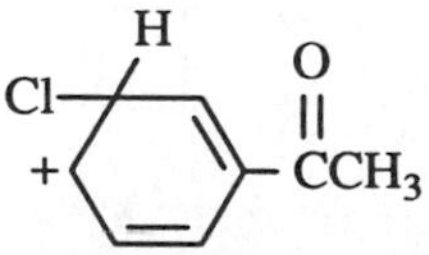

11.12 Draw resonance structures of the benzenonium intermediate for:

(a) the *para* nitration of anisole (methoxybenzene)

(b) the *meta* nitration of benzoic acid

11.13 Write the equations for the mononitration of the following compounds:

(a) *p*-toluenesulfonic acid (b) *p*-nitrophenol

(c) *p*-bromotoluene

11.14 Starting with aniline, benzene, or toluene, show how you could prepare the following compounds.

(a) CO_2H / NH_2 (para-substituted benzene ring)

(b) $\overset{O}{\overset{\|}{C}}CH_3$ / CH_3 (para-substituted benzene ring)

(c) CH_3 / SO_3H (meta-substituted benzene ring)

12

Substituted Benzenes

Some Important Features

This chapter includes the chemistry of alkylbenzenes, phenols, benzenediazonium salts and halobenzenes.

The benzylic position of an alkylbenzene has enhanced reactivity toward oxidation and free-radical reactions (Section 12.1). Recall from Chapter 5 that benzylic halides are highly reactive in S_N1 and S_N2 reactions.

Phenols are more acidic than alcohols, but usually less acidic than carboxylic acids.

$$ROH \xrightarrow[\text{extremely strong base}]{\text{Na, NaH, or}} RO^-$$

$$ArOH \xrightarrow{\text{NaOH}} ArO^-$$

$$RCO_2H \xrightarrow{\text{NaHCO}_3} RCO_2^-$$

Phenols can be esterified only with active derivatives of carboxylic acids (acid halides or anhydrides). Phenols undergo some unique reactions (see Section 12.2C-E).

Primary arylamines ($ArNH_2$) can be converted to diazonium salts, which can be converted, in turn, to a variety of other compounds (Section 12.3).

A ring substituted with an electron-withdrawing group is activated toward *nucleophilic* substitution.

O_2N–C_6H_4–Cl $\xrightarrow{:\ddot{O}H^-}$ [O_2N–C_6H_4(Cl)(OH)$^-$] → O_2N–C_6H_4–$\ddot{O}H$

Aromatic nucleophilic substitutions can also occur under "forcing" conditions through a benzyne intermediate (Section 12.4).

Additional Drill Problems

12.1 Complete the following equations:

(a) HO–C_6H_4–OH + 2 $CH_3C(=O)OC(=O)CH_3$ ⟶

(b) HO–C_6H_4–NO_2 + 2 Br_2 ⟶

(c) HO–C_6H_4–$CO_2CH_2CH_3$ + 2 Cl_2 ⟶

(d) C_6H_5–C_6H_4–OH + NaOH ⟶

(e) 2-naphthol (OH) + $CHCl_3$ + NaOH ⟶

12.2 Show how the following compounds could be prepared from benzene by a benzenediazonium procedure.

(a) HO–C_6H_4–$C(=O)CH_3$ (meta)

(b) 1,3,5-trichlorobenzene (Cl, Cl, Cl)

(c) CH_3–C_6H_4–N=N–C_6H_4–OH

12.3 What phenol and diazonium salt would be needed to prepare each of the following diazo compounds?

(a) [structure: 1-hydroxy-2-naphthyl–N=N–C6H4–SO3H]

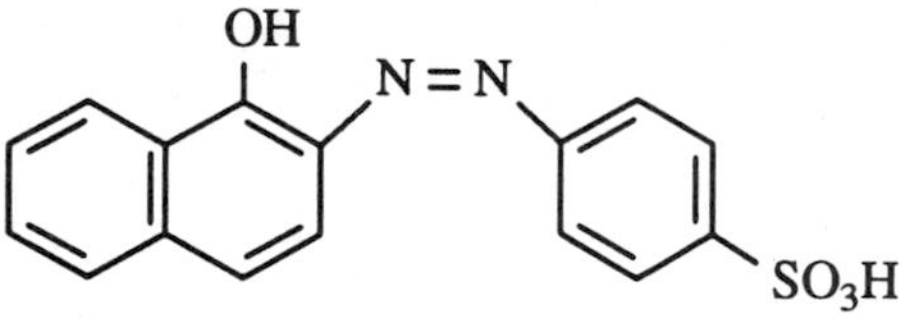

(b) [structure: 2-hydroxy-1-naphthyl–N=N–C6H4–SO3H]

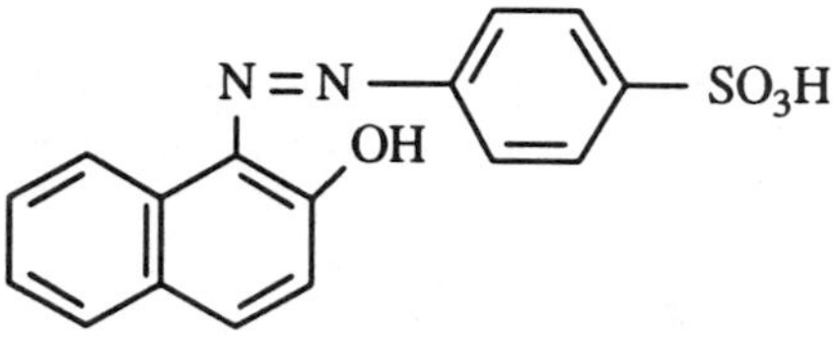

(c) [structure: NaO_3S, NH_2, OH, N=N, SO_3Na substituted naphthalene–N=N–biphenyl–N=N–naphthalene with OH, NH_2, SO_3Na, SO_3Na]

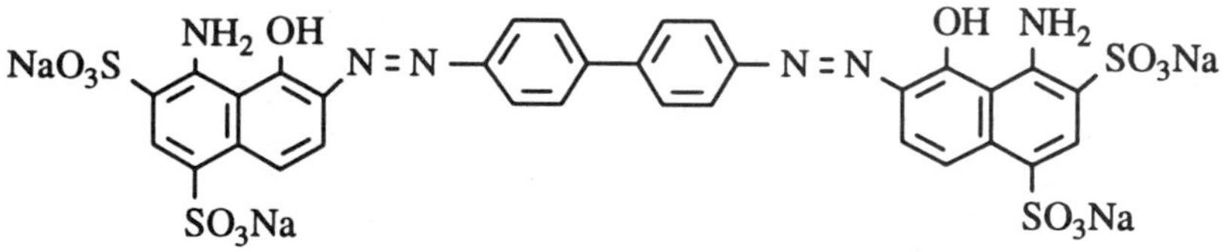

12.4 Write flow equations showing how the following three compounds can be synthesized from the same starting material.

(a) $HO-C_6H_4-NO_2$ (b) $H_2N-C_6H_4-NO_2$

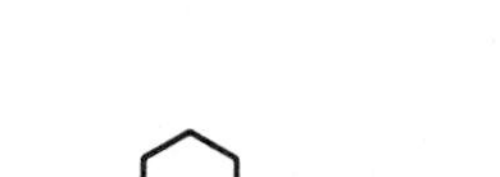

(c) $CH_3O-C_6H_4-NO_2$

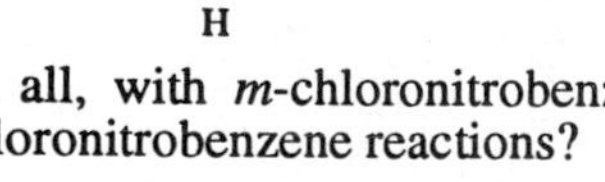

12.5 Piperidine ($C_5H_{10}NH$) reacts readily with *o*- and *p*-chloronitrobenzene but very slowly, if at all, with *m*-chloronitrobenzene. (a) What are the products of the *o*- and *p*-chloronitrobenzene reactions? (b) Why does the *m*-chloro isomer fail to react?

12.6 Complete the following equations:

(a) 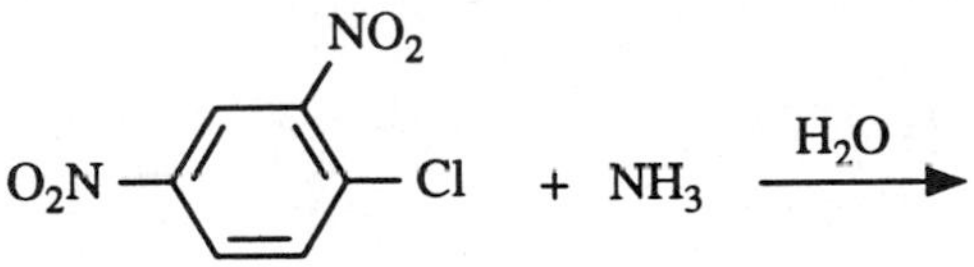 2-chloro-1,5-dinitrobenzene (O_2N–$C_6H_3(NO_2)$–Cl) + NH_3 $\xrightarrow{H_2O}$

(b) 2 O_2N–C_6H_4–Cl + S^{2-} $\longrightarrow$

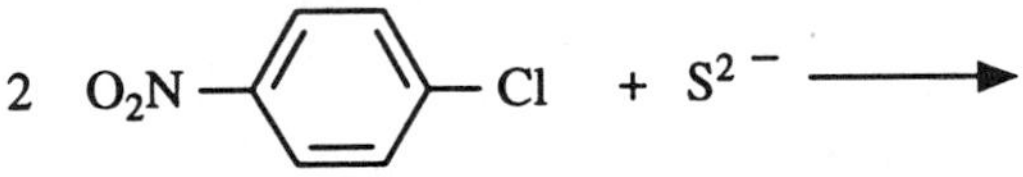

(c) O_2N–$C_6H_2(NO_2)_2$–Cl + NH_2NH_2 $\longrightarrow$

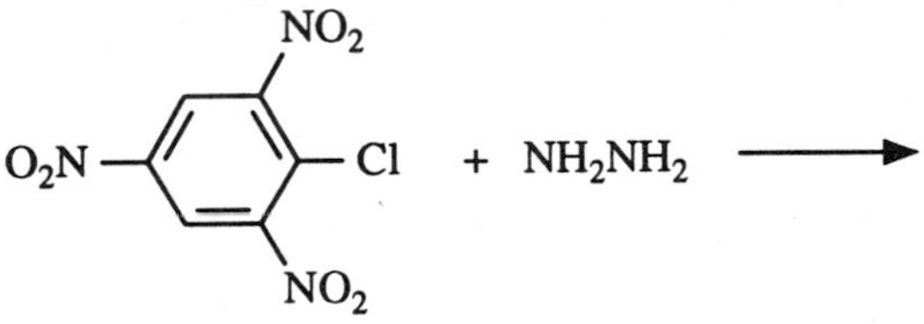

(d) 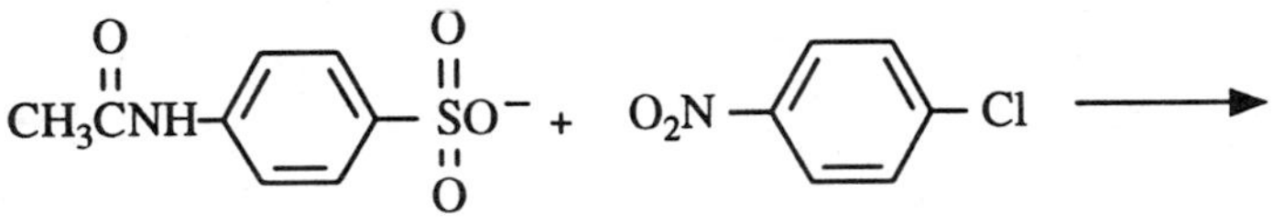 $CH_3C(=O)NH$–C_6H_4–SO_2O^- + O_2N–C_6H_4–Cl $\longrightarrow$

(e) 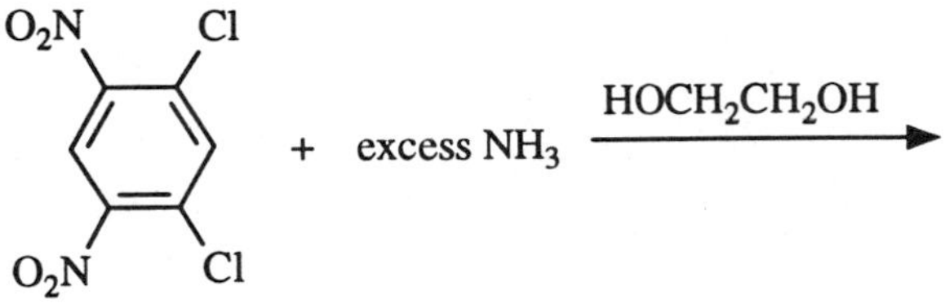 1,2-dichloro-4,5-dinitrobenzene + excess NH_3 $\xrightarrow{HOCH_2CH_2OH}$

12.7 Write flow equations for the preparation of the following disubstituted benzenes from benzene in two-step syntheses.

(a) 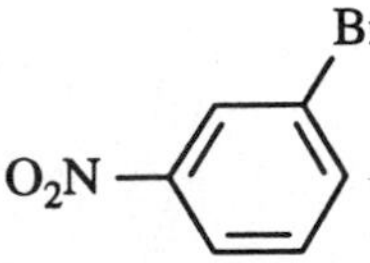 O_2N–C_6H_4–Br (meta)

(b) O_2N–C_6H_4–Br (para)

(c) CH_3–C_6H_4–SO_3H (meta)

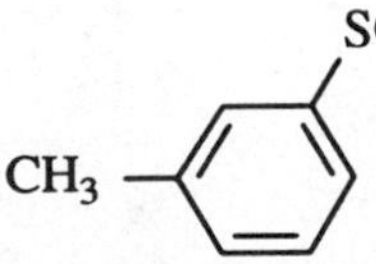

(d) 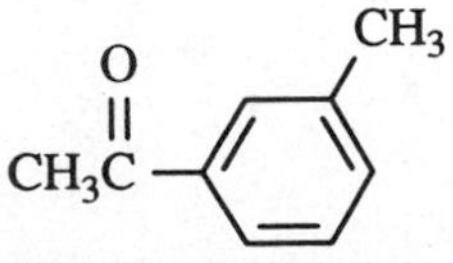 $CH_3C(=O)$–C_6H_4–CH_3 (meta)

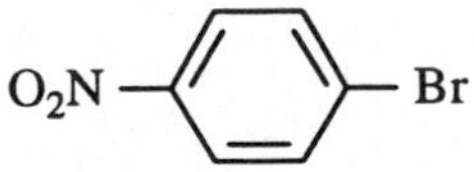

12.8 Which ring of phenyl benzoate would you expect to undergo electrophilic bromination? Explain your answer.

12.9 Show how the following compounds can be prepared from a monosubstituted benzene.

(a) Br–C_6H_4–NH_2 (b) CH_3O–C_6H_4–NH_2

(c) $CH_3\overset{O}{\overset{\|}{C}}O$–$C_6H_4$–$NH_2$ (d) CH_3O–C_6H_4–CN

12.10 Use a flow diagram to show how benzene can be converted to each of the following compounds:

(a) $C_6H_5CH_2Cl$ (b) $C_6H_5\underset{}{CH}(OH)–CH(OH)C_6H_5$ (c) Br–C_6H_4–CO_2H

Aldehydes and Ketones

Some Important Features

A carbonyl group (C=O) is polar and can be attacked by an electrophile such as H^+ or by a nucleophile.

In acidic solution, the oxygen is protonated:

$$>C=\ddot{O}: \; + \; H^+ \rightleftharpoons >C=\overset{+}{\ddot{O}}H$$

In base, the carbon is attacked:

$$-\underset{}{\overset{\ddot{O}:}{\overset{\|}{C}}}- \; + \; :Nu^- \rightleftharpoons -\overset{:\ddot{O}:^-}{\overset{|}{\underset{\underset{Nu}{|}}{C}}}-$$

Aldehydes and ketones can undergo many *addition reactions*. Some of these reactions are reversible; others, such as Grignard reactions, are irreversible.

$$\underset{\text{an aldehyde or ketone}}{\overset{\text{O}}{\overset{\|}{\text{RCH}}}} + \text{R'OH} \overset{\text{H}^+}{\rightleftharpoons} \underset{\text{a hemiacetal or hemiketal}}{\begin{matrix}\text{OH}\\ |\\ \text{RCH}\\ |\\ \text{OR'}\end{matrix}}$$

The carbonyl compound is favored by a less positive, more hindered C=O carbon, as in a ketone.

The addition product is favored by a more positive, less hindered C=O carbon, as in an aldehyde.

$$\underset{\text{a ketone}}{\overset{\delta-\ :\ddot{\text{O}}}{\overset{\|}{\text{RCR}}}\ \delta+} + \overset{\delta-\ \ \delta+}{\text{R'}-\text{MgX}} \longrightarrow \underset{\text{a magnesium alkoxide}}{\begin{matrix}:\ddot{\text{O}}:^-\ {}^+\text{MgX}\\ |\\ \text{RCR}\\ |\\ \text{R'}\end{matrix}}$$

Aldehydes and ketones can undergo *addition-elimination* reactions with many nitrogen compounds. The products are favored by resonance-stabilization or by electron-withdrawing groups on the nitrogen. Typical reagents are primary amines (RNH_2) and hydrazine derivatives ($RNHNH_2$).

$$\overset{:\ddot{\text{O}}}{\overset{\|}{\text{RCR}}} + \text{R}\ddot{\text{N}}\text{H}_2 \rightleftharpoons \begin{matrix}:\ddot{\text{O}}\text{H}\\ |\\ \text{RCR}\\ |\\ \text{NHR}\end{matrix} \overset{-\text{H}_2\text{O}}{\rightleftharpoons} \underset{\text{an imine}}{\begin{matrix}\text{RCR}\\ \|\\ \ddot{\text{N}}\text{R}\end{matrix}}$$

Aldehydes and ketones undergo catalytic hydrogenation with heat and pressure; however, these conditions also reduce carbon-carbon double bonds. Reduction to an alcohol with a metal hydride does not usually reduce a carbon-carbon double bond. The carbonyl group can be reduced to $-CH_2-$ or to -CHNHR (Section 13.6C and D). Aldehydes are also very easily oxidized to carboxylic acids.

A hydrogen atom α to a carbonyl group is slightly acidic. (For resonance structures of the product enolate ions, see Section 13.8.)

$$R_2\overset{\overset{O}{||}}{C}CR + {}^{-}\!:\!\ddot{O}R \rightleftharpoons R_2\ddot{C}\overset{\overset{O}{||}}{C}R^{-} + H\ddot{O}R$$

$$R\overset{\overset{O}{||}}{C}CH\overset{\overset{O}{||}}{C}R + {}^{-}\!:\!\ddot{O}R \rightleftharpoons R\overset{\overset{O}{||}}{C}\ddot{C}H\overset{\overset{O}{||}}{C}R^{-} + H\ddot{O}R$$

An aldehyde or ketone with an α hydrogen can undergo tautomerization. Unless the enol tautomer is stabilized relative to the keto tautomer (by hydrogen bonding, for example), the keto form is favored.

$$R_2CH\overset{\overset{O}{||}}{C}R \rightleftharpoons R_2C=\overset{\overset{OH}{|}}{C}R$$

keto *enol*

Reminders

In acidic solution the C=O is protonated, and a weak nucleophile can attack the C=O carbon.

$$\underset{\delta+}{R\overset{\overset{\delta-\ :O:}{||}}{C}R} \overset{H^+}{\rightleftharpoons} R\overset{\overset{+\ \ddot{O}H}{||}}{C}R \xrightarrow{:Nu^-} R\overset{\overset{:\ddot{O}H}{|}}{\underset{\underset{Nu}{|}}{C}}R$$

A strong nucleophile, such as CN^-, can attack the C=O carbon without prior protonation of the C=O group.

$$\underset{\delta+}{R\overset{\overset{\delta-\ :\ddot{O}}{||}}{C}R} + {}^{-}\!:\!CN \rightleftharpoons R\overset{\overset{:\ddot{O}:^-}{|}}{\underset{\underset{CN}{|}}{C}}R \overset{HCN}{\rightleftharpoons} R\overset{\overset{:\ddot{O}H}{|}}{\underset{\underset{CN}{|}}{C}}R + {}^{-}\!:\!CN$$

A strong base, such as ^{-}OR, can remove an α hydrogen.

Remember that an sp^2 carbon atom cannot be chiral.

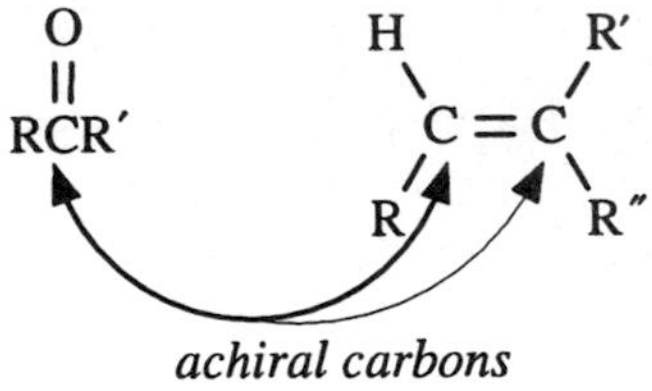

$$R\overset{O}{\overset{\|}{C}}R' \qquad \begin{matrix} H & & R' \\ & C=C & \\ R & & R'' \end{matrix}$$

achiral carbons

Additional Drill Problems

13.1 Write formulas for the following names:

(a) diethyl ketone
(b) α-chloropropionaldehyde
(c) ω-chlorobutyraldehyde
(d) dibenzyl ketone
(e) α,α′-dichloroacetone
(f) α,α-dichloroacetone

13.2 Write formulas for the following names:

(a) 2,2-diethylcyclopentanone
(b) *cis*-3-penten-2-one
(c) 1,4-cycloheptanedione
(d) 4-hydroxypentanal
(e) *p*-chlorobenzaldehyde
(f) propenal

13.3 By the use of flow equations, show how 2-butanone can be synthesized from each of the following starting materials:

(a) 2-butyne
(b) bromoethane
(c) 2-butanol
(d) 1-butene

13.4 Complete the following equations, showing only the major organic products:

(a) $C_6H_5\overset{CO_2H}{\overset{|}{C}}HCH_2\overset{OH}{\overset{|}{C}}HC_6H_5 \xrightarrow{H_2CrO_4}$

(b) $C_6H_5Cl \;+\; C_6H_5\overset{O}{\overset{\|}{C}}Cl \xrightarrow{AlCl_3}$

13.5 Using flow equations, show how each of the following conversions could be carried out:

(a) toluene (methylbenzene) ⟶ benzaldehyde

(b) propene ⟶ propanal

(c) propene ⟶ acetone

(d) benzene ⟶ (benzene ring bearing $\overset{O}{\overset{\parallel}{C}}CH_3$ and NO_2 in meta positions)

13.6 What are the structures of the following compounds?

(a) $C_9H_{10}O$, Spectra 13.1 (b) $C_9H_{10}O$, Spectra 13.2

(c) C_4H_7ClO, Spectra 13.3

Spectra 13.1

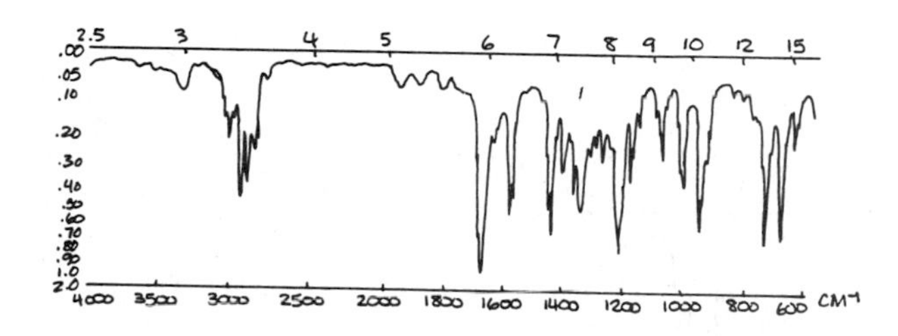

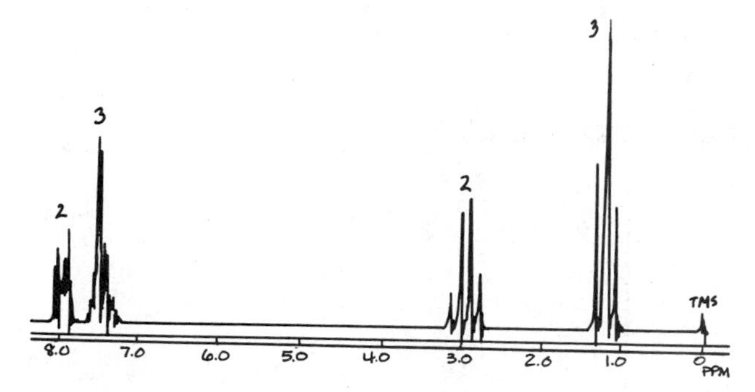

Spectra 13.2

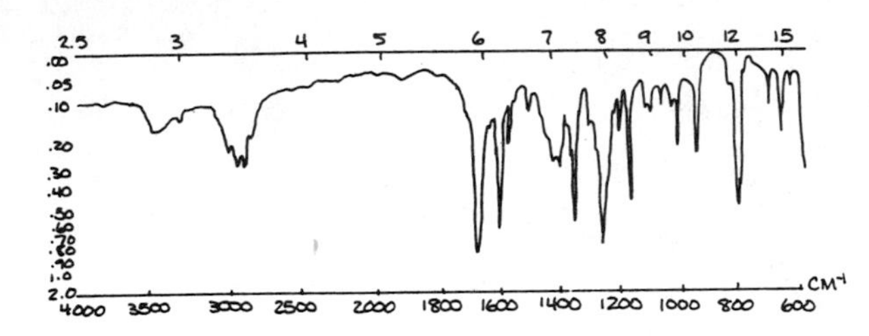

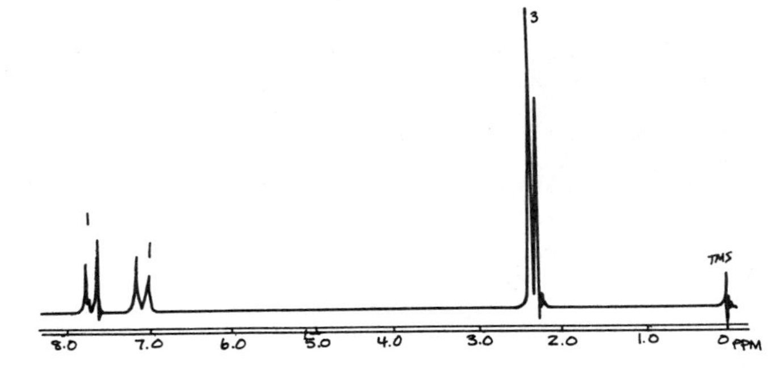

Spectra 13.3

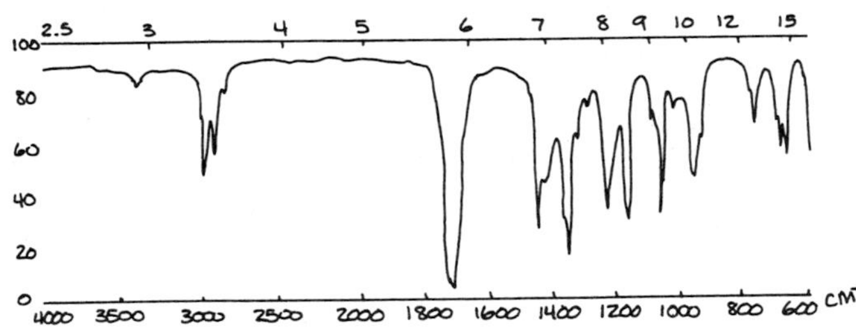

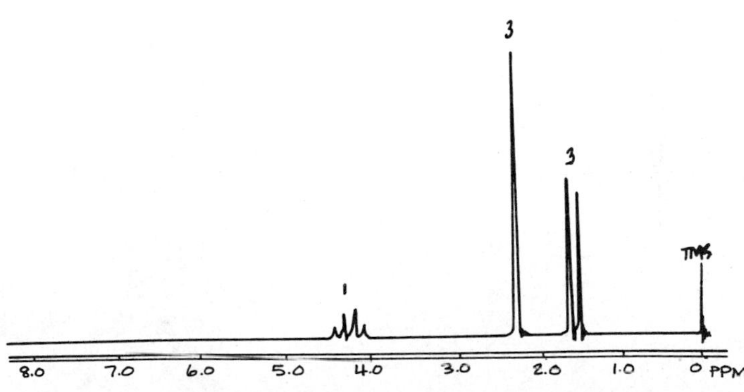

13.7 Which member of each of the following pairs will be more reactive towards a nucleophilic reagent?

(a) $CCl_3\overset{O}{\overset{\|}{C}}H$ or $CHCl_2\overset{O}{\overset{\|}{C}}H$ (b) $CH_3OCH_2\overset{O}{\overset{\|}{C}}H$ or $CH_3SCH_2\overset{O}{\overset{\|}{C}}H$

(c) $NCCH_2\overset{O}{\overset{\|}{C}}CH_3$ or $CH_3CH_2\overset{O}{\overset{\|}{C}}CH_3$

(d) 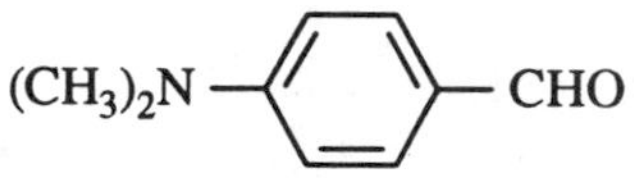 $(CH_3)_2N-C_6H_4-CHO$ or C_6H_5-CHO

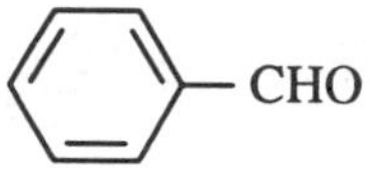

13.8 Arrange the following aldehydes and ketones in order of increasing reactivity towards a nucleophilic reagent (least reactive first):

(a) $H\overset{O}{\overset{\|}{C}}H$ (b) $C_6H_5CH_2\overset{O}{\overset{\|}{C}}CH_3$

(c) $CH_3\overset{O}{\overset{\|}{C}}H$ (d) $CH_3CH_2\overset{O}{\overset{\|}{C}}CH_2CH_3$

13.9 Write the formulas for the cyanohydrin products that could form when the following compounds are treated with an aqueous solution of HCN and NaCN. Show stereochemistry where appropriate.

(a) 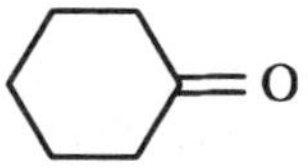 cyclohexanone (cyclohexane ring $=O$) (b)  $O_2N-C_6H_4-CHO$

(c) $H\overset{O}{\overset{\|}{C}}H$ (d) 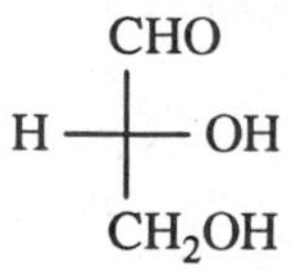 Fischer projection: CHO (top), H (left), OH (right), CH_2OH (bottom)

13.10 Write an equation showing the equilibrium of each of the following hydrates with a carbonyl compound:

(a) $CH_3CH_2CH_2CH(OH)OH$ (b) 1-methylcyclopentane with OH, OH (c) tetrahydropyran ring with HO and OH

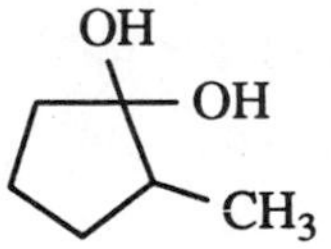

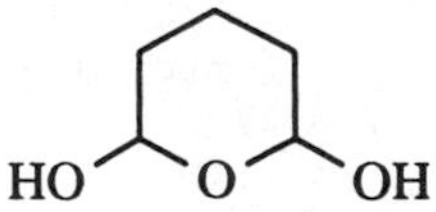

13.11 Complete the following equations showing both the hemiacetal and acetal products:

(a) $(CH_3)_2CHCH_2CHO + CH_3CH_2OH \xrightleftharpoons{H^+}$

(b) 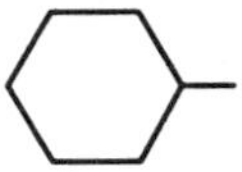 cyclohexyl–CHO $+ CH_2{=}CHCH_2OH \xrightleftharpoons{H^+}$

13.12 Predict the cyclic acetal (or ketal) product:

(a) $HOCH_2CH_2CH_2CH_2CH(=O) + CH_3OH \xrightleftharpoons{H^+}$

(b) $CH_3CH(=O) + HOCH_2CH_2OH \xrightleftharpoons{H^+}$

(c) 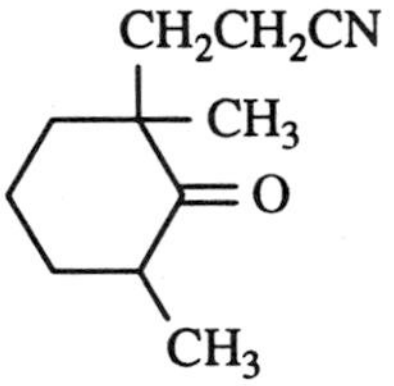 cyclohexanone bearing CH_2CH_2CN, CH_3, and CH_3 substituents $+ HOCH_2CH_2OH \xrightleftharpoons{H^+}$

13.13 What products, other than the structure shown, would be formed if each of the following compounds were treated with dilute aqueous acid?

(a) 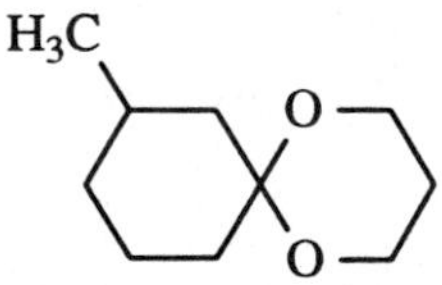 spiro ketal with H_3C, O, O (b) 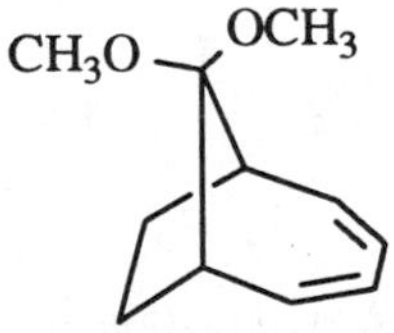 bicyclic dimethyl ketal with CH_3O, OCH_3

(c) 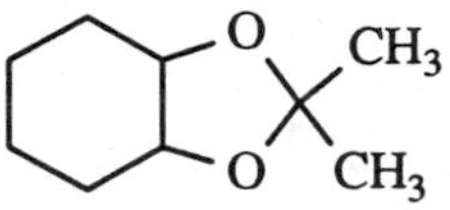 [cyclohexane ring fused to a 1,3-dioxolane; acetal carbon bearing two CH_3 groups]

13.14 Complete the following equations:

(a) 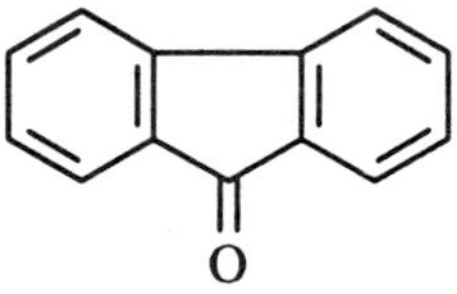 [fluorenone] $\xrightarrow[(2)\ H_2O,\ H^+]{(1)\ CH_3Li}$

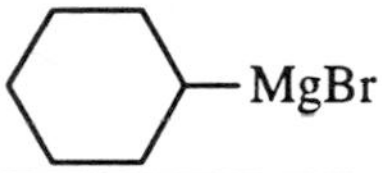

(b) (*R*)-3-phenylbutanal $\xrightarrow[(2)\ H_2O,\ H^+]{(1)\ CH_3MgBr}$

(c) HCH (O=) $\xrightarrow[(2)\ H_2O,\ H^+]{(1)\ \text{cyclohexyl}-MgBr}$

(d) 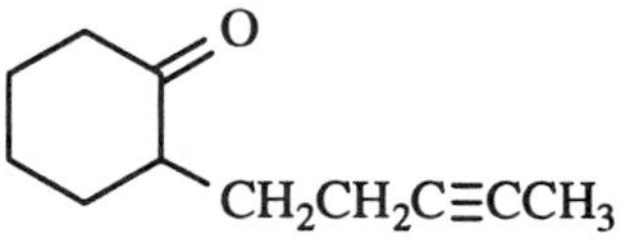 [cyclohexanone with $CH_2CH_2C{\equiv}CCH_3$ substituent at C-2] $\xrightarrow[(2)\ H_2O,\ H^+]{(1)\ CH_3CH_2MgBr}$

13.15 Write flow equations showing how each of the following alcohols could be prepared from aldehydes or ketones by Grignard reactions. If there is more than one method possible, show all methods.

(a) $(CH_3CH_2CH_2)_3COH$

(b) $C_6H_5CH(OH)CH_2CH_3$

(c) $CH_3CH_2CH_2CH_2CH_2OH$

(d) $C_6H_5C(OH)(CH_2CH_3)CH_3$

13.16 Show how the following conversions could be carried out by Grignard reactions:

(a) $ClCH_2CH_2CHO$ $\longrightarrow$ $CH_3CH(OH)CH_2CH_2CHO$

(b) 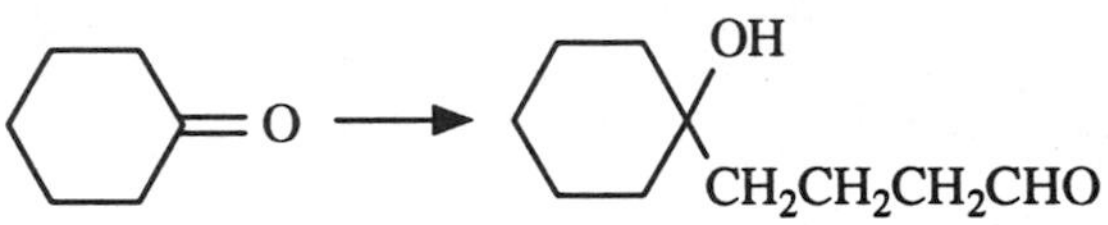 cyclohexanone $\longrightarrow$ 1-(OH)-1-($CH_2CH_2CH_2CHO$)cyclohexane

13.17 Complete the following equations:

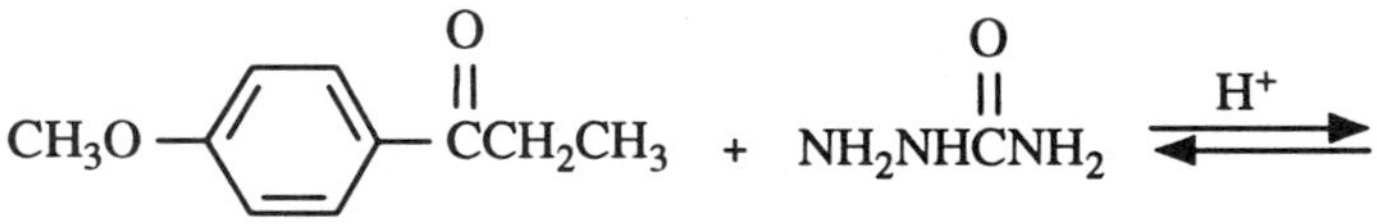

(a) 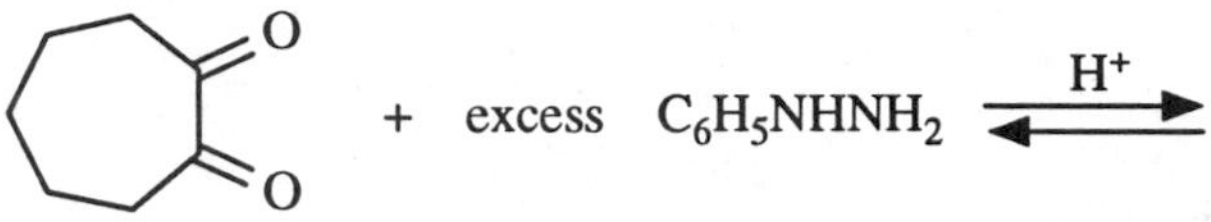 cycloheptane-1,2-dione + excess $C_6H_5NHNH_2$ $\xrightleftharpoons{H^+}$

(b) $CH_3O-C_6H_4-CCH_2CH_3$ (C=O) + NH_2NHCNH_2 (C=O) $\xrightleftharpoons{H^+}$

(c) 4,4-dimethylcyclohexanone (H_3C, H_3C) + morpholine (O, N–H) $\xrightleftharpoons{H^+}$

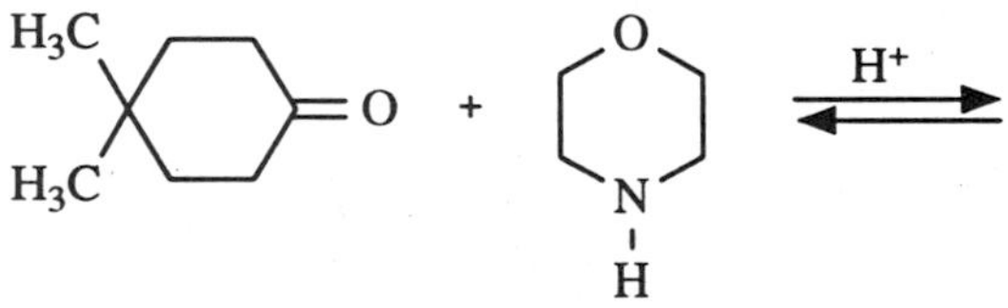

13.18 Write a complete reaction mechanism for each of the following reactions. Be sure to include all proton transfer steps.

(a) $CH_3CHO + 2\ CH_3OH \xrightleftharpoons{H^+} CH_3CH(OCH_3)OCH_3 + H_2O$

(b) $CH_3CHO + CH_3NH_2 \xrightleftharpoons{H^+} CH_3CH{=}NCH_3 + H_2O$

(c) $CH_3CHO + (CH_3)_2NH \xrightleftharpoons{H^+} H_2C{=}CHN(CH_3)_2 + H_2O$

13.19 Write equations that show how benzaldehyde can be converted into the following compounds:

(a) $C_6H_5CH{=}CH_2$ (b) $C_6H_5CH_2OH$ (c) $C_6H_5CH{=}NNH\overset{O}{\overset{\|}{C}}NH_2$

13.20 Predict the products:

(a) $(C_6H_5)_3P \xrightarrow[(2)\ CH_3(CH_2)_3Li]{(1)\ CH_3CH_2Br}$

(b) $(C_6H_5)_3P{=}CHCH_2OCH_3 + CH_3CH_2CHO \xrightarrow{THF}$

(c) $C_6H_5CHO + (C_6H_5)_3P{=}CHCH{=}CH_2 \xrightarrow{THF}$

13.21 Write equations showing how each of the following compounds could be synthesized from an aldehyde or a ketone and a Wittig reagent:

(a) 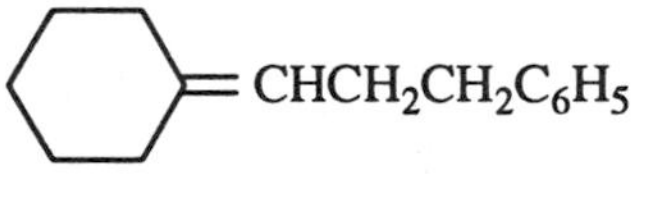 cyclohexane ring $=CHCH_2CH_2C_6H_5$ (b) 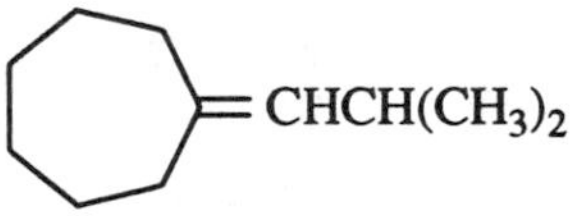 cycloheptane ring $=CHCH(CH_3)_2$

13.22 Write flow equations showing how the following conversions could be carried out:

(a) bromoethane ⟶ 3-ethyl-3-pentanol

(b) bromocyclohexane ⟶ cyclohexylmethanol (by a Wittig reaction)

(c) bromocyclohexane ⟶ cyclohexylmethanol (by a Grignard reaction)

13.23 Predict the $NaBH_4$ reduction products:

(a) 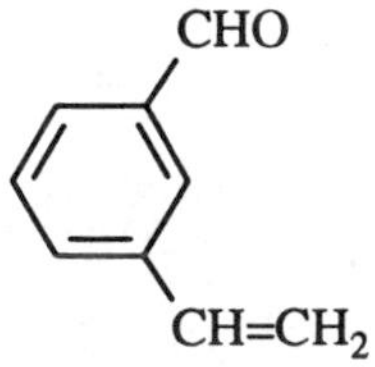 benzene ring with CHO and $CH{=}CH_2$ (b) 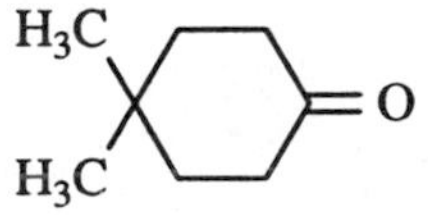 H_3C, H_3C cyclohexane ring $=O$ (c) 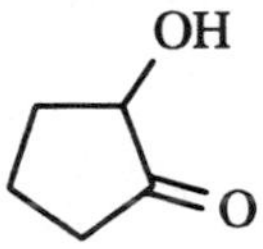 cyclopentane ring with OH and $=O$

13.24 Predict the principal organic products:

(a) $CH_3CH{=}CHCHO \xrightarrow[\text{heat, pressure}]{\text{excess } H_2\text{, catalyst}}$

(b) $CH_3CH_2\overset{O}{\overset{\|}{C}}C_6H_5 \xrightarrow[\text{(2) KOH}]{\text{(1) } NH_2NH_2\text{, } H^+}$

(c) $C_6H_5\overset{O}{\overset{\|}{C}}CH_2CH_3 \xrightarrow[\text{HCl}]{\text{Zn/Hg}}$

(d) $C_6H_5\overset{O}{\overset{\|}{C}}CH_2CH_3 \xrightarrow[H_2\text{, Ni}]{CH_3NH_2}$

13.25 Predict the principal organic products:

(a) cyclopentyl–CHO $\xrightarrow[H_2O]{Ag(NH_3)_2^+}$

(b) $H\overset{O}{\overset{\|}{C}}CH_2CH_2CH_2CH_2OH + H_2CrO_4 \xrightarrow{\text{heat}}$

(c) $C_6H_5CH{=}\underset{CH_2CH_3}{\underset{|}{C}}\overset{O}{\overset{\|}{C}}H \xrightarrow[H_2O]{Ag(NH_3)_2^+}$

(d) C_6H_5CH (cyclic acetal: CH bonded to two O atoms joined by CH_2CH_2 ring) $+ H_2CrO_4 \xrightarrow{\text{heat}}$

13.26 Draw formulas for all the enol tautomers:

(a) $N{\equiv}CCH_2\overset{O}{\overset{\|}{C}}OCH_3$

(b) $CH_3CH{=}CH\overset{O}{\overset{\|}{C}}CH_3$

13.27 Draw the keto structures for each of the following enols:

(a) 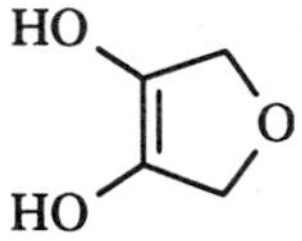 [structure: ring with HO, HO, O] (b) 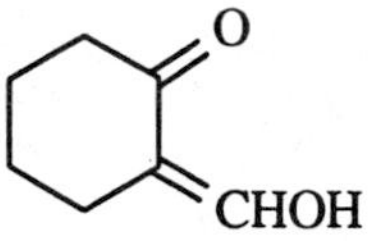 [structure: cyclohexane ring with =O and =CHOH] (c) 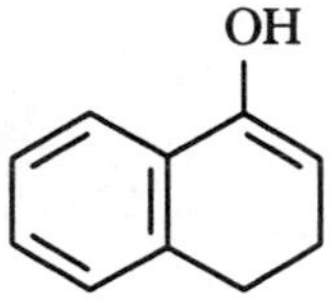 [structure: bicyclic ring with OH]

13.28 Complete the following equations, showing only the principal organic products:

(a) $C_6H_5\overset{O}{\overset{\|}{C}}CH_3$ + excess NaOI (I_2 + NaOH) $\longrightarrow$

(b) 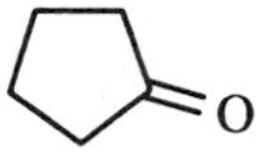 cyclopentanone + 2 Br_2 $\xrightarrow{OH^-}$

(c) 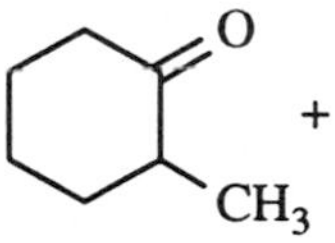 2-methylcyclohexanone (CH_3) + Br_2 $\xrightarrow{H^+}$

13.29 Write equations for reactions of acetophenone and the following reagents:

(a) (1) C_6H_5Li; (2) H_2O, H^+

(b) (1) $NaBH_4$; (2) H_2O, H^+

(c) $CH_3CH_2CH{=}P(C_6H_5)_3$

(d) (1) NaOI; (2) H_2O, H^+

(e) H_2, Ni

(f) NH_2OH, H^+

(g) excess CH_3OH, H^+

(h) NH_3, H_2, Ni

(i) Br_2, H^+

(j) Zn(Hg), HCl

(k) $NH_2NH\overset{O}{\overset{\|}{C}}NH_2$, H^+

(l) 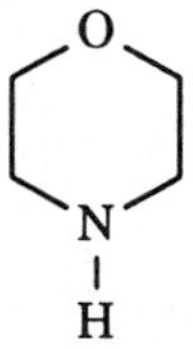 morpholine (O, N–H ring) , H^+

(m) HNO_3, H_2SO_4

(n) (1) NH_2NH_2; (2) $CH_3CH_2O^-$ Na^+

(o) (1) CH_3MgBr; (2) H_2O, H^+

(p) (1) HCN, NaCN; (2) H_2, Ni

(q) (1) $LiAlH_4$; (2) H_2O, H^+

(r) $CH_3CH_2NH_2$, H^+; (2) H_2, Ni

Carboxylic Acids

Some Important Features

Carboxylic acids (RCO_2H) are more acidic than alcohols, phenols, or carbonic acid. Carboxylic acids yield carboxylate salts when treated with any base stronger than the carboxylate ion itself.

$$RCO_2H + Na^+ \; HCO_3^- \rightleftharpoons RCO_2^- Na^+ + H_2O + CO_2$$

Factors affecting acid strength are discussed in detail in this chapter of the text. Among these factors are the *electronegativity* of the atom attached to the acidic hydrogen. A greater electronegativity means a stronger acid; therefore, alkanes are extremely weak acids, while amines (R_2NH), alcohols (ROH), and hydrogen halides (HX) are successively stronger acids.

Hybridization of the atom attached to H affects acidity: RC≡CH is more acidic than $R_2C{=}CHR$ or RH.

The *inductive effect* increases or decreases acid strength by withdrawing or releasing electron density.

$$ClCH_2CO_2H \rightleftharpoons ClCH_2CO_2 + H^+$$

Cl strengthens the acid by stabilizing the anion relative to the acid through dispersal of the negative charge.

$$Cl{-}C_6H_4{-}CO_2H \rightleftharpoons Cl{-}C_6H_4{-}CO_2^- + H^+$$

Resonance stabilization of the anion also increases acid strength. This resonance stabilization is the primary reason for the acidity of carboxylic acids. Resonance stabilization of the anion also explains why phenols are more acidic than alcohols.

$$\mathrm{R\overset{\overset{O}{\|}}{C}OH} \rightleftharpoons \left[\mathrm{R\overset{\overset{:\ddot{O}}{\|}}{C}-\ddot{O}:^-} \longleftrightarrow \mathrm{R\overset{\overset{:\ddot{O}:^-}{|}}{C}=\ddot{O}:} \right] + H^+$$

resonance structures

Other factors, such as hydrogen bonding in the anion, can also affect acid strength.

A carboxylic acid can undergo esterification with an alcohol. The rate of esterification decreases with increasing steric hindrance around the carboxyl groups. (Review the mechanism for this reaction in Section 14.6).

$$\mathrm{R\overset{\overset{O}{\|}}{C}OH} + R'OH \underset{}{\overset{H^+}{\rightleftharpoons}} \mathrm{R\overset{\overset{O}{\|}}{C}OR'} + H_2O$$

an ester

Decarboxylation of β-keto acids is important both in the laboratory and in biological systems.

$$\mathrm{R\overset{\overset{O}{\|}}{C}CH_2\overset{\overset{O}{\|}}{C}OH} \xrightarrow[\text{or enzymes}]{\text{heat}} \mathrm{R\overset{\overset{O}{\|}}{C}CH_3} + CO_2$$

a β-keto acid

Other reactions of carboxylic acids are reduction by $LiAlH_4$ (Section 14.7), and anhydride formation by diacids (Section 14.8B).

Reminders

For calculations of K_a and pK_a values, review Section 1.7F.

Remember that the conjugate base of a strong acid is a weak base, while the conjugate base of a weak acid is a strong base.

$$RCO_2H \rightleftharpoons RCO_2^- + H^+$$

strong acid *weak base*

$$ROH \rightleftharpoons RO^- + H^+$$

weak acid *strong base*

Because a carboxylic acid contains oxygen atoms with unshared electrons, it can be protonated by a strong acid or it can undergo hydrogen bonding.

$$RC(=O)OH \underset{}{\overset{H^+}{\rightleftharpoons}} \left[RC(=\overset{+}{O}H)OH \longleftrightarrow RC(OH)\overset{+}{O}H \longleftrightarrow RC(OH)=\overset{+}{O}H \right]$$

resonance structures for protonated acid

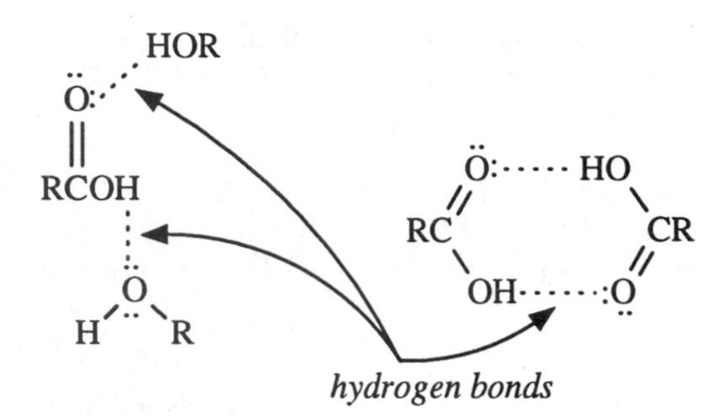

Additional Drill Problems

14.1 Name the following groups:

(a) $CH_3C(=O)-$ (b) $CH_3C(=O)O-$ (c) $C_6H_5C(=O)-$

14.2 What are the structures of the following compounds:

(a) $C_9H_{10}O_2$, Spectra 14.1

(b) $C_4H_8O_2$, Spectra 14.2

(c) $C_3H_6O_2$, Spectra 14.3

(d) $C_9H_{10}O_3$, Spectra 14.4

Spectra 14.1

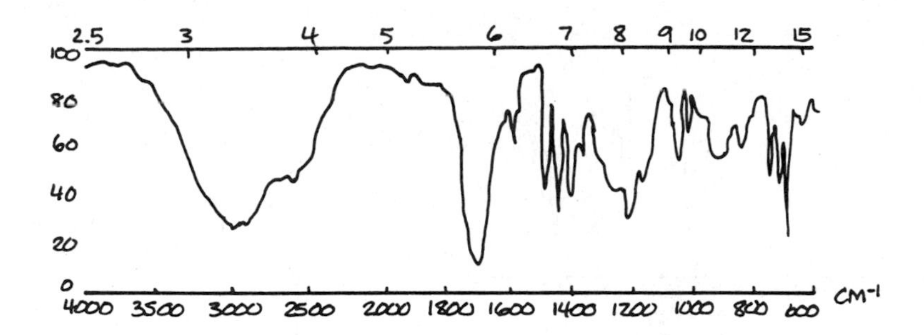
2.5
3
4
5
6
7
8
9
10
12
15
100
80
60
40
20
0
4000
3500
3000
2500
2000
1800
1600
1400
1200
1000
800
600
CM-1

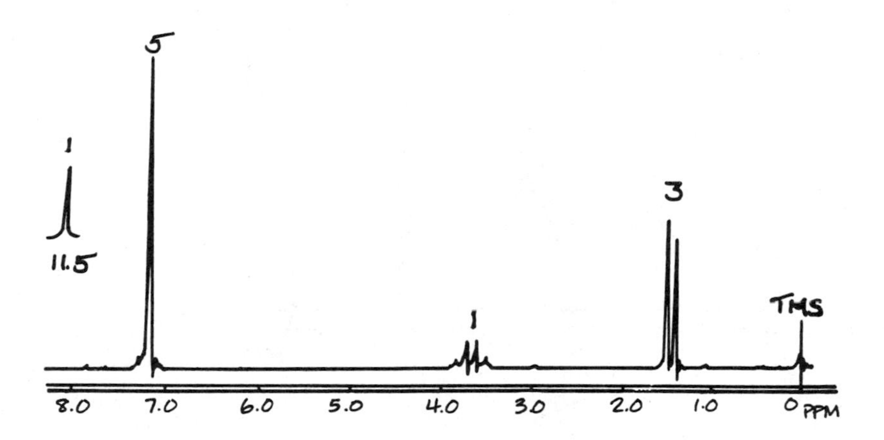
5
1
11.5
1
3
TMS
8.0
7.0
6.0
5.0
4.0
3.0
2.0
1.0
0
PPM

Spectra 14.2

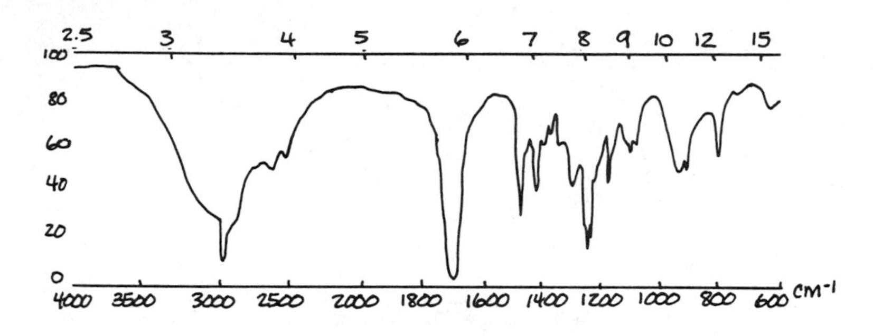
2.5 3 4 5 6 7 8 9 10 12 15
100
80
60
40
20
0
4000 3500 3000 2500 2000 1800 1600 1400 1200 1000 800 600 CM-1

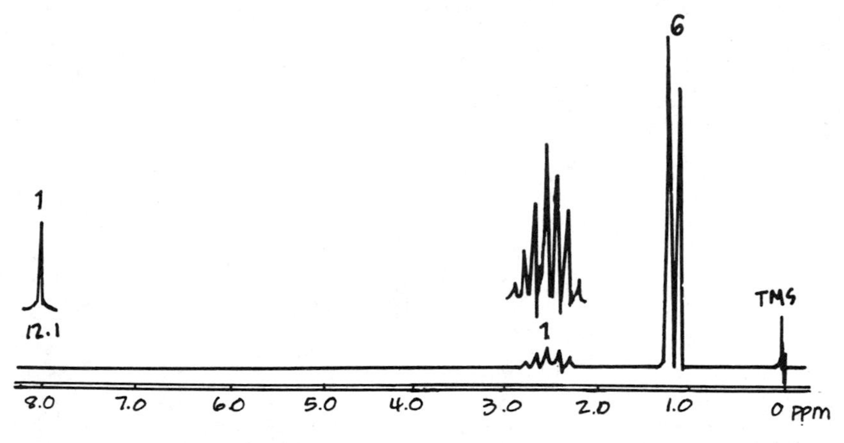
6
1
12.1
1
TMS
8.0 7.0 6.0 5.0 4.0 3.0 2.0 1.0 0 PPM

Spectra 14.3

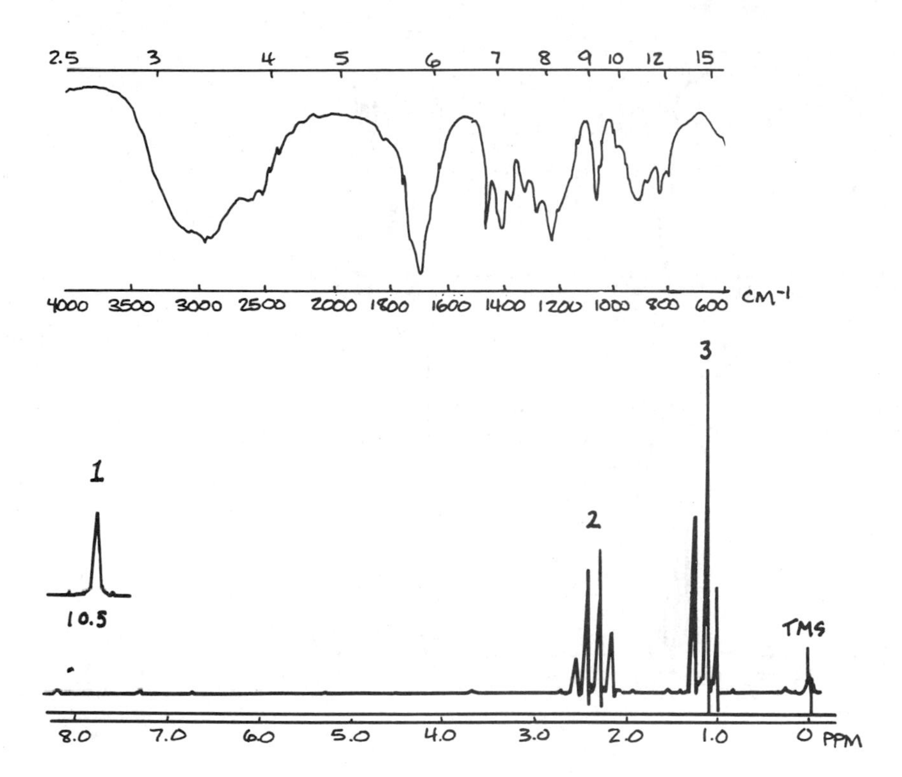

2.5
3
4
5
6
7
8
9
10
12
15
4000
3500
3000
2500
2000
1800
1600
1400
1200
1000
800
600
CM⁻¹
1
10.5
2
3
TMS
8.0
7.0
6.0
5.0
4.0
3.0
2.0
1.0
0
PPM

Spectra 14.4

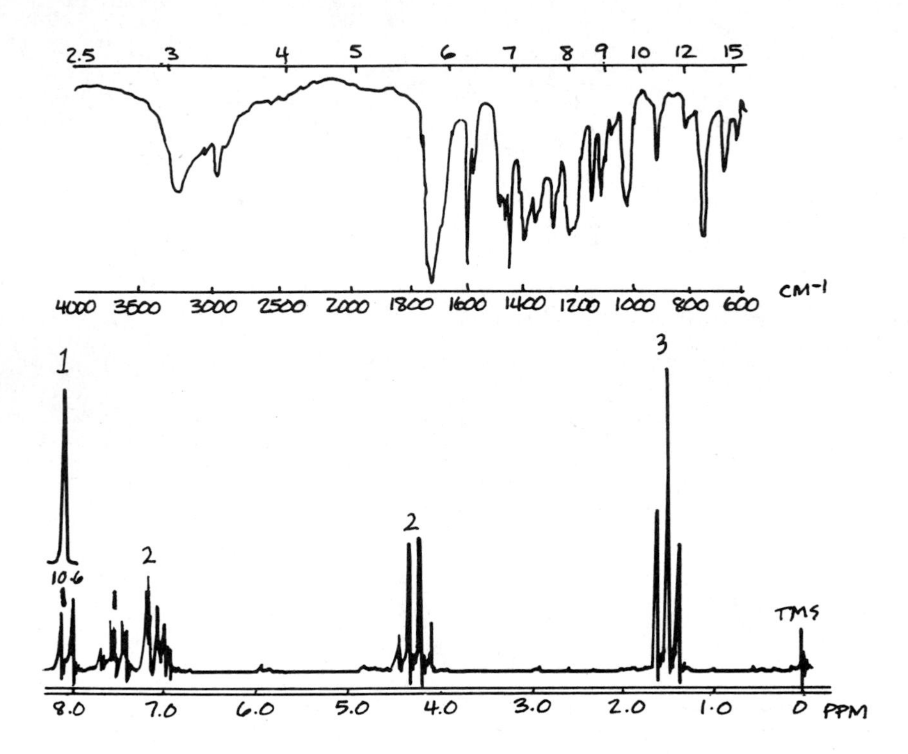

2.5 3 4 5 6 7 8 9 10 12 15
4000 3500 3000 2500 2000 1800 1600 1400 1200 1000 800 600
CM-1
1
10.6
2
2
3
TMS
8.0 7.0 6.0 5.0 4.0 3.0 2.0 1.0 0 PPM

14.3 Write equations showing how each of the following carboxylic acids could be prepared by a Grignard synthesis:

(a) $(CH_3)_3CCH_2CO_2H$

(b) 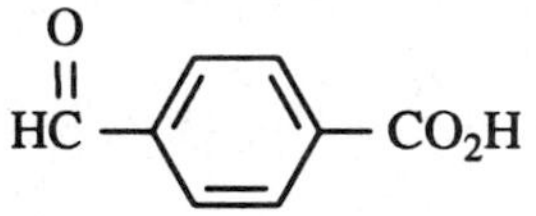 $HC(=O)-C_6H_4-CO_2H$ (para)

(c) cyclopentyl–CO_2H

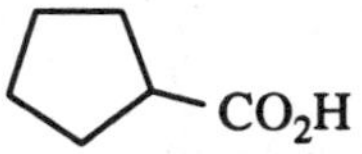

14.4 Write flow equations for *two* procedures by which each of the following conversions could be carried out:

(a) $C_6H_5CH_2Br$ to $C_6H_5CH_2CO_2H$

(b)  cyclohexyl–Br $\longrightarrow$ cyclohexyl–CO_2H

14.5 Write equations showing how each of the following compounds could be converted to benzoic acid:

(a) $C_6H_5CH(=O)$ (b) $C_6H_5CH_3$ (c) $C_6H_5CO_2CH_3$

(d) C_6H_5CN (e) *trans*-$C_6H_5CH{=}CHC_6H_5$

14.6 Would a Grignard synthesis or a nitrile synthesis be better for each of the following transformations? Write equations for and defend your answers.

(a) $CH_3CH_2CH(Br)CH(CH_3)_2 \longrightarrow CH_3CH_2CH(CO_2H)CH(CH_3)_2$

(b) $CH_3CH_2CH(OH)CH_2CH_2Cl \longrightarrow CH_3CH_2CH(OH)CH_2CH_2CO_2H$

(c) $C_6H_5Br \longrightarrow C_6H_5CO_2H$

14.7 Using flow diagrams, show how the following conversions could be carried out:

(a) $CH_3(CH_2)_6CH_2OH \longrightarrow CH_3(CH_2)_8CO_2H$

(b) $CH_3(CH_2)_6CH_2OH \longrightarrow CH_3(CH_2)_7CO_2H$ (two routes)

(c) $CH_3(CH_2)_6CH_2OH \longrightarrow CH_3(CH_2)_6CO_2H$

(d) $CH_3(CH_2)_6CH_2OH \longrightarrow CH_3(CH_2)_5CO_2H$

14.8 Using flow equations, outline a procedure for each of the following conversions:

(a) methylenecyclopentane ($=CH_2$) $\longrightarrow$ cyclopentyl–CH_2CO_2H

(b) methylenecyclopentane ($=CH_2$) $\longrightarrow$ cyclopentyl–CO_2H

(c) $CH_3CH_2CHO \longrightarrow CH_3CH_2CH(OH)COOH$

(d) $CH_3CH_2CH{=}CHCH_2CH_3 \longrightarrow CH_3CH_2CO_2H$

14.9 Write the formula for the conjugate base of each of the following acids:

(a) $C_6H_5SO_3H$ (b) C_6H_5OH (c) $CH_3CH_2CO_2H$

14.10 Calculate the pK_a values for the following compounds, given their K_a values:

(a) heptanoic acid, 1.28×10^{-5} (b) phenylacetic acid, 5.2×10^{-5}

14.11 If the dissociation constant of pentanoic acid is 1.51×10^{-5}, what is the approximate H^+ concentration in an 0.0300 *M* solution of the acid?

14.12 If a 0.200 *M* solution of a carboxylic acid has an H^+ concentration of 0.00184 *M*, what is the dissociation constant of the acid? What would be the H^+ concentration of a liter of 0.200 *M* acid to which 0.100 mol of NaOH has been added?

14.13 Which compound in each pair is the stronger acid?

(a) CH_3OH or CH_3SH (b) CH_3CH_2OH or $CH_3CH_2NH_2$

(c) $CH_3CH(Br)COOH$ or $CH_3CH(I)COOH$

(d) C_6H_5–OH or NO_2–C_6H_4–CH_2OH

14.14 Arrange the following compounds in order of increasing acidity (least acidic first):

(a) H_2SO_4 (b) C_6H_5OH (c) cyclopentanol (cyclopentane–OH)

(d) CH_3CH_2COOH (e) H_2O (f) $HC{\equiv}CH$

14.15 Which group in each of the following pairs would have the greater inductive electron-withdrawing power?

(a) $CH_3C(=O)$- or Cl- (b) F- or Cl- (c) CH_3S- or CH_3O-

(d) C_6H_5- or CH_3- (e) CH_3CH_2- or $CH_2{=}CH$- (f) Br- or H-

14.16 Write the steps in the mechanism of the acid-catalyzed esterification of formic acid with excess ethanol.

14.17 The following acids, 1.0 mol each, are treated with 1.0 mol of methanol and a trace of H_2SO_4. The solutions are heated until equilibrium is reached. List the acids in order of increasing amount of ester formed (least ester first). Explain your answer.

(a) $CH_3CH_2CO_2H$ (b) $CH_3CH_2CH(CH_3)CO_2H$ (c) $CH_3CH_2C(CH_3)_2CO_2H$

14.18 Write an equation for the mechanism of the thermal decarboxylation of the following compound:

$$CH_3\overset{O}{\overset{\|}{C}}CH_2CO_2H$$

14.19 Complete the following equations:

(a) C_6H_{11}—$CO_2H + HCO_3^- \longrightarrow$ (cyclohexane ring)

(b) $C_6H_5OH + HCO_3^- \longrightarrow$

(c) $CH_3CH_2CH_2CO_2H + OH^- \longrightarrow$

(d) $C_6H_5OH + OH^- \longrightarrow$

(e) C_6H_{11}—$OH + OH^- \longrightarrow$ (cyclohexane ring)

(f) C_6H_{11}—$OH + {}^-NH_2 \longrightarrow$ (cyclohexane ring)

(g) $CH_3CO_2H + CH_3O^- \longrightarrow$

(h) $C_6H_5OH + C_6H_{11}$—$O^- \longrightarrow$ (cyclohexane ring)

(i) $CH_3CH_2CO_2^- + C_6H_5OH \longrightarrow$

(j) $C_6H_5O^- + CH_3CH_2CO_2H \longrightarrow$

14.20 Complete the following equations:

(a) $C_6H_5CO_2H \xrightarrow[\text{(2) } H_2O,\ H^+]{\text{(1) } LiAlH_4}$

(b) $(CH_3)_3CCH_2CO_2H + CH_3OH \xrightarrow[\text{heat}]{H^+}$

(c) 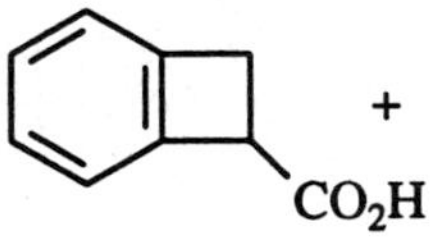 [benzocyclobutene-1-carboxylic acid: CO_2H] $+ C_6H_5MgBr \longrightarrow$

(d) 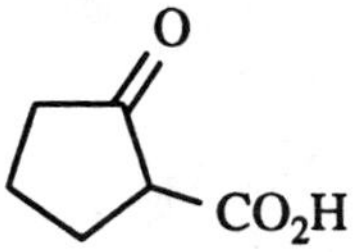 [2-oxocyclopentanecarboxylic acid: O, CO_2H] $\xrightarrow{\text{heat}}$

15

Derivatives of Carboxylic Acids

Some Important Features

The common derivatives of carboxylic acids are esters, amides, acid halides, acid anhydrides, and nitriles. Except for nitriles, these compounds all undergo nucleophilic acyl substitution (addition-elimination) reactions with reagents such as water, alcohols, ammonia, or amines. Nitriles undergo similar reactions. (See the chapter-end summary in the text.)

$$\mathrm{\overset{\overset{:\ddot{O}}{\|}}{R\overset{\delta+}{C}Cl}} + \mathrm{H_2\overset{\delta-}{\ddot{O}}:} \xrightarrow{\text{addition}} \left[\mathrm{RC(:\ddot{O}:^-)(Cl)(:\overset{+}{O}H_2)}\right] \xrightarrow[\text{and loss of } H^+]{\text{elimination of } Cl^-} \mathrm{\overset{\overset{:\ddot{O}}{\|}}{RC}{-}:\ddot{O}H}$$

The reactivity of the carboxylic acid derivatives depends partly on the leaving group (Section 15.1). For example, hydrolysis of an acid chloride proceeds more rapidly than hydrolysis of an ester partly because Cl^- is a better group than RO^- is.

easier:

$$\underset{\textit{an acid chloride}}{\mathrm{\overset{\overset{O}{\|}}{RCCl}}} + \mathrm{H_2O} \longrightarrow \mathrm{\overset{\overset{O}{\|}}{RCOH}} + \mathrm{HCl}$$

more difficult:

$$\underset{\textit{an ester}}{\mathrm{\overset{\overset{O}{\|}}{RCOR'}}} + \mathrm{H_2O} \underset{}{\overset{H^+,\ \text{heat}}{\rightleftharpoons}} \mathrm{\overset{\overset{O}{\|}}{RCOH}} + \mathrm{R'OH}$$

In your study of the reactions of the carboxylic acid derivatives, pay particular attention to the similarities of the reactions and their mechanisms. The reactions of esters and amides, for example, are quite similar.

$$RCOR' + {}^-{:}\ddot{O}H \xrightarrow{heat} \left[RC(O^-)(OR')(OH) \right] \xrightarrow{-R'\ddot{O}:^-} RC(=O)OH \xrightarrow[-H_2O]{^-:\ddot{O}H} RC(=O){-}\ddot{O}:^-$$

an ester

reacts with H_2O to yield $R'OH$ and ^-OH

$$RCNH_2 + {}^-{:}\ddot{O}H \xrightarrow{heat} \left[RC(O^-)(NH_2)(OH) \right] \xrightarrow{-{}^-\ddot{N}H_2} RC(=O)OH \xrightarrow[-H_2O]{^-:\ddot{O}H} RC(=O){-}\ddot{O}:^-$$

an amide

reacts with H_2O to yield NH_3 and ^-OH

All carboxylic acid derivatives contain unsaturation and therefore can be reduced by catalytic hydrogenation or by $LiAlH_4$.

$$RCOR' \xrightarrow[heat,\ pressure]{H_2,\ Ni} RCH_2OH + HOR'$$

an ester

Other topics covered in this chapter are α halogenation of acid halides, some other types of esters (lactones, polyesters, thioesters), some other types of amides (lactams, imides, carbamates, and others), and the reactions of nitriles.

Reminders

In writing mechanisms for the reactions of carboxylic acid derivatives, use electron dots for unshared electrons. They will help you see where protonation can occur and the direction of the electron shifts. (Remember that protonation can occur in acidic solution, but not in alkaline solution.)

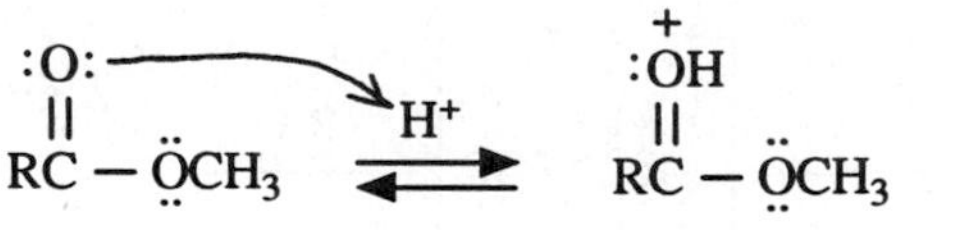

$$\mathrm{R\overset{\overset{\large :O:}{\|}}{C}-\ddot{O}CH_3 \;\underset{}{\overset{H^+}{\rightleftharpoons}}\; R\overset{\overset{\large :\overset{+}{O}H}{\|}}{C}-\ddot{O}CH_3}$$

a protonated ester

You have probably noticed that we are using more-complex structures now than we did earlier in the text. Do not be intimidated by a complex structure. For example, under saponification conditions, the only reactive functional group in the following compound is an ester group.

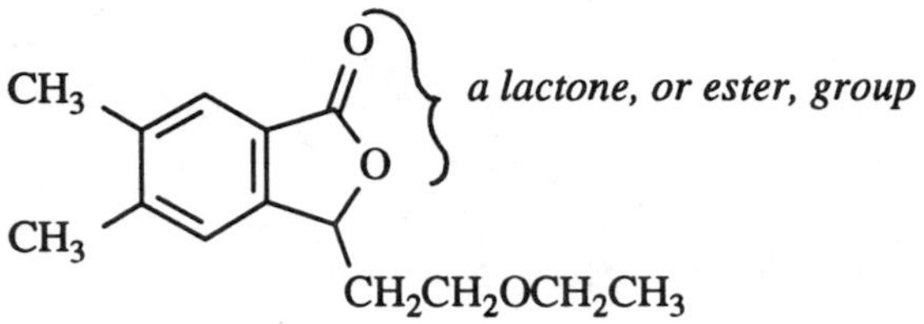

CH_3, CH_3, $CH_2CH_2OCH_2CH_3$ — *a lactone, or ester, group*

Additional Drill Problems

115.1 Name the following compounds:

(a) 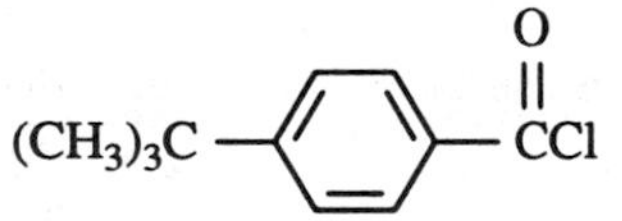 $(CH_3)_3C-C_6H_4-\overset{O}{\overset{\|}{C}}Cl$

(b) $C_6H_5-\overset{O}{\overset{\|}{C}}OCH_2CH(CH_3)_2$

(c) $CH_3CH_2O\overset{O}{\overset{\|}{C}}CH_2\overset{O}{\overset{\|}{C}}OCH_2CH_3$

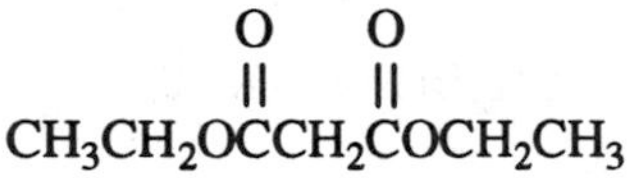

(d) 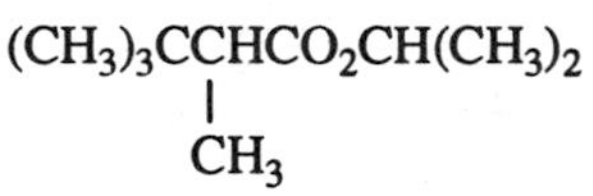 $(CH_3)_3C\underset{CH_3}{\underset{|}{C}}HCO_2CH(CH_3)_2$

(e) 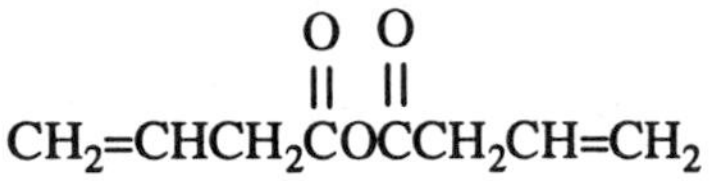 $CH_2{=}CHCH_2\overset{O}{\overset{\|}{C}}O\overset{O}{\overset{\|}{C}}CH_2CH{=}CH_2$

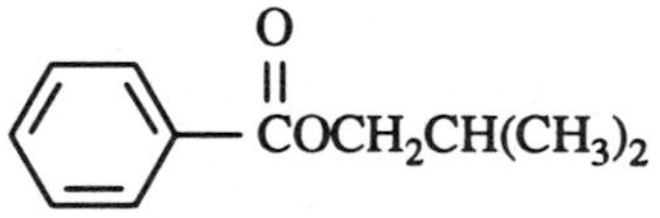

(f) 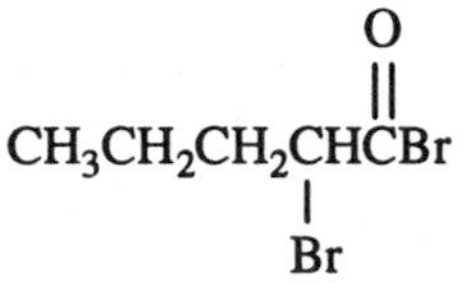 $CH_3CH_2CH_2\underset{Br}{\underset{|}{C}}H\overset{O}{\overset{\|}{C}}Br$

(g) $CH_3CH=CHCH_2CN$

(h) 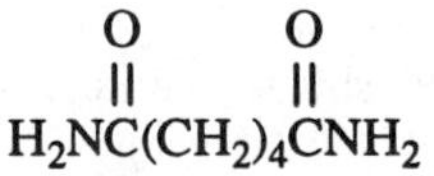 $H_2N\overset{O}{\overset{\|}{C}}(CH_2)_4\overset{O}{\overset{\|}{C}}NH_2$

15.2 Write IUPAC names for the following compounds:

(a) $CH_3CH_2CH_2\overset{O}{\overset{\|}{C}}O\overset{O}{\overset{\|}{C}}CH_3$

(b) $CH_2{=}CHCH_2\overset{O}{\overset{\|}{C}}Cl$

(c) $C_6H_{11}{-}O\overset{O}{\overset{\|}{C}}CH_3$ (cyclohexyl)

(d) $C_6H_{11}{-}\overset{O}{\overset{\|}{C}}OCH_3$ (cyclohexyl)

(e) $(CH_3)_3CCH_2CN$

(f) $CH_3\underset{CH_3}{\underset{|}{C}}H\overset{O}{\overset{\|}{C}}NHCH_3$

15.3 Draw formulas for the following compounds:

(a) methyl γ-bromobutyrate

(b) methyl chloroacetate

(c) chloromethyl acetate

(d) β-chloropropionitrile

(e) acetic formic anhydride

(f) *N,N*-dimethylformamide

15.4 Draw formulas of the following compounds:

(a) butanoic anhydride

(b) pentanoyl fluoride

(c) *N*-ethylbenzamide

(d) 3-methylbutanenitrile

(e) methyl 2-cyclopropanoate

(f) *p*-ethylbenzamide

(g) 2,4-dinitrobenzoyl chloride

15.5 What are the structures of the following compounds?

(a) C_3H_7NO, Spectra 15.1

(b) $C_6H_{11}BrO_2$, Spectra 15.2

(c) $C_{16}H_{15}N$, Spectra 15.3

Spectra 15.1

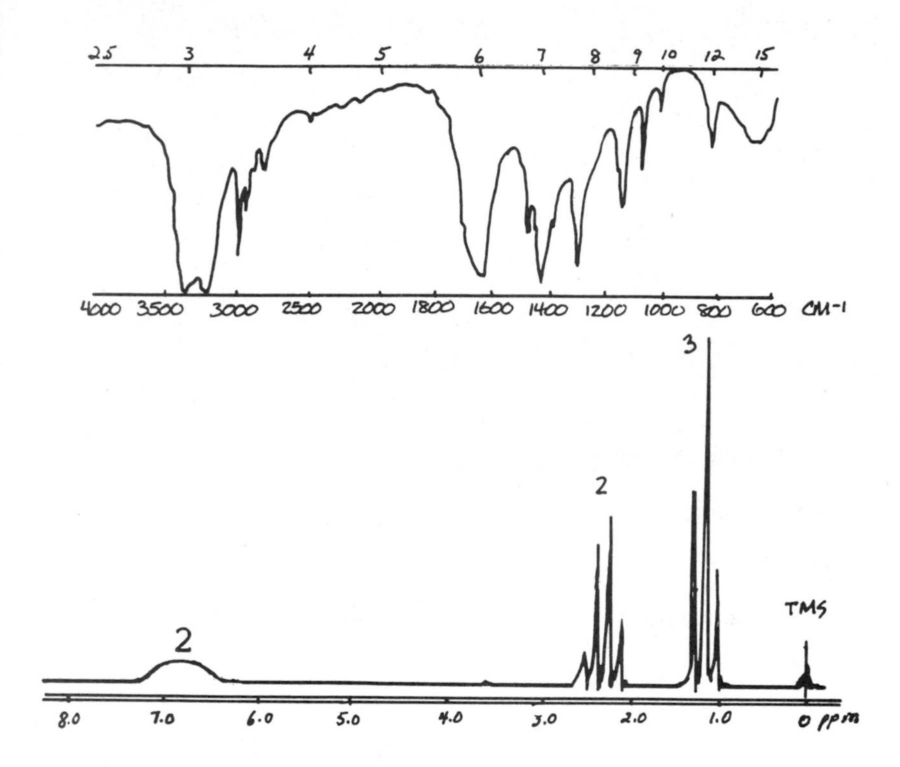

2.5 3 4 5 6 7 8 9 10 12 15
4000 3500 3000 2500 2000 1800 1600 1400 1200 1000 800 600 CM-1
3
2
2
TMS
8.0 7.0 6.0 5.0 4.0 3.0 2.0 1.0 0 ppm

Spectra 15.2

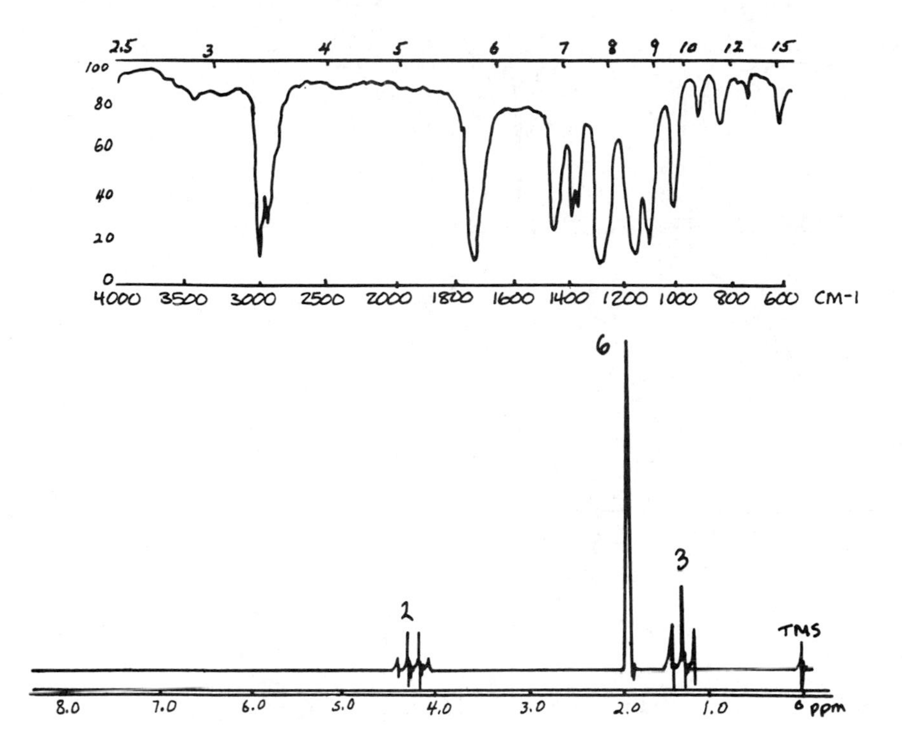
2.5
3
4
5
6
7
8
9
10
12
15
100
80
60
40
20
0
4000
3500
3000
2500
2000
1800
1600
1400
1200
1000
800
600
CM-1
6
3
2
TMS
8.0
7.0
6.0
5.0
4.0
3.0
2.0
1.0
0 ppm

Spectra 15.3

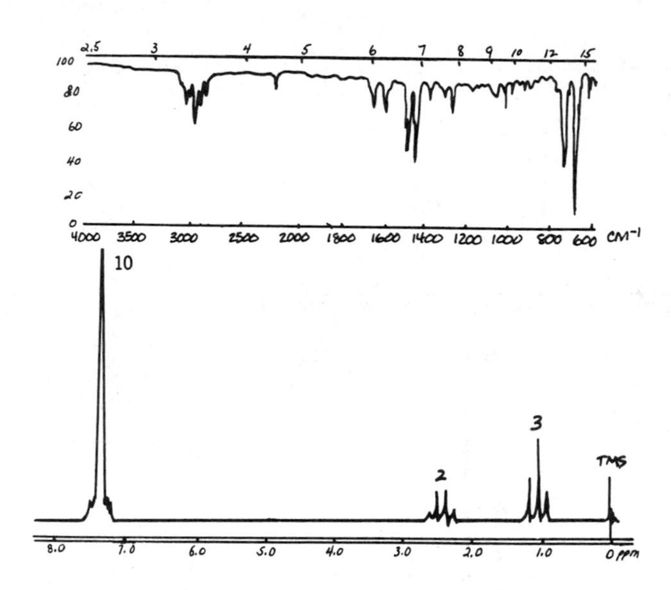

15.6 Complete the following equations, showing only the major organic products:

(a) $(CH_3)_2CHCH_2CO_2H + Br_2 \xrightarrow{PBr_3}$

(b) $CH_3\overset{O}{\overset{\|}{C}}Cl + (CH_3)_2CHCH_2OH \xrightarrow{\text{pyridine}}$

(c) $(CH_3)_3C\overset{O}{\overset{\|}{C}}Cl + (CH_3)_2CuLi \longrightarrow$

(d) O_2N (benzene ring) $-\overset{O}{\overset{\|}{C}}Cl \xrightarrow[\text{(2) } H_2O,\ H^+]{\text{(1) } LiAlH[OC(CH_3)_3]_3}$

(e) 3,5-$(H_3C)_2C_6H_3C(=O)Cl$ + C_6H_6 $\xrightarrow{AlCl_3}$

(f) o-$(C_6H_5CH_2)C_6H_4C(=O)Cl$ $\xrightarrow{AlCl_3}$

(g) CH_3CCl (C=O) + $CH_3C(=O)\ddot{C}HC(=O)CH_3$ Na^+ $\longrightarrow$

15.7 Write equations for three different procedures for synthesizing 1-phenyl-4-methyl-1-pentanone from an acid chloride.

15.8 What combination of reagents could produce each of the following compounds in a single chemical reaction:

(a) butanoic anhydride

(b) *p*-nitrobenzoyl chloride

(c) *N,N*-diethylacetamide

15.9 Starting with acetic acid, write equations showing how the following compounds could be prepared:

(a) propionitrile

(b) acetophenone

(c) *N,N*-dimethylethanamide

(d) acetic anhydride

(e) ethyl acetate

(f) chloroacetic acid

15.10 What would be the structure of the anhydride formed if each of the following carboxylic acids were treated with acetic anhydride?

(a) o-$C_6H_4(CO_2H)_2$

(b) O_2N–C_6H_4–CO_2H (para)

(c) 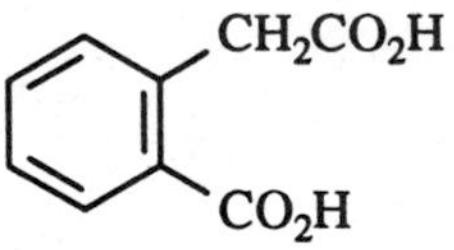 (benzene ring with CH_2CO_2H and CO_2H substituents)

15.11 Write equations showing how the following anhydrides could be prepared from carboxylic acids:

(a) $CH_3\overset{O}{\overset{\|}{C}}O\overset{O}{\overset{\|}{C}}H$ (b) 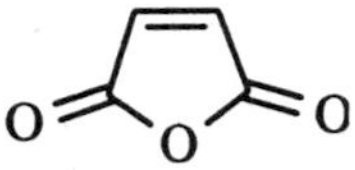 (maleic anhydride ring, O, O, O) (c) 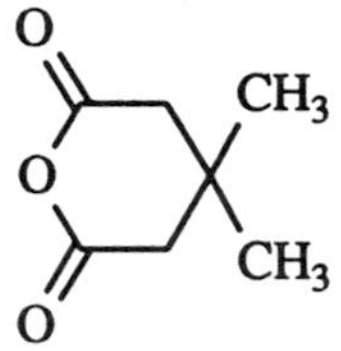 (six-membered cyclic anhydride with O, O, O and CH_3, CH_3)

15.12 Write formulas for the product(s) that would be formed when the following compounds are treated with excess acetic anhydride:

(a) benzyl alcohol (b) aniline

(c) benzoic acid (d) (*R*)-2,3-dihydroxypropanal

(e) sodium phenoxide (f) dimethylamine

15.13 Complete the following equations:

(a) excess CH_3CH_2O–(benzene ring)–NH_2 + $CH_3\overset{O}{\overset{\|}{C}}O\overset{O}{\overset{\|}{C}}CH_3$ ⟶

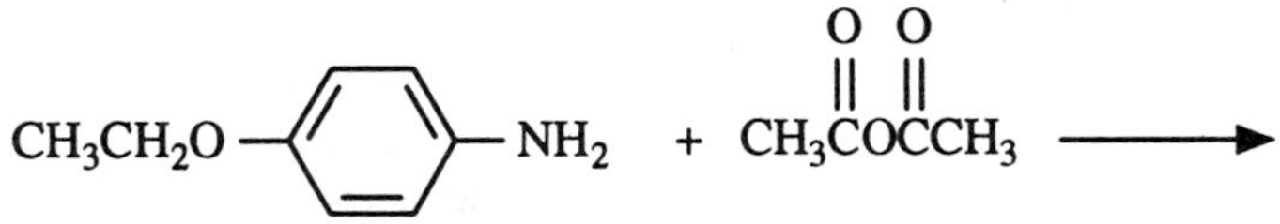

(b) (six-membered cyclic anhydride, O, O, O) + H_2O $\xrightarrow{H^+}$

(c) (five-membered cyclic anhydride, O, O, O) + NaOH $\xrightarrow{H_2O}$

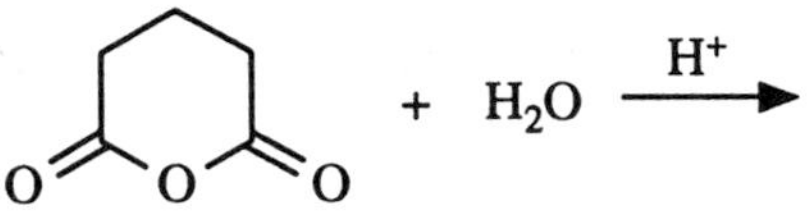

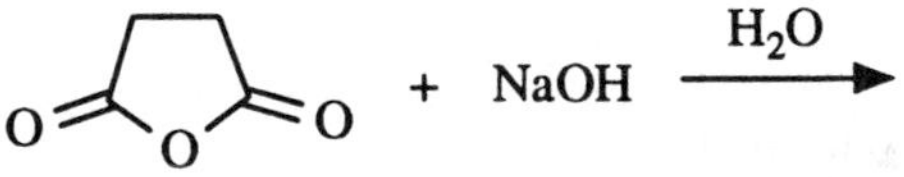

(d) 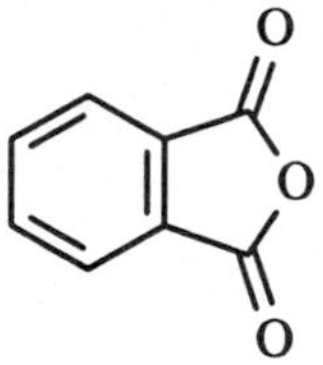 $C_6H_4(CO)_2O$ (phthalic anhydride: benzene ring fused to $\overset{O}{\overset{\|}{C}}-O-\overset{O}{\overset{\|}{C}}$) + excess $CH_3CH_2NH_2 \longrightarrow$

15.14 Write flow equations for the mechanisms of the following reactions:

(a) $CH_3\overset{O}{\overset{\|}{C}}Cl + 2\ C_6H_5NH_2 \longrightarrow CH_3\overset{O}{\overset{\|}{C}}NHC_6H_5 + C_6H_5NH_3^+\ Cl^-$

(b) $CH_3\overset{O}{\overset{\|}{C}}O\overset{O}{\overset{\|}{C}}CH_3 + C_6H_5NH_3 \longrightarrow CH_3\overset{O}{\overset{\|}{C}}NHC_6H_5 + CH_3\overset{O}{\overset{\|}{C}}OH$

15.15 Using flow equations, show how the following conversions could be carried out:

(a) 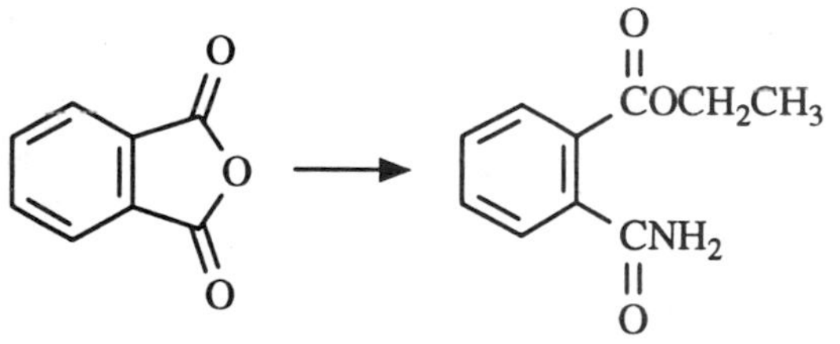 phthalic anhydride (benzene ring fused to $\overset{O}{\overset{\|}{C}}-O-\overset{O}{\overset{\|}{C}}$) $\longrightarrow$ benzene ring bearing $\overset{O}{\overset{\|}{C}}OCH_2CH_3$ and, ortho, $\underset{O}{\underset{\|}{C}}NH_2$

(b) phenol $\longrightarrow$ *p*-nitrophenyl acetate

(c) pentanedioic acid $\longrightarrow CH_3O\overset{O}{\overset{\|}{C}}CH_2CH_2CH_2\overset{O}{\overset{\|}{C}}N(CH_3)_2$

15.16 Show how butanoic acid can be converted to the following compounds:

(a) $CH_3CH_2CH_2\overset{O}{\overset{\|}{C}}OCH_3$

(b) $CH_3CH_2\underset{Br}{\underset{|}{C}}H\overset{O}{\overset{\|}{C}}OCH_3$

(c) $CH_3CH_2CH_2CH_2\overset{O}{\overset{\|}{C}}OCH_2CH_3$

(d) $CH_3CH_2CH_2\overset{O}{\overset{\|}{C}}OCH_2CH_2CH_2CH_3$ (three routes)

15.17 Complete the following equations:

(a) [bicyclic ketone bearing CO_2CH_3 and CH_3] $\xrightarrow[(2)\ H_2O,\ H^+]{(1)\ LiAlH_4}$

(b) $C_6H_5\overset{O}{\overset{\|}{C}}O\text{-}C_6H_4\text{-}NO_2$ + NaOH $\xrightarrow[heat]{H_2O}$

(c) $(CH_3)_3CCH_2\overset{O}{\overset{\|}{C}}OCH_2CH_3$ + excess cyclohexyl-OH $\xrightarrow[heat]{H^+}$

(d) $CH_2{=}CHCH_2\overset{O}{\overset{\|}{C}}OC_6H_5$ $\xrightarrow[heat,\ pressure]{H_2,\ catalyst}$

(e) $CH_2{=}CHCH_2\overset{O}{\overset{\|}{C}}OC_6H_5$ $\xrightarrow[(2)\ H_2O,\ H^+]{(1)\ excess\ C_6H_5MgBr}$

15.18 Explain the following observations:

(a) [cyclohexane with CO_2H and OH] $\xrightarrow[heat]{H^+}$ a lactone

(b) 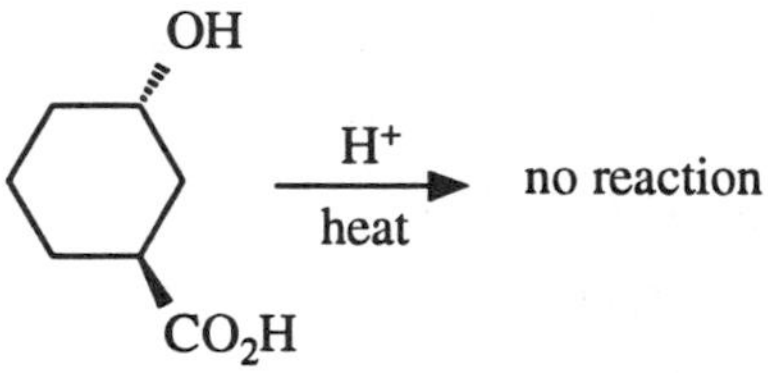 [cyclohexane with OH (hashed) and CO_2H (wedged)] $\xrightarrow[\text{heat}]{H^+}$ no reaction

15.19 Name three different reagents that might be used to synthesize heptanol from ethyl heptanoate.

15.20 Write equations for complete reaction mechanisms for the following reactions:

(a) $CH_3CH_2\overset{O}{\overset{\|}{C}}OCH_3 + H_2O \overset{H^+}{\rightleftharpoons}$

(b) 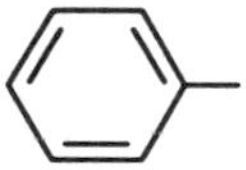 C_6H_5—$CO_2CH_3 + NaOH \xrightarrow[\text{heat}]{H_2O}$

15.21 What product(s) would be formed when the following compounds are treated with an excess of methylmagnesium iodide followed by hydrolysis?

(a) $C_6H_5CO_2CH_3$ (b) $C_6H_5\overset{O}{\overset{\|}{C}}C_6H_5$ (c) 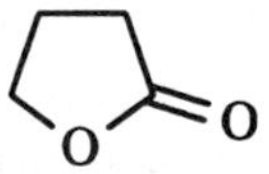 [five-membered lactone ring with ring O and C=O]

(d) $CH_3O\overset{O}{\overset{\|}{C}}OCH_3$ (e) $C_6H_5\overset{O}{\overset{\|}{C}}H$ (f) $C_6H_5\overset{O}{\overset{\|}{C}}Cl$

15.22 When $CH_3O_2CCH_2CH_2\overset{O}{\overset{\|}{C}}CH_3$ is reduced with $NaBH_4$, a lactone is formed.

(a) What is the structure of the lactone?

(b) How is the lactone formed?

(c) Would the lactone be formed if $LiAlH_4$ was the reducing agent?

15.23 Write equations showing the preparation of *N*-methylpentanamide from (a) an acid chloride, (b) an acid anhydride, and (c) an ester.

15.24 Complete the following equations:

(a) 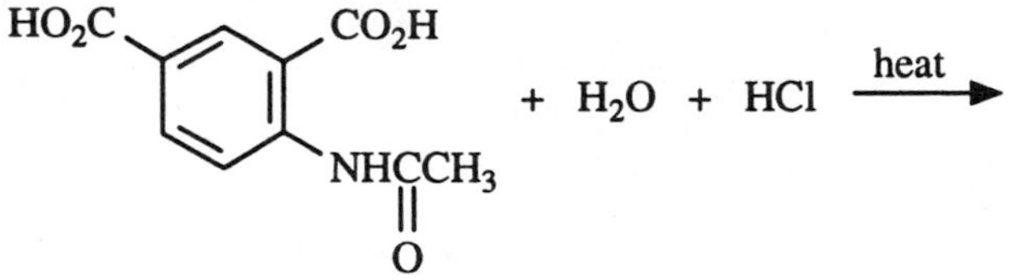 [2-acetamido-isophthalic acid: benzene ring bearing HO_2C, CO_2H, and $NHCCH_3$ (with C=O)] $+ H_2O + HCl \xrightarrow{\text{heat}}$

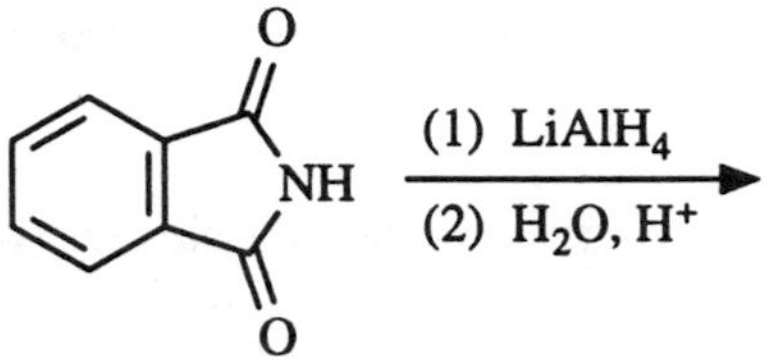

(b) [phthalimide] $\xrightarrow[\text{(2) } H_2O,\ H^+]{\text{(1) } LiAlH_4}$

(c) $C_6H_5\overset{O}{\overset{\|}{C}}N(CH_3)_2 + {}^{-}OH \xrightarrow[\text{heat}]{H_2O}$

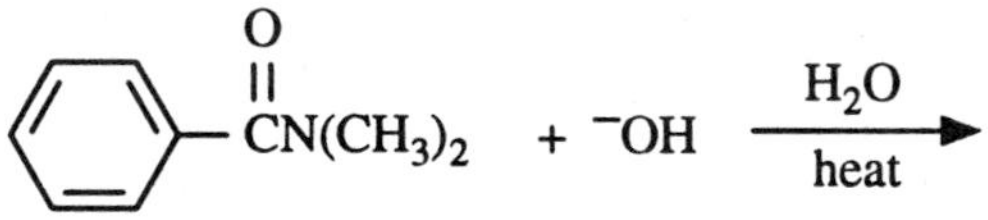

15.25 Write equations for the reactions of *N*-methylbenzamide with the following reagents:

(a) H_2O, H^+

(b) H_2O, OH^-

(c) (1) $LiAlH_4$; (2) H_2O, H^+

(d) CH_3OH, H^+

15.26 Write equations for the reactions of the following compounds with $LiAlH_4$ followed by hydrolysis:

(a) $(CH_3)_2CH\overset{O}{\overset{\|}{C}}CH_3$

(b) $(CH_3)_2CHCH_2\overset{O}{\overset{\|}{C}}H$

(c) $C_6H_5CO_2H$

(d) $C_6H_5\overset{O}{\overset{\|}{C}}Cl$

(e) $C_6H_5CO_2CH_3$

(f) $C_6H_5\overset{O}{\overset{\|}{C}}NH_2$

15.27 What is the structure of the nitrile (or dinitrile) that can be used to prepare each of the following carboxylic acids?

(a) 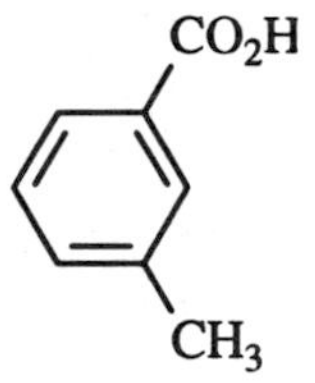

(b) $HO\overset{O}{\overset{\|}{C}}CH_2CH_2\overset{O}{\overset{\|}{C}}OH$

(c) 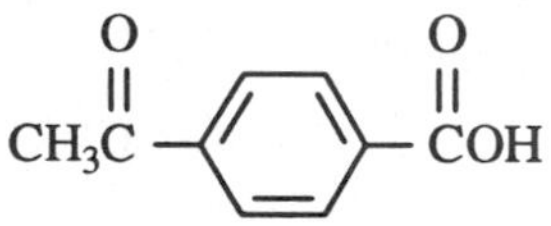

15.28 Write flow equations for the following conversions:

(a) 1-propanol ⟶ butanenitrile

(b) benzyl alcohol ⟶ phenylacetonitrile (phenylethanenitrile)

(c) benzene ⟶ benzonitrile

15.29 Write equations for the reactions of benzonitrile with the following reagents:

(a) H_2O, H^+, 100°

(b) H_2, Raney Ni

(c) (1) $LiAlH_4$; (2) H_2O, H^+

(d) HCl, H_2O, 40°

15.30 Write equations for two different synthetic procedures that can be used to carry out the following conversions:

(a) 

(b) $CH_3CH_2CH_2Br \longrightarrow CH_3CH_2CH_2\overset{O}{\overset{\|}{C}}OH$

15.31 Complete the following equations:

(a) $(CH_3\overset{O}{\overset{\|}{C}})_2O + CH_3CH_2OH \longrightarrow$

(b) $CH_3\overset{O}{\overset{\|}{C}}Cl + CH_3CH_2OH \longrightarrow$

(c) $CH_3C(=O)NHCH_3 + CH_3CH_2OH \longrightarrow$

(d) $C_6H_5C(=O)Cl \xrightarrow[(2)\ H_2O,\ H^+]{(1)\ \text{excess } CH_3CH_2MgBr}$

(e) $C_6H_5C(=O)Cl + (CH_3CH_2)_2CuLi \longrightarrow$

(f) 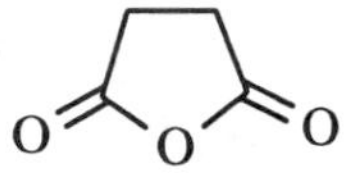 (cyclic anhydride: five-membered ring with O, two C=O) $+ CH_3CH_2OH \longrightarrow$

(g) 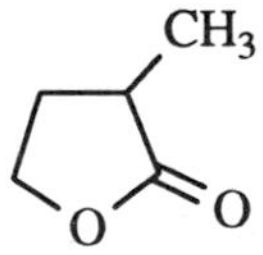 (lactone: five-membered ring with O, C=O, and CH_3 on the α-carbon) $\xrightarrow[(2)\ H_2O,\ H^+]{(1)\ LiAlH_4}$

(h) 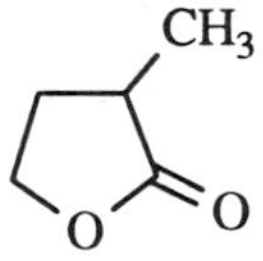 (lactone: five-membered ring with O, C=O, and CH_3 on the α-carbon) $\xrightarrow[(2)\ H_2O,\ H^+]{(1)\ \text{excess } CH_3MgI}$

15.32 Using flow equations, show how each of the following conversions could be carried out:

(a) $CH_3CH_2CH{=}CH_2 \longrightarrow CH_3CH_2CH_2C(=O)NH_2$

(b) 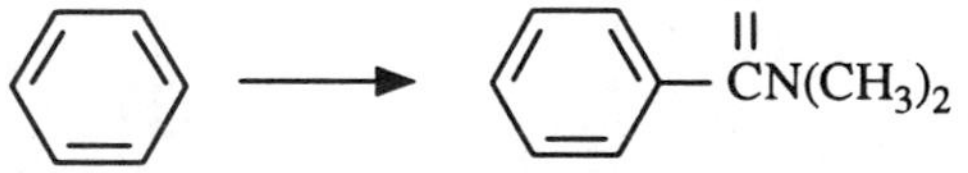 benzene $\longrightarrow C_6H_5C(=O)N(CH_3)_2$

(c) $CH_3CO_2H \longrightarrow CH_3CH_2NH_2$

16

Conjugate Addition

Some Important Features

The key feature to understanding conjugate addition is to look upon the conjugated pi bond system as a single functional group rather than as two. When viewed in this manner, it is easier to identify the two ends of the conjugated system that undergo substitution as well as the inner two carbon atoms at which the new double bond forms.

View as a single functional group:

$CH_2{=}CH{-}CH{=}CH\,CH_3$ $\qquad$ $CH_3\,CH{=}CH{-}C(=O){-}H$

Substitution occurs here

$$CH_3\,CH{=}CH{-}CH{=}CH{-}CH_3 + Br_2 \xrightarrow{\text{1,4-Addition}} CH_3CH(Br)CH{=}CHCH(Br)CH_3$$

double bond forms here

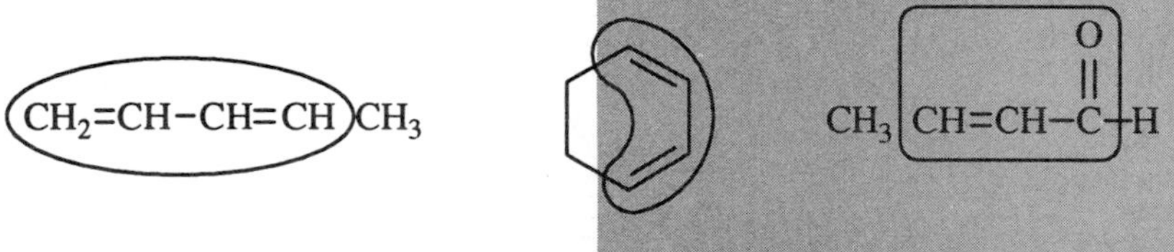

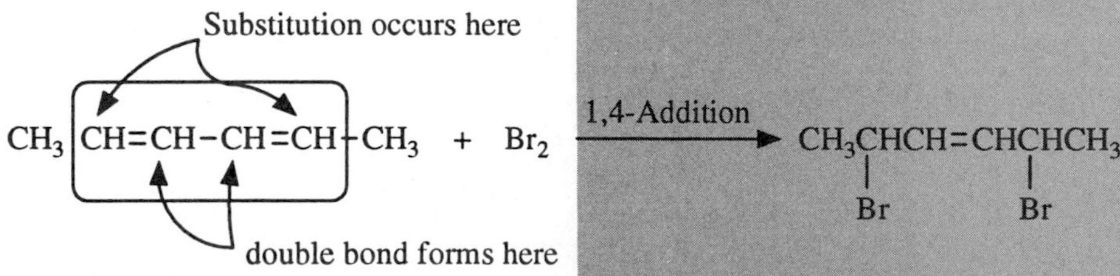

When conjugate addition takes place with a conjugated carbonyl compound, remember that there are two steps. The first step is a standard 1,4-addition reaction. This standard reaction, however, forms an enol. Therefore, the second step is the conversion of the enol to a keto form. The intermediate enol is never considered to be the product of the reaction.

$$CH_3CH{=}CH{-}\overset{\overset{O}{\|}}{C}{-}CH_3 \; + \; HCl$$

$$\Big\downarrow \text{ 1,4-addition}$$

$$\left[\begin{array}{c} CH_3\underset{\underset{Cl}{|}}{C}H{-}CH{=}\overset{\overset{OH}{|}}{C}{-}CH_3 \\ \textit{an enol} \end{array}\right]$$

$$\Big\updownarrow \text{ enol-keto equilibrium}$$

$$CH_3\underset{\underset{Cl}{|}}{C}H{-}CH_2\overset{\overset{O}{\|}}{C}CH_3$$

observed product

Additional Drill Problems

16.1 Complete the following equations, showing the most likely products in your answer:

(a) $CH_2{=}\overset{\overset{C_6H_5}{|}}{C}CH{=}CH_2 \; + \; 1\ HI \longrightarrow$

(b) $CH_3{-}C_6H_4{-}CH{=}CHCH{=}CH{-}C_6H_4{-}CH_3 \; + \; 1\ HBr \longrightarrow$ (para-substituted benzene rings)

(c) $CH_2{=}C\overset{\overset{CH_3}{|}}{\underset{\underset{CH_3}{|}}{C}}{=}CH_2 \; + \; 1\ Br_2 \longrightarrow$

16.2 What products would you expect if 1.0 mol of each of the following dienes were treated with (1) 1.0 mol of Br_2; (2) 1.0 mol of HBr?

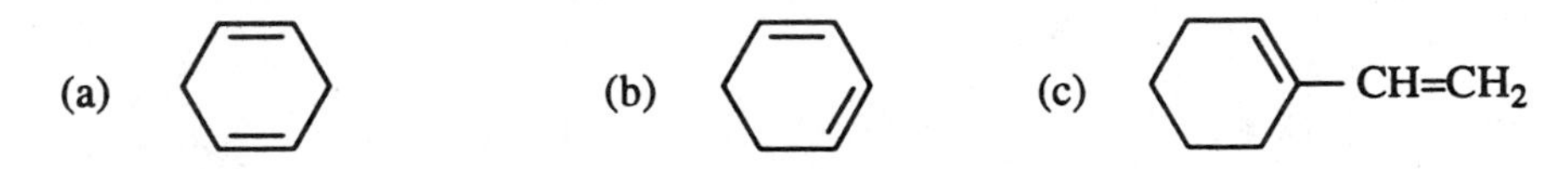

16.3 Predict all likely organic products of the reaction of 4-phenyl-3-buten-2-one with:

(a) H_2O, H^+ (b) CH_3MgI (c) HBr

(d) excess CH_3NH_2 (e) NaCN, H^+

16.4 What is wrong with the following equation?

$$\text{piperidine (N–H)} + CH_2{=}CH{-}\overset{O}{\overset{\|}{C}}{-}CH_3 \xrightarrow{H^+} CH_3\underset{\text{N (piperidinyl)}}{\underset{|}{C}H}{-}\overset{O}{\overset{\|}{C}}CH_3$$

16.5 Complete the following equations.

(a) 2-cyclohexen-1-one + $CH_3OH \xrightarrow{H^+}$

(b) $C_6H_5{-}\overset{O}{\overset{\|}{C}}CH{=}CH_2 + (CH_3)_2NH \xrightarrow{H^+}$

(c) $C_6H_5{-}SH + CH_2{=}CH{-}\overset{O}{\overset{\|}{C}}CH_3 \xrightarrow{CH_3O^-}$

(d) $C_6H_5{-}\overset{O}{\overset{\|}{C}}{-}CH{=}CH_2 + H_2O \xrightarrow{H^+}$

16.6 Hydrazine reacts with 2-pentenal to yield the following product. Predict the product of the following reaction:

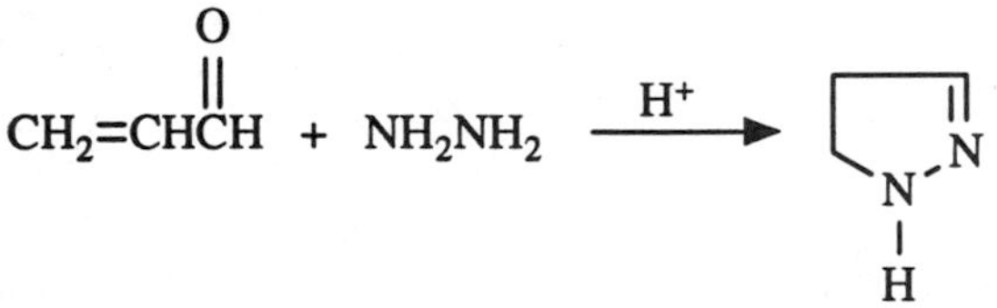

$$CH_2{=}CH\overset{O}{\overset{\|}{C}}H + NH_2NH_2 \xrightarrow{H^+}$$ (ring: N, N, N–H, H)

$$C_6H_5{-}CH{=}CH{-}\overset{O}{\overset{\|}{C}}{-}C_6H_5 + NH_2NH_2 \xrightarrow{H^+} ?$$

16.7 Predict the Diels-Alder reaction product of 1,3-butadiene and each of the following compounds:

(a) $CH_2{=}CH\overset{O}{\overset{\|}{C}}H$

(b) $(NC)_2C{=}C(CN)_2$

(c) $HC{\equiv}C\overset{O}{\overset{\|}{C}}OCH_2CH_3$

(d) $CH_2{=}CH\overset{O}{\overset{\|}{C}}OCH_2CH_3$

(e) *cis*-$CH_3O\overset{O}{\overset{\|}{C}}CH{=}CH\overset{O}{\overset{\|}{C}}OCH_3$

(f) *trans*-$CH_3O\overset{O}{\overset{\|}{C}}CH{=}CH\overset{O}{\overset{\|}{C}}OCH_3$

16.8 Complete each of the following equations showing the Diels-Alder product:

(a) 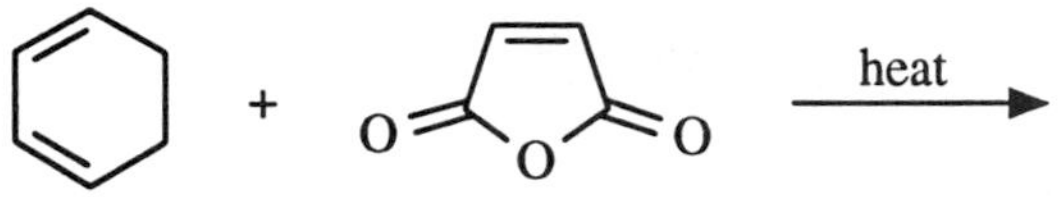 [1,3-cyclohexadiene] + [maleic anhydride: O, O, O] $\xrightarrow{heat}$

(b) 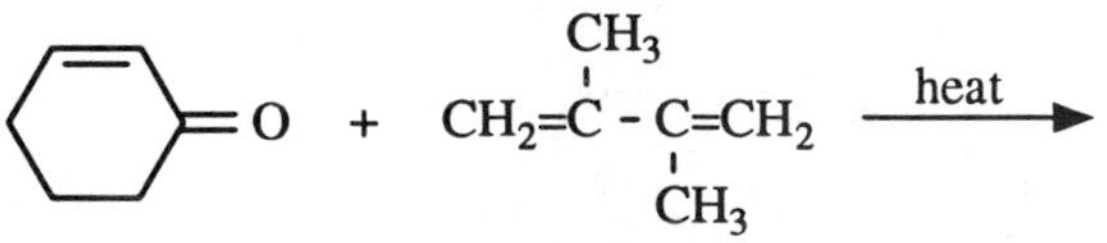 [2-cyclohexenone]=O + $CH_2{=}\underset{CH_3}{\overset{CH_3}{C}}$ - $\underset{CH_3}{C}{=}CH_2 \xrightarrow{heat}$

16.9 Draw the structures of the diene and the dienophile needed for the preparation of each of the following cyclohexenes:

(a) 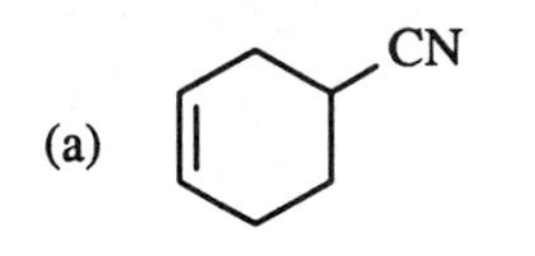

(b)

(c)

H

CHO

H

CH_2CH_3

(d)

O_2CCH_3

O_2CCH_3

O_2CCH_3

(e)

H

CO_2CH_3

(f)

O

CN

CN

17

Enolates and Carbanions: Building Blocks for Organic Synthesis

Some Important Features

A hydrogen atom α to a carbonyl group is slightly acidic and can be removed by a base. The reason for this acidity of carbonyl compounds is the resonance stabilization of the enolate ion. The extent of enolate formation is determined by the acidity of the carbonyl compound and by the strength of the base.

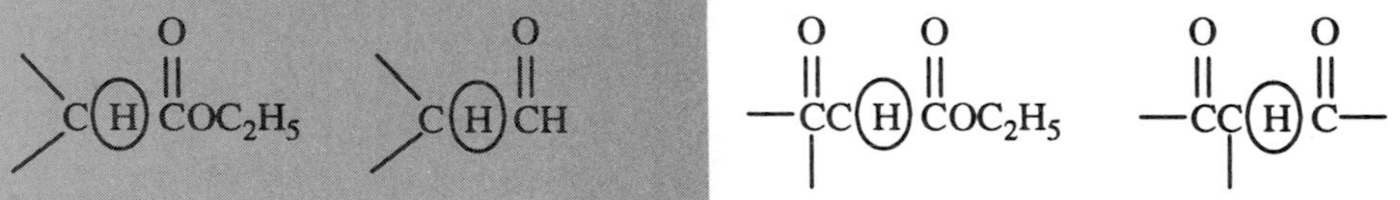

$$>C\text{(H)}-\overset{O}{\overset{\|}{C}}OC_2H_5 \qquad >C\text{(H)}-\overset{O}{\overset{\|}{C}}H$$

less acidic than ROH

$$-\overset{O}{\overset{\|}{C}}-\underset{|}{C}\text{(H)}-\overset{O}{\overset{\|}{C}}OC_2H_5 \qquad -\overset{O}{\overset{\|}{C}}-\underset{|}{C}\text{(H)}-\overset{O}{\overset{\|}{C}}-$$

more acidic than ROH

$$CH_3\overset{O}{\overset{\|}{C}}H + {}^{-}\!:\ddot{\underset{..}{O}}C_2H_5 \rightleftharpoons {}^{-}\!:CH_2\overset{O}{\overset{\|}{C}}H + H\ddot{\underset{..}{O}}C_2H_5$$

$$CH_3\overset{O}{\overset{\|}{C}}CH_2\overset{O}{\overset{\|}{C}}CH_3 + {}^{-}\!:\ddot{\underset{..}{O}}C_2H_5 \rightleftharpoons CH_3\overset{O}{\overset{\|}{C}}\overset{-}{\underset{..}{C}}H\overset{O}{\overset{\|}{C}}CH_3 + H\ddot{\underset{..}{O}}C_2H_5$$

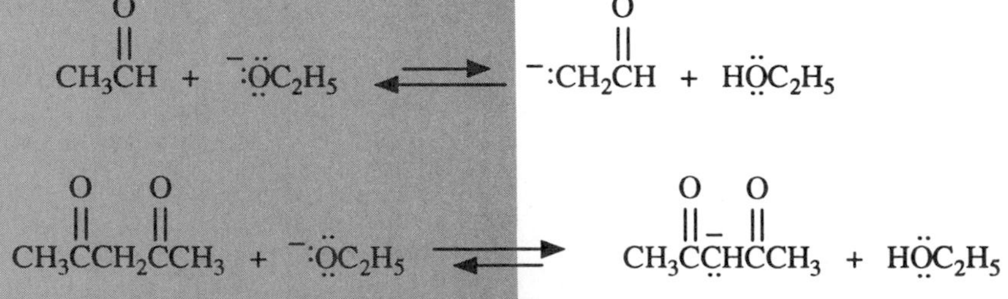

Enolate ions are useful as synthetic intermediates. For example, enolate ions can act as nucleophiles in substitution reactions with alkyl halides. (Enamine reactions are similar; see Section 17.5.)

An alkylation reaction:

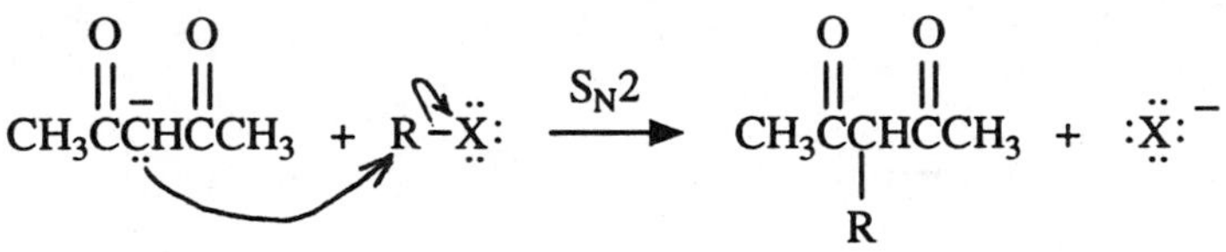

$$CH_3C(=O)\overset{-}{\ddot{C}}HC(=O)CH_3 + R{-}\ddot{X}: \xrightarrow{S_N2} CH_3C(=O)CH(R)C(=O)CH_3 + :\ddot{X}:^-$$

Enolates ions can also attack carbonyl groups.

An aldol condensation:

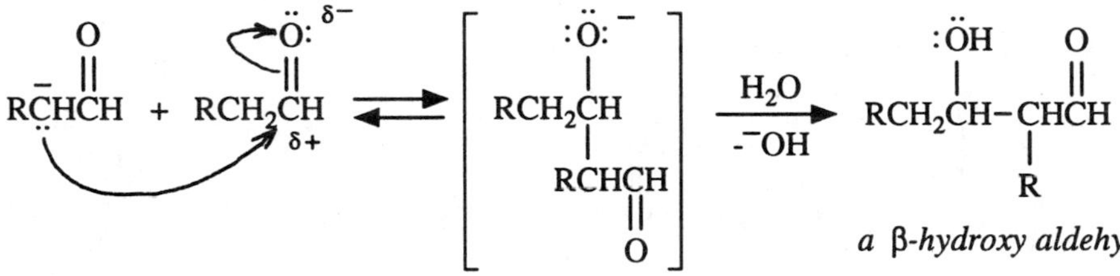

$$R\overset{-}{\ddot{C}}HC(=O)H + RCH_2\overset{\delta+}{C}H{=}\ddot{O}:^{\delta-} \rightleftharpoons \left[RCH_2CH(:\ddot{O}:^-)\text{–}RCHCH(=O) \right] \xrightarrow[-^-OH]{H_2O} RCH_2CH(:\ddot{O}H){-}CH(R)CH(=O)$$

a β-hydroxy aldehyde

An ester condensation:

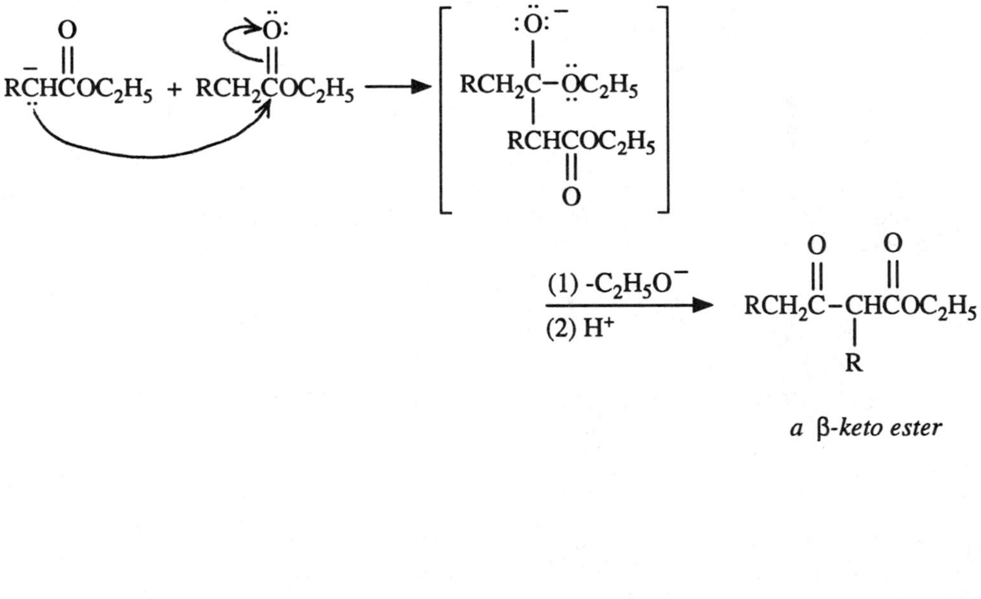

$$R\overset{-}{\ddot{C}}HC(=O)OC_2H_5 + RCH_2C(=\ddot{O}:)OC_2H_5 \longrightarrow \left[RCH_2C(:\ddot{O}:^-){-}\ddot{O}C_2H_5 \text{ – } RCHC(=O)OC_2H_5 \right]$$

$$\xrightarrow[(2)\ H^+]{(1)\ -C_2H_5O^-} RCH_2C(=O){-}CH(R)C(=O)OC_2H_5$$

a β-keto ester

Enolate ions can also attack an α,β-unsaturated carbonyl compound in a 1,4-addition reaction (Michael addition).

$$(C_2H_5OC)_2\bar{C}H + RCH{=}CH{-}\overset{O}{\overset{\|}{C}}R \longrightarrow$$

$$\left[(C_2H_5OC)_2CHCHCH{=}\overset{:\ddot{O}:^-}{\overset{|}{C}}R \longleftrightarrow (C_2H_5OC)_2CHCH\bar{C}H{-}\overset{\ddot{O}:}{\overset{\|}{C}}R\right]$$

$$\rightleftharpoons (C_2H_5OC)_2\bar{C}CHCH_2CR \xrightarrow{H^+} (C_2H_5OC)_2CHCHCH_2CR$$

more stable enolate

The products of enolate reactions may be subjected to further reactions, such as a second enolate reaction, hydrolysis, decarboxylation, or dehydration. These topics are discussed in the text.

Reminders

In predicting the products of an enolate or related reaction, look first for the most acidic hydrogen. Then look for the most likely target of nucleophilic attack--for example, a carbon-halogen bond or a carbonyl group.

$$CH_3CCH_2CCH_3 + NaOC_2H_5 \rightleftharpoons CH_3C\bar{C}HCCH_3\ Na^+ + HOC_2H_5$$

most acidic *nucleophilic*

$$CH_3C\bar{C}HCCH_3 + \overset{\delta+}{C}H_2{-}\overset{\delta-}{\ddot{C}l}:\ (CH_2CH_3) \longrightarrow CH_3CCHCCH_3\ (CH_2CH_2CH_3) + :\ddot{C}l:^-$$

attacking a partially positive carbon

When solving problems asking for a synthetic route, first determine the type of product. Is the product a substituted acetic acid? (Try a malonic ester alkylation.) A substituted acetone? (Try an acetoacetic ester alkylation.) An α,β-unsaturated aldehyde? (Try an aldol condensation.) A β-keto ester? (Try an ester condensation.) After you deduce a likely route, use your pencil to divide the structure into its pieces.

$$(CH_3CH_2)_2\,\boxed{CHCO_2H}$$

from 2 RX → $(CH_3CH_2)_2$; *from malonic ester* → $CHCO_2H$

At this point, you can work the problem backwards.

$$(CH_3CH_2)_2C(CO_2H)_2 \xrightarrow[-CO_2]{\text{heat}} \text{product}$$

$$\uparrow \; H_2O,\ H^+,\ \text{heat}$$

$$(CH_3CH_2)_2C(CO_2C_2H_5)_2$$

$$\uparrow \; (1)\ NaOC_2H_5 \quad (2)\ CH_3CH_2Br$$

$$CH_3CH_2CH(CO_2C_2H_5)_2 \xleftarrow{CH_3CH_2Br} {}^{-}CH(CO_2C_2H_5)_2 \xleftarrow{NaOC_2H_5} CH_2(CO_2C_2H_5)_2$$

Additional Drill Problems

17.1 Write equations showing all reactants, intermediates, and products in the decarboxylation of the following compound:

$(CH_3)_2C(CO_2C_2H_5)_2$

17.2 Complete the following flow equations:

(a) $$CH_3\overset{O}{\overset{\|}{C}}CH_2\overset{O}{\overset{\|}{C}}CH_3 \xrightarrow[CH_3CH_2OH]{Na^{+}\ {}^{-}OCH_2CH_3} \underline{\quad(1)\quad} \xrightarrow{C_6H_5CH_2Cl} \underline{\quad(2)\quad} \xrightarrow[\text{heat}]{H_2O,\ H^+} \underline{\quad(3)\quad}$$

(b) $CH_2(CO_2C_2H_5)_2 \xrightarrow[CH_3CH_2OH]{Na^+ \,^-OCH_2CH_3}$ (1) ___ $\xrightarrow{\text{(2-furyl)}CH_2Br}$ (2) ___

$\xrightarrow[heat]{H_2O,\ NaOH}$ (3) ___ $\xrightarrow{H_2O,\ H^+}$ (4) ___ $\xrightarrow{heat}$ (5) ___

(c) $C_2H_5O_2CCH_2CN \xrightarrow[CH_3CH_2OH]{Na^+ \,^-OCH_2CH_3}$ (1) ___ $\xrightarrow{BrCH_2CO_2C_2H_5}$ (2) ___

$\xrightarrow[heat]{H_2O,\ H^+}$ (3) ___

(d) morpholine (O, N–H ring) $\xrightarrow[H^+]{\text{cyclopentanone}}$ (1) ___ $\xrightarrow{C_6H_5CH_2Br}$ (2) ___ $\xrightarrow[heat]{H_2O,\ H^+}$ (3) ___

17.3 Which of the following organohalogen compounds could be used as an alkylating agent in an acetoacetic ester alkylation?

(a) $(CH_3)_3CBr$

(b) $CH_2{=}CHCH_2Cl$

(c) cyclohexyl–Br

(d) $BrCH_2CO_2CH_3$

(e) $C_6H_5CH_2I$

(f) (2-furyl)–CH_2Cl

(g) $C_6H_5\overset{O}{\overset{\|}{C}}Cl$

(h) cyclohexen-1-yl–Br

17.4 Write equations showing how each of the following compounds could be prepared from an aldehyde or a ketone in an enamine synthesis:

(a) $CH_2{=}CHCH_2CH_2CH{=}O$

(b) 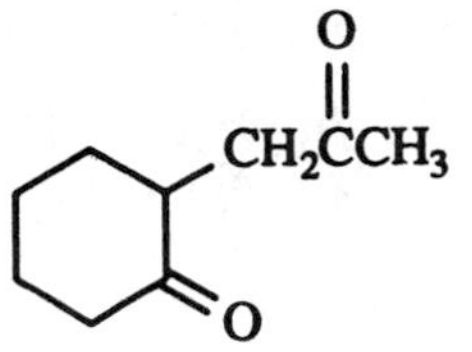 2-($CH_2C(=O)CH_3$)cyclohexanone

(c) 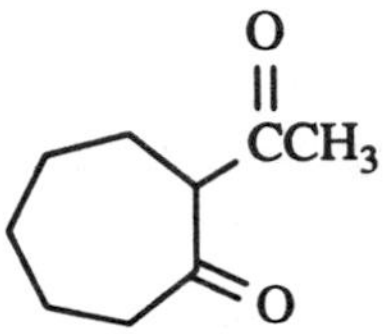 2-($C(=O)CH_3$)cycloheptanone

17.5 Write flow equations to show expected intermediates and aldol addition products of the following aldehydes:

(a) $(CH_3)_2CHCH_2CH{=}O$

(b) 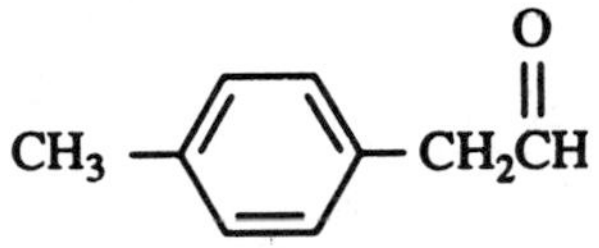 CH_3–C_6H_4–$CH_2CH{=}O$ (para)

17.6 Complete the following equations:

(a) 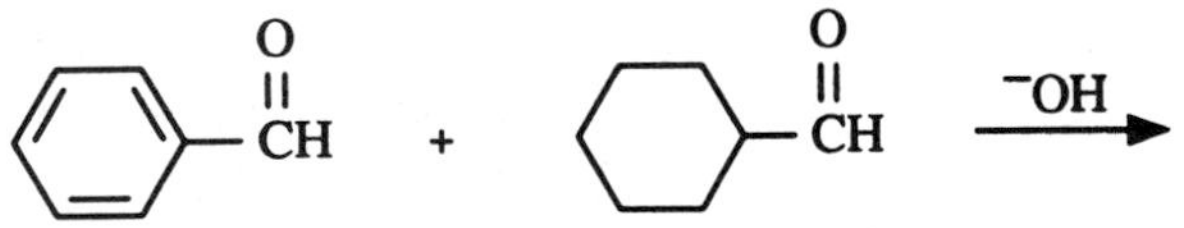 $C_6H_5CH{=}O$ + $C_6H_{11}CH{=}O$ $\xrightarrow{^{-}OH}$

(b) $C_6H_5C(=O)CH_3$ + 2-furyl–CHO $\xrightarrow{^{-}OH}$

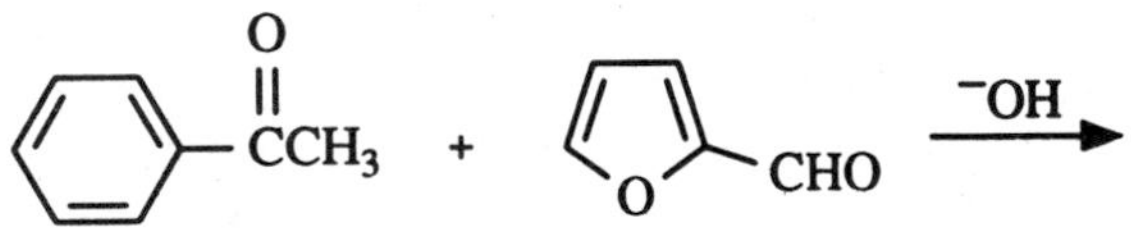

(c) $CH_3C(=O)(CH_2)_5CH{=}O$ $\xrightarrow{^{-}OH}$

17.7 Write the equation for the reaction that would be expected to occur between *p*-nitrobenzaldehyde and each of the following sets of reagents:

(a) $CH_3CH(=O) + NaOH$

(b) $CH_2(CO_2C_2H_5)_2 + NaOC_2H_5$

(c) $C_6H_5C(=O)CH_3 + NaOH$

17.8 Complete each of the following equations showing the Knoevenagel condensation product:

(a) $CH_3CH_2CH_2CHO + CH_2(CO_2C_2H_5)_2 \xrightarrow[\text{heat}]{\text{piperidine}}$

(b) $C_6H_5CH{=}CHCHO + CH_2(CO_2H)_2 \xrightarrow[\text{heat}]{NH_3}$

(c) $CH_2(CN)_2$ + furan-2-CHO $\xrightarrow[\text{heat}]{NH_3}$

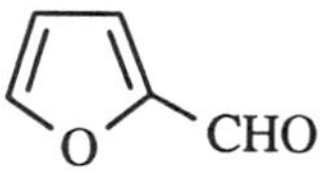

(d) $NCCH_2CO_2C_2H_5$ + $CH_3C(=O)$-quinoline $\xrightarrow[\text{heat}]{Na^+ \; {}^-O_2CCH_3}$

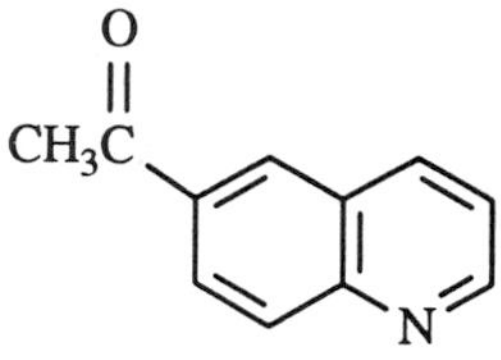

17.9 Which of the following esters could undergo an ester condensation reaction when treated with an alkoxide?

(a) $CH_3CH_2C(=O)OCH_3$

(b) $CH_3CH_2C(CH_3)(CO_2CH_2CH_3)CH_3$

(c) $CH_3C(=O)CH_2C(=O)OCH_3$

(d) $CH_3OC(=O)CH_2CH_2CH_2C(=O)OCH_3$

17.10 Write equations for the ester condensations in the preceding problem.

17.11 Write abbreviated mechanisms for the following reactions:

(a) $CH_3\overset{O}{\overset{||}{C}}CH_2CH_2CH(CO_2C_2H_5)_2 \xrightarrow[(2)\ H_2O,\ H^+,\ heat]{(1)\ C_2H_5O^-Na^+}$ 1,3-cyclohexanedione

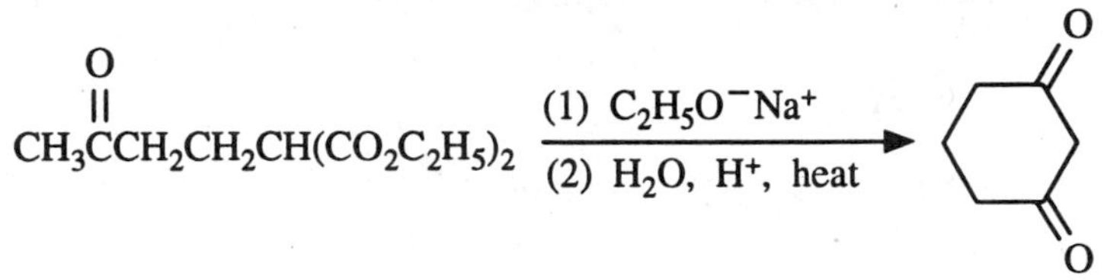

(b) 2-naphthyl–$\overset{O}{\overset{||}{C}}(CH_2)_4\overset{O}{\overset{||}{C}}OC_2H_5 \xrightarrow[(2)\ H_2O,\ H^+,\ heat]{(1)\ C_2H_5O^-Na^+}$ 2-naphthyl–$\overset{O}{\overset{||}{C}}$–(2-oxocyclopentyl)

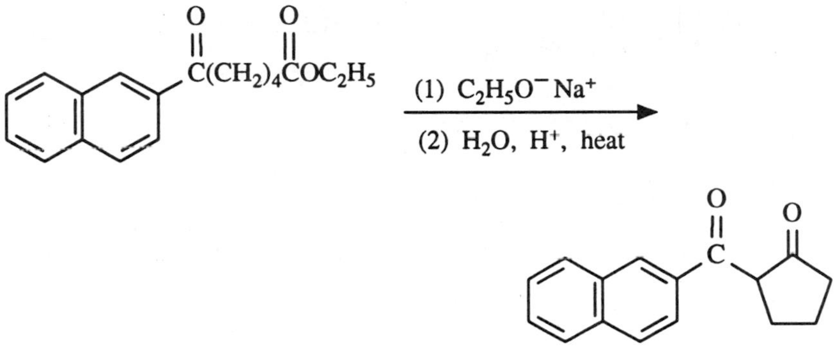

17.12 Complete the following equations showing the expected intermediates and Michael addition products.

(a) 2-cyclohexen-1-one $\xrightarrow[(2)\ H_2O,\ H^+,\ heat]{(1)\ Na^+\ ^-CH(CO_2C_2H_5)_2}$

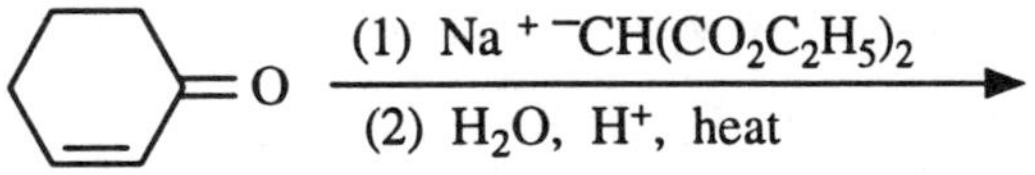

(b) bicyclic enone (octalone) $\xrightarrow[(2)\ H_2O,\ H^+,\ heat]{(1)\ Na^+\ ^-CH(CO_2C_2H_5)_2}$

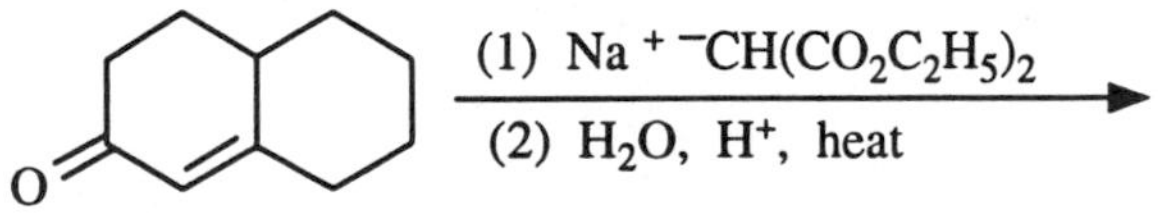

(c) $CH_3CH_2\overset{O}{\overset{||}{C}}CH{=}CH_2 \xrightarrow[(2)\ H_2O,\ H^+,\ heat]{(1)\ CH_3\overset{O}{\overset{||}{C}}\overset{-}{C}H\overset{O}{\overset{||}{C}}OC_2H_5\ Na^+}$

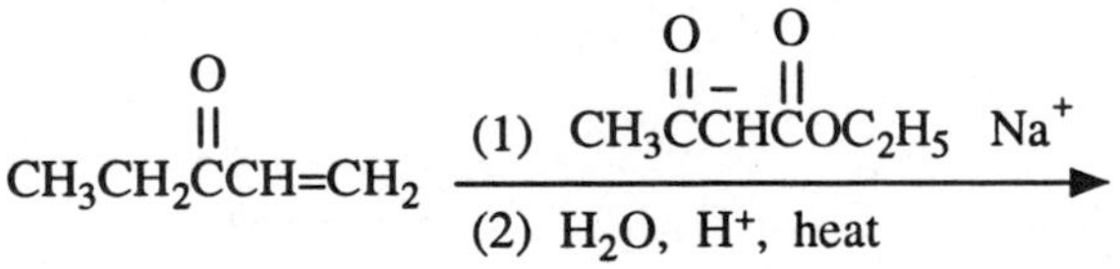

17.13 Complete the following equations:

(a) $C_6H_5CH_2CO_2CH_3 + CH_3O^- \longrightarrow$

(b) $CH_3CH_2\overset{\overset{\displaystyle O}{\|}}{C}H + OH^- \longrightarrow$

(c) 1-(3,4-dihydronaphthalen-2-yl)piperidine $+ CH_3I \longrightarrow$

(d) 2-allylcyclohexylidene morpholinium ion ($CH_2CH{=}CH_2$ on the ring carbon next to $C{=}N^+$) $+ H_2O \xrightarrow{H^+}$

(e) 5,5-dimethylcyclohexane-1,3-dione (H_3C, H_3C, O, O) $\xrightarrow[\text{(2) } CH_3CH_2I]{\text{(1) } Na^+ \ ^-OCH_3}$

17.14 Complete the following equations:

(a) $CH_3\overset{\overset{\displaystyle O}{\|}}{C}CH_2CO_2CH_3 \xrightarrow[\text{(2) } C_6H_5CH_2Cl]{\text{(1) } NaOCH_3}$

(b) $CH_2(CO_2C_2H_5)_2 + C_6H_5CO_2C_2H_5 \xrightarrow{NaOC_2H_5}$

(c) [bicyclo[2.2.2]octane ring bearing C=O and an adjacent =CH_2 group]
$\xrightarrow[\text{(2) } H_2O,\ H^+,\ \text{cold}]{\text{(1) } Na^+\ ^-CH(CO_2C_2H_5)_2}$

(d) $C_6H_5CHO + CH_3NO_2 \xrightarrow{NaOH}$

17.15 Show how pentanal can be converted into the following compounds:

(a) $CH_3CH_2CH_2CH_2CH(OH)CH(CH_2CH_2CH_3)CH{=}O$

(b) $CH_3CH_2CH_2CH(CH_2C_6H_5)CH{=}O$

(c) $CH_3(CH_2)_3C(=O)OC_2H_5$

(d) $CH_3(CH_2)_3C(=O)CH(CH_2CH_2CH_3)CO_2C_2H_5$

18

Amines

Some Important Features

Amines are compounds in which N is bonded to three other groups (or hydrogen). The nitrogen of an amine contains an unshared pair of electrons; therefore, an amine is basic and can act as a nucleophile.

$$R_3N{:} + H-\ddot{O}H \rightleftharpoons R_3\overset{+}{N}H + {:}\ddot{O}H^-$$

an amine

$$R_3N{:} + H-Cl \rightleftharpoons R_3\overset{+}{N}H\ Cl^-$$

an amine salt

Although amines can react as nucleophiles with alkyl halides, this reaction often leads to mixtures of products.

$$R_3N{:} + CH_3-I \xrightarrow{S_N2} R_3\overset{+}{N}CH_3\ I^-$$

only product

$$RNH_2 + CH_3I \longrightarrow R\overset{+}{N}H_2CH_3\ I^- + R\overset{+}{N}H(CH_3)_2\ I^- + R\overset{+}{N}(CH_3)_3\ I^-$$

For this reason, other methods for synthesizing amines have been developed (see Section 18.4).

Amines are *weak* bases. Their basicity is determined by the relative stabilization of amine versus amine salt.

$$R_2\ddot{N}H + H_2O \rightleftharpoons R_2\overset{+}{N}H_2 + OH^-$$

R is electron-releasing and strengthens base

$$C_6H_5-\ddot{N}H_2 + H_2O \rightleftharpoons C_6H_5\overset{+}{N}H_3 + OH^-$$

resonance stabilization of amine weakens base

Basicity is also affected by the amount of solvation (which stabilizes the conjugate acid) and the hybridization of the N atom (*sp^3* N more basic than *sp^2* N).

Reminders

When determining the relative basicities of amines, always ask yourself about the availability of the unshared electrons for donation to H^+. For example, the electrons of aniline are *less* available than those of alkylamine; therefore, aniline is less basic.

Additional Drill Problems

18.1 Classify each of the following amines or ammonium ions as 1°, 2°, or 3° or as a quaternary ammonium salt.

(a) $C_6H_5CH_2NH_2$ (b) $(CH_3CH)_2NH$ with CH_3 on each CH (c) $(CH_3)_4N^+\ Br^-$

(d) $CH_2{=}CHCH_2NH_2$ (e) $H_2N(CH_2)_{10}NH_2$

18.2 Which of the following compounds would you expect to be capable of resolution into isolable enantiomers?

(a) $CH_3NCH_2CH_3$ with $CH_2CH_2CH_3$ on N

(b) $(CH_3)_2\overset{+}{N}CH_2CH_2\overset{+}{N}CH_3\ 2\,Br^-$ with C_6H_5 on each N and CH_2CH_3 on the second N

18.3 Determine the structures of the following compounds:

(a) $C_8H_9NO_2$, Spectra 18.1 (b) $C_9H_{13}N$, Spectra 18.2

Spectra 18.1

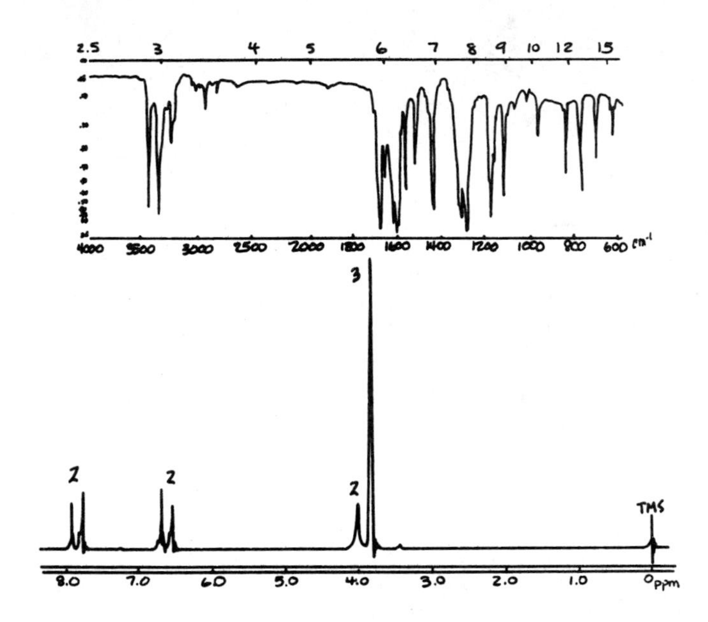

Spectra 18.2

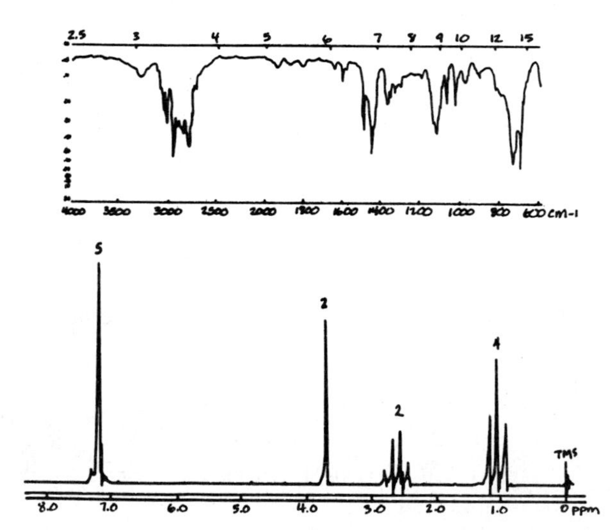

18.4 What reagent(s) would be needed to carry out the following conversions?

(a) Cyclohexyl–$\overset{O}{\overset{\|}{C}}NH_2$ $\longrightarrow$ Cyclohexyl–CH_2NH_2

(b) O_2N–(benzene ring with NO_2)–Cl $\longrightarrow$ O_2N–(benzene ring with NO_2)–NH_2

(c) Cyclohexyl–$\overset{O}{\overset{\|}{C}}NH_2$ $\longrightarrow$ Cyclohexyl–NH_2

(d) Cyclohexyl–$\overset{O}{\overset{\|}{C}}H$ $\longrightarrow$ Cyclohexyl–CH_2NHCH_3

(e) $CH_3CH_2CH_2CN \longrightarrow CH_3CH_2CH_2CH_2NH_2$

18.5 Complete the following equations:

(a) (bicyclic ring with NCH_3) + CH_3I $\longrightarrow$

(b) (benzene ring with CN and CH_2Br) + $(CH_3)_2NH$ $\longrightarrow$

18.6 Which path would be better for the synthesis of each of the following amines--the Gabriel phthalimide synthesis or the Hofmann rearrangement? Explain.

(a) $CH_3CH_2\overset{CH_3}{\overset{|}{\underset{CH_3}{\underset{|}{C}}}}NH_2$ (b) $C_6H_5\underset{NH_2}{\underset{|}{C}}HCO_2H$

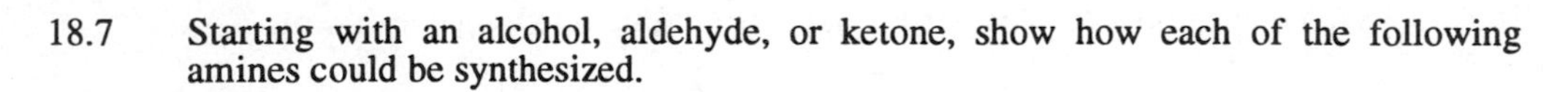

18.7 Starting with an alcohol, aldehyde, or ketone, show how each of the following amines could be synthesized.

(a) $(CH_3)_2CH-C_6H_{10}-NH_2$

(b) $CH_3CH_2CHCH_2NH_2$ (with OH on the CH)

(c) $CH_3(CH_2)_4N(CH_3)_2$

18.8 Write formulas for the conjugate acids of the following compounds:

(a) $CH_3CH_2NH_2$

(b) HN O (ring)

(c) $CH_3CHCO_2^-$ with $^+NH_3$ on the CH

18.9 Complete the following table. (The K_a shown is for the conjugate acid.)

Amine	K_a	pK_a	K_b	pK_b
$C_4H_9NH_2$	?	10.77	5.89×10^{-4}	?
$C_5H_{11}NH_2$	2.34×10^{-11}	?	?	?
$(CH_3CH_2)_2NH$	?	?	3.09×10^{-4}	?
N H (piperidine ring)	?	?	?	2.88

18.10 Which of the following compounds would be the most basic? Explain.

(a) $C_6H_5NH_2$

(b) $CH_3CH_2NH_2$

(c) $C_6H_5NHCCH_3$ (with C=O)

(d) N H (ring)

18.11 In the preceding problem, which compound would be the least basic? Explain.

18.12 Which member of each of the following pairs of ions is the more acidic? Explain.

(a) $CH_3C(=O)\overset{+}{N}H_3$ or $CH_3CH_2\overset{+}{N}H_3$

(b) $C_6H_5\overset{+}{N}H_3$ (benzene ring) or $C_6H_{11}\overset{+}{N}H_3$ (cyclohexane ring)

(c) piperidinium ion (ring $\overset{+}{N}H_2$) or $(CH_3CH_2)_2\overset{+}{N}H_2$

(d) $O_2N-C_6H_4-\overset{+}{N}H_3$ or $Cl-C_6H_4-\overset{+}{N}H_3$

18.13 Predict the major organic products:

(a) $C_6H_5CH_2CH_2N(CH_3)_2 \;+\; OH^- \xrightarrow{\text{heat}}$

(b) $(CH_3)_3CCH_2\overset{+}{N}(CH_3)_2CH_2CH_3 \;+\; OH^- \xrightarrow{\text{heat}}$

(c) N-ethyl-N-methylpiperidinium ion (ring $\overset{+}{N}$ bearing H_3C and CH_2CH_3) $+\; OH^- \xrightarrow{\text{heat}}$

18.14 Complete the following equations:

(a) $C_6H_5NH_2 \;+\; NaNO_2 \;+\; HCl \xrightarrow{0^\circ}$

(b) pyridine (ring N) $+\; HCl \longrightarrow$

(c) N-acetylpiperidine ($NCCH_3$ with C=O) $\xrightarrow[\text{(2) } H_2O,\ H^+]{\text{(1) } LiAlH_4}$

(d) cyclohexene $\xrightarrow[\text{(3) } NH_3;\ \text{(4) } Br_2,\ NaOH]{\text{(1) } H_2CrO_4;\ \text{(2) } SOCl_2}$

(e) $(CH_3CH_2)_2NH$ + HNO_2 $\longrightarrow$

(f) phthalimide (NH) $\xrightarrow[\text{(2) } BrCH_2CO_2C_2H_5]{\text{(1) } KOH}$

(g) $CH_3CH_2NH_2$ + C_6H_5CHO $\xrightarrow{H^+}$

(h) $CH_3CH_2NH_2$ + $CH_3\overset{O}{\overset{\|}{C}}O\overset{O}{\overset{\|}{C}}CH_3$ $\longrightarrow$

19

Polycyclic and Heterocyclic Aromatic Compounds

Some Important Features

The polycyclic aromatic compounds, such as naphthalene and anthracene, are not symmetrical as is benzene. Consequently, some carbon-carbon bonds of the polycyclic aromatic compounds have more double-bond character than others.

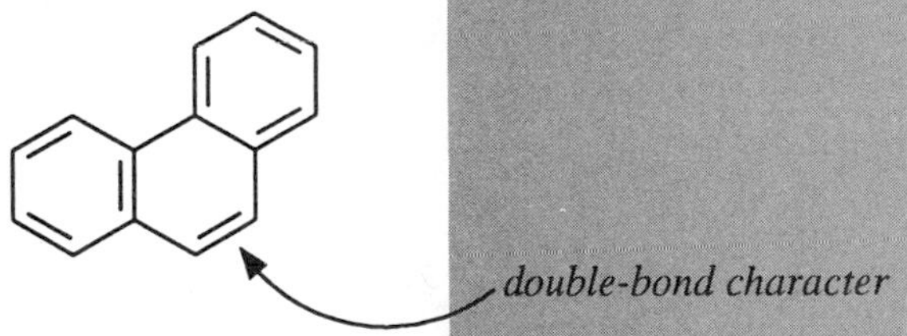

double-bond character

The aromatic polycyclic compounds are more reactive toward electrophiles, oxidizing agents, and reducing agents than is benzene because the intermediates still contain one or more rings with aromatic character.

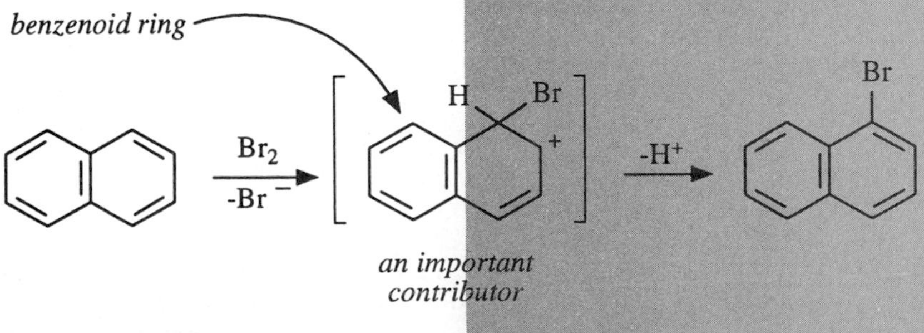

benzenoid ring

Br_2, $-Br^-$; $-H^+$

an important contributor

Sulfonation of naphthalene, like the sulfonation of benzene, is reversible. Thus, the sulfonation of naphthalene can lead to substitution at the 1- or 2-position, depending on the reaction conditions (see Section 19.5A).

The position of the second substitution on naphthalene is determined, in part, by the first substituent.

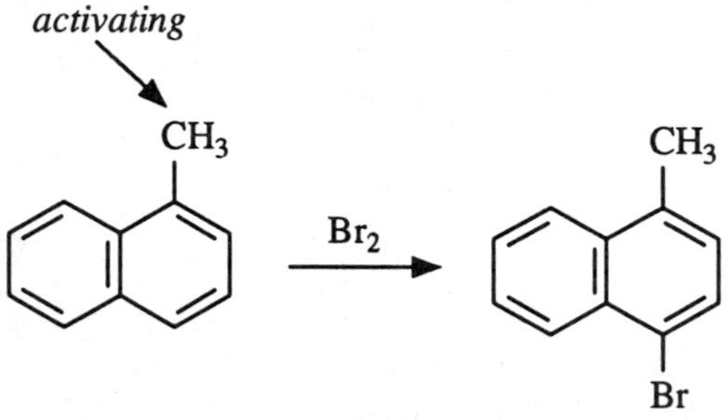

activating

CH_3 $\xrightarrow{Br_2}$ CH_3 Br

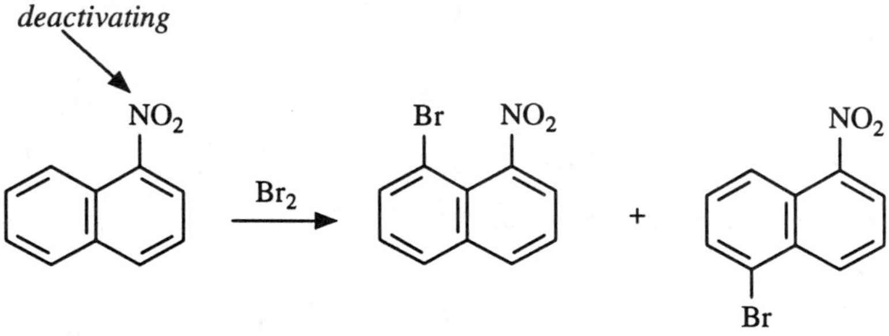

deactivating

NO_2 $\xrightarrow{Br_2}$ Br NO_2 + NO_2 Br

Although pyridine is a weaker base than a tertiary amine (because the N is sp^2 hybridized), pyridine still reacts with acids or with alkyl halides to yield salts.

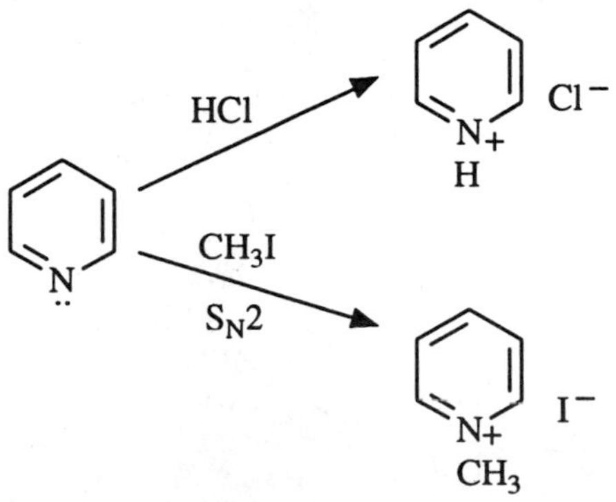

HCl N+ H Cl^-

N CH_3I S_N2 N+ CH_3 I^-

Compared to benzene, pyridine is deactivated to electrophilic substitution and activated to nucleophilic substitution because the nitrogen withdraws electron density from the rest of the ring.

δ+ N —E⁺, more difficult→ (E) N   N —Nu:⁻, easier→ N Nu

Pyrrole is much less basic than pyridine because it has no unshared valence electrons; the "extra" pair of electrons on the nitrogen are both contributed toward the aromatic pi cloud.

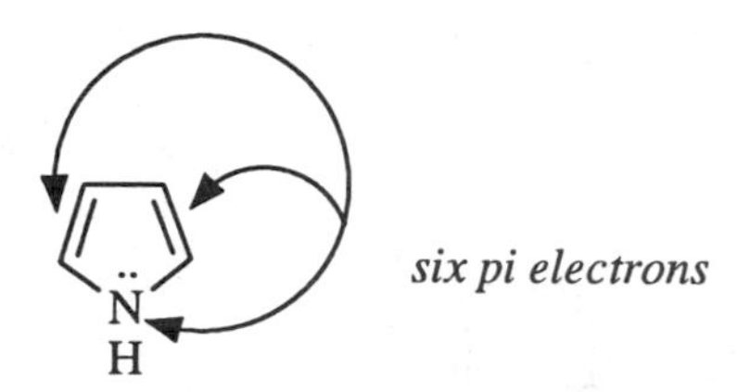

Because the nitrogen is electron-deficient, the rest of the pyrrole ring is electron-rich and undergoes electrophilic substitution more easily than benzene.

δ - N H —E⁺, easy→ N H E

Other important topics in this chapter are quinoline and isoquinoline (Section 19.8).

Reminders

Review Chapter 11 if aromaticity and aromatic substitution reactions are not clear to you. Review Chapter 18 if the structural features controlling the relative basicities of amines are not clear.

Electrophilic substitution occurs under *acidic* conditions (or neutral conditions for activated rings). Keep in mind the relative reactivities of aromatic compounds toward electrophilic substitution, such as nitration or bromination.

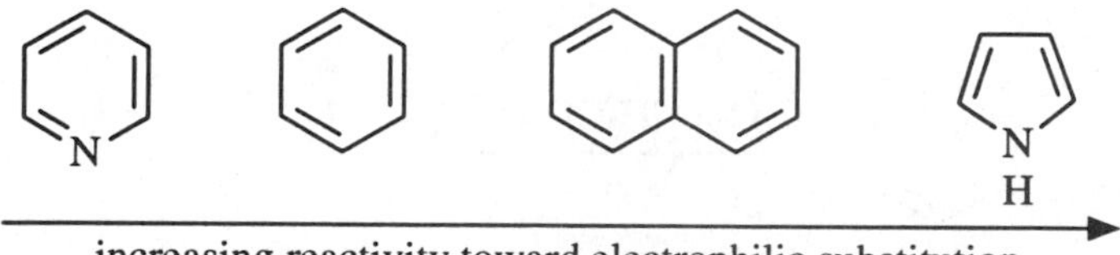

increasing reactivity toward electrophilic substitution

Nucleophilic substitution occurs only with a strong base under any circumstances.

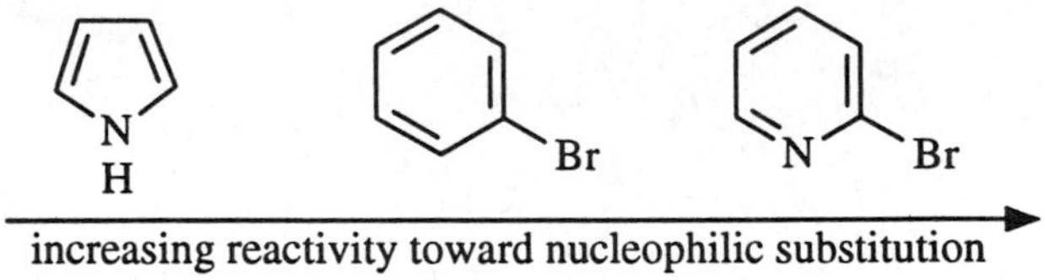

N H   Br   N Br

increasing reactivity toward nucleophilic substitution

Additional Drill Problems

19.1 What is the total number of pi electrons in each of the following polynuclear hydrocarbons?

(a)

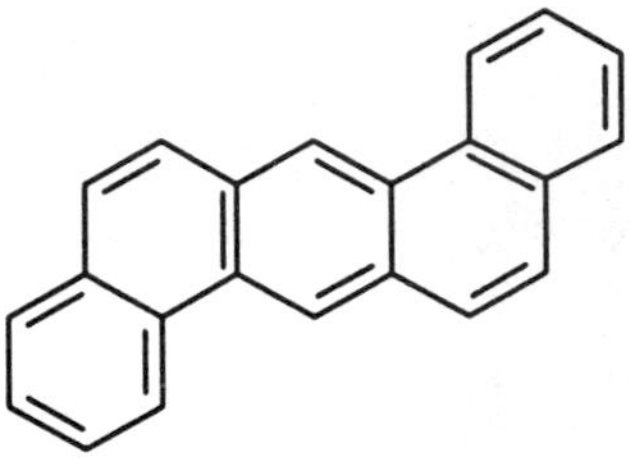

(b)

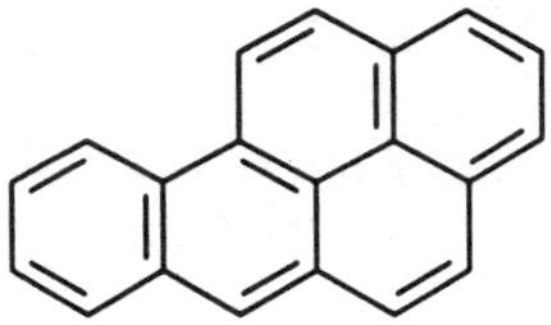

19.2 Complete the following equations showing only the principal organic products:

(a) 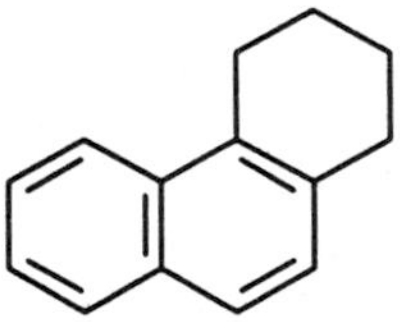 + HNO_3 $\xrightarrow{H_2SO_4}$

(b) 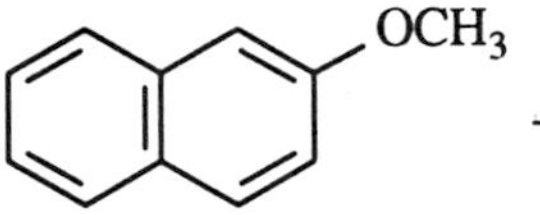 OCH_3 + Br_2 $\xrightarrow{FeBr_3}$

(c) 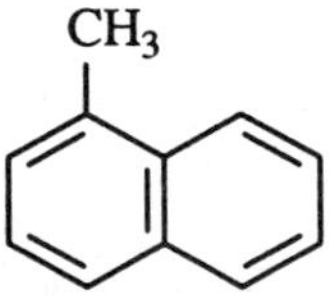 CH_3 $\xrightarrow[\text{heat}]{CrO_3,\ CH_3CO_2H}$

(d) 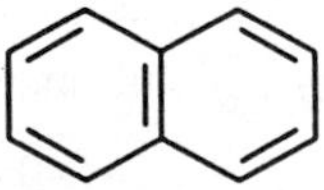 $\xrightarrow[\text{heat}]{CrO_3,\ H_2SO_4}$

(e) 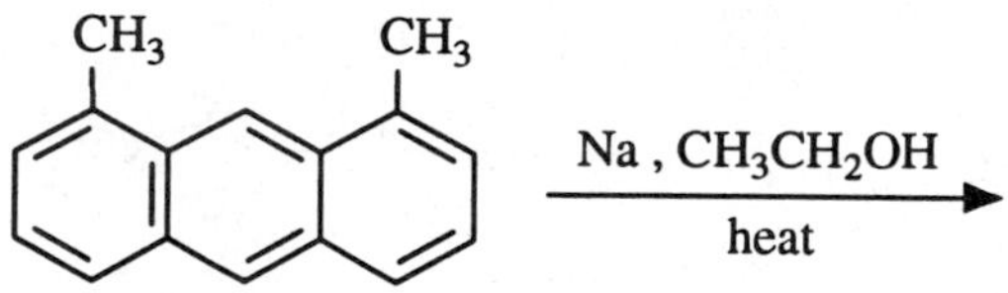 [1,8-dimethylanthracene] $\xrightarrow[\text{heat}]{Na,\ CH_3CH_2OH}$

(f) 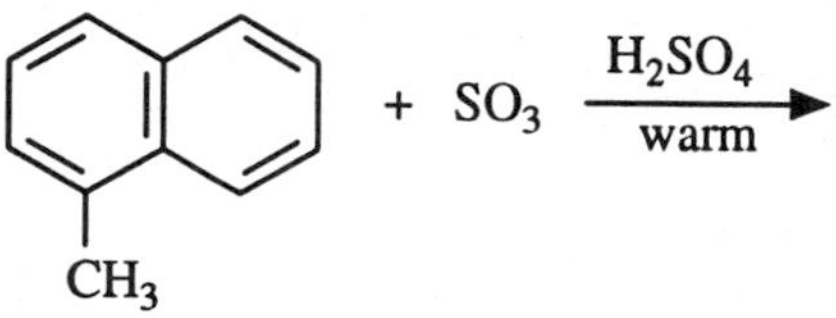 [1-methylnaphthalene] + SO_3 $\xrightarrow[\text{warm}]{H_2SO_4}$

19.3 Starting with a polynuclear hydrocarbon, write flow equations showing how you could prepare each of the following compounds:

(a) 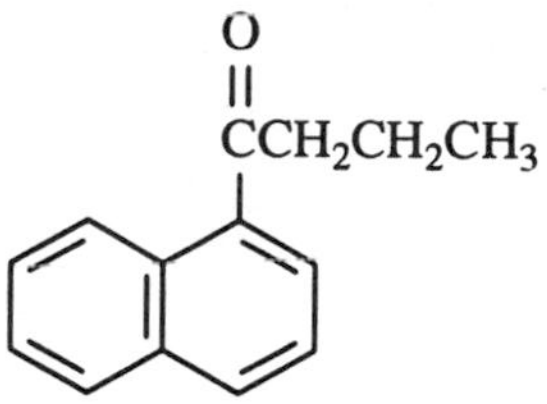 [1-naphthyl with $CCH_2CH_2CH_3$ (C=O)]

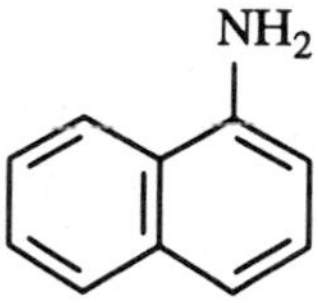

(b) [1-naphthyl with NH_2]

(c) [1-naphthyl with $C(C_6H_5)_2$ bearing OH]

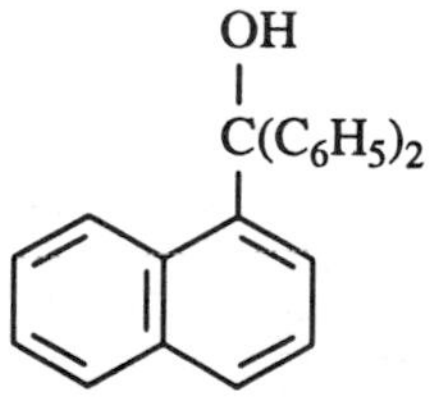

(d) 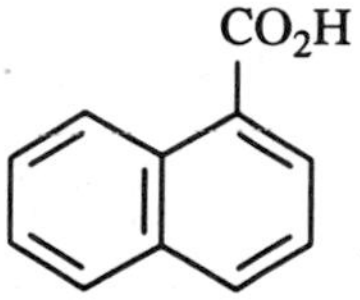 [1-naphthyl with CO_2H]

19.4 Draw formulas for the following names.

(a) 3-bromoquinoline

(b) 3-aminopyridine

(c) 4-hydroxypyridine

(d) 2-nitrofuran

19.5 Write names for the following compounds:

(a) 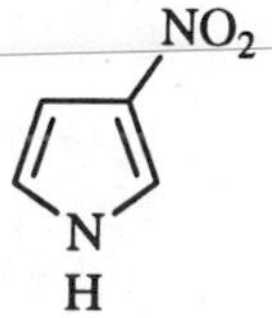 [pyrrole (N–H) with NO_2]

(b) 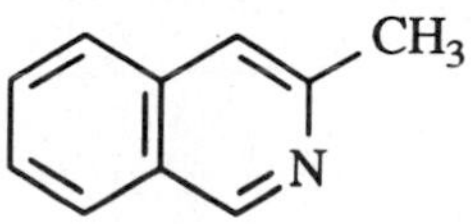 [isoquinoline with CH_3]

(c) 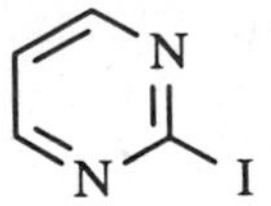 (d) 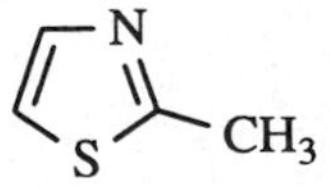 CH_3

N, N, I, S

(e) 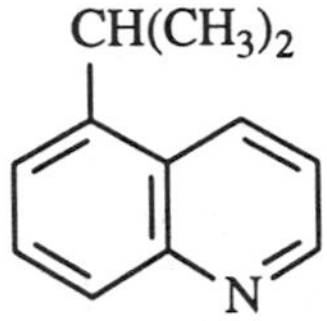 $CH(CH_3)_2$ N

(f) 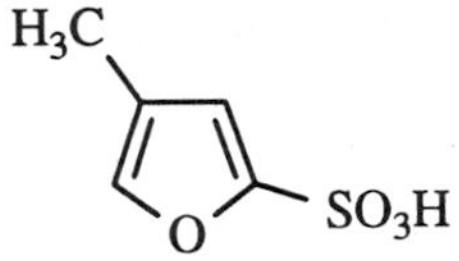 H_3C O SO_3H

19.6 Draw the five resonance structures of pyrrole. Which resonance structure would be the major contributor?

19.7 Which of the following compounds would you expect to undergo a coupling reaction with benzenediazonium chloride?

(a) 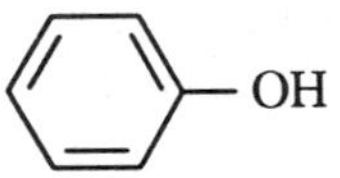 OH

(b) 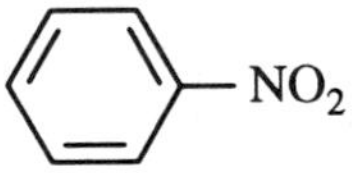 NO_2

(c)  N H

(d) 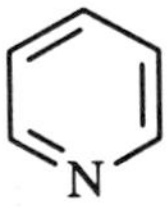 N

Natural Products: Studies in Organic Synthesis

Some Important Features

The types of natural products, naturally occurring organic compounds, discussed here are terpenes, pheromones, and alkaloids.

A terpene is composed of head-to-tail isoprene units.

$CH_3CO_2CH_2CH_2CH(CH_3)_2$

A pheromone is a communication compound.

N, N, CH_3

An alkaloid contains one or more basic nitrogen atoms.

A number of synthetic techniques are introduced or reiterated in this chapter. The abscisic acid synthesis shows the use of a ketal blocking group, standard permanganate oxidation reactions, nucleophilic addition to a carbonyl group by an acetylide ion, a reduction, and a hydrolysis reaction.

The pheromone discussion introduces asymmetric induction (the introduction of chiral centers) by the use of chiral catalysts, solvents, or reagents. It is not easy to predict the stereochemistry of these reactions without knowledge of the reactions of other related compounds.

The alkaloid synthesis shown in Figure 20.3 of the text shows how classical organic reactions and resolution can sometimes be used to synthesize a chiral natural product.

Additional Drill Problems

20.1 Which of the following compounds belongs to the class of terpenes or terpenoids?

(a)

(b) $CH_3C(CH_3)=CHCH_2C(CH_3)=CHCH_2OH$

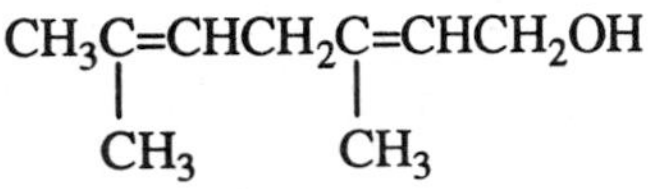

(c) CH_3 OH CH_3 CH_3 CH_3

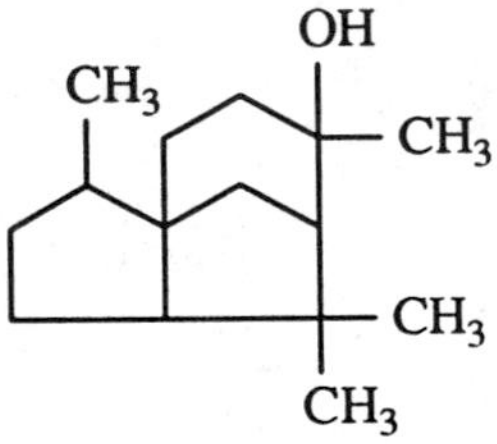

(d) CH_3 CH_3 O CH_3

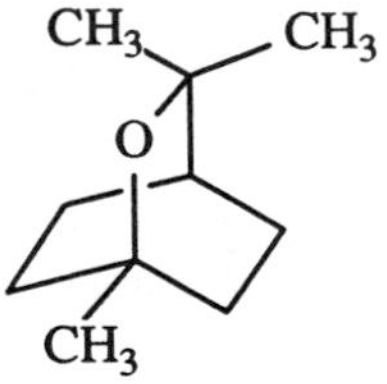

20.2 Circle the isoprene units in the following monoterpenes.

(a) $CH_3-C(CH_3)=CH-CH_2CH_2-C(=CH_2)-CH=CH_2$

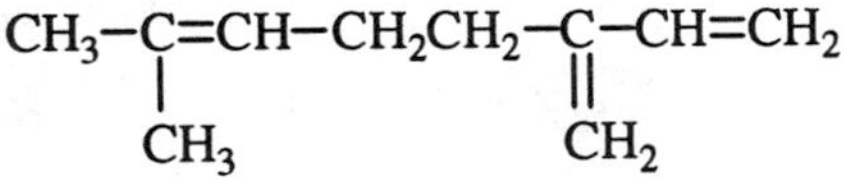

(b) $CH(CH_3)_2$ CH_3 HOC(=O) CH_3

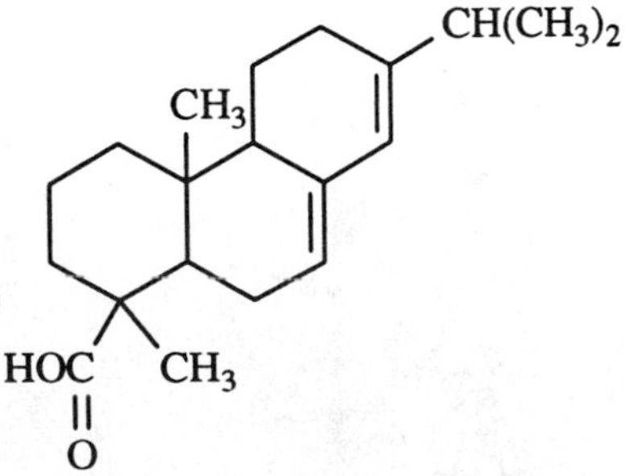

20.3 Identify each of the following as a monoterpene, sesquiterpene, or higher terpene.

(a)

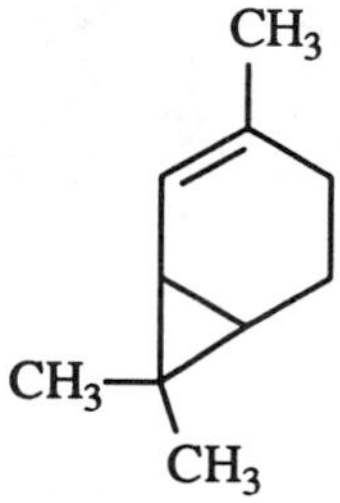

CH_3
CH_3
CH_3

(b)

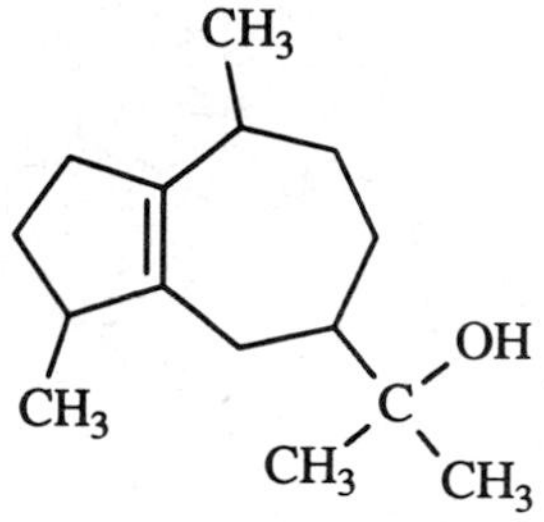

CH_3
CH_3
OH
C
CH_3 CH_3

(c)

CH_3
$CH_2{=}C$
CH_3 CH_3 CH_3
CH_3
HO
CH_3 CH_3

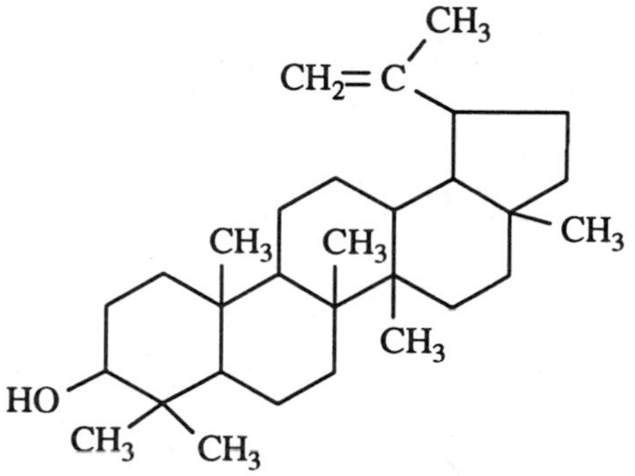

(d)

O
CH_3 O
O
CH_3
CH_3

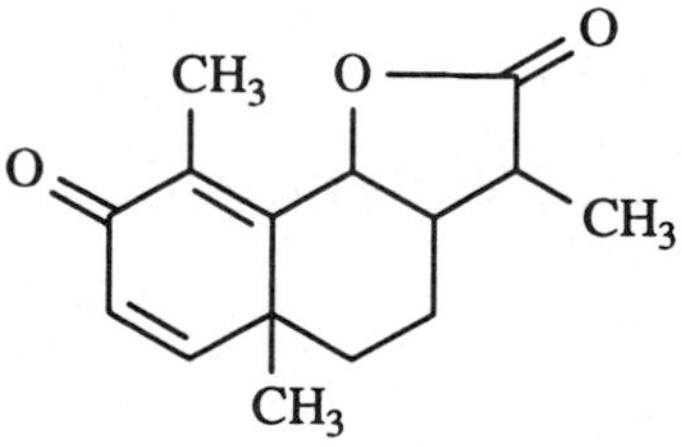

20.4 Which of the following natural products are alkaloids? terpenes?

(a)

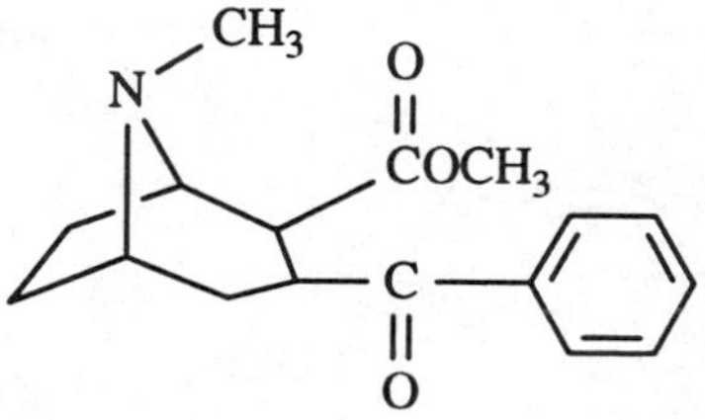

CH_3
N
O
‖
$COCH_3$
C
‖
O

(b)

CH_3
CH_3
CH_2

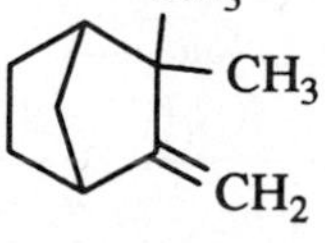

(c) 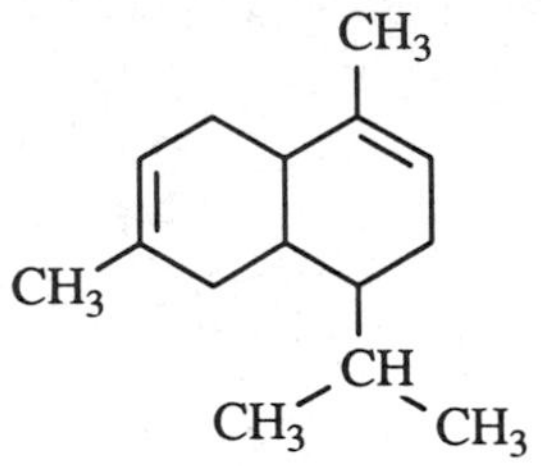

(d) 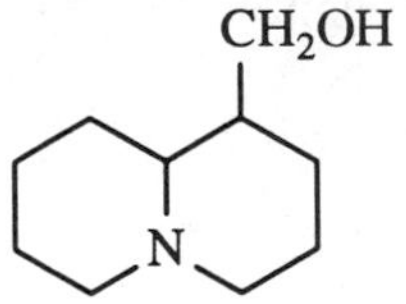

20.5 If a compound is isolated in 70% yield and has an enantiomeric excess of 90%, what is the percent of each enantiomer in the mixture?

20.6 What chemical reaction do all alkaloids have in common?

20.7 Write an equation showing the starting material and reagents to prepare the following compounds.

(a) an aldehyde ⟶ 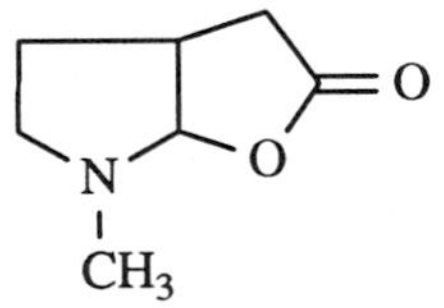

(b) an alkaloid ⟶ 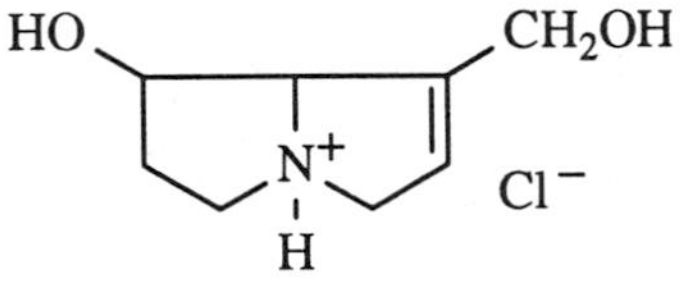

(c) an alcohol ⟶ 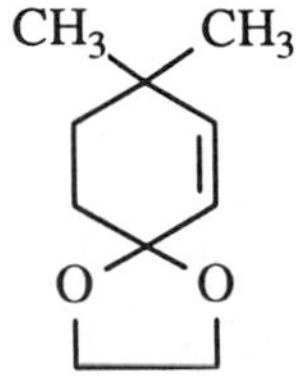

(d) a ketone ⟶ 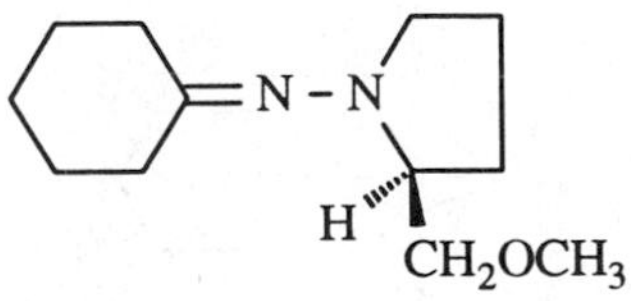

21

Pericyclic Reactions

Some Important Features

Pericyclic reactions are concerted reactions with cyclic transition states involving pi orbitals. The most important pericyclic reactions are *cycloaddition reactions, electrocyclic reactions,* and *sigmatropic rearrangements.*

Pericyclic reactions are often stereospecific. Photo-induction and thermal induction of these reactions often yield different products. The frontier orbital approach is one technique used to account for these observations and to predict the course of such reactions. In this technique, the phases of the *p*-orbital components of the HOMO and LUMO are considered. A pericyclic reaction does not proceed readily unless it is symmetry-allowed -- that is, the *p*-orbital components of the pertinent molecular orbitals must be of the same phase to undergo overlap and form a new bond.

Cycloaddition reactions are reactions in which the HOMO of one pi system overlaps with the LUMO of another pi system (in the same molecule or in different molecules) to form sigma bonds. The result is a cyclization. A [2 + 2] cycloaddition (two $\pi\ e^-$ + two $\pi\ e^-$) is photo-induced and results in a four-membered ring. A [4 + 2] cycloaddition is thermally induced and results in a six-membered ring.

R R R R R R R R R R R R $h\nu$ [2 + 2] +

O RC R heat [4 + 2] O CR R

Electrocyclic reactions are reversible cyclizations of conjugated polyenes. These reactions may be either thermally induced or photo-induced. In properly selected compounds, the stereochemistry of the product is determined by the method of induction used. To predict the products, we consider the HOMO of the polyene (not of the cyclic product) and determine whether conrotatory or disrotatory motion brings the in-phase *p*-orbital components together. Table 21.1 in the text summarizes the types of electrocyclic reactions so that you need not consider the molecular orbitals each time you are confronted with an electrocyclic problem.

Cyclization of a (4n + 2) polyene:

CH_3 heat (disrotatory) CH_3
CH_3 CH_3

CH_3 $h\nu$ (conrotatory) CH_3
CH_3 CH_3

Sigmatropic rearrangements are usually thermally induced and involve migration of groups from one end of a pi system to another portion of the same molecule. To predict whether a sigmatropic reaction is symmetry-allowed, we examine the HOMO of the hypothetical free radical that would be formed if the migrating group were homolytically cleaved. Common symmetry-allowed migrations are those that are [1,5]; [1,7]; and [3,3]. (These classifications are discussed in Section 21.4A.)

Reminders

Keep in mind that electrons are promoted to a higher energy level by light, but not by ordinary heating. When a compound is heated, the electrons of the molecule remain in the ground state.

Models may be helpful in determining the stereochemistry of a pericyclic reaction, especially a Diels-Alder reaction. (You may wish to review the discussion of the stereochemistry of Diels-Alder reactions in Section 16.3.)

Additional Drill Problems

21.1 Identify each of the following reactions as being (1) a cycloaddition; (2) an electrocyclic reaction; or (3) a sigmatropic rearrangement.

(a) [structure: bicyclic tosylate with TsO and H] ⇄ [structure: bicyclic tosylate with H and OTs]

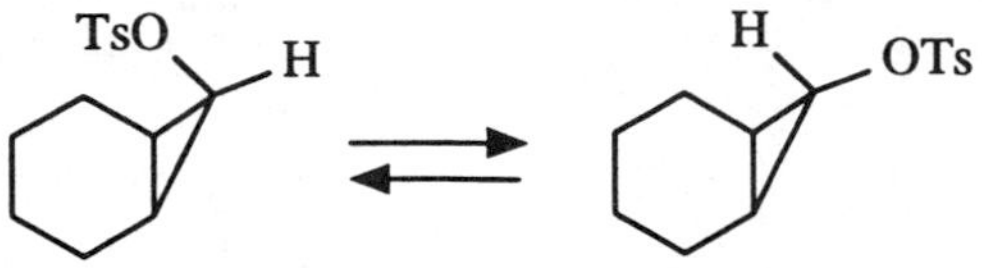

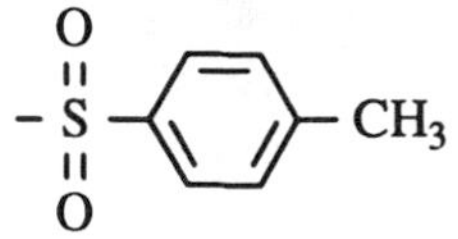

where Ts = $-SO_2-C_6H_4-CH_3$ (tosylate)

(b) [structure: cyclopropane with CH_3, H, H and vinyl substituent] $\xrightarrow{\Delta}$ [structure: cyclopentene with CH_3, H, H, CH_3]

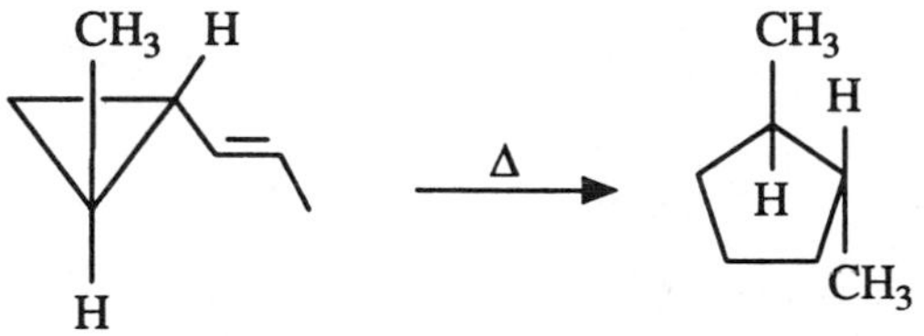

(c) $CH_3CH{=}CHCH_3$ + O_3 ⟶ [structure: ozonide ring with two CH–CH_3 groups and three O atoms]

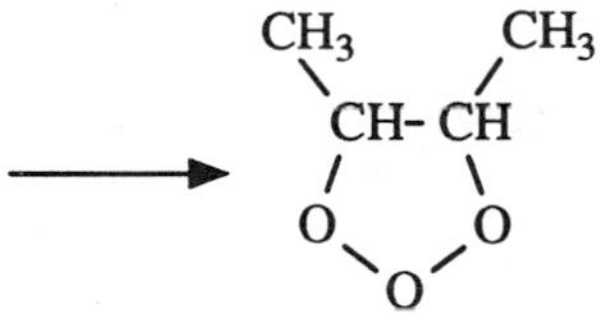

21.2 Draw orbital pictures that represent (1) the HOMO and (2) the LUMO for each of the following pi systems:

(a) $CH_3CH{=}CHCH_3$ (b) 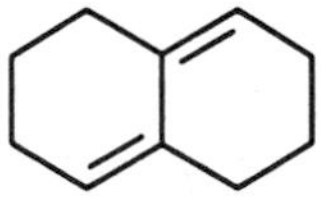 [structure: bicyclic diene] (c) 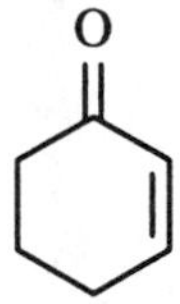 [structure: cyclohexenone]

21.3 Construct a diagram of the array of bonding and antibonding pi molecular orbitals for each of the following structures:

(a) *cis*-$CH_3CH{=}CH\bar{C}HCH_2CH_3$ (b) $CH_2{=}CHCH{=}O$ (c) [structure: cyclohexenyl cation, +]

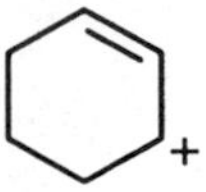

21.4 Suggest syntheses for the following compounds by cycloaddition reactions:

(a) 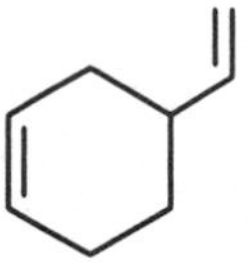 (b)

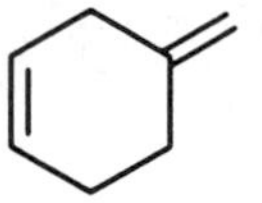

(c) 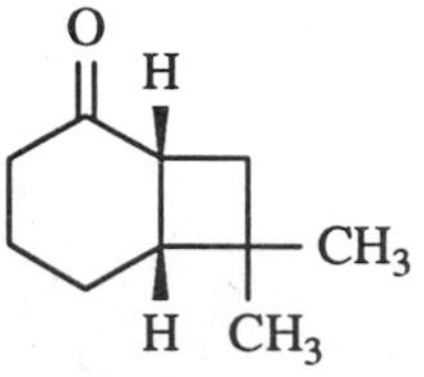 (d)

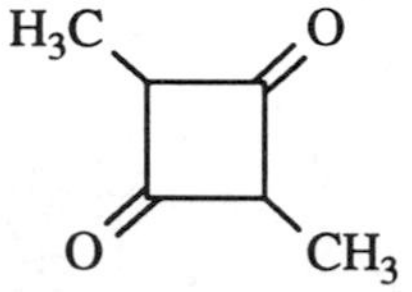

21.5 Predict the structures and stereochemistry of the major organic products:

(a) $CH_2{=}C(CH_3)CH{=}CH_2$ + $C_6H_5CH{=}CH_2$ $\xrightarrow{\text{heat}}$

(b) (*E*,*E*)-$CH_3OCH{=}CHCH{=}CHOCH_3$ + $CH_3O_2CC{\equiv}CCO_2CH_3$ $\xrightarrow{\text{heat}}$

(c) 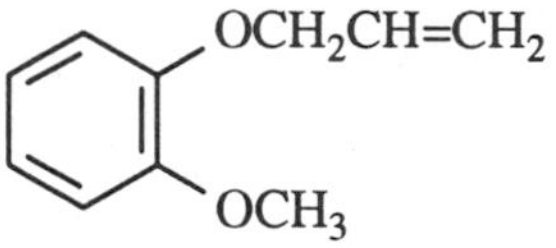 $\xrightarrow{\text{heat}}$

(d) *trans*-$C_6H_5CH{=}CHC_6H_5$ + *trans*-$CH_3CH{=}CHCO_2CH_3$ $\xrightarrow{h\nu}$

(e) 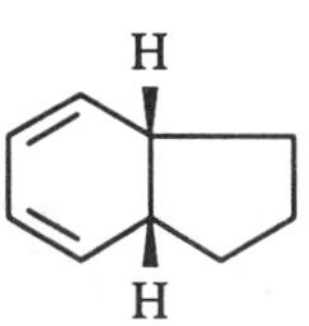 $\xrightarrow{h\nu}$

(f) [the product from (e)] $\xrightarrow{\text{heat}}$

21.6 Did conrotatory or disrotatory motion take place in each of the following electrocyclic reactions?

(a) [structure] $\xrightarrow[0°]{H_2SO_4}$ [structure]

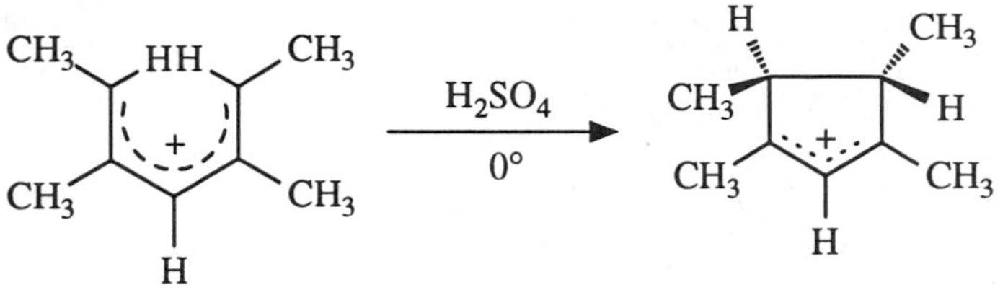

(b) [structure] $\xrightarrow[SO_2ClF,\ -100°]{SbF_5}$ [structure]

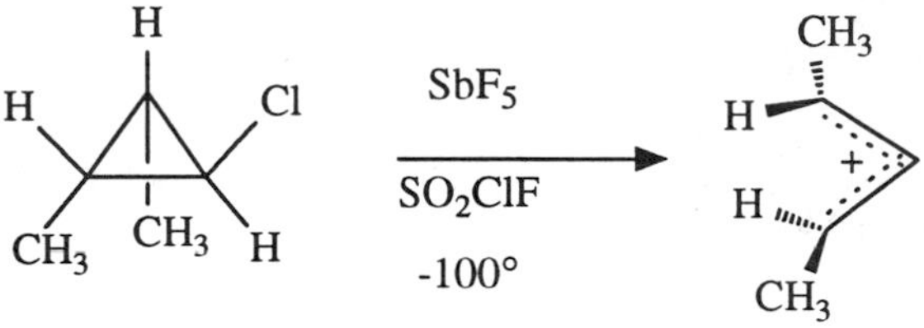

21.7 Suggest mechanisms for the following reactions:

(a) [structure with CO_2CH_3, CO_2CH_3] $\xrightarrow{heat}$ [structure with CH_3O_2C, CH_3O_2C]

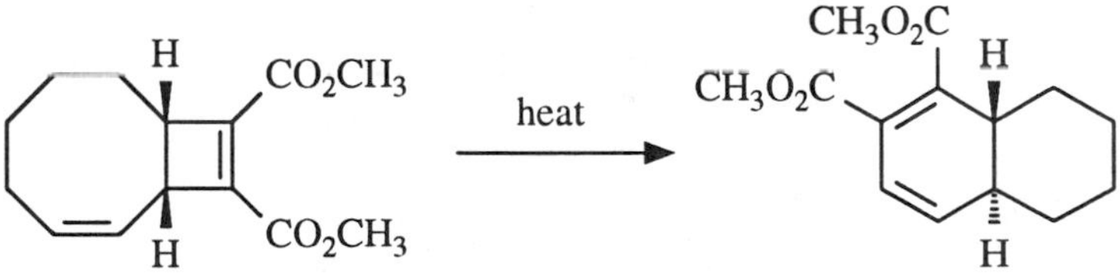

(b) 2,6-dimethylphenyl–$O^{14}CH_2CH{=}CH_2$ $\xrightarrow{heat}$ $CH_2{=}CH^{14}CH_2$–(3,5-dimethyl-4-hydroxyphenyl)

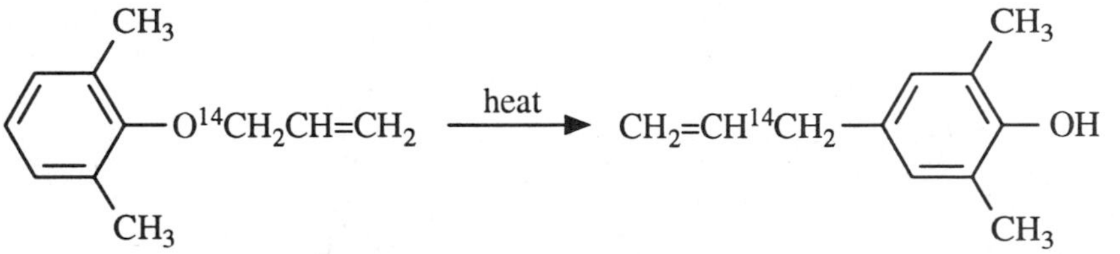

(c) $CH_3CH_2CH_2CO_2H$ + $CH_3\overset{O}{\overset{\|}{C}}C{\equiv}CCH_2CH_2CH_3$ $\xrightarrow{h\nu}$ $CH_3CH_2CH_2\overset{O}{\overset{\|}{C}}\underset{CH_3C=O}{CH}\overset{O}{\overset{\|}{C}}CH_2CH_2CH_3$

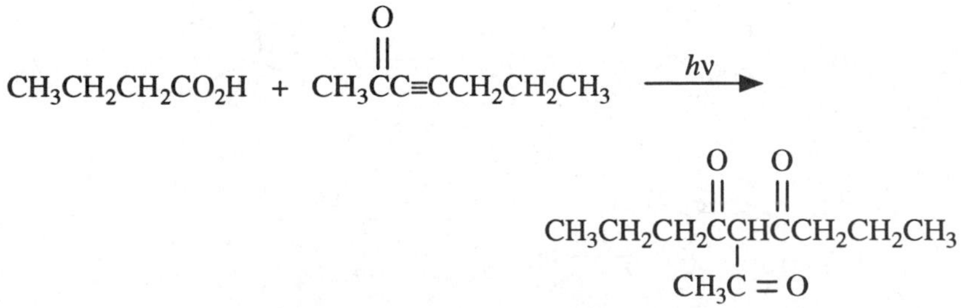

21.8 Complete the following equations, including stereochemistry where appropriate:

(a) 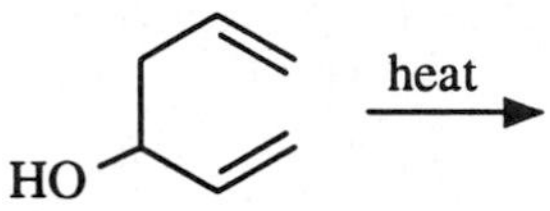 HO — heat →

(b) 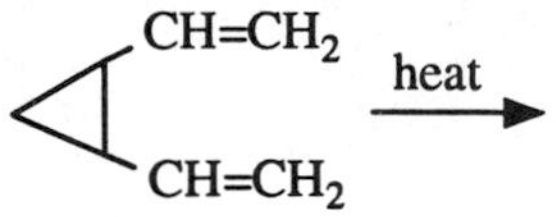 $CH{=}CH_2$, $CH{=}CH_2$ — heat →

(c) 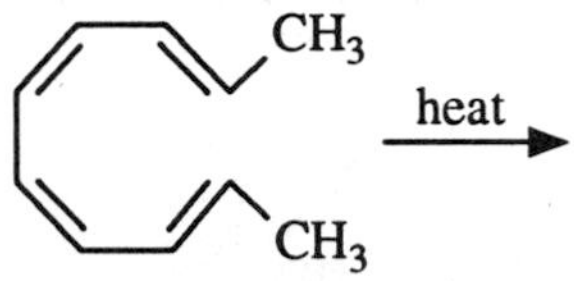 CH_3, CH_3 — heat →

21.9 The mechanism for the Fischer indole synthesis is given below. Identify the [3,3] sigmatropic rearrangement in this mechanism.

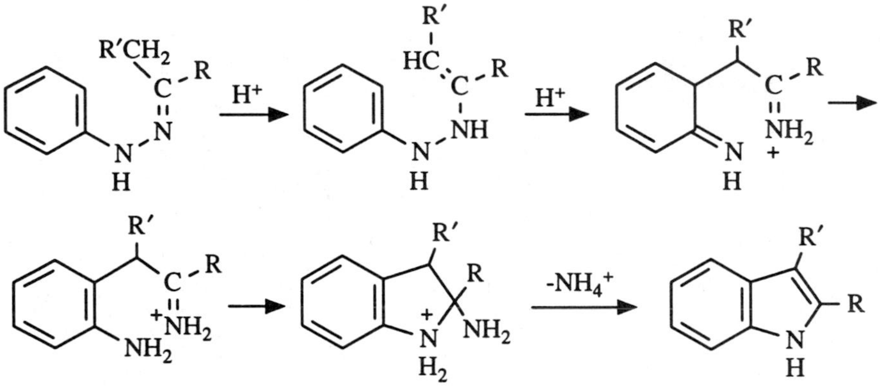

$R'CH_2$, R, N, N, H — H^+ → R′, HC, R, NH, N, H — H^+ → R′, R, NH_2, N +, H →

R′, R, $^+NH_2$, NH_2 → R′, R, NH_2, $\overset{+}{N}$ H_2 — $-NH_4^+$ → R′, R, N, H

21.10 Explain the following reactions:

(a) 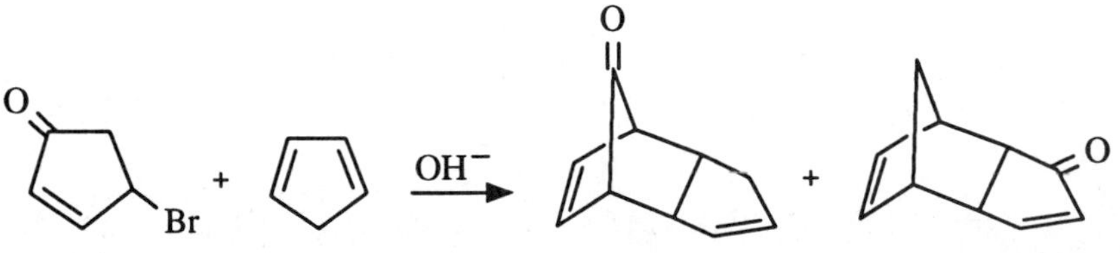 O, Br + — OH^- → O + O

(b) 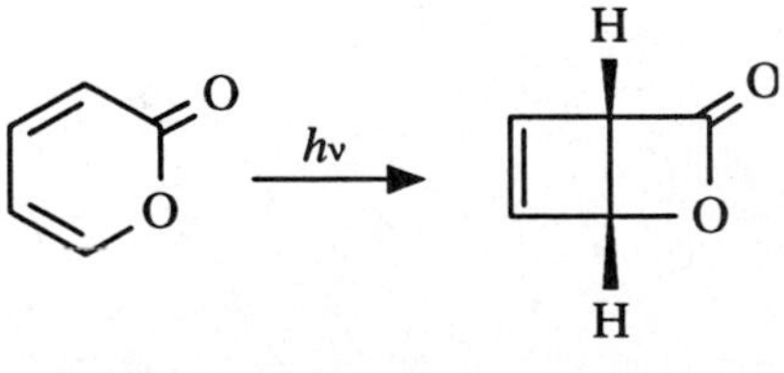 O, O — $h\nu$ → H, O, O, H

21.11 Claisen rearrangements are usually thought of as rearrangements of aryl allyl ethers to yield phenols. Yet, Claisen originally reported the rearrangement in 1912 with allyl vinyl ethers. Predict the product of the following Claisen rearrangement:

$$\begin{array}{c} OCH_2CH{=}CH_2 \\ | \\ CH_3C{=}CH_2 \end{array} \xrightarrow{\text{heat}}$$

Spectroscopy II: Ultraviolet Spectra, Color and Vision, Mass Spectra

Some Important Features

Ultraviolet and visible spectra arise from the promotion of pi electrons, especially pi electrons that are part of a conjugated system. In general, the larger the conjugated system, the longer is the wavelength of light absorbed. If the conjugated system is sufficiently long, the compound appears colored.

$$\mathrm{C{=}C} \qquad \mathrm{C{=}C{-}C{=}C} \qquad \mathrm{C{=}C{-}C{=}C{-}C{=}C} \longrightarrow$$

Absorbed light of increasing λ (decreasing energy)

The quantity of radiation absorbed (*the molar absorptivity* ϵ) is calculated from the formula $\epsilon = A/c\ l$ (Section 22.2). Transitions of the $\pi \rightarrow \pi^*$ type generally have a high value for ϵ compared with $n \rightarrow \sigma^*$ or $n \rightarrow \pi^*$ transitions.

Color usually arises from the absorption of specific wavelengths from the full visible range (white light). Reflection of the remaining unabsorbed wavelengths to the eye results in color vision.

Some colored organic compounds are discussed in Section 22.4B. Of particular interest are the indicators that change color depending on the pH of the solution. An acid-base reaction of an indicator molecule results in a change in the length of the conjugated system and a change in the wavelength of absorption.

Mass spectra arise from cleavage of molecules into ions and ion-radicals when the molecules are bombarded with high-energy electrons. Loss of a single electron gives rise to the *molecular ion*, often the farthest peak to the right in a mass spectrum.

$$CH_3\overset{\overset{\ddot{O}}{\|}}{C}CH_3 \xrightarrow{-e^-} \left[CH_3\overset{\overset{\dot{O}^{\bullet}}{\|}}{C}CH_3\right]^{\dot{+}}$$

molecular ion
m/e = 58

Fission of the molecular ion often occurs at a branch or α to an electronegative atom.

$$\left[R-\overset{\overset{CH_3}{|}}{CH}-R\right]^{\dot{+}} \qquad \left[R-\overset{\overset{O}{\|}}{C}-R\right]^{\dot{+}}$$

Small molecules, such as H_2O, can be lost:

$$[RCH_2CH_2OH]^{\dot{+}} \xrightarrow{-H_2O} [RCH-CH_2]^{\dot{+}}$$

An alkene molecule may be lost in a McLafferty rearrangement (Section 22.9D).

Reminders

In a problem concerning the color change of a compound with a change in pH: (1) look for the most acidic or basic group in the molecule; (2) write the acid-base reaction; (3) write resonance structures for reactant and product.

In mass spectral problems, *m/e* is the *mass of the ion or ion-radical* divided by the *ionic charge* (usually +1).

To determine fragmentation patterns, write the structure of the molecular ion. Mark all likely positions of cleavage, note any small molecules that could be lost, and then write formulas for the fragments, keeping track of electrons.

$$CH_3\overset{\overset{\ddot{O}}{\|}}{C}CH_3 \xrightarrow{-e^-} CH_3\overset{\overset{:\overset{+}{O}\bullet}{\|}}{C}-CH_3 \xrightarrow{-\dot{C}H_3} [CH_3\overset{+}{C}\equiv O: \longleftrightarrow CH_3\overset{+}{C}=\ddot{O}:]$$

usually written:

$$CH_3\overset{\overset{O}{\|}}{C}CH_3 \xrightarrow{-e^-} \left[CH_3\overset{\overset{O}{\|}}{C}CH_3\right]^{\dot{+}} \xrightarrow{-\dot{C}H_3} [CH_3\equiv O \longleftrightarrow CH_3C=O\,]^{+}$$

the odd electron was lost with $\bullet CH_3$

$$CH_3CH_2\ddot{O}H \xrightarrow{-e^-} [CH_2(H)-CH_2\dot{\ddot{O}}H^+] \xrightarrow{-H_2\ddot{O}:} \dot{C}H_2-\overset{+}{C}H_2$$

usually written:

$$CH_3CH_2OH \xrightarrow{-e^-} [CH_3CH_2OH]^{+\bullet} \xrightarrow{-H_2O} [CH_2-CH_2]^{+\bullet}$$

the odd electron was not lost with H_2O

Additional Drill Problems

22.1 Fill in the blanks:

(a) A photon of light with a shorter wavelength contains (more/less) energy than a photon of light with a longer wavelength.

(b) Electromagnetic radiation with a higher frequency contains (more/less) energy than radiation with a lower frequency.

22.2 What is the change in energy (in ergs) when light of the following wavelengths is absorbed by a compound?

(a) 230 nm (b) 6.2 μm

22.3 Given the ultraviolet spectrum of acetone in cyclohexane (Spectrum 22.1):

(a) what compound is absorbing the ultraviolet radiation?

(b) what is the λ_{max} of the compound?

(c) what is the molar absorptivity of the compound?

(d) what type of electronic transitions are taking place? (Identify them on the spectrum.)

Spectrum 22.1

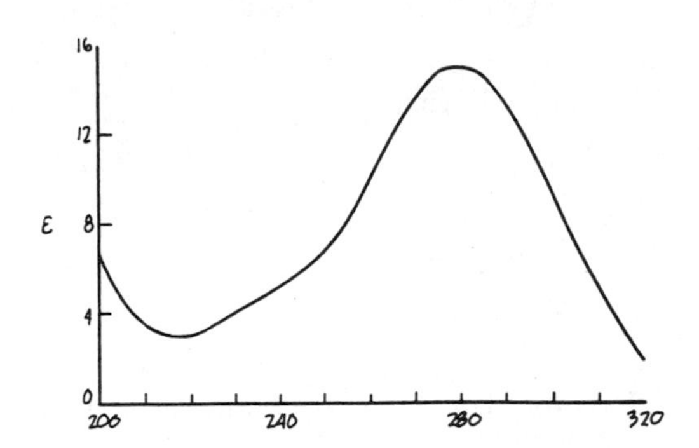

22.4 In Spectrum 22.2, identify (a) the base peak and (b) the molecular ion peak.

Spectrum 22.2

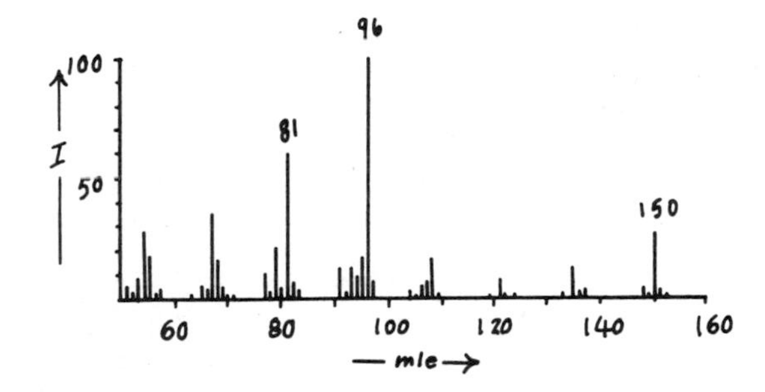

22.5 Write equations that show the molecular ion and the most likely mass spectrometry fragmentation products of the following compounds. Use curved arrows to show how these could be formed.

(a) $(CH_3CH_2CH_2)_2C{=}O$

(b) CH_3CH_2CHOH with CH_3 on the CH carbon ($CH_3CH_2CH(CH_3)OH$)

(c) $CH_3CH_2CHCH_2CH_2CH_3$ with CH_3 on the CH carbon ($CH_3CH_2CH(CH_3)CH_2CH_2CH_3$)

(d) $(CH_3)_2CHNHCH_2CH_2CH_3$

22.6 The mass spectrum of ethanolamine ($H_2NCH_2CH_2OH$) shows two peaks at m/e 30 and m/e 31. What are the structures of the ions giving rise to these peaks?

22.7 Predict the fragmentation pattern for each of the following:

(a) $CH_3CH_2\overset{O}{\overset{\|}{C}}OCH_3$

(b) 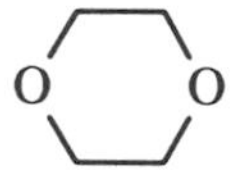 1,4-dioxane ring (six-membered ring with O at positions 1 and 4)

22.8 What type of spectroscopy (infrared, ^1H NMR, UV, or mass) would be best for distinguishing between the compounds in each of the following pairs? Explain your answers.

(a) $CH_2{=}CHCH{=}CH_2$ and $CH_2{=}CHCH_2CH{=}CH_2$

(b) 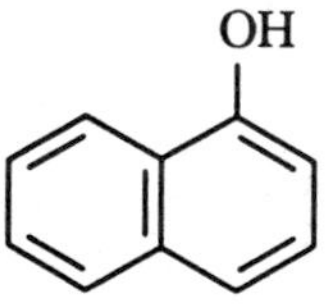 1-naphthol (OH on aromatic naphthalene ring) and the corresponding saturated decalol (OH on decalin ring)

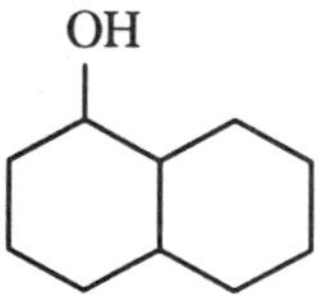

(c) $C_6H_5CH_3$ and $C_6H_5CH_2CH_3$

(d) $CH_3CH_2\overset{CH_3}{\overset{|}{C}}H(CH_2)_7CH_3$ and $CH_3\overset{CH_3}{\overset{|}{C}}H(CH_2)_8CH_3$

(e) $CH_3\overset{O}{\overset{\|}{C}}CH_3$ and $CH_3\overset{O}{\overset{\|}{C}}OCH_3$

22.9 Match the mass spectra in the figure labeled Spectra 22.3 with the following structures:

(a) CH_3CH_2OH (b) CH_2BrCH_2Cl (c) CH_3CH_2F

(d) C_4H_9Br (e) $CH_3CH_2CH_2I$ (f) CH_3CH_2Cl

(g) CH_3CH_2Br (h) CH_3CH_2I

Spectra 22.3

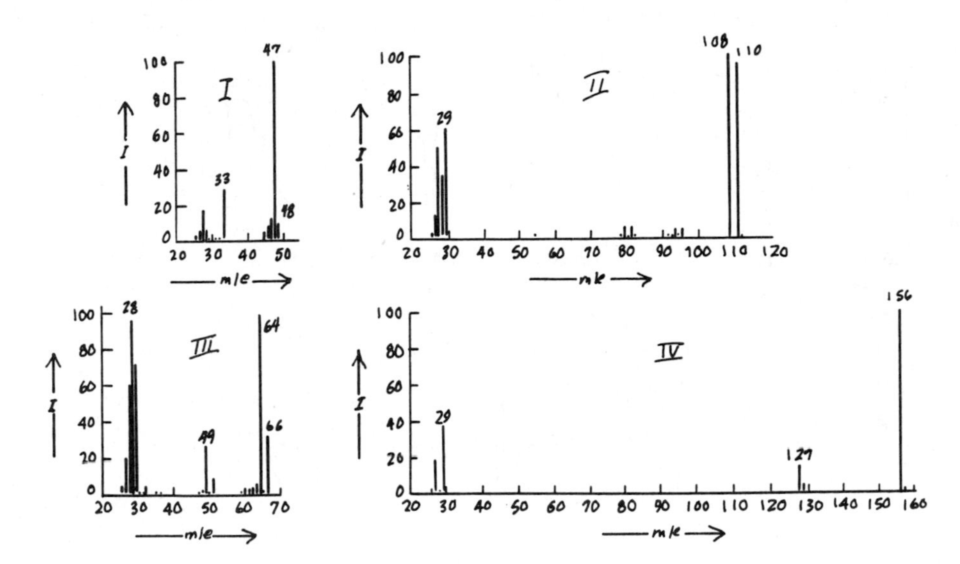

22.10 For each of the following compounds, the infrared, ^{1}H NMR, and mass spectra are shown. Determine the structures.

(a) $C_6H_{10}O$, Spectra 22.4

(b) $C_3H_6Cl_2$, Spectra 22.5

(c) $C_6H_{13}N$, Spectra 22.6

(d) $C_6H_{12}O$, Spectra 22.7

(e) $C_4H_4O_3$, Spectra 22.8

Spectra 22.4

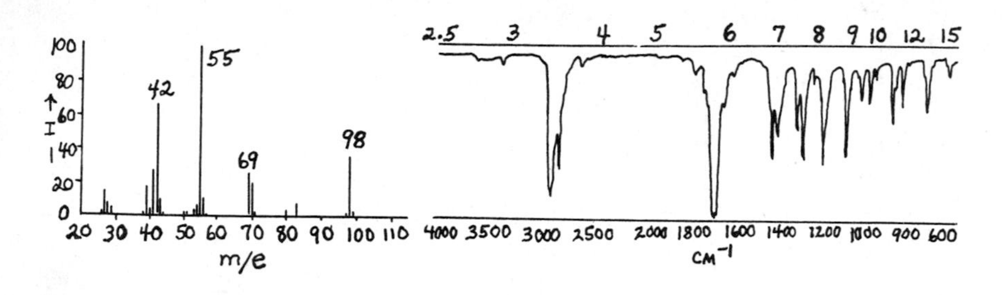

100
80
60
40
20
0
I
55
42
69
98
20 30 40 50 60 70 80 90 100 110
m/e
2.5 3 4 5 6 7 8 9 10 12 15
4000 3500 3000 2500 2000 1800 1600 1400 1200 1000 900 600
CM^{-1}

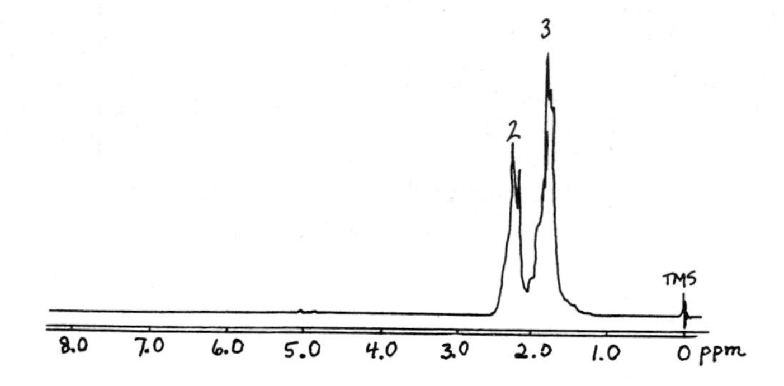

3
2
TMS
8.0 7.0 6.0 5.0 4.0 3.0 2.0 1.0 0 ppm

Spectra 22.5

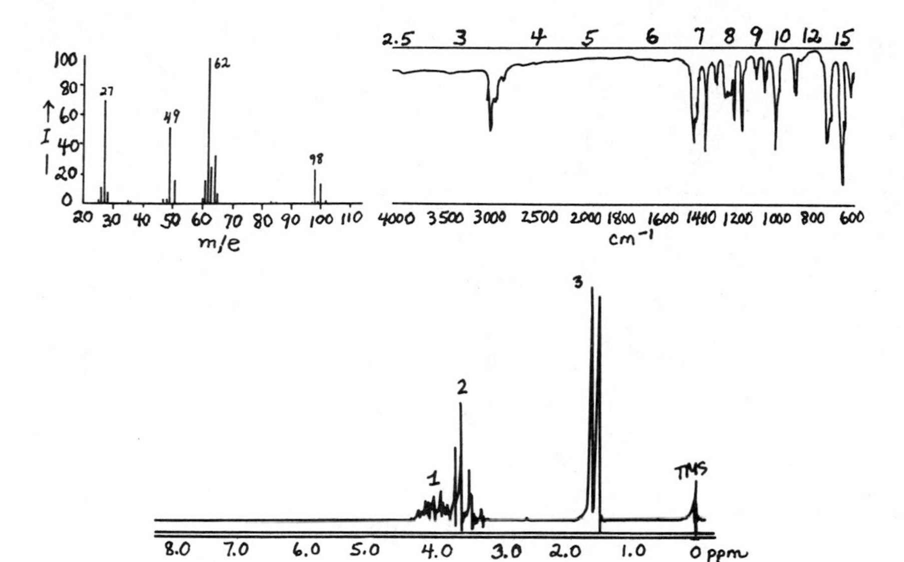
100
80
60
40
20
0
I
27
49
62
98
20 30 40 50 60 70 80 90 100 110
m/e
2.5 3 4 5 6 7 8 9 10 12 15
4000 3500 3000 2500 2000 1800 1600 1400 1200 1000 800 600
cm⁻¹
1
2
3
TMS
8.0 7.0 6.0 5.0 4.0 3.0 2.0 1.0 0 ppm

Spectra 22.6

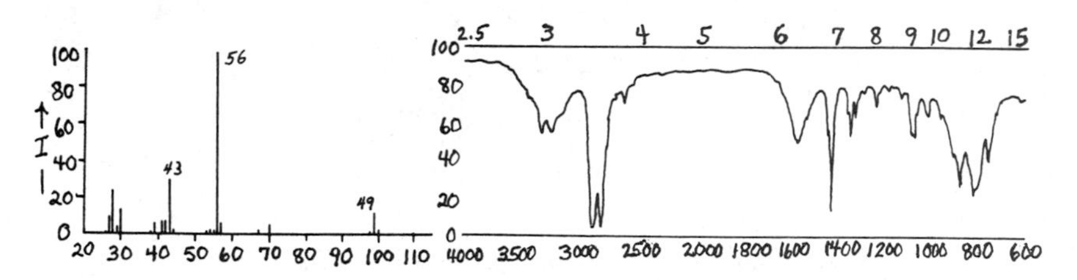
100
80
60
40
20
0
I
56
43
49
20 30 40 50 60 70 80 90 100 110
2.5 3 4 5 6 7 8 9 10 12 15
4000 3500 3000 2500 2000 1800 1600 1400 1200 1000 800 600

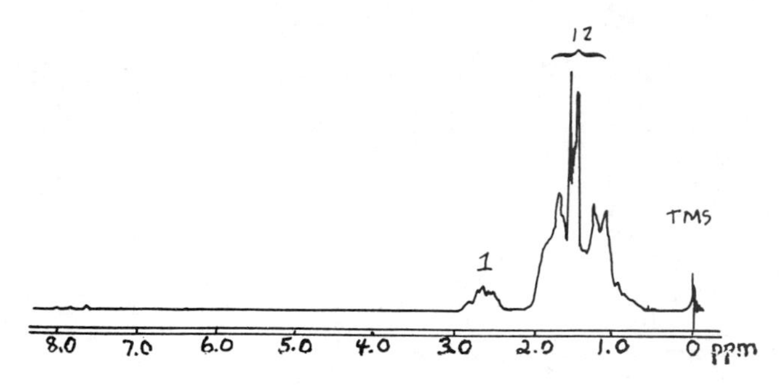
12
1
TMS
8.0 7.0 6.0 5.0 4.0 3.0 2.0 1.0 0 ppm

Spectra 22.7

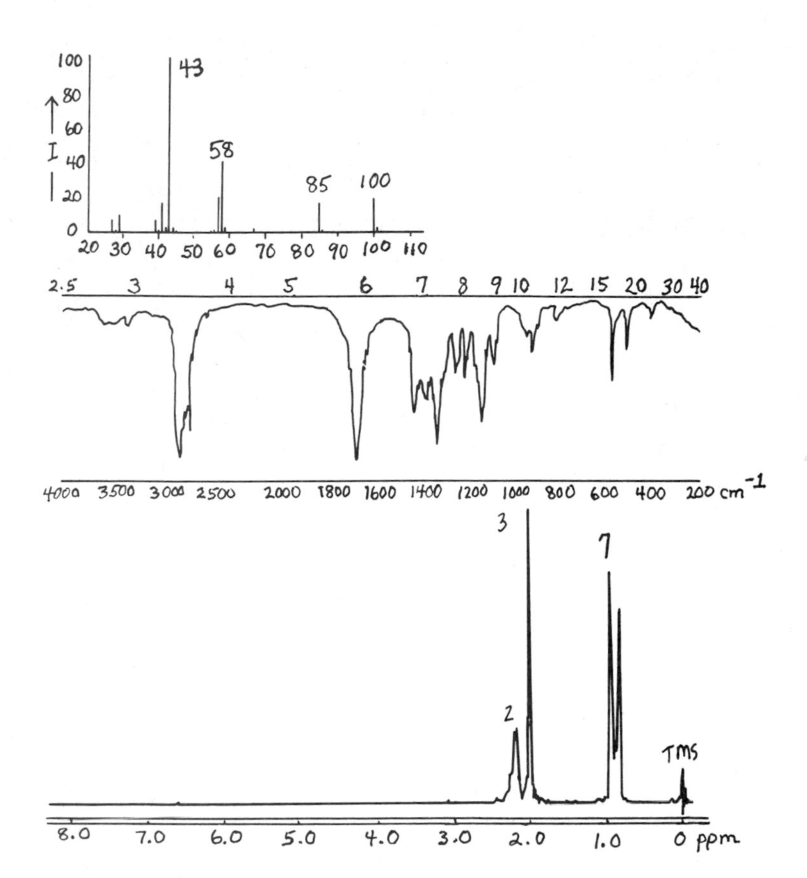
100
80
60
I
40
20
0
43
58
85
100
20 30 40 50 60 70 80 90 100 110
2.5 3 4 5 6 7 8 9 10 12 15 20 30 40
4000 3500 3000 2500 2000 1800 1600 1400 1200 1000 800 600 400 200 cm-1
3
7
2
TMS
8.0 7.0 6.0 5.0 4.0 3.0 2.0 1.0 0 ppm

Spectra 22.8

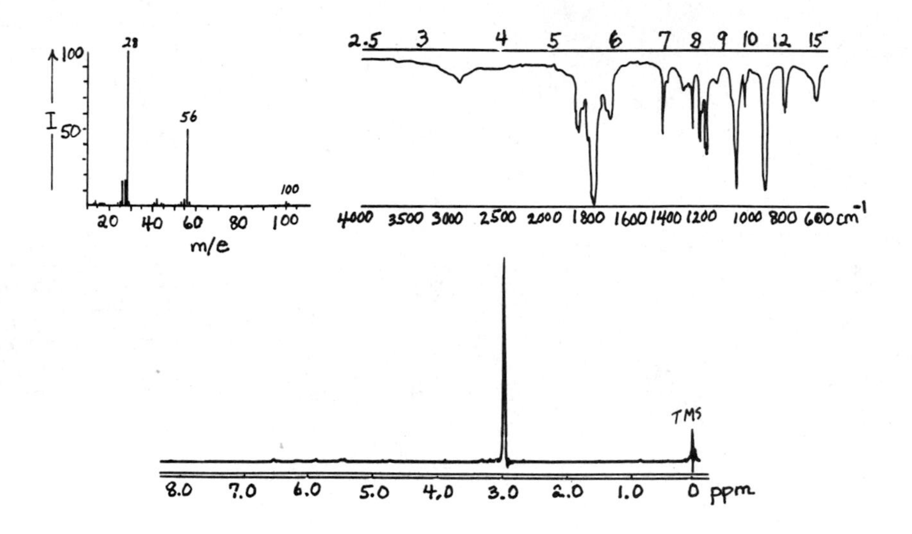
100
I
50
28
56
100
20
40
60
80
100
m/e
2.5 3 4 5 6 7 8 9 10 12 15
4000 3500 3000 2500 2000 1800 1600 1400 1200 1000 800 600 cm^{-1}
TMS
8.0 7.0 6.0 5.0 4.0 3.0 2.0 1.0 0 ppm

22.11 What are the structures of (a) the compound with the spectra labeled 22.9 and (b) the compound with the spectra labeled 22.10?

Spectra 22.9

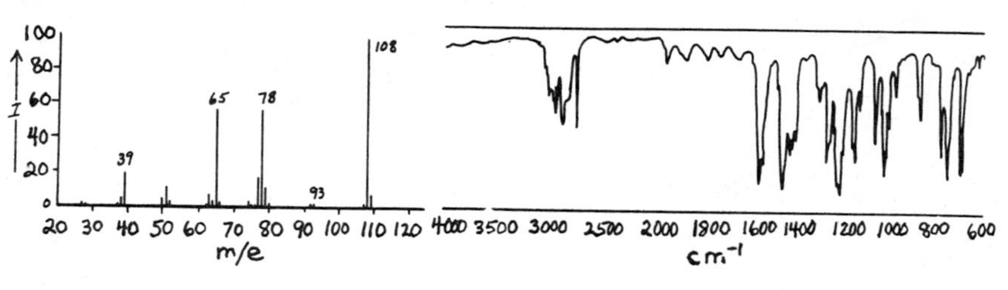

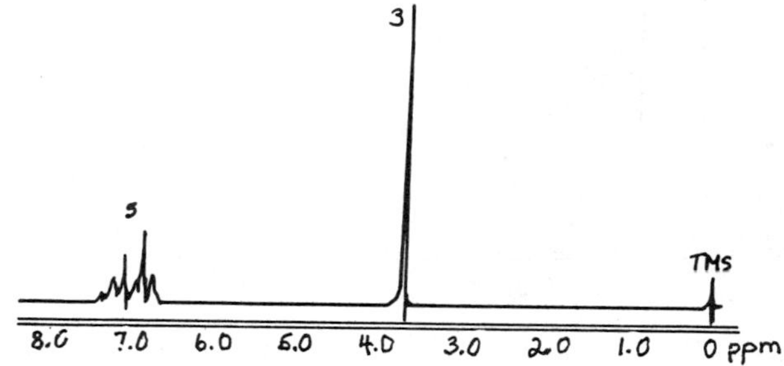

Spectra 22.10

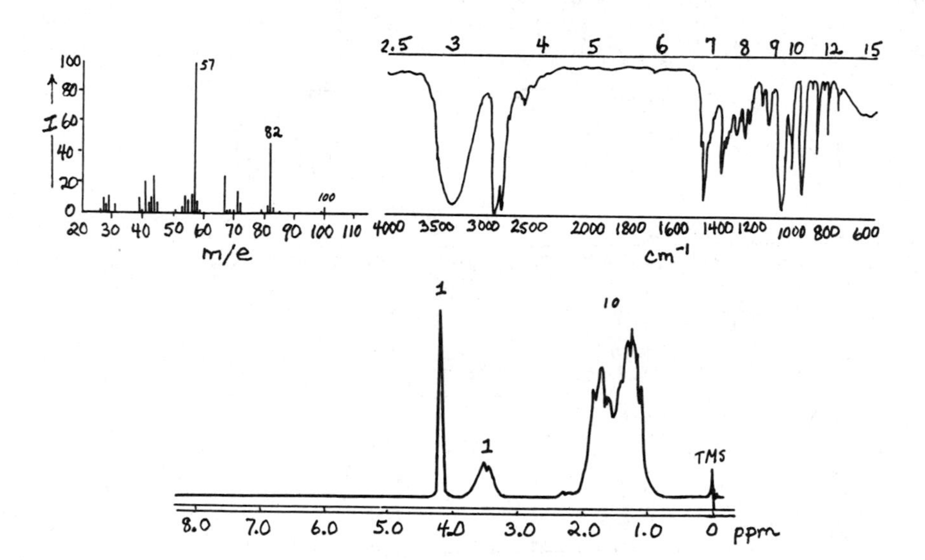

Carbohydrates

Some Important Features

Glucose and the other monosaccharides exist primarily as pairs of cyclic hemiacetals (anomers), which are in equilibrium with the open-chain carbonyl form in solution. Hemiacetals can undergo aldehyde reactions because of this equilibrium.

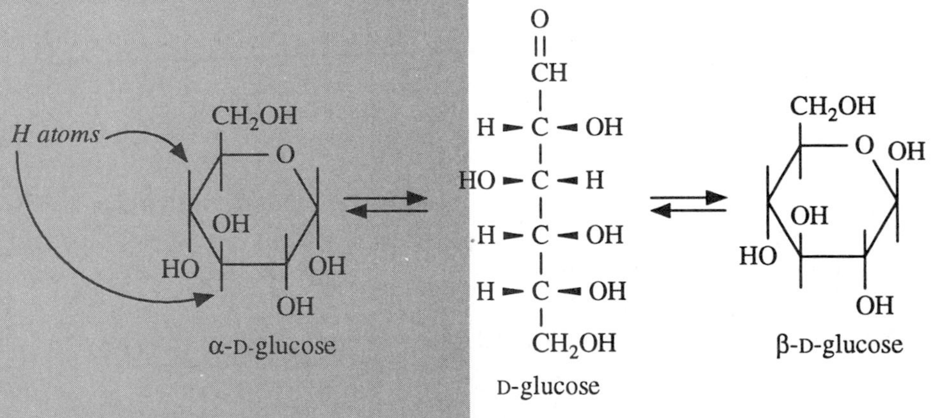

H atoms

α-D-glucose ⇌ D-glucose *open chain* ⇌ β-D-glucose

The monosaccharides form acetals (*glycosides*) when treated with an alcohol. A glycoside is not in equilibrium with the carbonyl form in neutral or alkaline solutions; therefore, glycosides do not undergo aldehyde reactions. Glycoside links can be hydrolyzed in acidic solution or with appropriate enzymes.

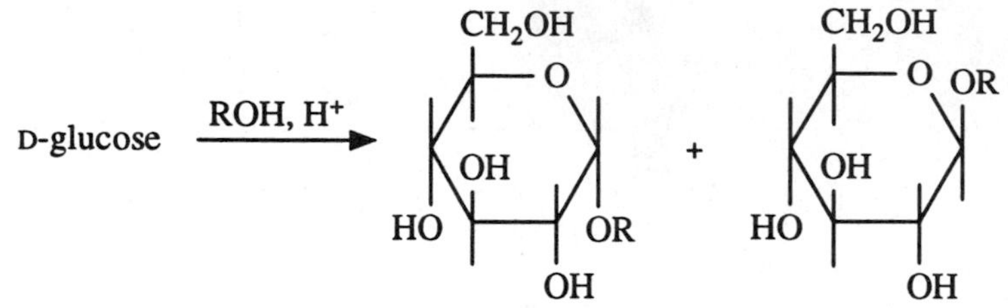

D-glucose $\xrightarrow{ROH,\ H^+}$ an α-D-glucoside + a β-D-glucoside

The relative configurations of sugars can be determined by synthesis.

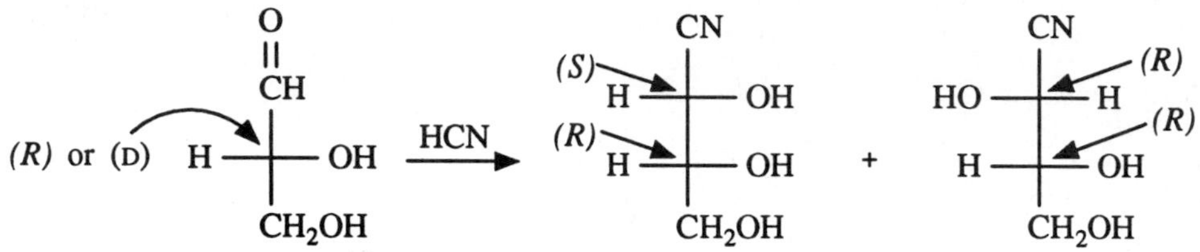

(R) or (D) D-glyceraldehyde $\xrightarrow{HCN}$ a pair of diastereomers

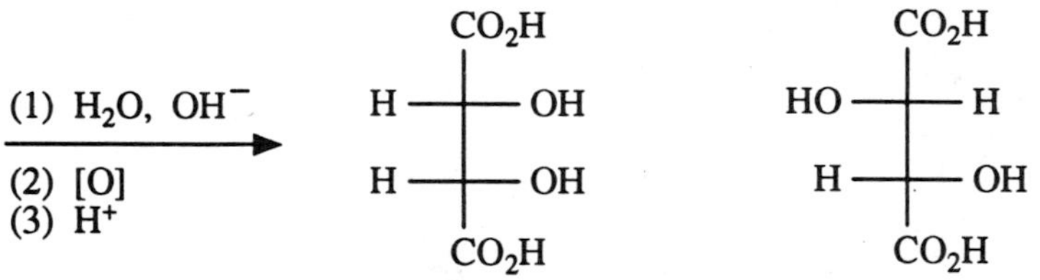

$\xrightarrow[(2)\ [O]\ (3)\ H^+]{(1)\ H_2O,\ OH^-}$ *meso*-tartaric acid + D-(-)-tartaric acid

Both have the same configuration as D-glyceraldehyde at carbon 3.

Some important reactions of monosaccharides are the oxidation to *aldonic acids* (carbon 1 oxidized), *aldaric acids* (both carbon 1 and the last carbon oxidized), and *uronic acids* (only the last carbon oxidized). Oxidative cleavage with HIO_4 is useful in structure determinations. These oxidations are discussed in Section 23.6. The monosaccharides also can be reduced to *alditols* (Section 23.7). The hydroxyl groups can be esterified or converted to alkoxyl groups (Section 23.8). They can also react with aldehydes or nonhindered ketones to form cyclic acetals or ketals (Section 23.8C).

Disaccharides are formed from two monosaccharide units joined by a glycoside link.

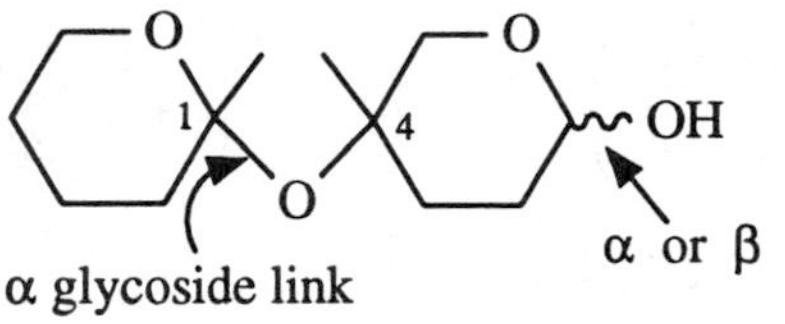

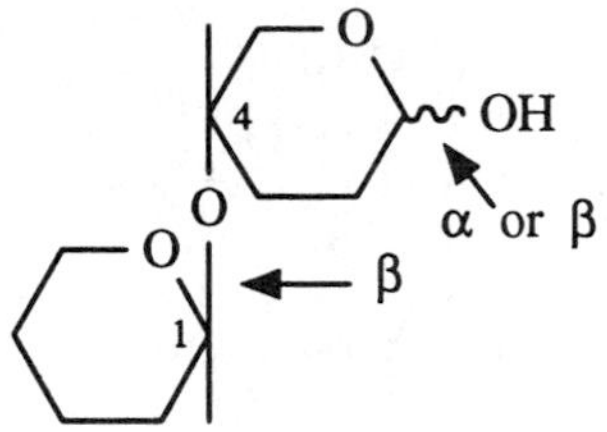

α glycoside link — α or β — α or β — β

Polysaccharides are composed of many monosaccharide units joined by glycoside links.

Reminders

Review Fischer projections in Section 4.6C. Memorize the Fischer projections and Haworth formulas for α- and β-D-glucose. Then, interconversions for other sugars are simplified.

If you forget the configurations of the carbon atoms of glucose, draw the favored chair form. β-D-Glucose has all substituents in equatorial positions.

Be sure you know what a *meso* compound is and how to assign *(R)* and *(S)* configurations. (See Chapter 4.)

Additional Drill Problems

23.1 Name the monosaccharides represented by the following Fischer projections:

(a)
CHO
H — OH
H — OH
H — OH
CH_2OH

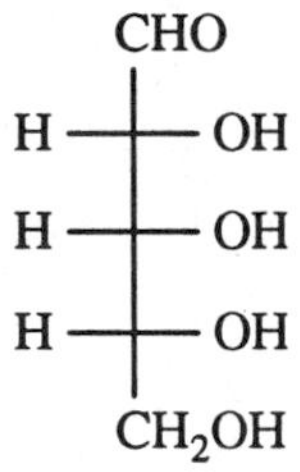

(b)
CHO
H — OH
HO — H
H — OH
H — OH
CH_2OH

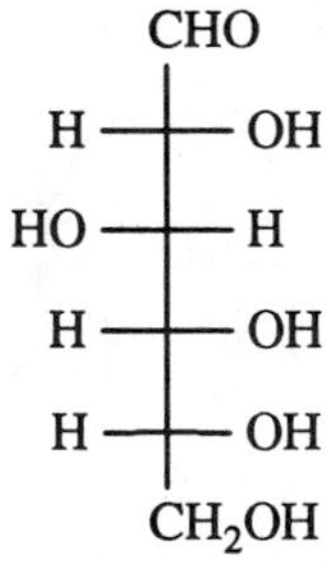

(c)
CH_2OH
C=O
HO — H
H — OH
H — OH
CH_2OH

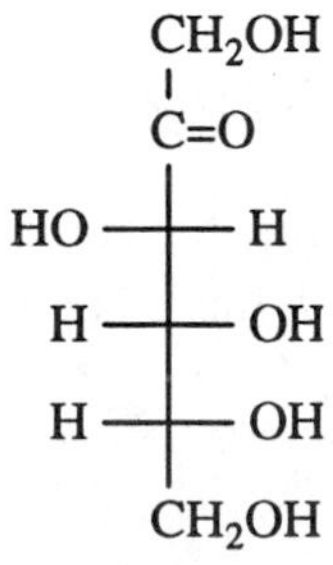

(d)
CHO
H — OH
HO — H
HO — H
H — OH
CH_2OH

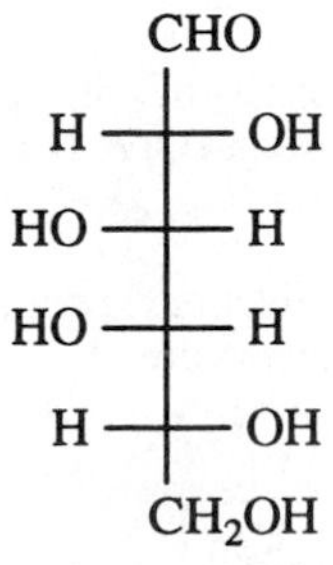

23.2 Label each of the following saccharides as a pyranose or as a furanose:

(a) 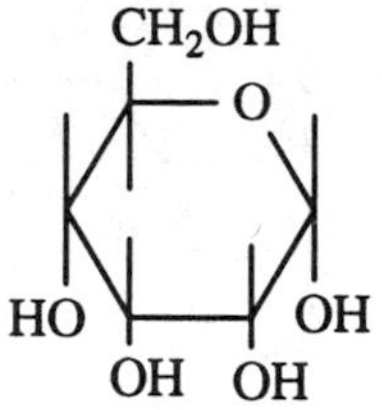 (b) 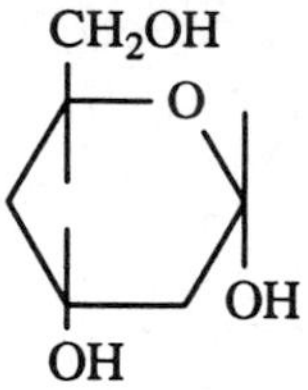 (c)

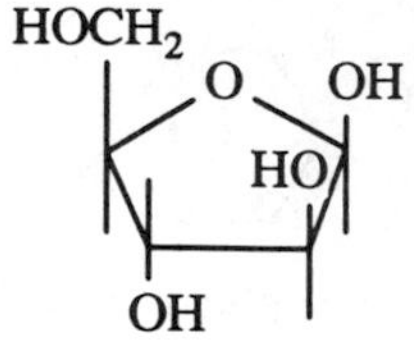

23.3 Draw formulas for (1) the anomer and (2) the mirror image for the following:

(a) 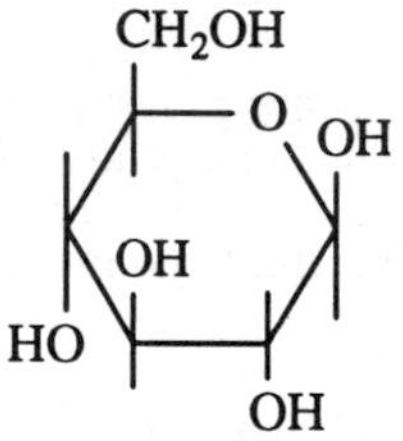 (b)

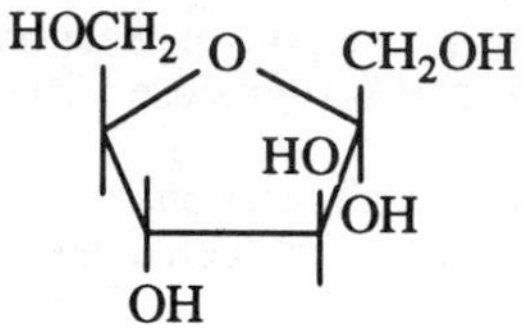

23.4 Name the following saccharides:

(a) 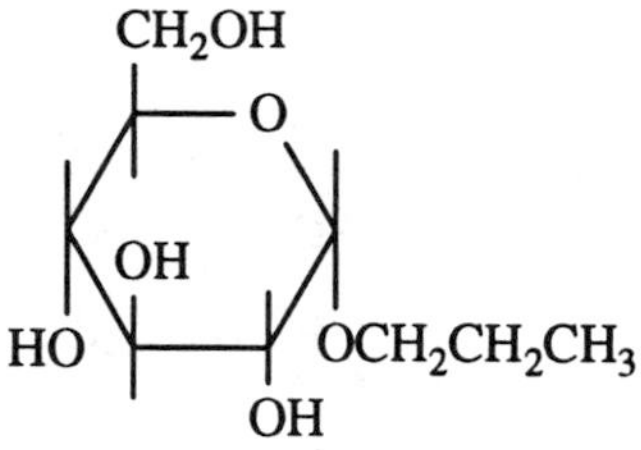 (b)

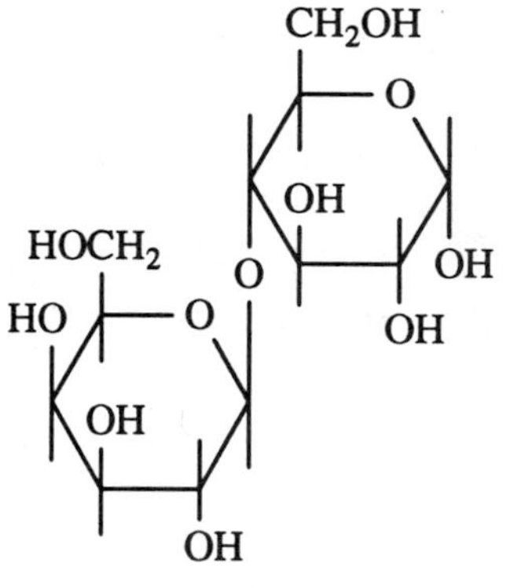

23.5 State which of the following compounds are reducing sugars and which are nonreducing. Give reasons for your answers.

23.6 Write Fischer projections for the open-chain forms of the following monosaccharides:

(a) 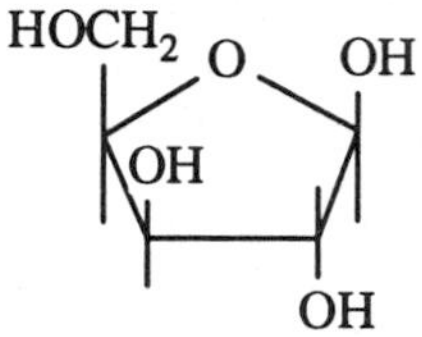 $HOCH_2$ O OH / OH / OH (b) 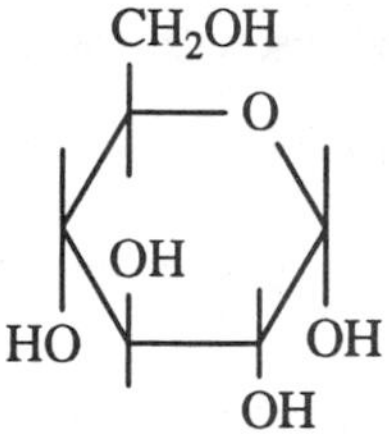 CH_2OH O / OH / HO OH / OH (c) 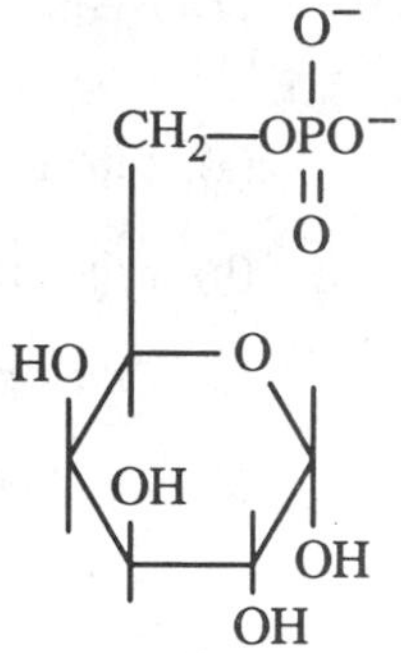 CH_2—OPO^- (O^-, =O) / HO O / OH / OH / OH

23.7 What products are expected from the periodic acid oxidation of the following compounds:

(a) 1,2-ethanediol

(b) 1,2-propanediol

(c) 2-hydroxypropanal

(d) 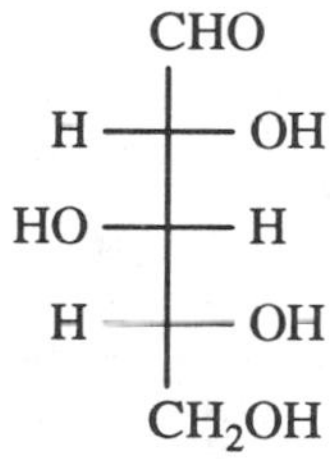 CHO / H—OH / HO—H / H—OH / CH_2OH

(e) 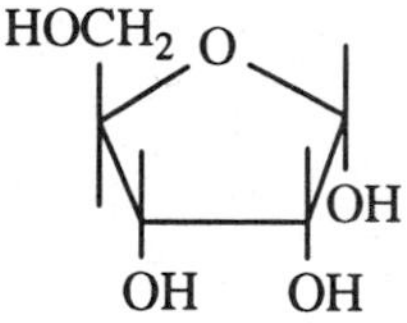 $HOCH_2$ O / OH / OH OH

23.8 What are the structures of A and B?

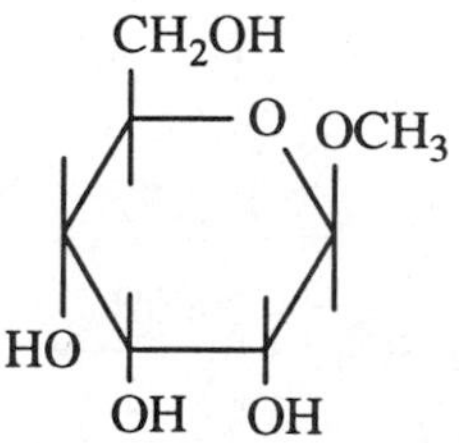

CH_2OH O OCH_3 / HO / OH OH $\xrightarrow{1.0\text{ mol } HIO_4}$ HCO_2H + A

B (a methyl glycoside) $\xrightarrow{1.0\text{ mol } HIO_4}$ A (+ no HCO_2H)

23.9 Write equations showing how each of the following transformations could be carried out:

(a) the acylation of D-arabinose

(b) the oxidation of D-idose to an aldaric acid

(c) the oxidation of D-idose to an aldonic acid

(d) the methylation of α-maltose (all hydroxyls)

23.10 Draw the structures of the products of the following reactions:

(a)

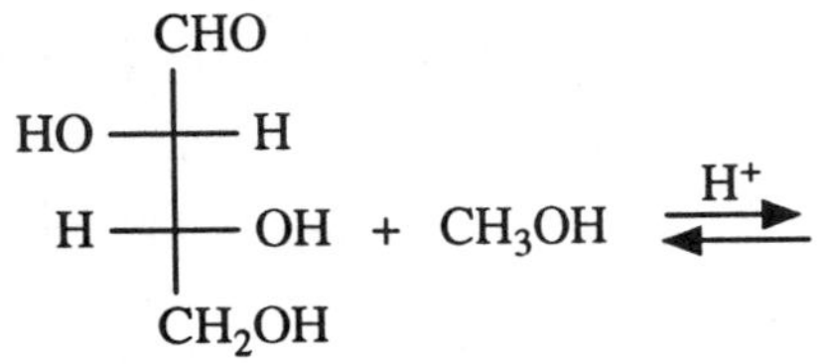

(b)

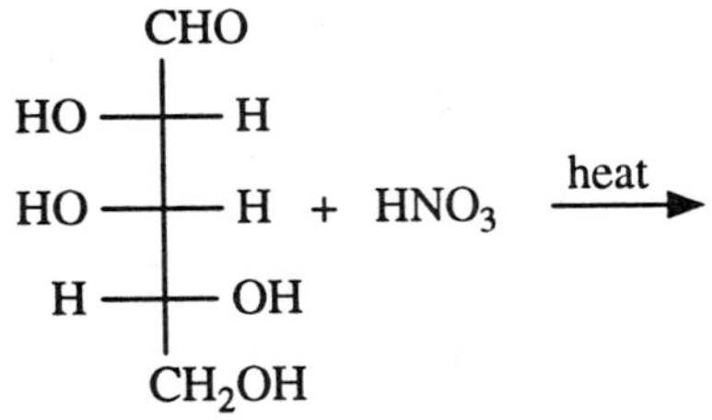

(c)

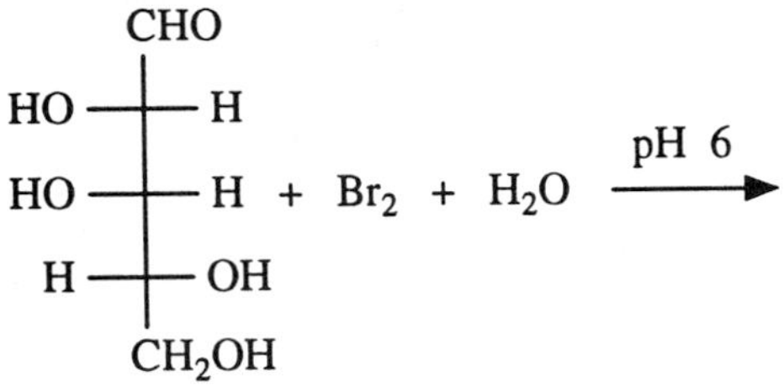

(d)

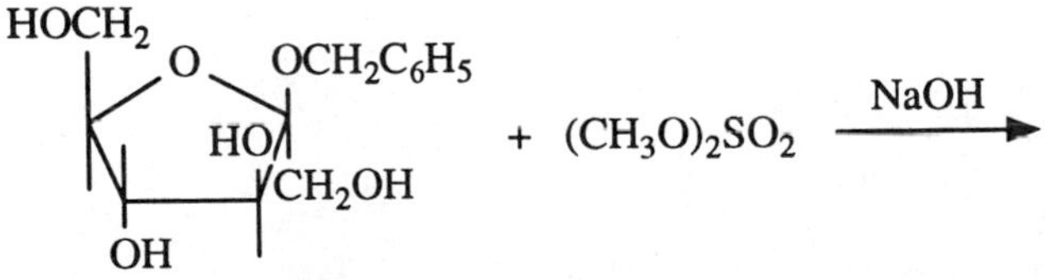

23.11 Predict the major organic product for the reaction of D-mannose with:

(a) $Br_2 + H_2O$

(b) HNO_3

(c) ethanol + H^+

(d) [the product from (c)] + excess dimethyl sulfate and NaOH

(e) [the product from (c)] + CH_3I, Ag_2O

(f) acetic anhydride

(g) acetyl chloride + pyridine

(h) $NaBH_4$

(i) (1) HCN; (2) H_2O, HCl

(j) (1) $LiAlH_4$; (2) H_2O, H^+

(k) H_2, Ni catalyst

(l) [the product from (c)] + H_2O, HCl

23.12 Fill in the blanks (give *all* possibilities):

(a) ______________ (a D-tetrose) $\xrightarrow{\text{hot } HNO_3}$ *meso*-tartaric acid

(b) ______________ $\xrightarrow[\text{cold}]{\text{dil. HCl}}$ [five-membered lactone ring: $HOCH_2$, ring O, =O, OH, OH]

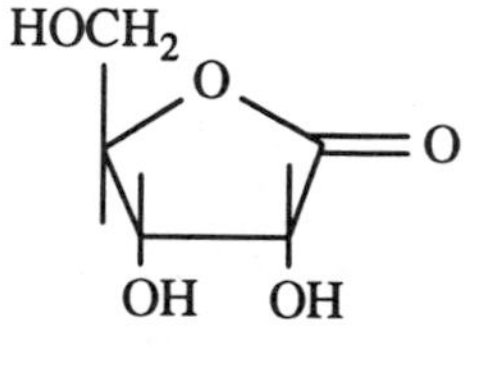

(c) ______________ (a D-aldohexose) $\xrightarrow[(2)\ H_2O,\ H^+]{(1)\ NaBH_4}$ a *meso*-alditol

23.13 Give the structure and the name of the organic product obtained when D-galactose is treated with:

(a) aqueous Br_2

(b) hot dilute HNO_3

(c) Tollens reagent

(d) (1) $NaBH_4$; (2) aqueous HCl

23.14 Compound A ($C_5H_{10}O_5$) is water soluble and sweet. Upon oxidation with Br_2 + H_2O, an optically active carboxylic acid B ($C_5H_{10}O_6$) is obtained. When compound A is oxidized with HNO_3, an optically inactive dicarboxylic acid C ($C_5H_8O_7$) is obtained.

(a) What are the two possible structures for C?

(b) If compound A could be degraded by a sequence of reactions starting at the carbonyl end to D-(-)-tartaric acid, what are the structures of A, B, and C?

23.15 (a) Write the sequence of reactions by which D-ribose is converted to D-allose.

(b) Show how you would determine that your product in (a) is indeed D-allose and not D-altrose

23.16 Predict the major organic product of the treatment of amylose with dimethyl sulfate and NaOH, followed by hydrolysis with dilute HCl.

23.17 How many possible disaccharides could be formed from just D-glucopyranose? (The α and β links form *different* disaccharides.)

Lipids

Some Important Features

Some important lipids are the edible fats and oils, terpenes, and steroids. Fats are triglycerides, or triacylglycerols, of fatty acids containing few carbon-carbon double bonds, while the edible oils are triglycerides of fatty acids containing a greater number of carbon-carbon double bonds. Saponification of fats or oils yields soaps, RCO_2Na, where R is a long alkyl or alkenyl chain. Twenty-carbon fatty acids are used to biosynthesize prostaglandins (Section 24.4).

Phospholipids are compounds with hydrocarbon chains and a dipolar phosphate-amine group. These compounds form part of cell walls.

Steroids are compounds containing the following ring system:

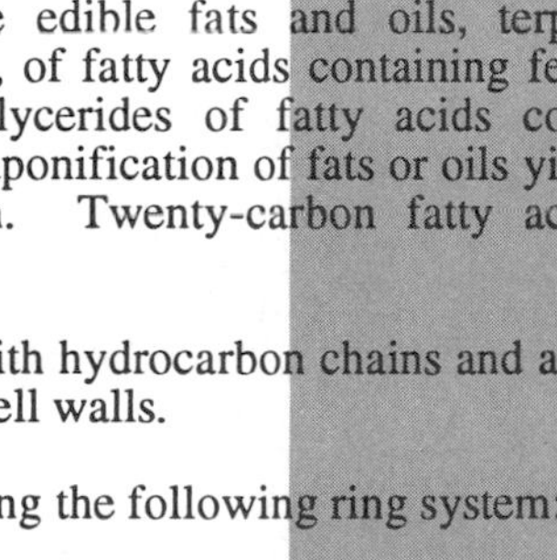

The stereochemistry of steroids is discussed in Section 24.5A. Most steroids have *trans* ring junctures; the bile acids are a notable exception.

Some steroids act as hormones. A few hormonal steroids, along with cholestrol, are discussed in the text.

Reminders

Although many lipids have complex structures, keep in mind that the reactions of lipids are predictable from their functional groups. For example, triglycerides are esters and may also contain carbon-carbon double bonds. Their reactions are typical of these functional groups. Sphingomyelin (Section 24.3) contains a carbon-carbon double bond, a hydroxyl group, an amide group, an inorganic ester group, and a quaternary nitrogen. Its reactions are typical of these functional groups.

Additional Drill Problems

24.1 Which of the following would be partly soluble or principally insoluble in water?

(a) 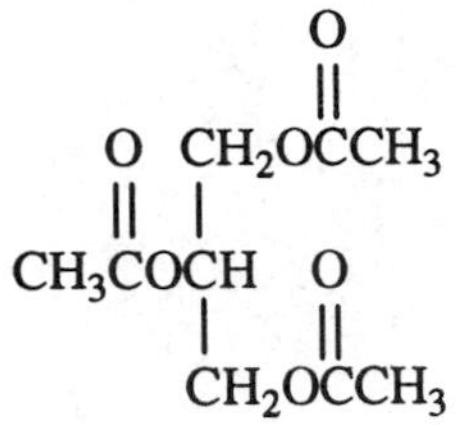
$$
\begin{array}{l}
\qquad\quad\;\; O \\
\qquad\quad\;\; \| \\
O \;\; CH_2OCCH_3 \\
\| \quad | \\
CH_3COCH \;\; O \\
\quad\;\; | \quad \| \\
\quad\;\; CH_2OCCH_3
\end{array}
$$

(b) 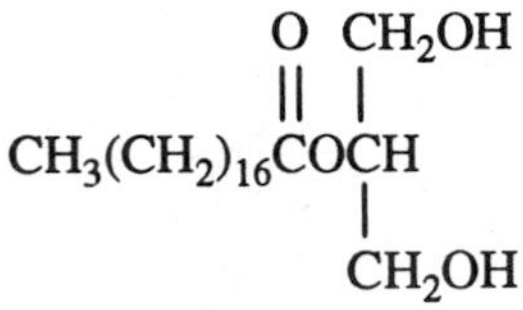
$$
\begin{array}{l}
\qquad\qquad\quad O \;\; CH_2OH \\
\qquad\qquad\quad \| \quad | \\
CH_3(CH_2)_{16}COCH \\
\qquad\qquad\qquad | \\
\qquad\qquad\qquad CH_2OH
\end{array}
$$

(c) 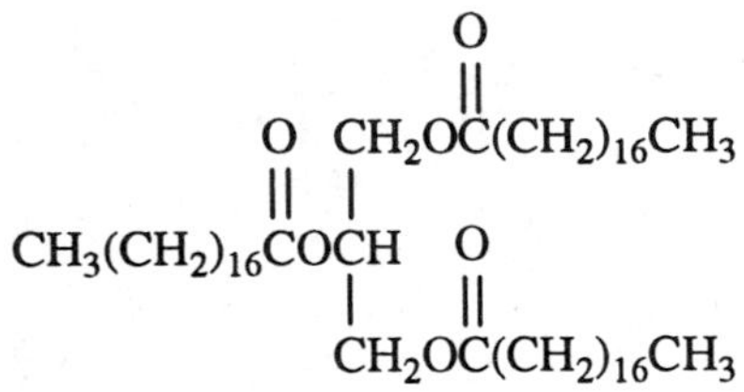
$$
\begin{array}{l}
\qquad\qquad\qquad\quad\;\; O \\
\qquad\qquad\qquad\quad\;\; \| \\
\qquad\qquad O \;\; CH_2OC(CH_2)_{16}CH_3 \\
\qquad\qquad \| \quad | \\
CH_3(CH_2)_{16}COCH \;\; O \\
\qquad\qquad\quad\;\; | \quad \| \\
\qquad\qquad\quad\;\; CH_2OC(CH_2)_{16}CH_3
\end{array}
$$

(d) 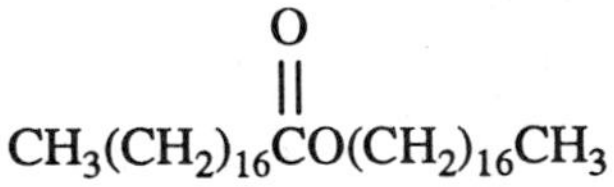
$$
\begin{array}{l}
\qquad\qquad\quad O \\
\qquad\qquad\quad \| \\
CH_3(CH_2)_{16}CO(CH_2)_{16}CH_3
\end{array}
$$

24.2 Which structure(s) shown below would not be expected to be found in a naturally occurring fat or oil?

(a) $CH_3(CH_2)_{16}\overset{\overset{\displaystyle O}{\|}}{C}OH$ (b) $CH_3(CH_2)_{17}\overset{\overset{\displaystyle O}{\|}}{C}OH$ (c) $CH_3(CH_2)_{18}\overset{\overset{\displaystyle O}{\|}}{C}OH$

24.3 Draw a formula for linolenic acid emphasizing the *cis* geometry of each carbon-carbon double bond.

24.4 Write the formula for a pair of enantiomers of a triglyceride that contains two stearic acid residues and one oleic acid residue.

24.5 Label each of the following compounds as a cephalin, lecithin, a sphingomylein, or a cerebroside (see Text Problem 24.17).

(a) 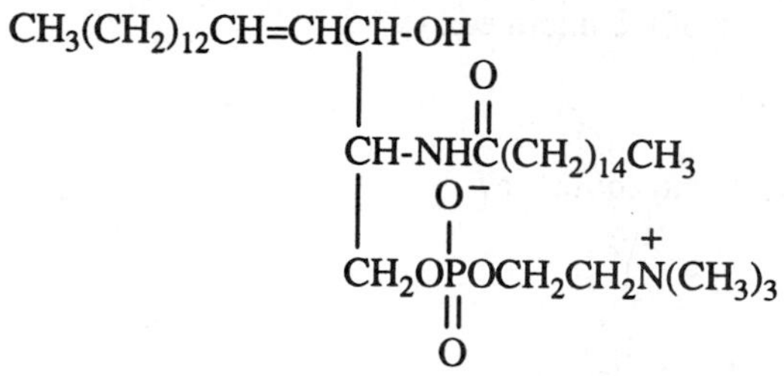 $CH_3(CH_2)_{12}CH{=}CHCH\text{-}OH$

$\quad\quad\quad\quad\quad\quad\quad\quad\;|\quad\quad\;\; O$

$\quad\quad\quad\quad\quad\quad\quad\quad\;|\quad\quad\;\; \|$

$\quad\quad\quad\quad\quad\quad\quad CH\text{-}NHC(CH_2)_{14}CH_3$

$\quad\quad\quad\quad\quad\quad\quad\quad\;|\quad\; O^-$

$\quad\quad\quad\quad\quad\quad\quad\quad\;|\quad\;\; |\quad\quad\quad\quad\quad +$

$\quad\quad\quad\quad\quad\quad\quad CH_2OPOCH_2CH_2N(CH_3)_3$

$\quad\quad\quad\quad\quad\quad\quad\quad\quad\;\; \|$

$\quad\quad\quad\quad\quad\quad\quad\quad\quad\; O$

(b)

$\quad\quad\quad\quad\quad\quad\quad\quad O$

$\quad\quad\quad\quad\quad\quad\quad\quad \|$

$CH_3(CH_2)_{14}COCH_2$

$\quad\quad\quad\quad\quad\quad\quad\quad\;\;|\quad\; O$

$\quad\quad\quad\quad\quad\quad\quad\quad\;\;|\quad\; \|$

$\quad\quad\quad\quad\quad\quad\quad\; CHOC(CH_2)_{14}CH_3$

$\quad\quad\quad\quad\quad\quad\quad\quad\;\;|\quad O^-$

$\quad\quad\quad\quad\quad\quad\quad\quad\;\;|\quad\; |\quad\quad\quad\quad\quad +$

$\quad\quad\quad\quad\quad\quad\quad\; CH_2OPOCH_2CH_2N(CH_3)_3$

$\quad\quad\quad\quad\quad\quad\quad\quad\quad\;\; \|$

$\quad\quad\quad\quad\quad\quad\quad\quad\quad\; O$

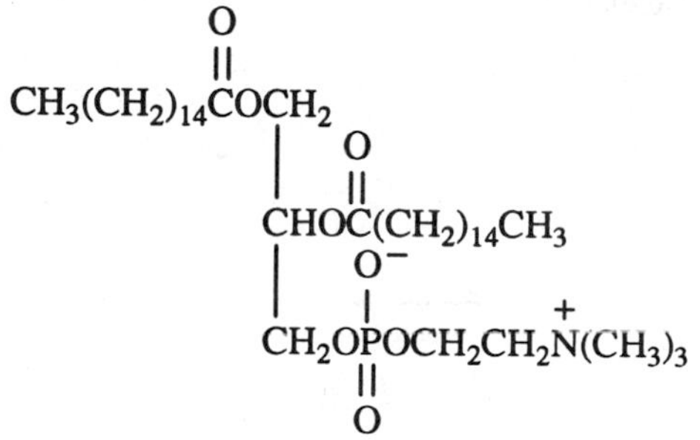

24.6 What distinguished the following compound classes from each other?

(a) a sphingomylein

(b) a steroid

(c) a cerebroside (see Text Problem 24.17)

(d) a glyceride

24.7 Would you expect the following steroids to show androgenic or estrogenic activity?

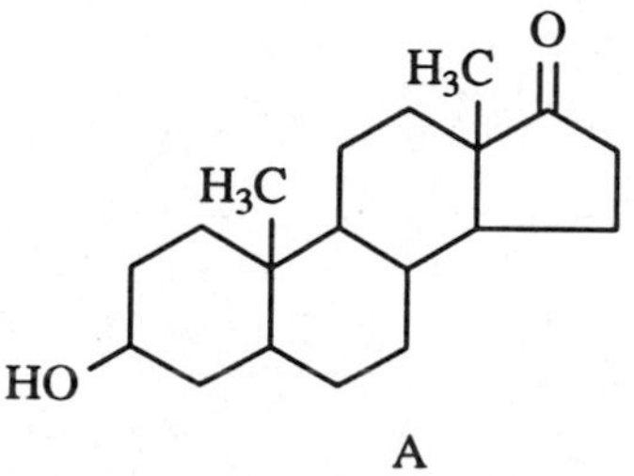

H_3C O $\quad$ H_3C $\quad$ HO

A

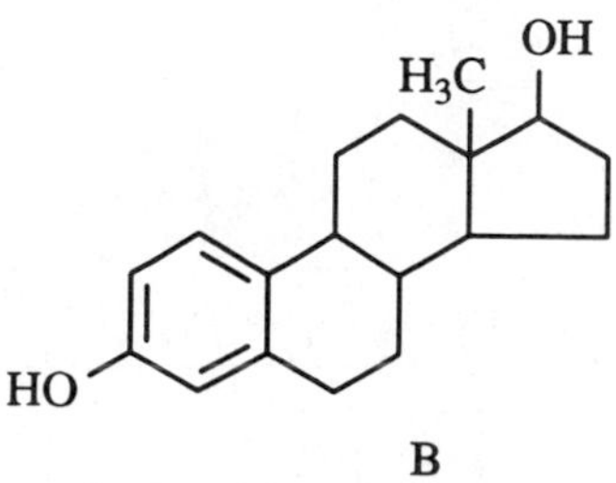

H_3C OH $\quad$ HO

B

24.8 Equilenin, a female sex hormone isolated from horses, was synthesized in 1939. The last step in the synthesis is shown, in part below.

(a) What is the structure of the intermediate **A**?

(b) How was it formed?

(c) How was it converted to equilenin?

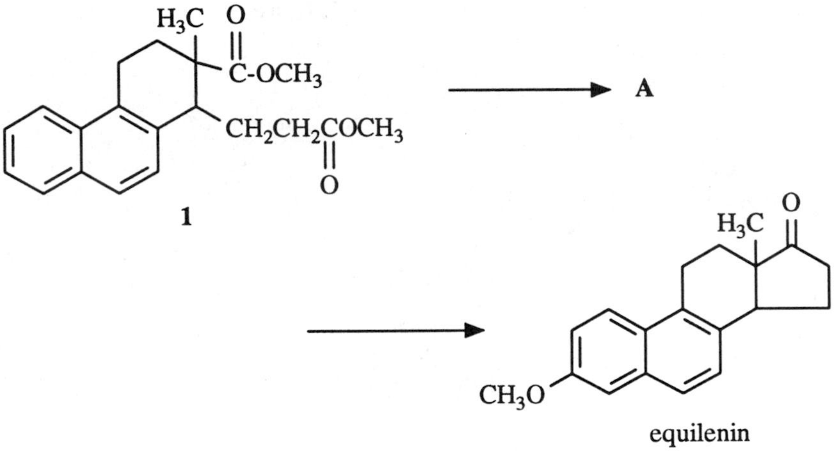

H_3C O
C-OCH_3
$CH_2CH_2COCH_3$
O
1
→ **A**

→

H_3C O
CH_3O
equilenin

24.9 Write equations for the reaction of cholesterol with the following reagents, showing the principal organic products:

(a) H_2CrO_4

(b) Br_2 in CCl_4

(c) NBS, *hv*

(d) NaH

24.10 Identify the changes in structure of the compound on the left that need be made if it is to be converted to the compound on the right.

(a)

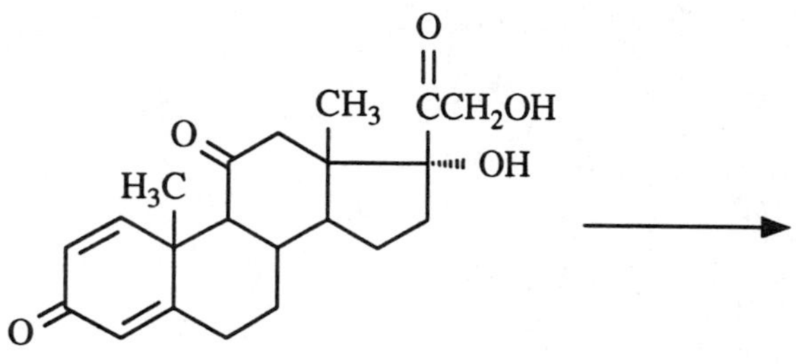

prednisone 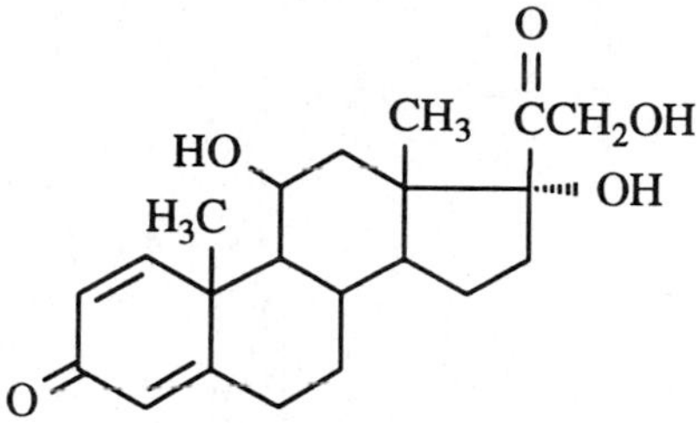 → prednisolone

(b)

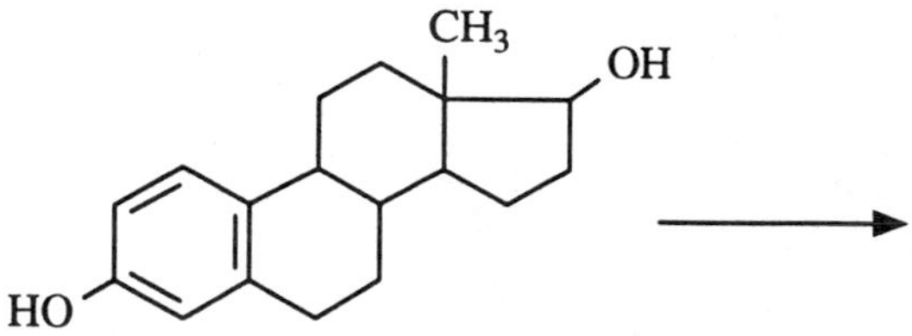

estradiol 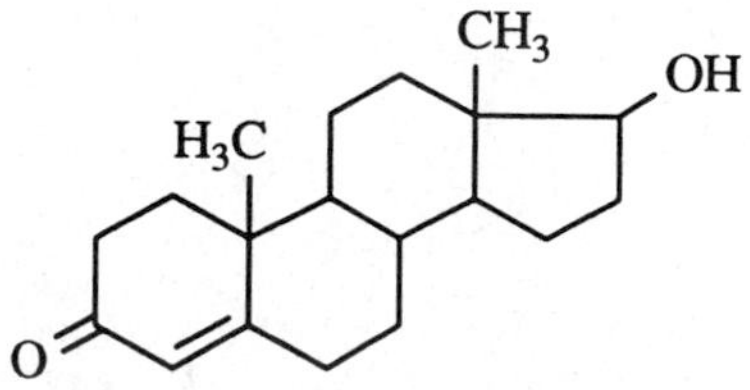 → testosterone

Amino Acids and Proteins

Some Important Features

Proteins are polyamides formed from L-α-amino acids. These acids may be acidic (acidic side chain), basic (basic side chain), or neutral (side chain neither acidic nor basic).

	neutral	*acidic*	*basic*
	$H_2N-C(CO_2H)(H)-CH_3$	$H_2N-C(CO_2H)(H)-CH_2CH_2CO_2H$	$H_2N-C(CO_2H)(H)-(CH_2)_4NH_2$
isoelectric point:	6	3	9

Amino acids exist as dipolar ions:

$$H_2\ddot{N}-C(CO_2H)(H)(CH_3) \rightleftharpoons H_3\overset{+}{N}-C(CO_2^-)(H)(CH_3)$$

Some techniques for synthesizing amino acids are discussed in Section 25.2.

A small protein molecule is called a *peptide*. The synthesis of peptides can be accomplished by blocking some reactive groups within the amino acids and then allowing other functional groups to react (Section 25.7). The carboxyl group may be activated toward amide formation by treatment with $ClCO_2C_2H_5$ or dicyclohexylcarbodiimide. In biological systems, amino acids are joined into protein chains enzymatically by mRNA, ribosomes, and tRNA.

The sequence of amino acids (the *primary structure*) in peptides and proteins can be determined by partial hydrolysis and terminal residue analysis (Section 25.6). The Edman reagent ($C_6H_5N{=}C{=}S$) and Sanger reagent (1-fluoro-2,4-dinitrobenzene) are used for *N*-terminal analysis. Enzymes that catalyze specific cleavages can be used in partial hydrolysis.

Hydrogen bonding between NH and C=O groups and side-chain interactions allow a protein to assume a *secondary structure* (the shape of a chain) and possibly a *tertiary structure* or *quaternary structure* (interactions between different parts of a chain or between two or more chains). Thus, a chain can form a helix that can fold into a globule or interact with other helices. The higher structures can add strength or solubility to a protein. When these higher structures are disrupted by a change in environment (such as a change in pH), the protein is said to be *denatured.*

Enzymes are proteins that act as biological catalysts. Their specific catalytic action depends on the unique protein surface presented to substrates and on the active site, which may be a *coenzyme* (a nonprotein organic molecule or a metal ion). Many vitamins are coenzymes.

Reminders

In predicting acid-base reactions of amino acids or peptides, look for the most acidic and most basic sites in the molecule.

Remember that any synthesis from an achiral molecule leads to an achiral or racemic product; however, a racemic mixture of enantiomers can be separated by procedures outlined in Chapter 4.

Additional Drill Problems

25.1 Write the abbreviations for the following amino acids:

(a) lysine (b) cysteine

(c) asparagine (d) tyrosine

(e) $HOCH_2CH(\overset{+}{N}H_3)CO_2^-$ (f) $C_6H_5CH_2CH(\overset{+}{N}H_3)CO_2^-$

(g) pyrrolidine ring (N–H) with CO_2H at C-2 (h) $HO_2CCH_2CH_2CH(NH_2)CO_2H$

25.2 Would you expect an aqueous solution of each of the following amino acids to be acidic, basic, or neutral? Explain.

(a) lysine (b) glutamine

(c) aspartic acid (d) leucine

25.3 Write flow equations showing how the following conversions could be carried out.

(a) $CH_3CH(CH_3)CH_2CH_2CO_2H \longrightarrow$ leucine

(b) $CH_3CH(Br)CH_2CH_3 \longrightarrow$ isoleucine

(c) $(CH_3)_2CHCH{=}O \longrightarrow$ valine

25.4 Predict the major organic products:

(a) $CH_3CH_2CO_2H \xrightarrow[\text{(2) excess } NH_3]{\text{(1) } Br_2,\ PBr_3}$

(b) Potassium phthalimide (phthalimide–NK) $\xrightarrow[\text{(2) HCl, } H_2O\text{, heat}]{\text{(1) } ClCH_2CO_2C_2H_5}$

(c) Phthalimide–$NCH(CO_2C_2H_5)_2$ $\xrightarrow{\text{(1) } Na^{+}\ {}^{-}OC_2H_5}$
(2) $C_6H_5CH_2Cl$
(3) HCl. H_2O, heat

(d) $(CH_3)_2CHCH_2CH{=}O + NH_3 + HCN \longrightarrow$

(e) $C_6H_5CH_2CH(NH_2)CO_2H + CH_3OH + HCl \xrightarrow{\text{heat}}$

25.5 Predict the major products:

(a) $1\ H_2N\underset{\displaystyle CH(CH_3)_2}{\underset{|}{C}}HCO_2^- + 1\ HCl \longrightarrow$

(b) $1\ H_2N\underset{\displaystyle (CH_2)_4NH_2}{\underset{|}{C}}HCO_2^- + 1\ HCl \longrightarrow$

(c) $1\ H_3\overset{+}{N}\underset{\displaystyle CH(CH_3)_2}{\underset{|}{C}}HCO_2H + 1\ NaOH \longrightarrow$

(d) $1\ H_3\overset{+}{N}\underset{\displaystyle (CH_2)_4NH_3^+}{\underset{|}{C}}HCO_2H + 1\ NaOH \longrightarrow$

25.6 Complete the following equations:

(a) $H_2N\overset{O}{\overset{||}{C}}CH_2CH_2\underset{\displaystyle NH_2}{\underset{|}{C}}HCO_2H + 2$ (ninhydrin: indane-1,3-dione with OH, OH on C-2) $\longrightarrow$

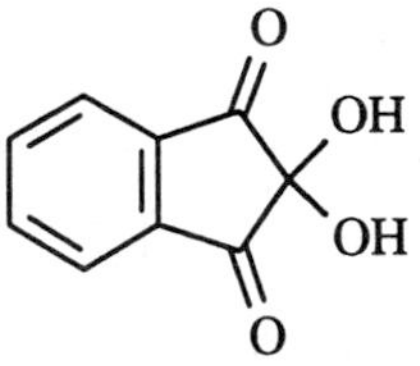

(b) $H_2N\overset{O}{\overset{||}{C}}CH_2CH_2\underset{\displaystyle NH_2}{\underset{|}{C}}HCO_2H + CH_3\overset{O}{\overset{||}{C}}O\overset{O}{\overset{||}{C}}CH_3 \longrightarrow$

(c) tyrosine + excess acetic anhydride $\longrightarrow$

(d) serine + excess acetic anhydride $\longrightarrow$

(e) alanine + ninhydrin $\longrightarrow$

25.7 Write the complete names for the following peptides:

(a) leu-lys

(b) lys-leu

(c) tyr-phe-ser

(d) ser-phe-tyr

25.8 Draw formulas for the following peptides:

(a) alanylglutamine (b) prolylthreonine

(b) glycylglutamine (d) lysyllysyltryptophan

25.9 Write the full name for each of the following peptides:

(a) $H_3\overset{+}{N}CH(CH_2C_6H_4OH)\overset{O}{\overset{\|}{C}}NHCH(CH_2OH)\text{-}\overset{O}{\overset{\|}{C}}NHCH(CH_2C_6H_5)\overset{O}{\overset{\|}{C}}O^-$

(b) $H_3\overset{+}{N}CH(CH_2\text{-indolyl})\overset{O}{\overset{\|}{C}}NHCH_2\overset{O}{\overset{\|}{C}}NHCH(CH_2\text{-indolyl})\overset{O}{\overset{\|}{C}}O^-$

25.10 Complete the following equations:

(a) $(CH_3)_3C\overset{O}{\overset{\|}{C}}\!\!\!\!\!\!\;\;O\overset{O}{\overset{\|}{C}}NHCH(CH_3)\overset{O}{\overset{\|}{C}}OC(=N\text{-}C_6H_{11})NH\text{-}C_6H_{11}$ + $H_2NCH_2CO_2H$ ⟶

(b) $C_6H_5CH_2O\overset{O}{\overset{\|}{C}}NHCH(CH_2C_6H_5)\overset{O}{\overset{\|}{C}}OCO_2C_2H_5$ + $H_2NCH_2CO_2CH_2CH_3$ ⟶

(c) the product from (b) $\xrightarrow{H_2,\ Pd}$

25.11 Write equations for the following reactions:

(a) alanine + $(CH_3)_3CO\overset{O}{\overset{\|}{C}}Cl \longrightarrow$

(b) acetic acid + dicyclohexylcarbodiimide $\longrightarrow$

(c) acetic acid + $ClCO_2CH_2CH_3 \longrightarrow$

(d) alanine + $C_6H_5CH_2O\overset{O}{\overset{\|}{C}}Cl \longrightarrow$

25.12 Consider the following peptide:

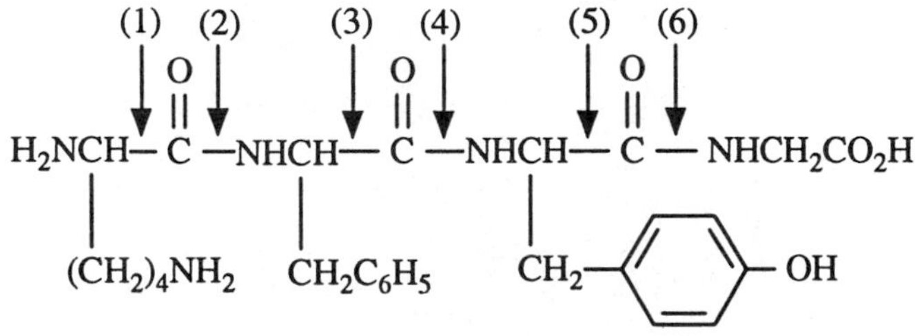

Which bonds, if any, would be hydrolyzed when the catalyst is:

(a) trypsin (b) pepsin (c) chymotrypsin

(d) BrCN (e) thermolysis (f) carboxypeptidase

24.13 Each of the following peptides is treated with trypsin. What fragments will be formed?

(a) lys-asp-gly-ala-ala-glu-ser-gly

(b) trp-cyS-lys-ala-arg-arg-gly

(c) ala-ala-his-arg-glu-lys-phe-ile-gly-glu-gly-glu

Nucleic Acids

Some Important Features

Nucleic acids are composed of repetitive units of nucleotides (base-sugar-phosphate). The phosphate group is mainly ionized at pH values near 6 or 7. Deoxyribose is the sugar in DNA, and ribose is the sugar in RNA. Be sure you know the structures of these sugars and of the principal bases (adenine, guanine, thymine, cytosine, and uracil).

The double-stranded DNA helix is held together primarily by hydrogen bonding by base pairs (T and A, C and G). DNA replicates by unwinding and directing the nucleotide sequence of a new complementary chain.

The sequence of nucleotides in mRNA is complementary to a portion of a DNA nucleotide sequence (a gene). The synthesis of mRNA according to the base sequence in a gene is called *transcription* of the genetic code.

The mRNA and ribosomes direct the synthesis of protein (*translation* of the genetic code). The genetic code is carried by particular series of three bases in a row called *codons*. Each codon is specific for a particular amino acid.

Reminders

The base T (or U in RNA) is always paired with A. The base C is always paired with G.

In problems, keep track of the strands--two complementary DNA strands; an mRNA strand with *codons* complementary to a portion of DNA strand; and a tRNA molecule with an *anticodon* complementary to a codon of the mRNA strand.

Additional Drill Problems

26.1 Draw a structure for each of the following:

(a) deoxyadenosine (b) adenosine

(c) guanine (d) deoxyguanosine

26.2 The following compound, called acyclovir, is an antiviral drug active against herpes viruses. To which nucleoside does it most closely correspond?

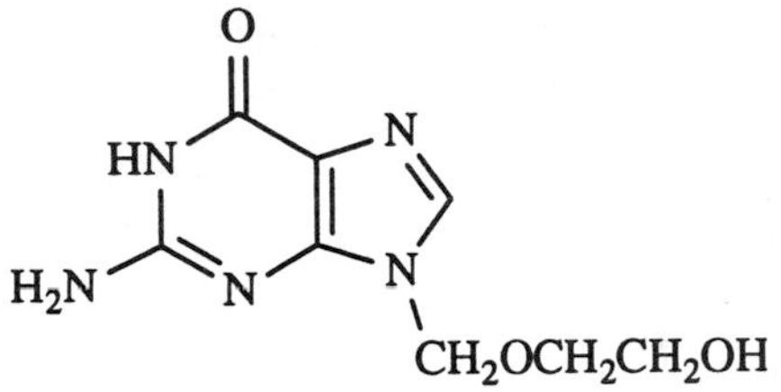

acyclovir

26.3 A sample of DNA is found to have 19% thymidine. What would be the expected percentages of guanine and cytosine in this DNA? (Assume the hydrogen-bonding ratio of the bases to be 1.0.)

26.4 Methylating agents, such as CH_3Br, can react with a guanosine residue in DNA to form the O-6 methyl derivative. Write an equation that illustrates this reaction. Does this reaction prevent the methylated guanosine residue from hydrogen bonding with cytosine?

26.5 The carcinogenic hydrocarbon benzo[a]pyrene is converted to a diol epoxide in the liver. The epoxide is also carcinogenic and has been shown to react with bases in DNA. Write an equation that illustrates the reaction between the epoxide and a guanosine in DNA.

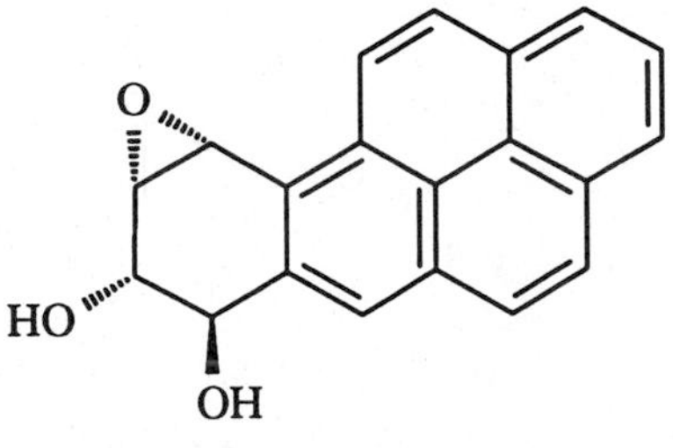

diol epoxide of benzo[a]pyrene

26.6 Ultraviolet light is known to cause skin cancer. One reason that has been suggested why is the fact that ultraviolet light causes a [2+2] cycloaddition reaction of pyrimidine bases. Write a reaction illustrating this reaction between two thymine residues in DNA to form a thymine dimer.

26.7 What are the major structural differences between DNA and RNA?

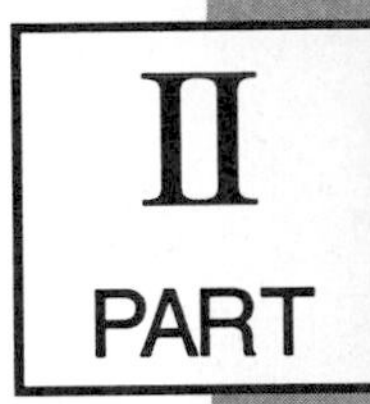

Solutions to the Additional Drill Problems

1

Atoms and Molecules — A Review

Answers to Additional Problems

1.1 (a) beryllium (Be) (b) sulfur (S)

(c) magnesium (Mg) (d) potassium (K)

1.2 (a) Na $1s^22s^22p^63s^1$
Na^+ $1s^22s^22p^63s^0$

(b) O $1s^22s^22p^4$
O^{2-} $1s^22s^22p^6$

(c) S $1s^22s^22p^63s^23p^4$
S^{2-} $1s^22s^22p^63s^23p^6$

(d) H $1s^1$
H^+ $1s^0$
H^- $1s^2$

1.3 (a) Al: $1s^22s^22p^63s^23p^1$ (b) Ca: $1s^22s^22p^63s^23p^64s^2$

(c) P: $1s^22s^22p^63s^23p^3$ (d) Ge: $1s^22s^22p^63s^23p^63d^{10}4s^24p^2$

1.4 (a) Mg^{2+} 2 :C̤̈l:$^-$

(b)
```
         :Ö:⁻
  ..     ..2+..
H:O: S :O:H
  ..     ..  ..
        :O:⁻
         ..
```

(c)
```
  H  +
     ..
H:C:O:H
  ..  ..
  H  H
```

(d)
```
        :O:
 –  ..   ..  ..
   :O:C:O:H
    ..      ..
```

(e)
```
     :O:
      ..
      ..
      N⁺
 ⁻:O.   .O:⁻
```

(f)
```
 –  ..     ..   .
   :O:N::O.
    ..
```

1.5 (a)

```
    H       H
    ..      ..
H : C : C ::: C : C : H
    ..      ..
    H       H
```

(b)

```
H.   H      .H
  C :: C : C :: C
H'       H      'H
```

(c)

```
    H
    ..   ..
H : C :  N : H
    ..   ..
    H    H
```

(d)

```
       :O:
        ::    ..
H :  C  :  O : H
              ..
```

(e)

```
      :O:
    H  ..
    ..      ..
H : C : C : O : H
    ..      ..
    H
```

(f)

```
    H    H    H
    ..   ..   ..
H : C :  C :  C : H
    ..   ..   ..
    H   :O:   H
         ..
         H
```

(g)

```
H.    H
      ..    ..
  C :: C : Cl:
H'          ..
```

(h)

```
         ..
        :Cl:
    ..   ..
:Cl :  C : H
    ..   ..
        :Cl:
         ..
```

1.6 (a)

```
    H     H
    |    /
H - C - N
    |    \
    H     H
```

C, 4; N, 3

(b)

```
H            H
 \          /
  C = C = C
 /          \
H            H
```

each C, 4

(c)

```
    H   O
    |   ||
H - C - C - O - H
    |
    H
```

each C, 4; each O, 2

(d)

```
    H   O   H
    |   ||  |
H - C - C - C - H
    |       |
    H       H
```

each C, 4; O, 2

(e)

```
H       H
 \     /
  C = C
 /     \
H       Cl
```

each C, 4; Cl, 1

(f)

```
    H       H
    |       |
H - C - O - C - H
    |       |
    H       H
```

each C, 4; O, 2

1.7 (a)
$$\begin{array}{ccccccc} & & & H & & & \\ & & & | & & & \\ & H & & O & & H & \\ & | & & | & & | & \\ H- & C & - & C & - & C & -H \\ & | & & | & & | & \\ & H & & H & & H & \end{array}$$

(b)
$$\begin{array}{ccccccccc} & H & & H & & H & & H & \\ & | & & | & & | & & | & \\ H- & C & - & C & -O- & C & - & C & -H \\ & | & & | & & | & & | & \\ & H & & H & & H & & H & \end{array}$$

(c)
$$\begin{array}{ccccccc} & H & & H & & O & \\ & | & & | & & \| & \\ H- & C & - & C & - & C & -H \\ & / & & | & & & \\ & H & H- & C & -H & & \\ & & & | & & & \\ & & & H & & & \end{array}$$

(d)
$$\begin{array}{cccccccc} & H & & H & & & H & \\ & | & & | & & & / & \\ H- & C & - & C & - & C & & \\ & | & & | & & \| & & \\ & H & & H & & C & -H & \\ & & & & / & & & \\ & & & & H & & & \end{array}$$

1.8 (a) CH_3OCH_3 or $(CH_3)_2O$

(b) $CH_3CH_2O^-$

(c) $Cl_2C{=}CHCl$ or $CCl_2{=}CHCl$

(d) $CH_2{=}CHNHCH{=}CH_2$ or $(CH_2{=}CH)_2NH$

(e) H_2SO_4 or $HOSO_2OH$

(f) HNO_3 or $HONO_2$

(g) $(CH_3)_2CHCO_2CH_3$ or $(CH_3)_2CH\overset{\overset{O}{\|}}{C}OCH_3$ or $CH_3\underset{\underset{CH_3}{|}}{C}H\overset{\overset{O}{\|}}{C}OCH_3$

(h) $CH_3\overset{\overset{O}{\|}}{C}NH_2$

(i) $(CH_3)_2CH\overset{\overset{O}{\|}}{C}CH(CH_3)_2$ or $[(CH_3)_2CH]_2C{=}O$ or $CH_3\underset{\underset{CH_3}{|}}{C}H\overset{\overset{O}{\|}}{C}\text{-}\underset{\underset{CH_3}{|}}{C}HCH_3$

1.9 (a) $CH_3\overset{\overset{\ddot{O}:}{\|}}{C}CH_3$ (b) $:\ddot{\underset{..}{C}}lCH_2\overset{\overset{\ddot{O}:}{\|}}{C}\ddot{N}HCH_3$ (c) $CH_3\overset{\overset{\ddot{O}:}{\|}}{C}\ddot{\underset{..}{O}}H$

1.10 (a)

```
     H   H
     |   |
 H — C — C — H
     |   |
 H — C — C — H
     |   |
     H   H
```

(b)

```
     H
     |      O
 H — C — C ⸗
     |   |
   ⸗ C — C — H
 O       |
         H
```

(c)

```
        H       H
        |       |
  H     C       C      H
    C ⸗   C   ⸗   C ⸗
    |     ||      |
    C ⸗   C   ⸗   C
  H     C       C      H
        |       |
        H       H
```

1.11 (a) CH_3 (b) 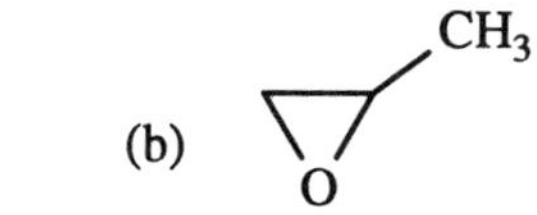

(c) S

1.12 (a) C (b) C

(c) each has the same value (d) Cl

1.13 (a) O (b) neither (c) C (d) Cl

(e) C (f) O (g) O (h) N

1.14 C-H < N-O < C-O < C-Mg

1.15 (a) $\overset{\delta+}{CH_3}-\overset{\delta-}{Br}$ (b) $\overset{\delta+}{CH_3}-\overset{\delta-}{OCH_3}$ (c) $\overset{\delta+}{CH_3}-\overset{\delta-}{NH_2}$

1.16 (a) $(CH_3)_2\ddot{N}H\cdots\cdot:\ddot{O}CH_3$ (with H on O) and $(CH_3)_2N:\cdots\cdot H\ddot{O}CH_3$ (with H on N)

(b) $CH_3CH_2\ddot{O}H\cdots\cdot:\ddot{O}CH_3$ (with H on O) and $CH_3CH_2\ddot{O}:\cdots\cdot H\ddot{O}CH_3$ (with H on O)

1.17 (a) and (b)

1.18 (a) $CH_3{-}H \longrightarrow \cdot CH_3 + H\cdot$

(b) $CH_3O{-}OCH_3 \longrightarrow CH_3O\cdot + \cdot OCH_3$

(c) $H{-}I \longrightarrow H\cdot + I\cdot$

1.19 (a) $(CH_3)_2CH{-}\overset{+}{O}H_2 \longrightarrow (CH_3)_2\overset{+}{CH} + OH_2$

(b) $(CH_3)_3C{-}Br \longrightarrow (CH_3)_3C^{+} + Br^{-}$

(c) $H{-}OSO_3H \longrightarrow H^{+} + {}^{-}OSO_3H$

(d) $CH_3CH_2{-}Li \longrightarrow CH_3CH_2^{-} + Li^{+}$

1.20 (a) $HOCH_2CH_2CH_2CH_2OH$ (b) CH_3CO_2H

(c) $HOCH_2CH_2OH$ (d) $CH_3CH_2NH_2$

In each case, the compound that can form the greater number of hydrogen bonds has the higher boiling point.

1.21 (a) CH_3CH_2OH (b) CH_3CO_2H

The compound that can form the greater number of hydrogen bonds with water is the more water-soluble compound of the pair.

1.22 (a) ${}^{-}OH$ (b) ${}^{-}OH$ (c) HSO_4^{-}

(d) $CH_3C(={}^{+}OH)CH_3$ (e) HSO_4^{-} (f) $CH_3\overset{+}{O}H_2$

1.23 Compounds (a), (c), and (d), because these compounds contain empty orbitals that can accept electrons. Compound (b) is not usually classified as a Lewis acid; however, under the proper reaction conditions, it can act as an electron acceptor in a Lewis acid-base reaction.

```
  ..                                 ..
CH3O—H  +  ¯:NH2   <=====>   CH3O:¯  +  :NH3
  ..         ..                  ..
```

1.24 (a) $-\log(5.9 \times 10^{-2}) = -\log(10^{-2}) - \log(5.9) = 2 - 0.77 = 1.23$

(b) $14 - \log(1.5) = 13.82$

(c) $6 - \log(8.5) = 5.07$

1.25 (a) + (b) − (c) −

1.26

(a)
```
      ..
 H H  O:               H   H   O
 .. .. ::              |   |   ||
H:C:C:C:O:H        H - C - C - C - O - H
 .. ..   ..            |   |
 H H                   H   H
```

(b)
```
  ..
  O:              O
  ::              ||
H:C:H         H - C - H
```

(c)
```
  ..
  O:   H              O       H
  ::.. ..             ||      |
H:C:O:C:H         H - C - O - C - H
    .. ..                     |
       H                      H
```

(d)
```
 H H   H              H   H    H
 .. .. .              |   |   /
H:C:C::C          H - C - C = C
 .. ..  .             |        \
 H      H             H         H
```

(e)
```
H        H          H         H
 .       .           \       /
 C::C::C              C = C = C
 .       .           /       \
H        H          H         H
```

or H:C:C⋮⋮C:H (with H above and below the first C) $H-CH_2-C\equiv C-H$

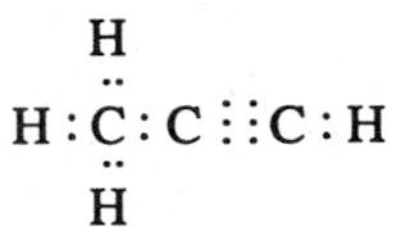

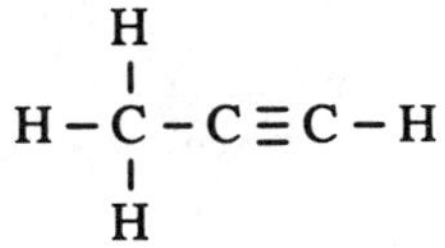

or H_2C ring with C::C (Lewis structure of cyclopropene)   △ (cyclopropene)

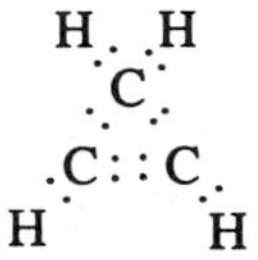

(f) H:C⋮⋮C:H $H-C\equiv C-H$

1.27 (b) < (a) < (c)

1.28 (a) $-\log(7.5 \times 10^{-4}) = -\log(10^{-4}) - \log(7.5) = 4 - 0.88 = 3.12$

(b) $pK_a = -\log(K_a)$ $-2.8 = \log(K_a)$ $\text{antilog}(-2.8) = K_a$ $\text{antilog}(-3 + 0.2) = K_a$ $\text{antilog}(-3) \times \text{antilog}(0.2) = K_a$ $K_a = 1.6 \times 10^{-3}$

(c) 4.82

1.29 (a) 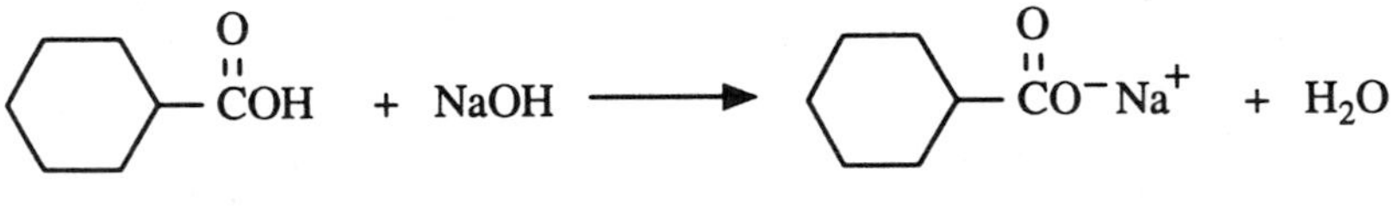 $C_6H_{11}COOH + NaOH \longrightarrow C_6H_{11}CO_2^-Na^+ + H_2O$ (cyclohexyl)

(b) $CH_3SO_2-OH + NaOH \longrightarrow CH_3SO_2-O^-Na^+ + H_2O$

(c) $C_6H_{11}-\overset{+}{O}H_2 + NaOH \longrightarrow C_6H_{11}-O-H + Na^+ \; H_2O$

(d) $CH_3CH_2CH_2\overset{+}{N}H_3 + NaOH \longrightarrow CH_3CH_2CH_2NH_2 + Na^+ + H_2O$

(e) $CH_3CH(^+NH_3)COOH$ + NaOH (excess) ⟶ $CH_3CH(NH_2)CO^-Na^+$ + Na^+ + H_2O

1.30 (a) $C_6H_5NH_2$ + HCl ⟶ $C_6H_5\overset{+}{N}H_3$ Cl^-

(b) $(CH_3CH_2)_3COH$ + HCl ⟶ $(CH_3CH_2)_3C\overset{+}{O}H_2$ Cl^-

(c) piperidine (N–H) + HCl ⟶ piperidinium ($\overset{+}{N}H_2$) Cl^-

(d) $C_6H_5CO_2^-$ + HCl ⟶ C_6H_5COOH + Cl^-

2

Orbitals and Their Role in Covalent Bonding

Answers to Additional Problems

2.1 (a)

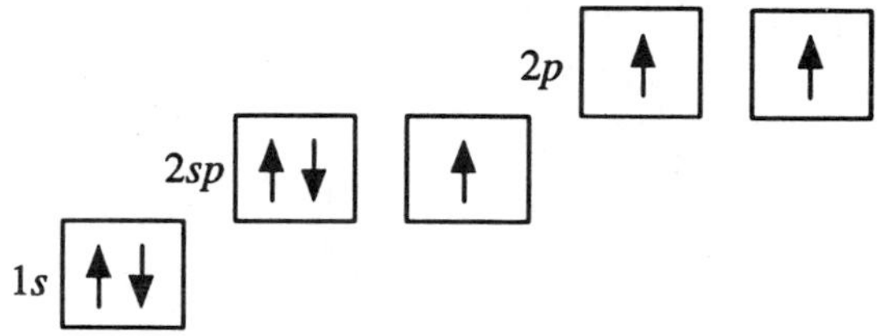

$1s$ ↑↓ ; $2sp$ ↑↓ ↑ ; $2p$ ↑ ↑

(b)

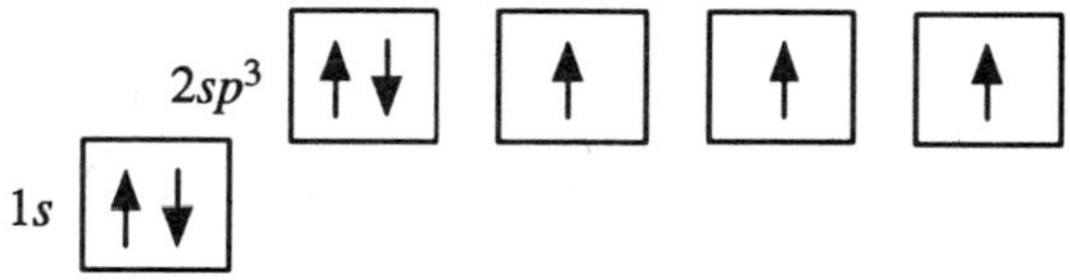

$1s$ ↑↓ ; $2sp^3$ ↑↓ ↑ ↑ ↑

(c)

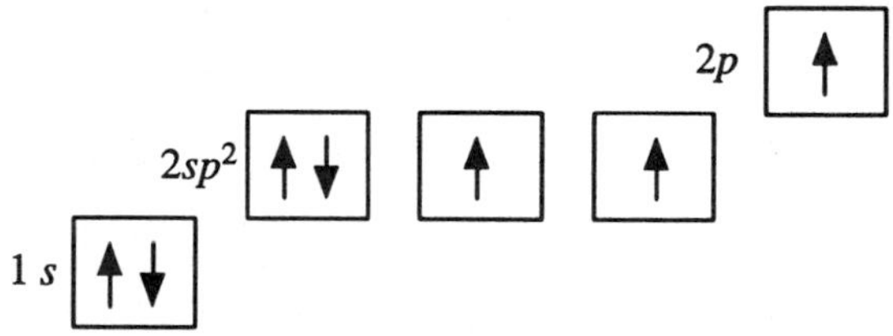

$1s$ ↑↓ ; $2sp^2$ ↑↓ ↑ ↑ ; $2p$ ↑

2.2 (a) cyclohexanone (C=O) + $HC{\equiv}N$ $\xrightarrow{^{-}C\equiv N}$ 1-hydroxycyclohexanecarbonitrile (OH, $C\equiv N$)

carbon, sp^2

oxygen, sp^2

oxygen, sp^3

carbon, sp^3

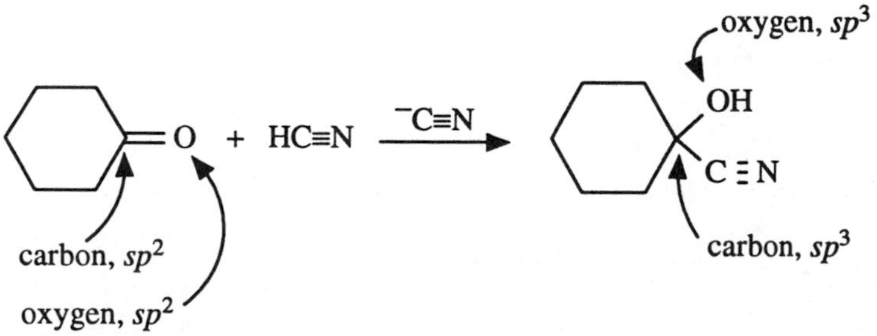

(b) $(CH_3)_3CCl$ + CH_3O^- ⟶ $(CH_3)_2C{=}CH_2$ + Cl^- + CH_3OH

all carbons, sp^3 (for $(CH_3)_3CCl$)

two carbons, sp^2 (for $C{=}CH_2$)

(c) $HC{\equiv}CH + H_2O \xrightarrow{H^+,\ Hg^{2+}} CH_3\overset{\overset{\displaystyle O}{\|}}{C}H$

- HC≡CH: two carbons, *sp*
- H_2O: oxygen, sp^3
- $CH_3CH=O$: oxygen, sp^2; carbon (C=O), sp^2; carbon (CH_3), sp^3

(d)

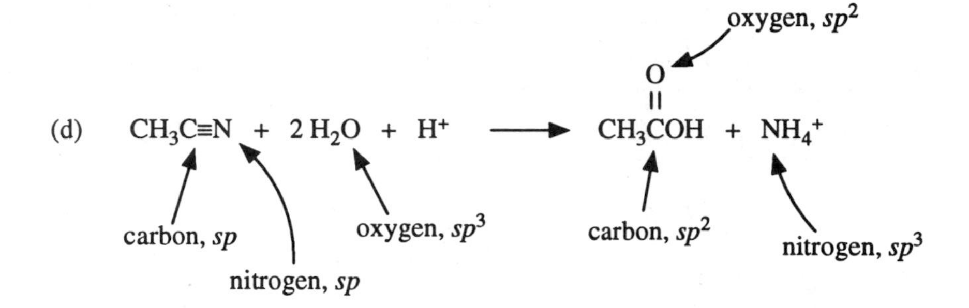

2.3 (a) O=C=O; each double bond contains one sigma bond and one pi bond

(b)

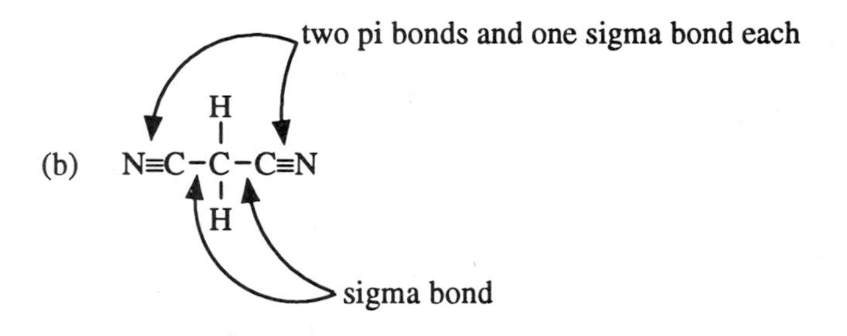

Each C-H bond is also a sigma bond.

(c) $CH_3\overset{\overset{\displaystyle CH_3}{|}}{C}{=}N{-}OH$

(C=N bond): one sigma bond and one pi bond; all other bonds are sigma bonds

2.4 (a) 180°: $H-C\equiv C-CH=CH-CH_2-\overset{O}{\overset{\|}{C}}-CH_3$

120° and 120°: $H-C\equiv C-\overset{H}{C}=\overset{H}{C}-CH_2-\overset{O}{\overset{\|}{C}}-CH_3$

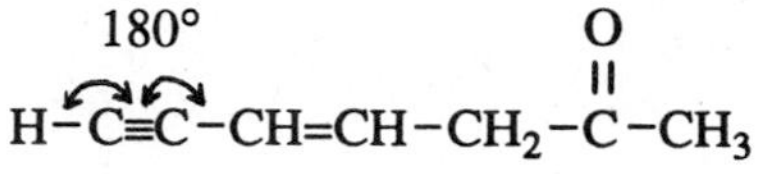

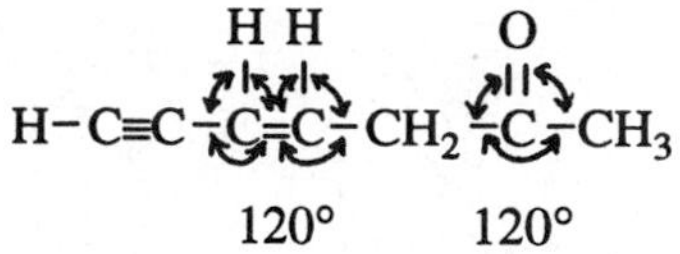

All other bond angles are approximately 109°.

(b) All sigma bond angles are approximately 120°.

2.5 (a) an sp^3 carbon; 25% *s* character

(b) (1) an sp^2 carbon; 33% *s* character (2) an *sp* carbon; 50% *s* character

(c) an sp^2 carbon; 33% *s* character

2.6 (a) all tetrahedral

(b) $C_6H_5-C\equiv C-C_6H_5$

C≡C: linear

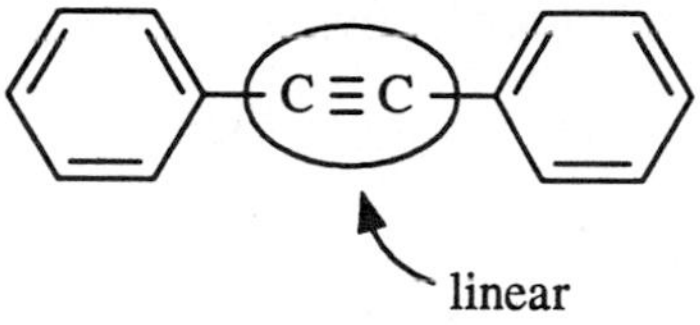

all others, trigonal

(c)

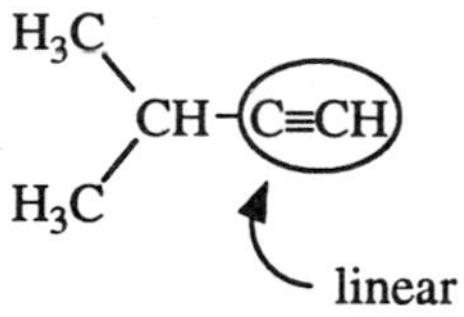

$$\begin{array}{l} H_3C \\ \quad\;\; CH-C\equiv CH \\ H_3C \end{array}$$

C≡CH: linear

all others, tetrahedral

2.7 (a) (1) because it is formed from sp^3-*s* orbital overlap

(b) (2) because a C-C bond is longer than a C-H bond

2.8 (a)

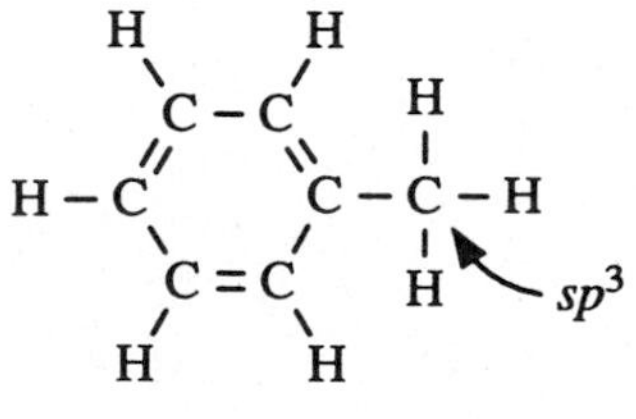

$$\begin{array}{ccccccc} & H & & H & & & \\ & & C-C & & & H & \\ H-C & & & & C-C-H & & \\ & & C=C & & & H & \\ & H & & H & & & \end{array}$$

CH_3 carbon: sp^3

all ring C atoms sp^2

all H atoms *s*

(b)

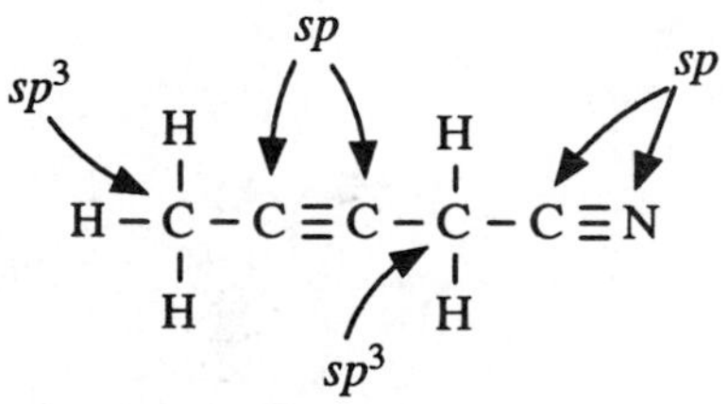

$$\begin{array}{ccccccc} & H & & & H & & \\ H- & C & -C\equiv C- & & C & -C\equiv N \\ & H & & & H & & \end{array}$$

first C: sp^3; C≡C carbons: *sp*; second CH_2 carbon: sp^3; C≡N: *sp*

(c) 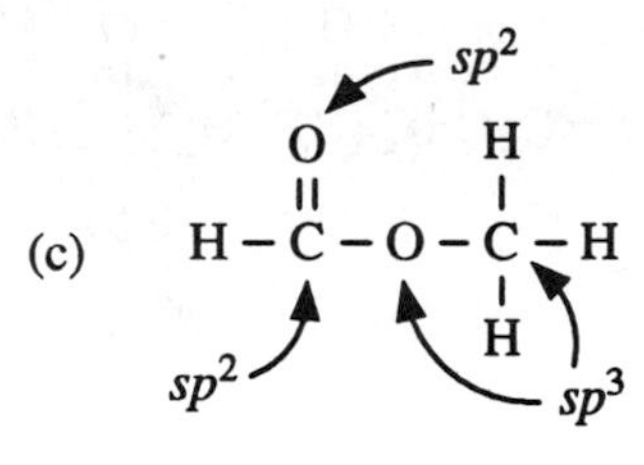

all H atoms *s*

2.9 (a) Both nitrogens are sp^2. (b) *sp*

(c) Both nitrogens are sp^2.

2.10 (a) (1) sp-sp^2; (2) sp^2-s (b) (1) sp^3-sp^2; (2) sp^2-s

(c) (1) sp^3-sp^3; (2) sp^2-sp^3 (d) (1) sp^3-sp^3; (2) sp^2-sp^2 and p-p

2.11 (a) 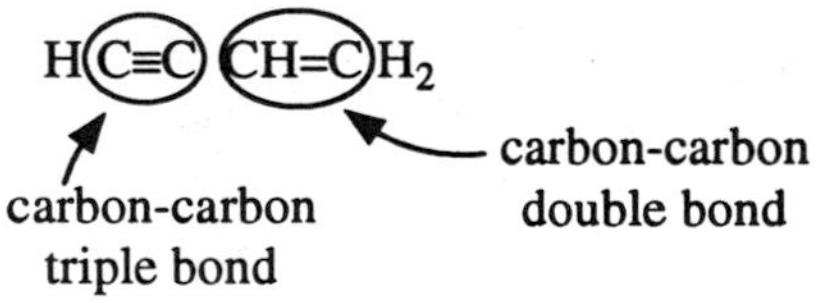 $HC{\equiv}CCH{=}CH_2$

carbon-carbon triple bond; carbon-carbon double bond

(b) $CH_3CH_2CH_2OH$ ← hydroxyl (alcohol)

(c) 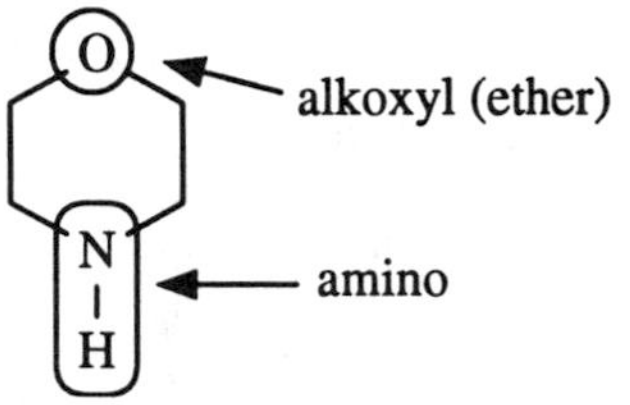 O ← alkoxyl (ether)

N–H ← amino

(d) 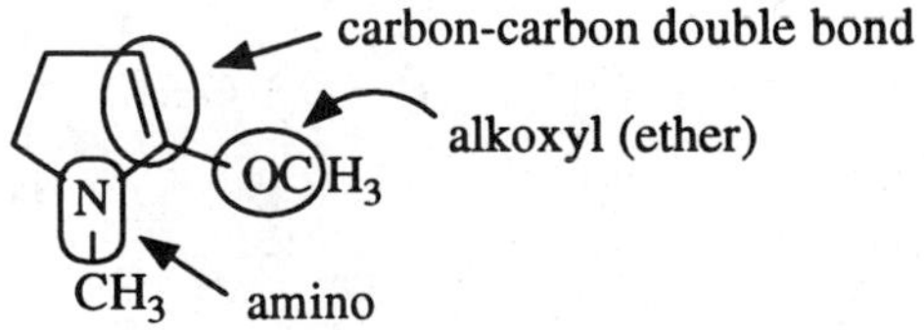 carbon-carbon double bond

OCH_3 ← alkoxyl (ether)

N–CH_3 ← amino

2.12 (a) CH_3CH_2NH—(cyclohexyl)

(b) $(CH_3CH_2)(C_6H_{11})C=C(C_6H_{11})(CH_2CH_3)$ or $(CH_3CH_2)(C_6H_{11})C=C(CH_2CH_3)(C_6H_{11})$

(c) $CH_3CH_2\overset{O}{\overset{\|}{C}}O$—(cyclohexyl)

(d) N-ethylpyrrolidine: N—CH_2CH_3

2.13 (a) ROH, alcohol (b) RNH_2, amine

(c) CH_2=CHR or RCH=CH_2, alkene

(d) ROR′, ether (e) RC≡CR′, alkyne

2.14 (a) nitrogen; *sp* (b) oxygen; sp^3

(c) carbon; *sp* (d) nitrogen; sp^2

2.15 (a) CH_3CH=CH-CH=$CHCH_3$ or CH_2=CH-CH=$CHCH_2CH_3$

(b) CH_2=C(CH_3)-CH=C(CH_3)-CH_3 (c) (1,3-cyclohexadiene)

2.16 Pairs (b), (c), and (d) are resonance structures because they differ only by the position of electrons. Pair (a) differ by the position of a hydrogen.

2.17 (a) (tetralin resonance structures) ⟷

(b) 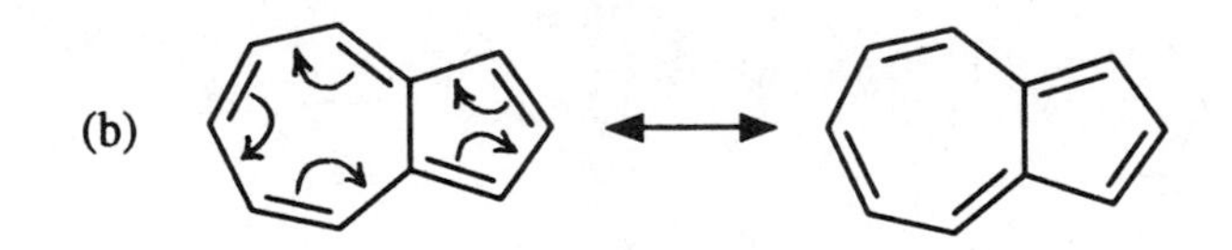

(c) cyclopropenyl cation $\longleftrightarrow$ cyclopropenyl cation $\longleftrightarrow$ cyclopropenyl cation (positive charge on each carbon in turn)

2.18 (a) $HC(=\ddot{O}:)-\ddot{N}H_2 \longleftrightarrow HC(-\ddot{O}:^-)=\overset{+}{N}H_2$

(b) $CH_3-C(=\ddot{O}:)-CH_3 \longleftrightarrow CH_3-\overset{+}{C}(-\ddot{O}:^-)-CH_3$

(c) $CH_2=CH-CH(=\ddot{O}:) \longleftrightarrow CH_2=CH-\overset{+}{C}H(-\ddot{O}:^-) \longleftrightarrow \overset{+}{C}H_2-CH=CH(-\ddot{O}:^-)$

(d) $CH_3-C(=\ddot{O}:)-\ddot{O}CH_3 \longleftrightarrow CH_3-C(-\ddot{O}:^-)=\overset{+}{O}CH_3$

(e) $CH_3-\ddot{N}=C=\ddot{O}: \longleftrightarrow CH_3-\ddot{N}=\overset{+}{C}-\ddot{O}:^- \longleftrightarrow CH_3-\overset{+}{N}\equiv C-\ddot{O}:^-$

(f) $:\overset{-}{C}H_2-\overset{+}{N}\equiv N: \longleftrightarrow :\overset{-}{C}H_2=\overset{2+}{N}=\ddot{N}:^- \longleftrightarrow CH_2=\overset{+}{N}=\ddot{N}:^-$

$\longleftrightarrow \overset{+}{C}H_2-\ddot{N}=\ddot{N}:^-$

(g) phenoxide $(C_6H_5-\ddot{O}:^-) \longleftrightarrow C_6H_5-\ddot{O}:^- \longleftrightarrow$ cyclohexadienone with $=\ddot{O}:$ and carbanion ($\ddot{}^-$) on ring

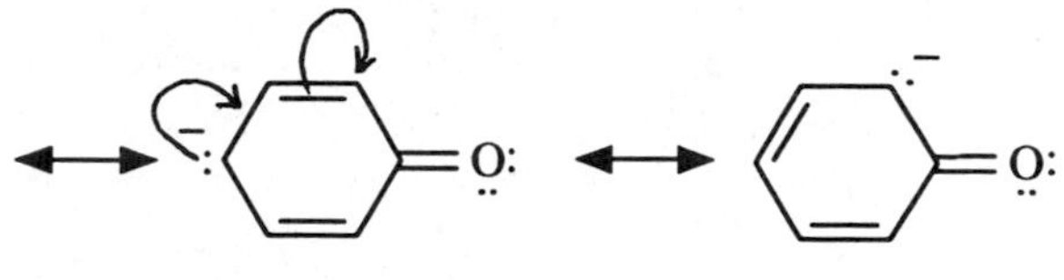

$\longleftrightarrow$ O: $\longleftrightarrow$ O:

2.19 (a) Ö• $\longleftrightarrow$ Ö• $\longleftrightarrow$ Ö:

$\longleftrightarrow$ Ö: $\longleftrightarrow$ Ö:

(b) $\overset{+}{C}H_2$ $\longleftrightarrow$ $\overset{+}{C}H_2$ $\longleftrightarrow$ CH_2

$\longleftrightarrow$ CH_2 $\longleftrightarrow$ CH_2

(c) $CH_3\overset{\overset{\ddot{O}:}{||}}{C}-\overset{-}{\ddot{C}}H-\overset{\overset{:\ddot{O}}{||}}{C}CH_3 \leftrightarrow CH_3\overset{\overset{:\ddot{O}:^-}{|}}{C}=CH-\overset{\overset{:\ddot{O}}{||}}{C}CH_3 \leftrightarrow CH_3\overset{\overset{\ddot{O}:}{||}}{C}-CH=\overset{\overset{:\ddot{O}:^-}{|}}{C}CH_3$

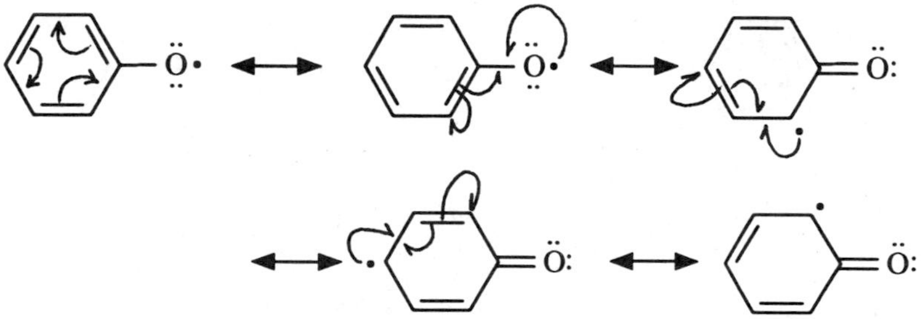

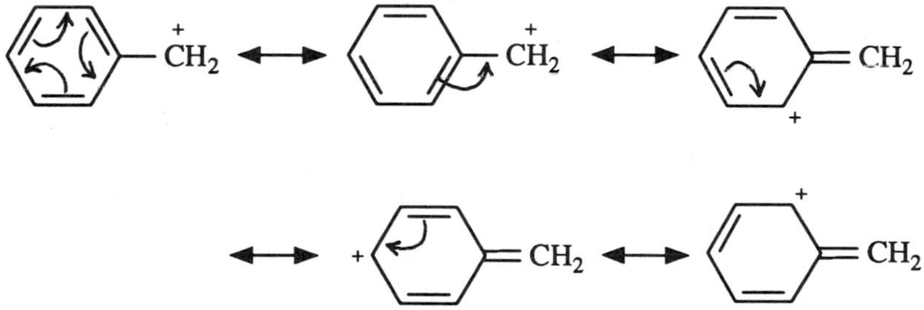

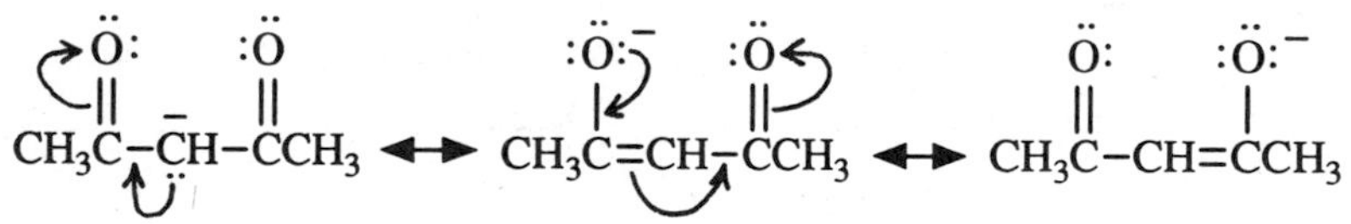

(d) H, Br $\longleftrightarrow$ H, Br $\longleftrightarrow$ H, Br

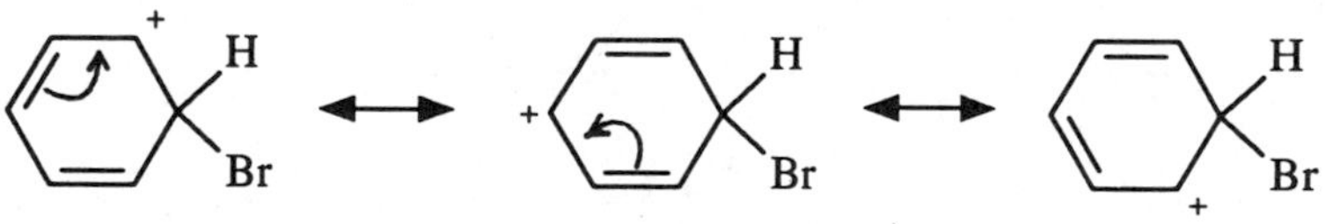

(e) $\overset{+}{C}H_2-CH=CH_2 \longleftrightarrow CH_2=CH-\overset{+}{C}H_2$

(f) $:N\equiv C-\overset{-}{\ddot{C}}H-C\equiv N: \leftrightarrow {}^{-}:\ddot{N}=C=CH-C\equiv N: \leftrightarrow :N\equiv C-CH=C=\ddot{N}:^{-}$

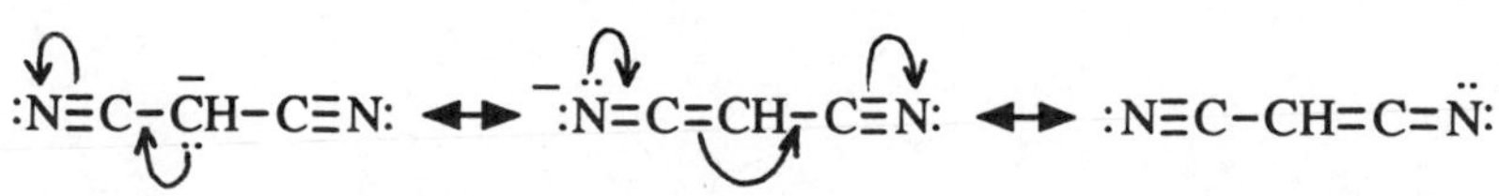

(g) 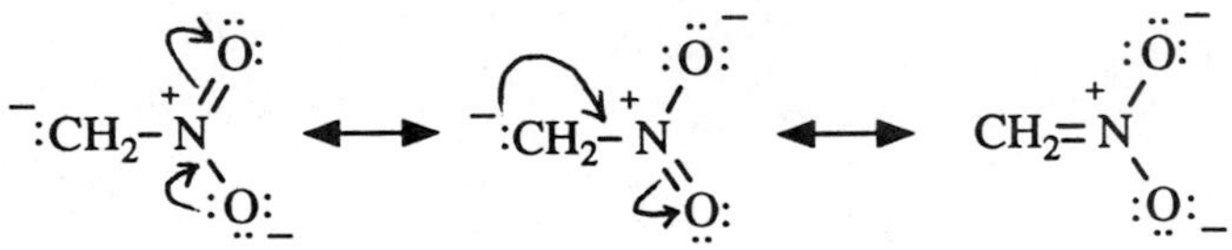 $^{-}:CH_2-\overset{+}{N}(=\ddot{O}:)(-\ddot{O}:^{-})$ ⟷ $^{-}:CH_2-\overset{+}{N}(-\ddot{O}:^{-})(=\ddot{O}:)$ ⟷ $CH_2=\overset{+}{N}(-\ddot{O}:^{-})(-\ddot{O}:^{-})$

(h) 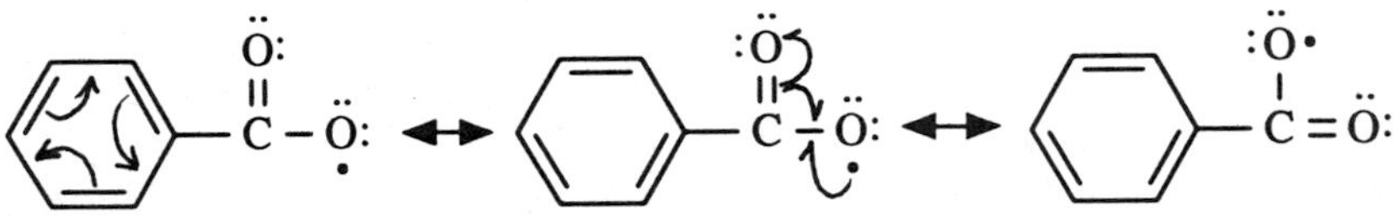 $C_6H_5-C(=\ddot{O}:)-\ddot{O}\cdot$ ⟷ $C_6H_5-C(=\ddot{O}:)-\ddot{O}\cdot$ ⟷ $C_6H_5-C(-\ddot{O}\cdot)=\ddot{O}:$

2.20 (a) 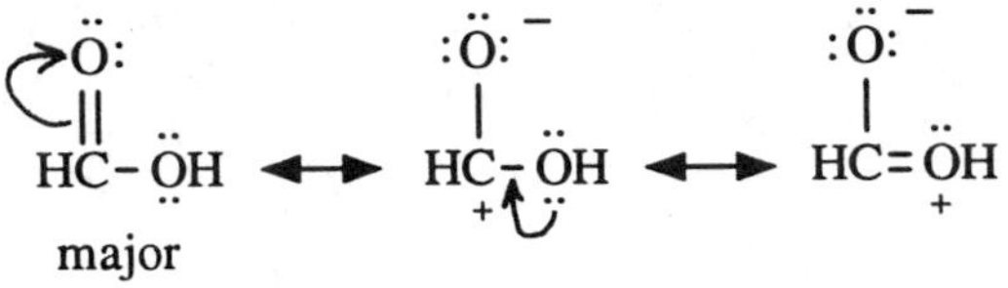 $HC(=\ddot{O}:)-\ddot{O}H$ ⟷ $H\overset{+}{C}(-\ddot{O}:^{-})-\ddot{O}H$ ⟷ $HC(-\ddot{O}:^{-})=\overset{+}{\ddot{O}}H$

major

(b) none

(c) 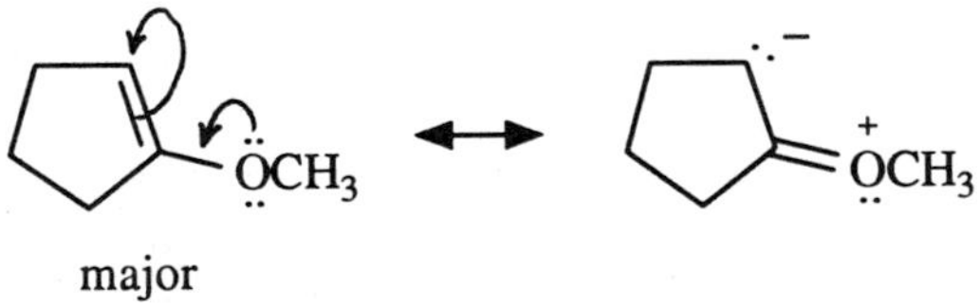 1-methoxycyclopentene ($\ddot{O}CH_3$) ⟷ cyclopentyl carbanion $=\overset{+}{O}CH_3$

major

2.21 (a) 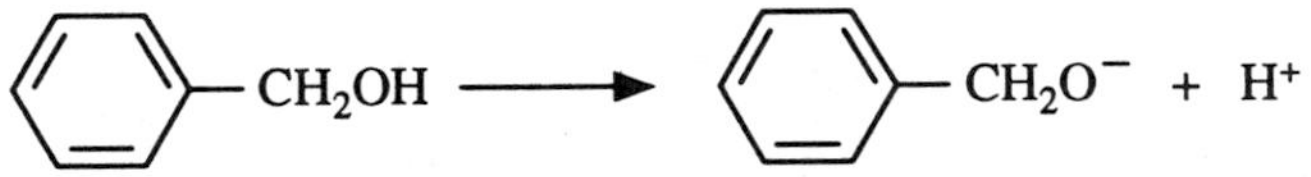 $C_6H_5CH_2OH$ ⟶ $C_6H_5CH_2O^-$ + H^+

(b) $C_6H_{11}COOH$ ⟶ $C_6H_{11}COO^-$ + H^+

(c) 1-naphthol (OH) ⟶ 1-naphthoxide (O^-) + H^+

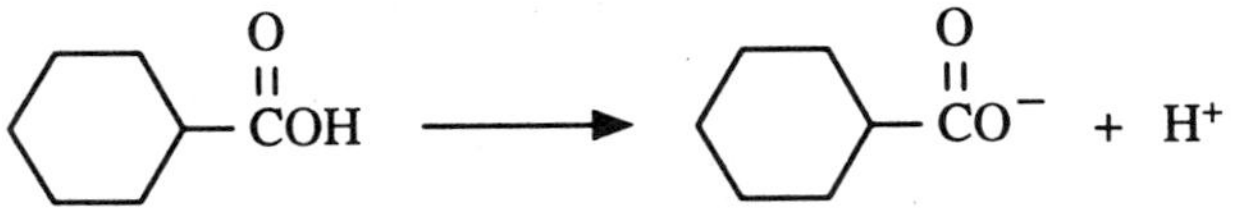

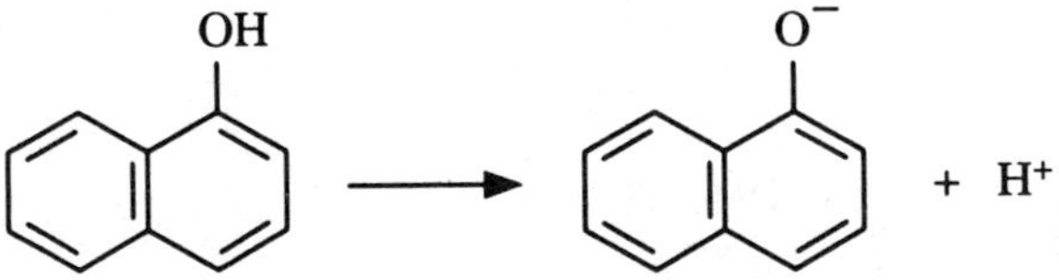

2.22 (a)  [structure: p-nitrophenol — OH on benzene ring, NO_2 para] because the negative charge of the conjugate base can be delocalized into the nitro group; i.e.

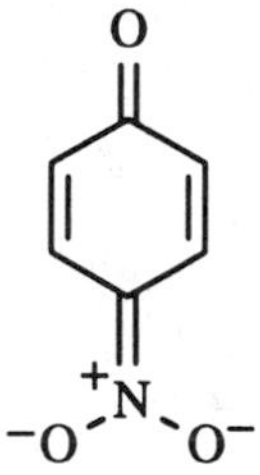

[structure: quinoid form, O= ring =$\overset{+}{N}(O^-)_2$]

(b) 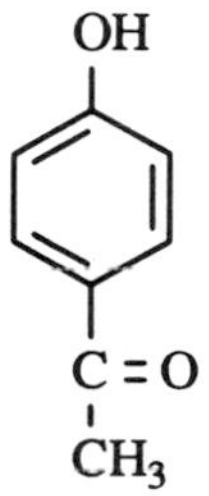 [structure: OH on benzene ring, para C=O with CH_3] because of the same reason as (a): [structure: O= ring =C(CH_3)(O^-)]

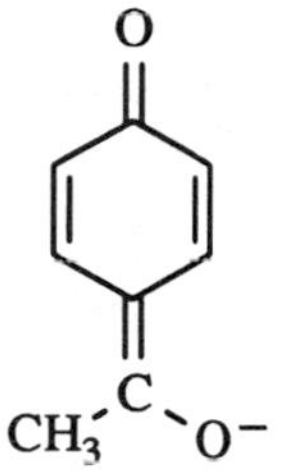

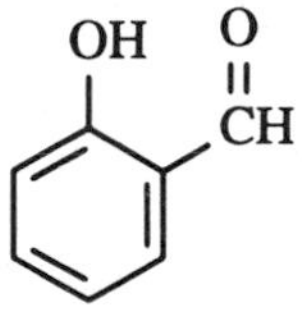

(c) [structure: OH on benzene ring, ortho CH=O] because of the same reason as (a): [structure: O= ring =CH(O^-)]

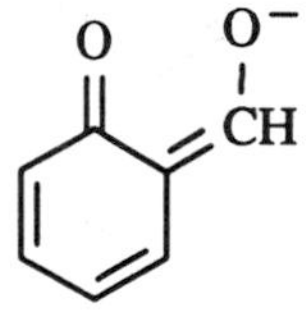

2.23 (a) The conjugate base is $O_2NCH_2CH_2O^-$; not resonance stabilized.

(b) The conjugate base is O_2N–[cyclohexane]–O^-; not resonance stabilized.

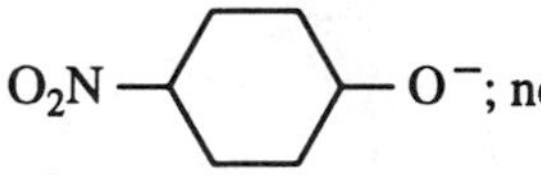

(c) The conjugate base is $O_2NCH_2\overset{O}{\overset{\|}{C}}O^-$; which is resonance stabilized.

$$O_2NCH_2\overset{O}{\overset{\|}{C}}O^- \longleftrightarrow O_2NCH_2\overset{O^-}{\overset{|}{C}}{=}O$$

2.24 a > b > c. The carboxylic acid is the strongest acid because the negative charge of its conjugate base can be delocalized onto two electronegative oxygens. The phenol is next in acidity because the negative charge of its conjugate base can be resonance stabilized. The alcohol is last in this list because the negative charge of its conjugate base cannot be delocalized.

2.25 (a)

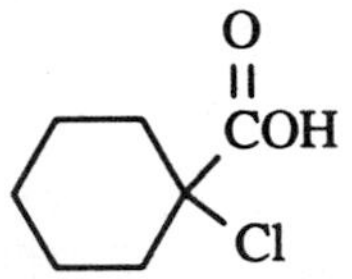

O
||
COH
Cl

(b)

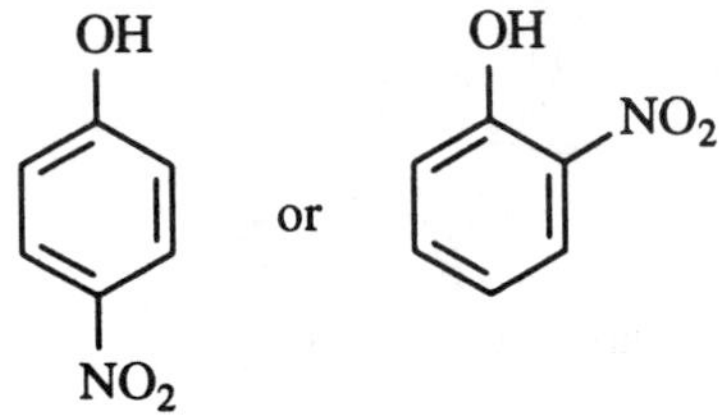

OH
NO$_2$

or

OH
NO$_2$

(c)

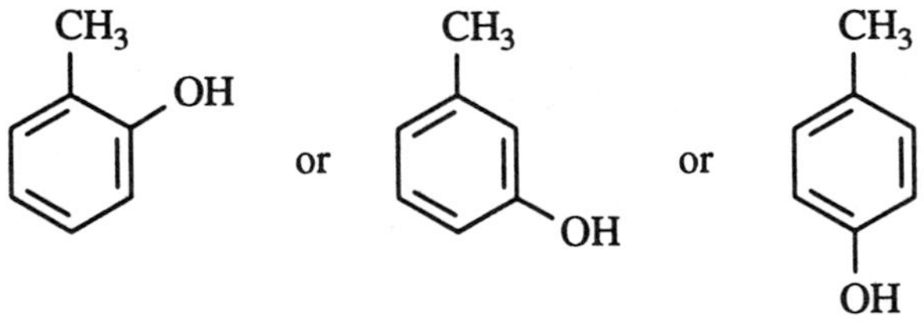

CH$_3$
OH

or

CH$_3$
OH

or

CH$_3$
OH

3

Structural Isomerism, Nomenclatu and Alkanes

Answers to Additional Problems

3.1 Compound (a) is saturated; compounds (b), (c), and (d) are unsaturated.

3.2 Compounds (b) and (c) are aliphatic. Compound (a) is aromatic.

3.3 (1) (a), (e), (g) (2) (b), (c), (d), (f), (h)

3.4 (a) $CH_2{=}C{=}CH_2$ (b) $CH_3OCH_2CH_3$ (c) $CH_3\overset{O}{\overset{\|}{C}}CH_2OH$

(d) $CH_3CH_2\overset{O}{\overset{\|}{C}}OH$ (e) $CH_3CH_2CH_2CH_2\overset{O}{\overset{\|}{C}}H$ (f) $C_6H_5OCH_2\overset{O}{\overset{\|}{C}}H$

In (f), the C_6H_5- group was not considered to be a functional group. If you consider it to be a functional group, then one correct answer would be:

$$HC{\equiv}CCH{=}CHCH{=}CHOCH_2\overset{O}{\overset{\|}{C}}H$$

3.5 (a) If the $C_{16}H_{10}$ compound were completely saturated, the compound would contain 2n+2, or 34, hydrogens. Therefore, there are (34-10), or 24, hydrogens missing. Since the insertion of each double bond (or ring) requires removing 2 hydrogens, the unsaturation number is 24/2, or 12. The other unsaturation numbers are:

(b) 2 (c) 4 (d) 3

3.6 $CH_3CH_2CH{=}CH_2$ *$CH_3CH{=}CHCH_3$ $(CH_3)_2C{=}CH_2$

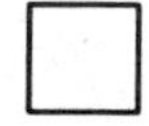

CH_3

The starred compound can exist as a pair of isomers as well:

$$\begin{matrix} H_3C & & CH_3 \\ & C=C & \\ H & & H \end{matrix} \quad \text{and} \quad \begin{matrix} H & & CH_3 \\ & C=C & \\ H_3C & & H \end{matrix}$$

These are not structural isomers, but geometric isomers, which will be discussed in Chapter 4.

3.7 $CH_3CH_2CH{=}CHCl$ $CH_3CH_2CCl{=}CH_2$ $CH_3CH(Cl)CH{=}CH_2$

$ClCH_2CH_2CH{=}CH_2$ $CH_3CH{=}CHCH_2Cl$ $CH_3CH{=}CClCH_3$

$ClCH_2C(CH_3){=}CH_2$ $(CH_3)_2C{=}CHCl$ (chlorocyclobutane: Cl on a cyclobutane ring)

(cyclopropane ring)–CH_2Cl (cyclopropane ring with Cl and CH_3 on the same carbon) (cyclopropane ring with CH_3 and Cl on adjacent carbons)

3.8 (1) (c) (2) (b) (3) (a)

3.9 (a) $CH_3\overset{+}{N}(O^-){=}O$ and $CH_2{=}\overset{+}{N}(O^-){-}OH$ or $CH_3{-}O{-}N{=}O$

(b) $CH_3CH_2CH_2Cl$ and $CH_3CH(Cl)CH_3$

3.10 $CH_3CH_2CH_2CH_2CH_2CH(CH_3)CH_3$ $\quad$ $CH_3CH_2CH_2CH_2CH(CH_3)CH_2CH_3$

$CH_3CH_2CH_2CH(CH_3)CH_2CH_2CH_3$ $\quad$ $CH_3CH_2CH_2CH(CH_2CH_3)CH_2CH_3$

3.11 (a)

ester

O O

O

alkoxyl (ether)

(b) $CH_3OCH_2CH_2CH(NH_2)COOH$

carboxyl

alkoxyl (ether) amino

(c)

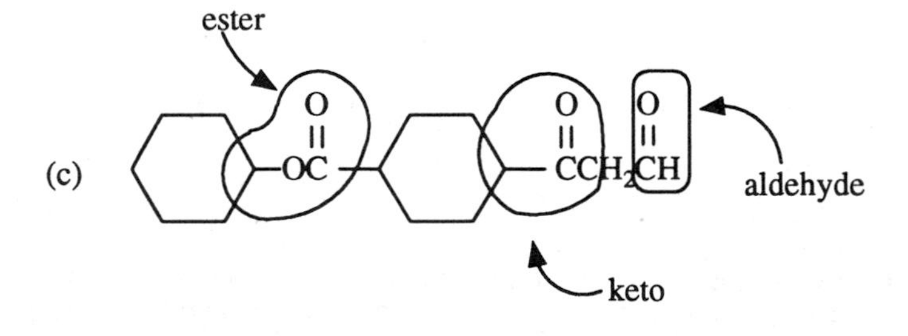

3.12 (a) Cyclobutane–COOH

(b) $CH_2{=}C(CH_3){-}CH_3$

(c) $HO{-}CH_2CH_2CH_2CH_2CH_2{-}Cl$

(d) $CH_3\overset{O}{\overset{||}{C}}CH_2CH_2CH_2\overset{O}{\overset{||}{C}}H$ (There are other correct answers.)

3.13 (a) $CH_3CH_2CH_2CH_2CH_3$; $CH_3CH_2CH(CH_3)_2$; $(CH_3)_4C$

(b) [cyclopentane] [methylcyclobutane, CH_3] [ethylcyclopropane, CH_2CH_3]

[1,2-dimethylcyclopropane, H_3C, H_3C] [1,1-dimethylcyclopropane, CH_3, CH_3]

(c) [bicyclo[2.1.0]pentane, CH_2] [bicyclo[1.1.1]pentane] [spiro[2.2]pentane]

[methylbicyclo[1.1.0]butane, CH_3] [1-methylbicyclo[1.1.0]butane, CH_3]

3.14 Molecular formula (c) represents only one structure. At least two structures, not necessarily representing stable compounds, can be drawn for the other formulas. A few of the possible examples for each molecular formula follow:

(a) $CH_3CH_2N{=}O$, $CH_3ON{=}CH_2$, $H_2NCH_2\overset{O}{\overset{||}{C}}H$, [ring: O, NH]

(b) $H_2NCH_2CH_2OH$, CH_3CH_2NHOH, CH_3ONHCH_3

3.15 (a) $CH_3CH{=}CHNH_2$ $CH_2{=}CHCH_2NH_2$ $CH_3CH_2CH{=}NH$

$CH_2{=}CHNHCH_3$ $CH_3CH{=}NCH_3$ $CH_2{=}NCH_2CH_3$

$CH_3C(NH_2){=}CH_2$ $CH_3C(CH_3){=}NH$

(b) $CH_3CH_2CH_2CH_2CH_3$ $CH_3CH_2CH(CH_3)_2$ $(CH_3)_4C$

3.16 (a) 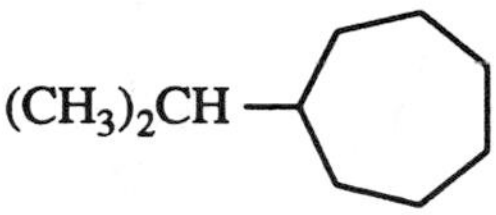 $(CH_3)_2CH$–cycloheptyl

isopropylcycloheptane

(b) 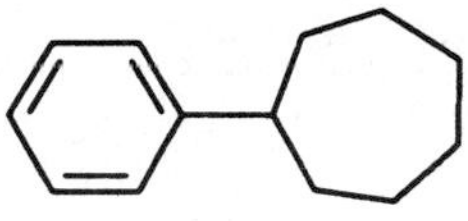 phenyl–cycloheptyl

phenylcycloheptane

(c) 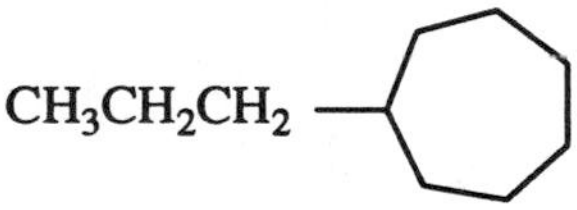 $CH_3CH_2CH_2$–cycloheptyl

propylcycloheptane

(d) 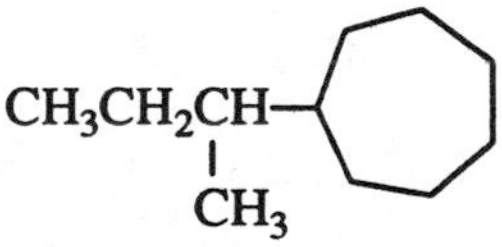 $CH_3CH_2CH(CH_3)$–cycloheptyl

sec-butylcycloheptane

(e) 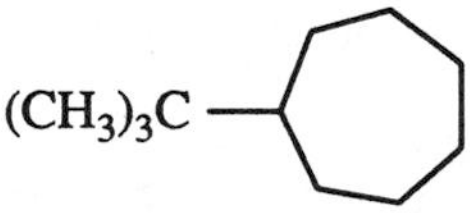 $(CH_3)_3C$–cycloheptyl

tert-butylcycloheptane

(f) 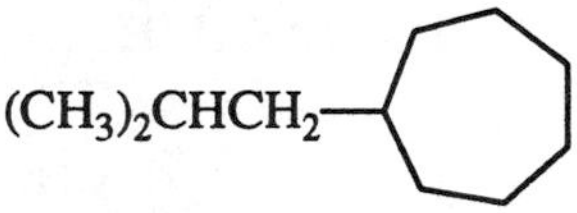 $(CH_3)_2CHCH_2$–cycloheptyl

isobutylcycloheptane

(g) 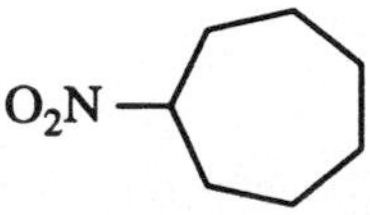 O_2N–cycloheptyl

nitrocycloheptane

(h) 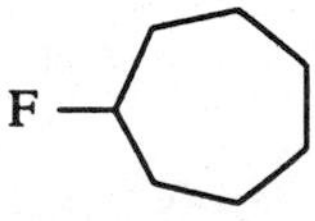 F–cycloheptyl

fluorocycloheptane

3.17 (a) $(CH_3)_2CHCH_2CH_2Cl$, 1-chloro-3-methylbutane

$(CH_3)_2CHCHCH_3$, 2-chloro-3-methylbutane
$\quad\quad\quad\quad\quad |$
$\quad\quad\quad\quad\quad Cl$

$(CH_3)_2CCH_2CH_3$, 2-chloro-2-methylbutane
$\quad\quad\quad\quad |$
$\quad\quad\quad\quad Cl$

$ClCH_2CHCH_2CH_3$, 1-chloro-2-methylbutane
$\quad\quad\quad\quad |$
$\quad\quad\quad\quad CH_3$

(b) $(CH_3)_3CCHCH_2Cl$, 1-chloro-2,3,3-trimethylbutane
$\quad\quad\quad\quad\quad |$
$\quad\quad\quad\quad\quad CH_3$

$(CH_3)_3CC(CH_3)_2$, 2-chloro-2,3,3-trimethylbutane
$\quad\quad\quad\quad\quad |$
$\quad\quad\quad\quad\quad Cl$

$\quad\quad\quad\quad CH_3$
$\quad\quad\quad\quad |$
$ClCH_2CCH(CH_3)_2$, 1-chloro-2,2,3-trimethylbutane
$\quad\quad\quad\quad |$
$\quad\quad\quad\quad CH_3$

3.18 (a) $(CH_3CH_2)_2CHCH_2CH_3$

(b) $(CH_3)_2CHCH(CH_3)_2$

(c) 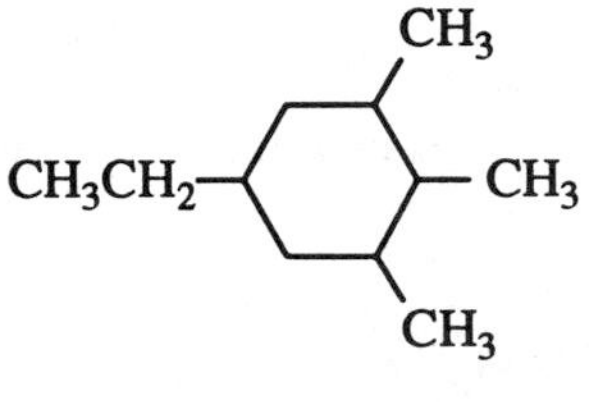 CH_3CH_2–cyclohexane ring bearing CH_3, CH_3, CH_3

(d) 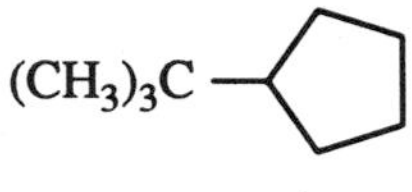 $(CH_3)_3C$–cyclopentane ring

(e) $CH_3CH_2CH_2C(CH_3)_2CH_2C(CH_3)_3$

(f) $[(CH_3)_3C]_2CHCH_3$

3.19 (a) 2-*sec*-butyl-1,3-dimethylcyclohexane

(b) 9-methyl-1,3,7-decatriene

(c) 6-*tert*-butyl-3,3-dimethyl-1,4-cyclohexadiene

(d) 1,3-dicyclopentylcyclopentane

3.20 (a) 2-iodo-3-pentanol

(b) 5-chloro-3-methyl-1-hepten-6-yne

(c) 3,3-diisopropyl-2,4-dimethylpentane

(d) 2,2,3,4,4-pentamethylpentane

(e) 2,2-dimethylpropanoic acid

(f) 2-cyclopentylpropanoic acid

(g) 3,4-dimethylheptane

3.21 (a) $CH_3CH_2CH(CH_3)CH_2CH(CH_3)CH_2OH$

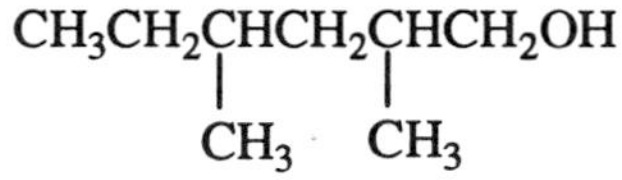

(b) benzene ring bearing three ethyl groups: CH_3CH_2, CH_2CH_3, CH_3CH_2 (1,3,5-positions)

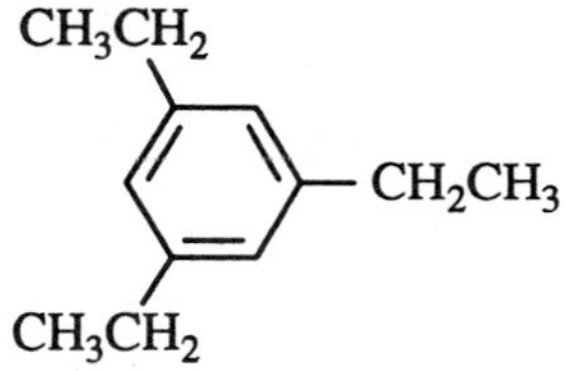

(c) $C_6H_5CH_2CH_2OH$, where C_6H_5 = phenyl

(d) $CH_3CH_2CH_2CCl_2-CH(=O)$

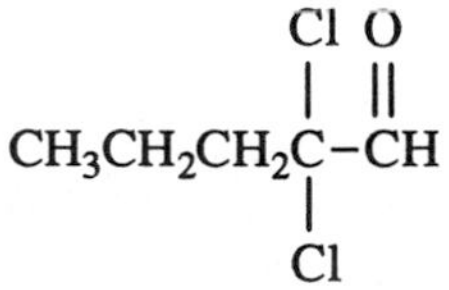

(e) $(CH_3)_2CH$– cyclopentane ring =O

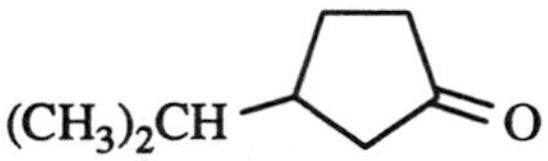

(f) cyclooctene ring – $CH(CH_3)CH_2CH_3$

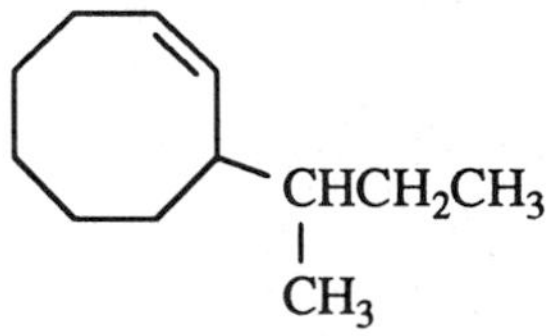

(g) $(CH_3)_3C$– benzene ring –$C(CH_3)_3$

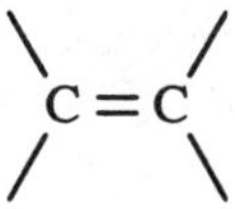

3.24 (a) $>C=C<$ (b) $>C=C<$ (c) $-\overset{O}{\overset{\|}{C}}OH$

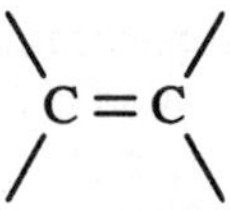

(d) $-\overset{O}{\overset{\|}{C}}H$ (e) $-\overset{O}{\overset{\|}{C}}H$ (f) $-\overset{O}{\overset{\|}{C}}H$

3.23 (a) CH_3CHCl_2, 1,1-dichloroethane $ClCH_2CH_2Cl$, 1,2-dichloroethane

(b) $CH_2{=}CHCH{=}CH_2$, 1,3-butadiene $CH_2{=}C{=}CHCH_3$, 1,2-butadiene

(c) $CH_3CH_2CH_2CH_2CH_2Cl$, 1-chloropentane

$CH_3CH_2CH_2CH(Cl)CH_3$, 2-chloropentane

$CH_3CH_2CH(Cl)CH_2CH_3$, 3-chloropentane

$CH_3CH_2CH(CH_3)CH_2Cl$, 1-chloro-2-methylbutane

$CH_3CH_2C(Cl)(CH_3)_2$, 2-chloro-2-methylbutane

$CH_3CH(Cl)CH(CH_3)_2$, 2-chloro-3-methylbutane

$ClCH_2CHCH(CH_3)_2$, 1-chloro-3-methylbutane

$ClCH_2C(CH_3)_3$, 1-chloro-2,2-dimethylpropane

(d) $CH_3CH{=}CHCHO$, 2-butenal $CH_2{=}CHCH_2CHO$, 3-butenal

$CH_2{=}C(CH_3)CHO$, 2-methylpropenal

(cyclopropyl)—CHO, cyclopropanecarbaldehyde (The nomenclature for this type of compound has not been discussed.)

3.24 (a) 2,2-dimethylpropanal

(b) hexafluoropropanone (numbers not necessary)

(c) 2-nitrobutane

(d) 2-chloro-3-methylbutanal

(e) propylbenzene

(f) 4-*tert*-butylcyclohexanol

3.25 Catalytic cracking is the process where large alkane molecules are broken into lower formula-weight alkanes and alkenes. Isomerization also occurs when catalytic cracking is carried out.

$$CH_3(CH_2)_6CH_3 \longrightarrow CH_2{=}CH_2 + CH_3\underset{\displaystyle CH_3}{\underset{|}{C}}HCH_3$$

In catalytic reforming, alkanes are concerted to aromatic compounds

$$CH_3(CH_2)_4CH_3 \longrightarrow C_6H_6 \text{ (benzene)} + 4\,H_2$$

3.26 (a) The octane rating would be 90.

(b) No. Market gasolines are ranked in antiknock characteristics against a test mixture that contains heptane-isooctane. The gasoline itself does not contain the mixture of heptane and isooctane.

4

Stereochemistry

Answers to Additional Problems

4.1 (a) and (b). Geometric isomers, or *cis-trans* isomers, are isomers (different compounds with the same molecular formula) that differ by groups being on the same side (*cis*) or opposite sides (*trans*) of a double bond or ring.

cis: $ClHC{=}CHCl$ (Cl and Cl on same side)

also (*Z*)

trans: $ClHC{=}CHCl$ (Cl and Cl on opposite sides)

also (*E*)

(c) Structural isomers are isomers that differ in structure or sequence of atoms, including the carbons skeleton and the position of a functional group.

$CH_3C(CH_3)_2CH_3$ and $CH_3CH_2CH_2CH_3$ $\quad$ $CH_3CH_2CH_2OH$ and $CH_3CH(OH)CH_3$

(d) Stereoisomers are isomers that differ only in the arrangement of their atoms in space, including diastereomers and enantiomers.

cis and *trans* $CH_3CH{=}CHCH_3$

$CH_3CH_2C(CH_3)(H)(OH)$ and $H(HO)C(CH_3)CH_2CH_3$

(e) Enantiomers are a pair of stereoisomers that are nonsuperposable mirror images of each other, such as the two butanols in (d).

(f) Diastereomers are two or more stereoisomers that are not enantiomers.

cis and *trans* $CH_3CH{=}CHCH_3$

```
       CO2H                      CO2H
  HO ───┼─── H              H ───┼─── OH
        │           and          │
   H ───┼─── OH             H ───┼─── OH
      CH2OH                     CH2OH
```

4.2 only (b) and (d).

```
(b)    H        CH2Cl          ClCH2       CH2Cl
        \      /                    \     /
         C = C                       C = C
        /      \                    /     \
   ClCH2        H                  H       H

          trans                       cis

(d)   Cl     H                 Cl        CH2CH2CH2Cl
        \   /                    \      /
         C = C                    C = C
        /     \                  /     \
       H       CH2CH2CH2Cl      H       H

          trans                       cis
```

4.3 Compounds (a) and (b) are *cis*. Compound (c) is *trans*.

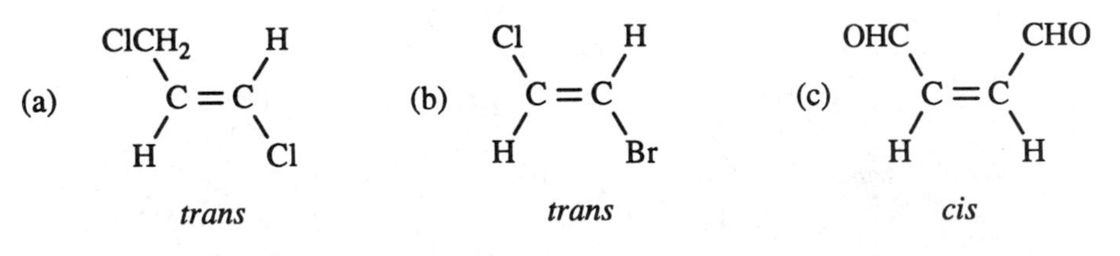

4.4 (a) -SH (b) $-CH_2CH_2CH_3$ (c) $-\overset{\overset{\text{O}}{\|}}{\text{C}}-Cl$ (d) $-CH(CH_3)_2$

4.5 (a)

$HOCH_2$ CH_3
C=C
H CH_2CH_2OH

(*E*) -3-methyl-2-pentene-1,5-diol

$HOCH_2$ CH_2CH_2OH
C=C
H CH_3

(*Z*) -3-methyl-2-pentene-1,5-diol

(b)

H CO_2H
C=C
H_3C H

(*E*) -2-butenoic acid (*trans*)

H_3C CO_2H
C=C
H H

(*Z*) -2-butenoic acid (*cis*)

(c)

H_3C CHO
C=C
H CH_2Cl

(*E*) -2-chloromethyl-2-butenal

H CHO
C=C
H_3C CH_2Cl

(*Z*) -2-chloromethyl-2-butenal

4.6 (a)

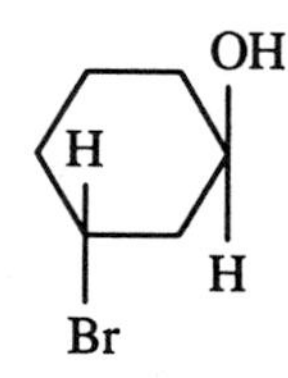

trans

H
H
OH
Br

cis

(b)

OH
H H O
CH_2OH

trans

H
H HO O
CH_2OH

cis

(c)

CH_3
CH_3
H H
CH_3
OH

trans

H
CH_3
HH_3C
CH_3
OH

cis

(d)

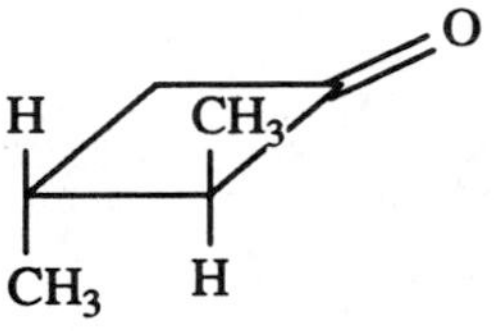

H CH$_3$ O

CH$_3$ H

trans

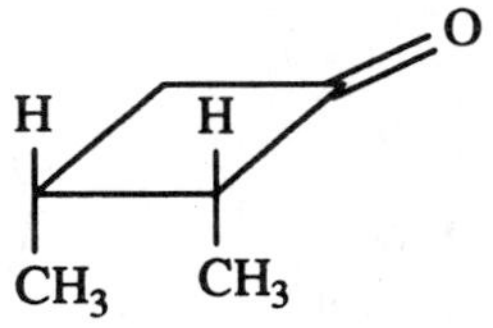

H H O

CH$_3$ CH$_3$

cis

4.7 (a)

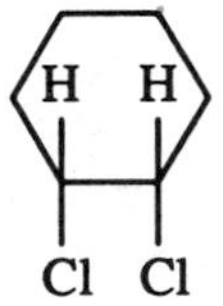

H H

Cl Cl

(b)

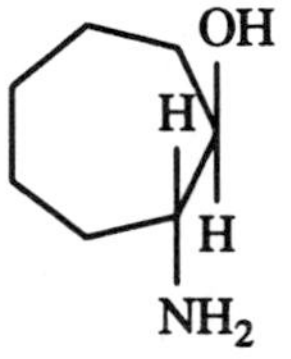

OH

H

H

NH$_2$

(c)

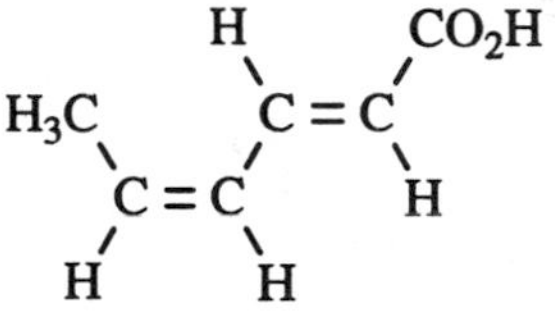

H CO$_2$H

H$_3$C C=C

C=C H

H H

(d)

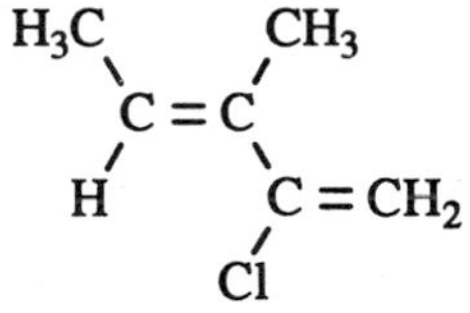

H$_3$C CH$_3$

C=C

H C=CH$_2$

Cl

(e)

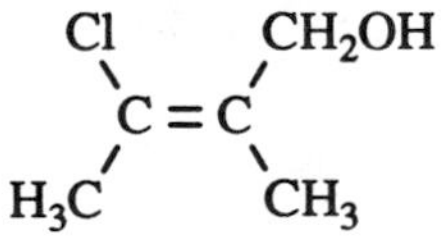

Cl CH$_2$OH

C=C

H$_3$C CH$_3$

(f)

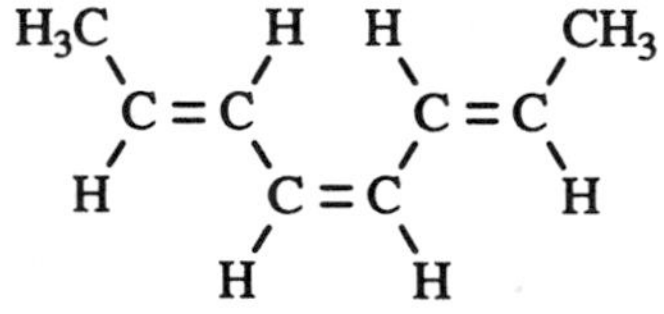

H$_3$C H H CH$_3$

C=C C=C

H C=C H

H H

4.8 (a)

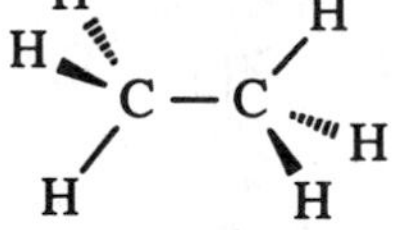

H H

H C–C H

H H

or

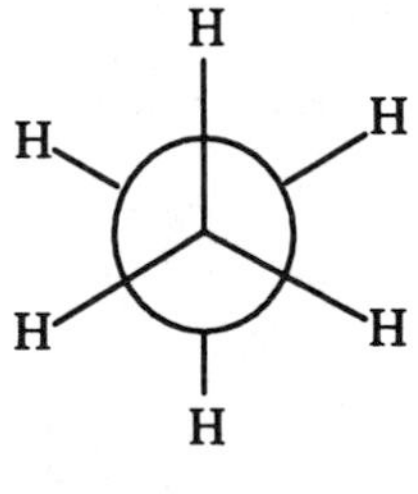

H

H H

H H

H

(b)

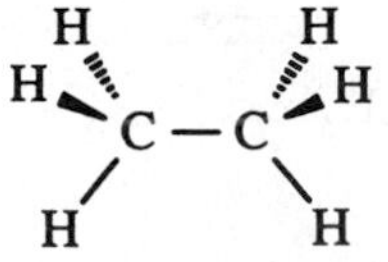

H H

H C–C H

H H

or

H H

H

H H

H

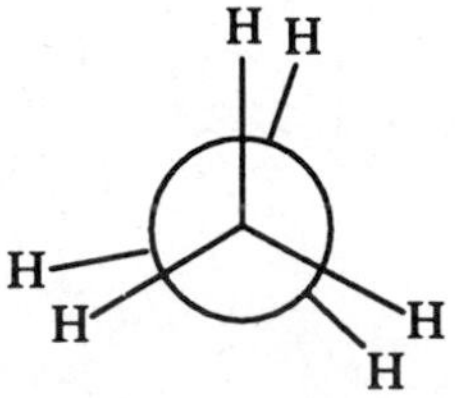

(c)

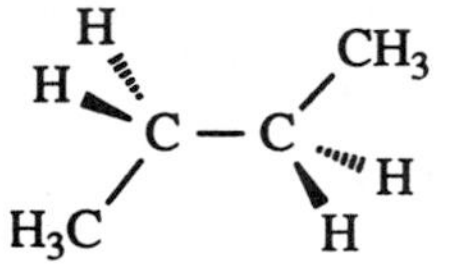

H H CH_3
C – C H
H_3C H

or

CH_3
H H
H H
CH_3

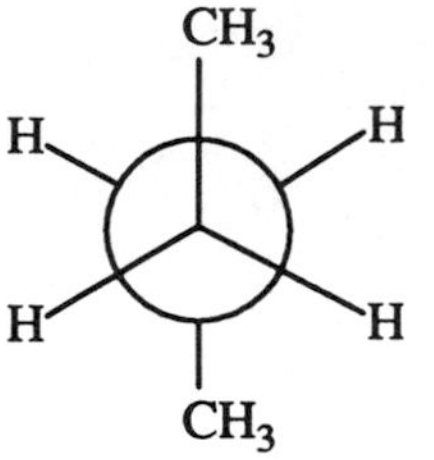

(d)

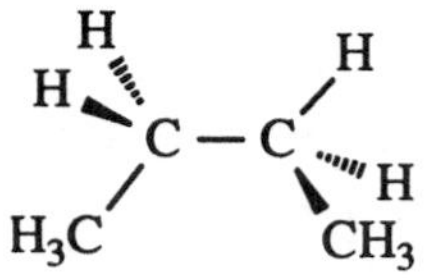

H H H
C – C H
H_3C CH_3

or

CH_3
H CH_3
H H
H

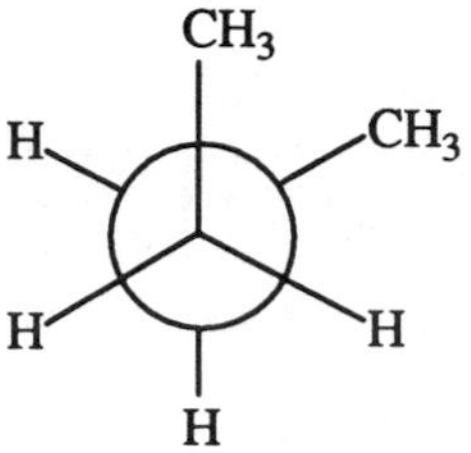

(e)

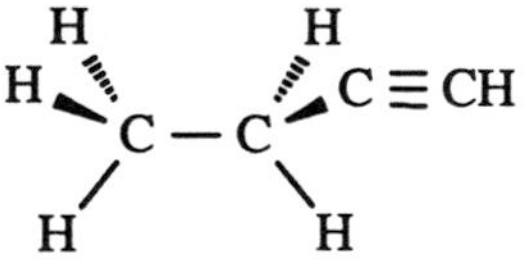

H H H
C – C C ≡ CH
H H

or

H H
H
H C≡CH
H

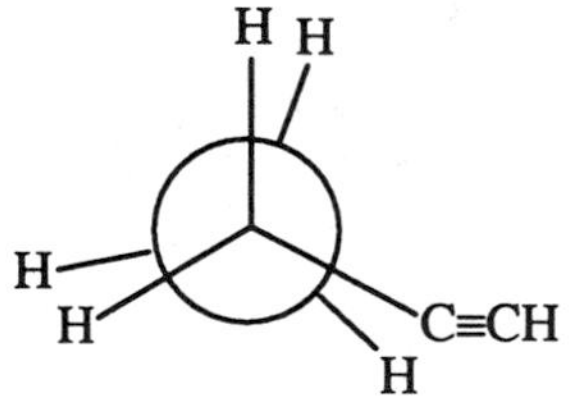

4.9 (a)

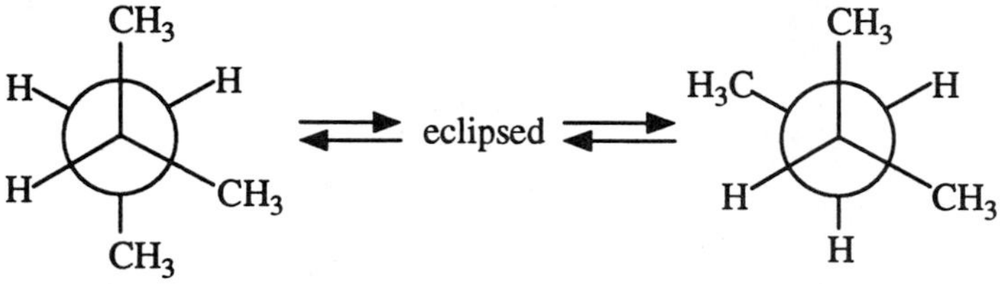

CH_3
H H
H CH_3
CH_3

anti/gauche

eclipsed

CH_3
H_3C H
H CH_3
H

anti/gauche

eclipsed

CH_3
H CH_3
H CH_3
H

eclipsed

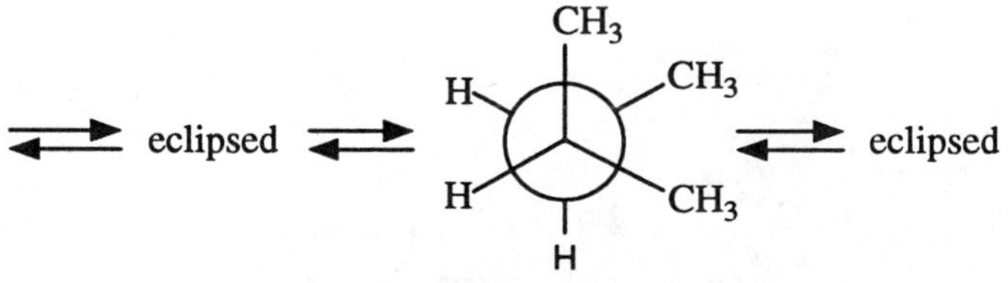

gauche/gauche

(Note crowding.)

(b) ⇄ eclipsed ⇄

⇄ eclipsed ⇄ ⇄ eclipsed

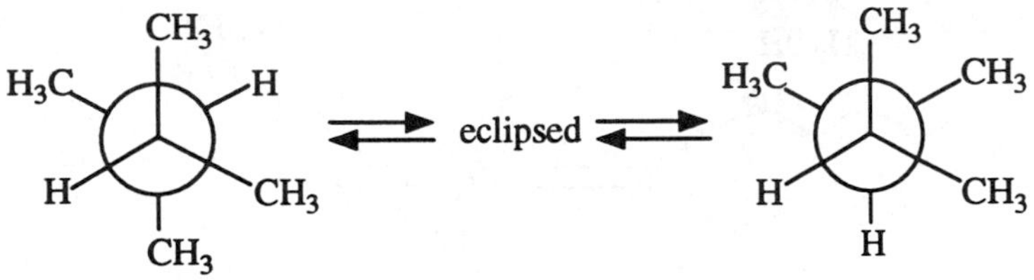

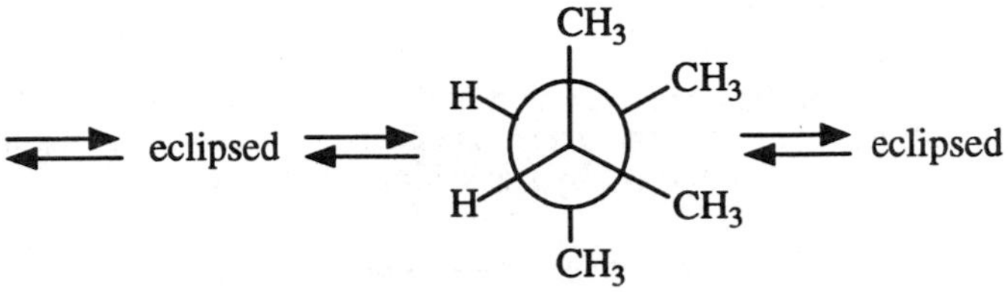

(Note the crowding of methyl groups in the second and third conformations shown).

4.10 (a) ⇄

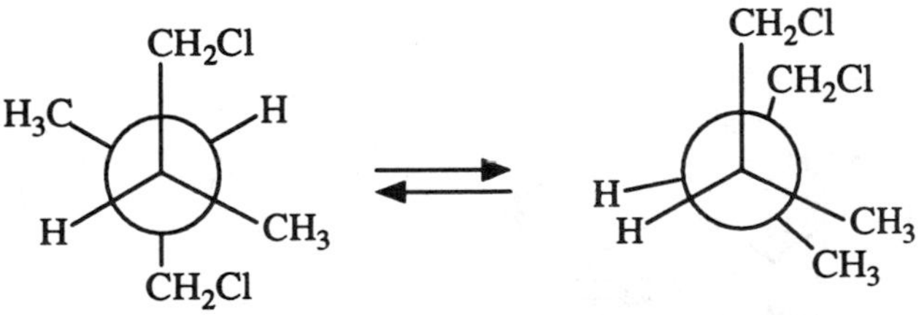

most stable
meso conformation

least stable

⇄

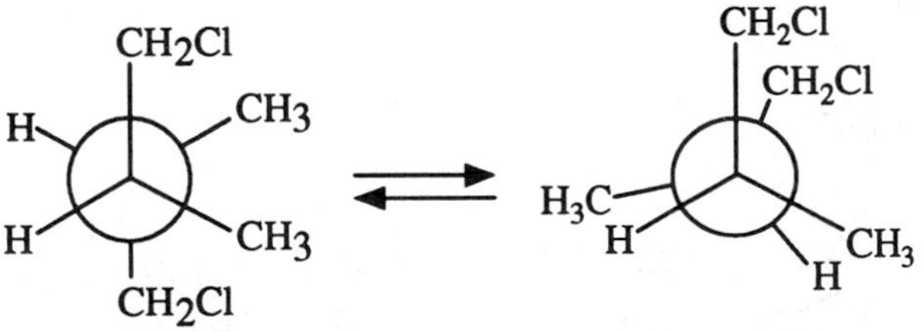

most stable
(2*R*,3*R*) conformation

least stable

In (a), you might have drawn the *meso*, the (2*R*,3*R*), or the (2*S*,3*S*) isomer. In each case, the eclipsed chloromethyl conformation is the least stable and the *anti* chloromethyl isomer is the most stable.

(b)

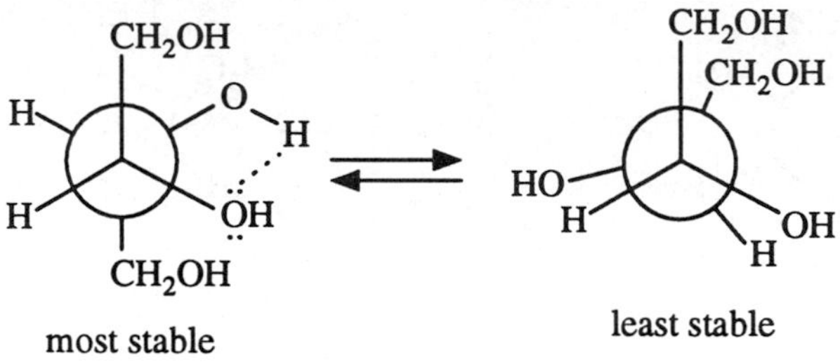
CH2OH
H
O
H
H
OH
CH2OH
most stable
CH2OH
CH2OH
HO
H
OH
H
least stable

In this stereoisomer, hydrogen bonding can help stabilize the most stable, *anti* configuration. In the *meso* isomer, the stabilization gained by hydrogen bonding might be offset by steric hindrance.

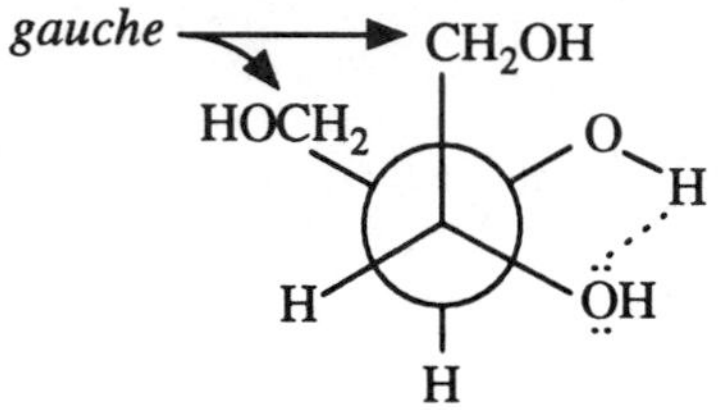
gauche
CH2OH
HOCH2
O
H
H
OH
H

(c)

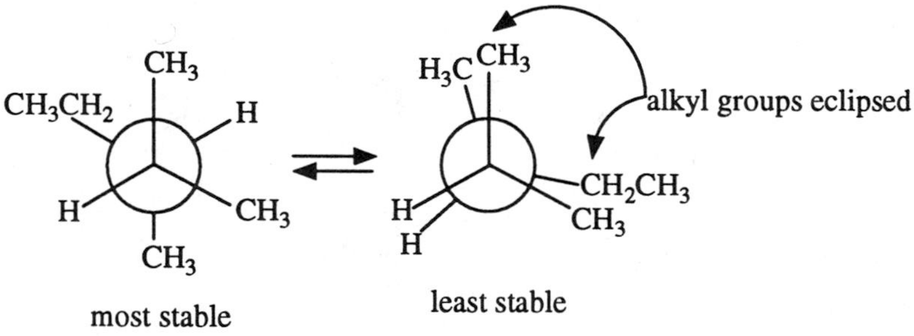
CH3
CH3CH2
H
H
CH3
CH3
most stable
H3C
CH3
alkyl groups eclipsed
CH2CH3
CH3
H
H
least stable

(d)

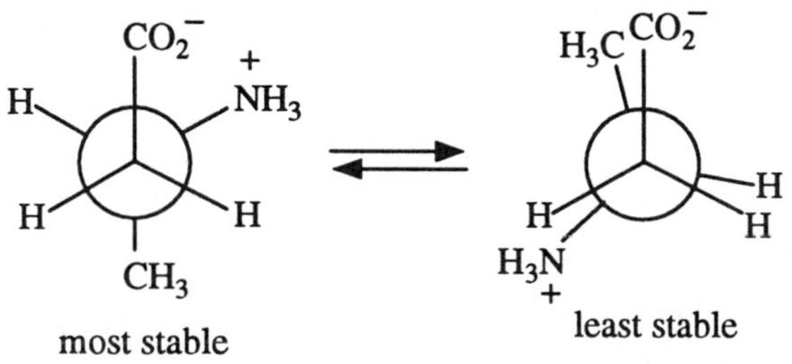
CO2
NH3
H
H
H
CH3
most stable
H3C
CO2
H
H
H
H3N
least stable

4.11 (a)

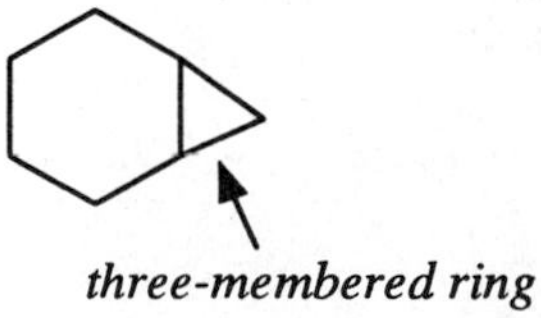
three-membered ring

(c)

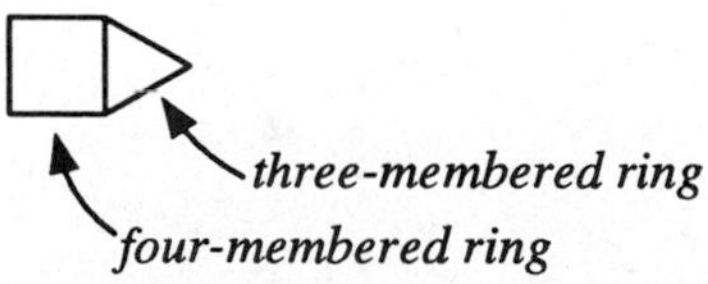
three-membered ring
four-membered ring

(d) 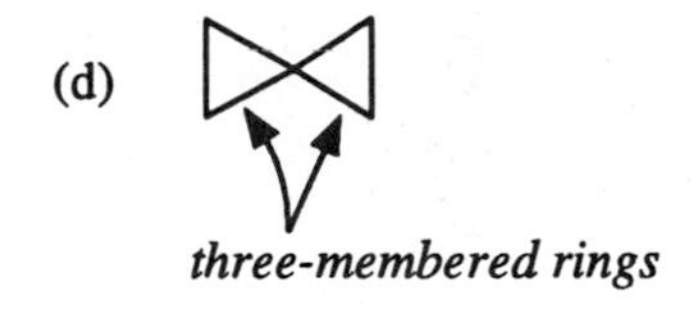

4.12 (a) (b)

H_3C $CH(CH_3)_2$
H
H H
OH

OH
H
H H
CH_3
$CH(CH_3)_2$

4.13 1,3-*cis*:

H H
H (H) (H) Cl (H) (H)
Cl H H H

1,3-*trans*:

H H
Cl (H) (H) Cl (H) (H)
H (H) H H

1,4-*cis*:

H Cl
H (H) (H) (H) Cl (H)
(H) H H H

1,4-*trans*:

H H
H (H) (H) Cl Cl (H)
(H) H H H

Equatorial H atoms are circled; axial H atoms are not.

4.14 (a) *e* (b) *a* (c) *a* on right-hand ring, *e* on left-hand ring

(d) *a* (e) *e*

4.15 (a) H_3C $CH(CH_3)_2$
H H

(b) $(CH_3)_2CH$ CH_3
H
CH_3

4.16 (a) H, Cl, H, CH_2CH_3 ⇌ CH_3CH_2, H, Cl, H ⇌ H, CH_3CH_2, Cl, H

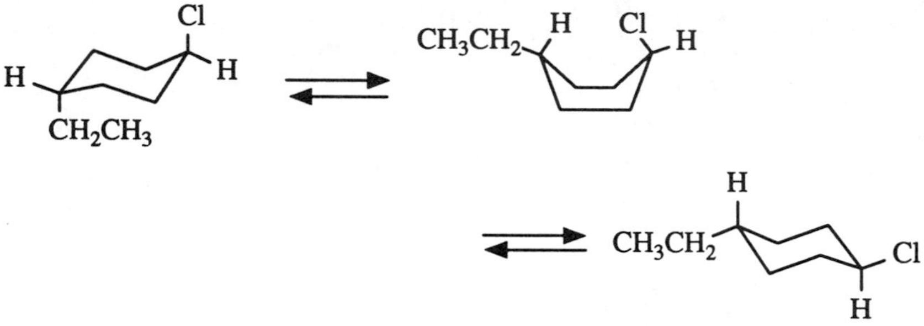

(b) CO_2H, H, H, OH ⇌ HO_2C, H, H, OH ⇌ H, CO_2H, HO, H

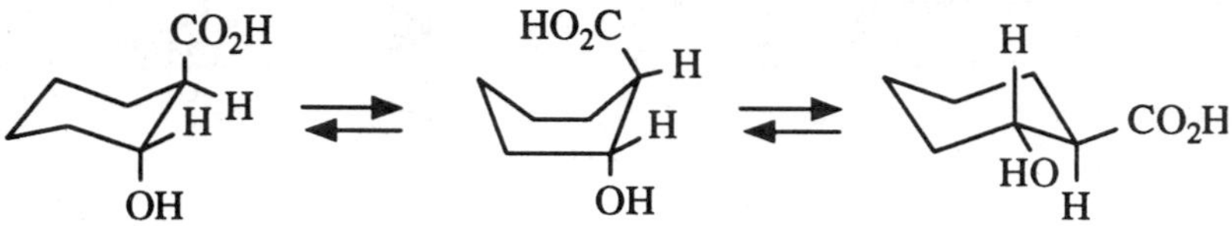

(c) H, NO_2, H, H_2N ⇌ O_2N, H, H, NH_2 ⇌ H, NO_2, NH_2, H

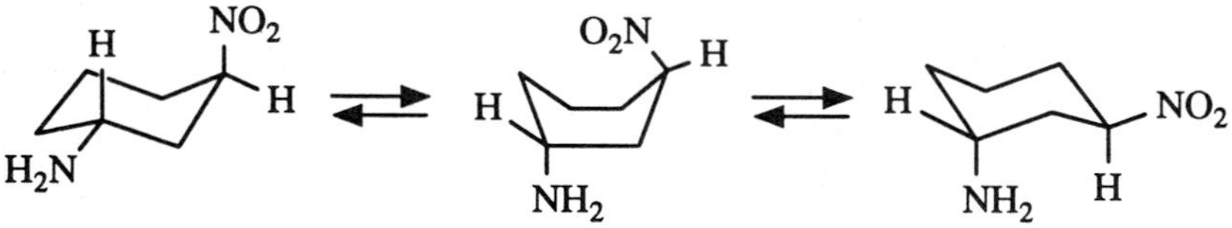

(d) NH_2, NO_2, H, H ⇌ O_2N, H, H_2N, H ⇌ H_2N, NO_2, H, H

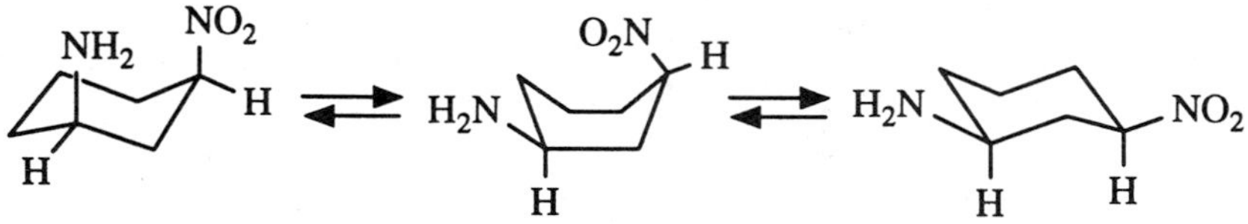

4.17 The best procedure for solving this type of problem is to draw all possible isomers and then to determine which ones are chiral. The answers given below show only those structures that have one or more chiral carbons. The achiral isomers are not drawn.

(a) $CH_3CH_2CH_2\overset{*}{C}HCH_2CH_3$ (with CH_3 on the starred carbon)

$CH_3CH_2\overset{*}{C}HCH(CH_3)_2$ (with CH_3 on the starred carbon)

(b) $CH_3CH_2\overset{*}{C}HCH_2Br$ (with Br on the starred carbon)

$CH_3\overset{*}{C}HCH_2CH_2Br$ (with Br on the starred carbon)

$CH_3\overset{*}{C}H\text{-}\overset{*}{C}HCH_3$ (with Br on each starred carbon)

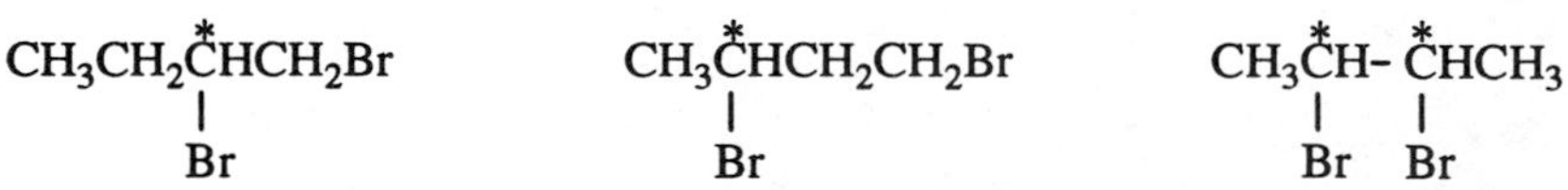

(c) $HOCH_2\overset{*}{C}HCHO$ (with OH on the starred carbon) $\qquad CH_3\overset{*}{C}HCO_2H$ (with OH on the starred carbon) $\qquad$ cyclopropane ring bearing HO, OH on one carbon and H, OH (starred carbon) on another

4.18 (a) $CH_3\overset{*}{C}HCHCH_2CH(CH_3)_2$, with OH on C-2 and Cl on C-3 (both starred) $\qquad$ (b) $CH_2{=}CH\overset{*}{C}HCH_2CH{=}CH_2$, with OH on the starred carbon

(c) Redraw in an open-chain form $Cl\text{-}\overset{*}{C}H\text{-}CH\text{-}CH_2CH_3$, with CH_3 on the first CH (starred) and CH_3 on the second CH (starred)

(d) achiral

4.19 The structures (a), (b), (c), and (f) represent the same compound. The structures (d) and (e) represent enantiomers. One way to solve a problem like this one is to determine the (*R*) or (*S*) configurations.

4.20 (a) Fischer projections: top CO_2H, H left, CH_3 right, bottom CH_2CH_3; and top CO_2H, H_3C left, H right, bottom CH_2CH_3

(b) Fischer projections: top CO_2H, H left, NH_2 right, bottom CH_3; and top CO_2H, H_2N left, H right, bottom CH_3

(c) Fischer projections: top CH_2OH, H left, OH right, bottom CH_2OCH_3; and top CH_2OH, HO left, H right, bottom CH_2OCH_3

(d)

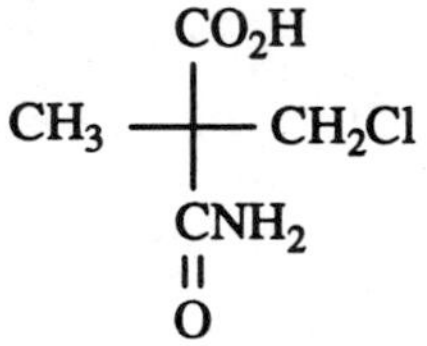

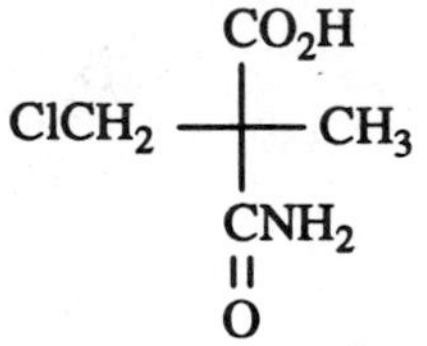

$$\begin{array}{c} CO_2H \\ CH_3 - \!\!\!+\!\!\! - CH_2Cl \\ \overset{\|}{C}NH_2 \\ O \end{array} \qquad \begin{array}{c} CO_2H \\ ClCH_2 - \!\!\!+\!\!\! - CH_3 \\ \overset{\|}{C}NH_2 \\ O \end{array}$$

4.21 (a) 2 chiral carbons, $n = 2$; $2^2 = 4$ (b) 4

(c) $n = 0$, $2^0 = 1$ (the one shown) (d) 8

4.22 $[\alpha]_D = \dfrac{\alpha}{l \times c}$ where α = observed rotation
c = concentration in g/mL
l = length of the tube in dm
(where 1.0 dm = 10 cm)

(a) $\alpha = +2.5°$
$c = 1.3$ g/25 mL = 0.052 g/mL
$l = 2.0$ dm

$$[\alpha]_D^{25} = \frac{+2.5°}{(0.052)\ (2.0)} = +24.0°$$

(b) $\alpha = -1.60°$; $c = 0.080$ g/mL; $l = 0.50$ dm; $[\alpha]_D^{20} = -40.0°$

4.23

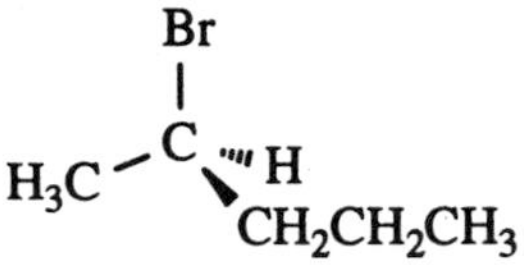

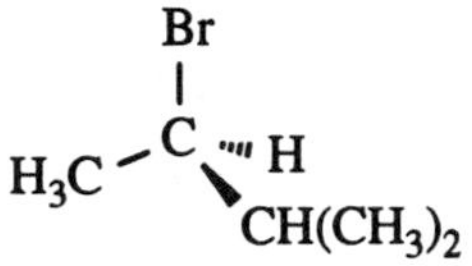

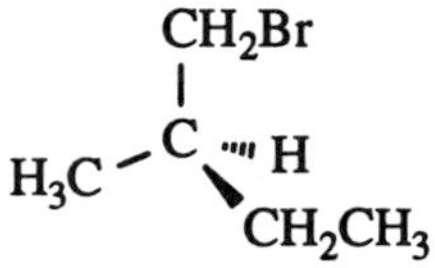

Br, H_3C, H, $CH_2CH_2CH_3$ on C; Br, H_3C, H, $CH(CH_3)_2$ on C; CH_2Br, H_3C, H, CH_2CH_3 on C

4.24 (a) 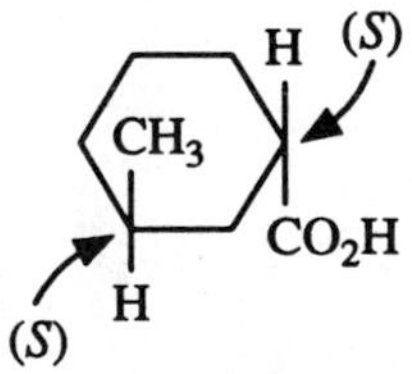 ring with CH_3, H (S), CO_2H, H (S) (b) (S)

(c) (*R*)

(d) 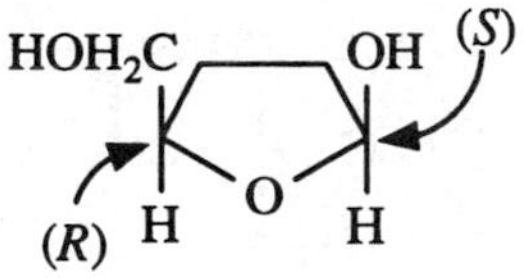

4.25 (a) 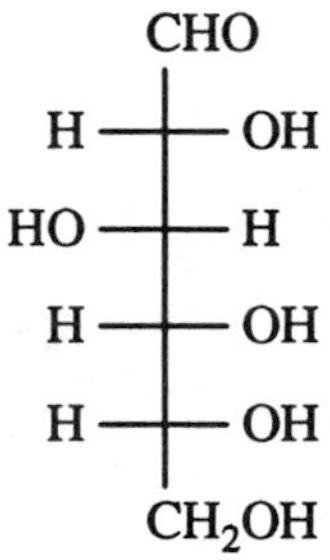

(b) 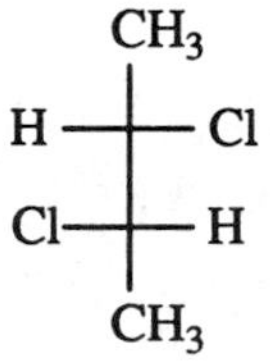

(c) 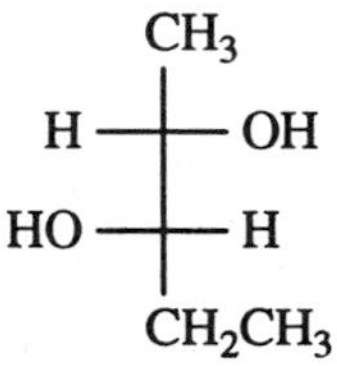

(d) 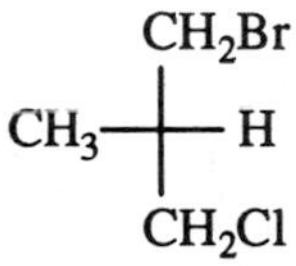

4.26 (a) 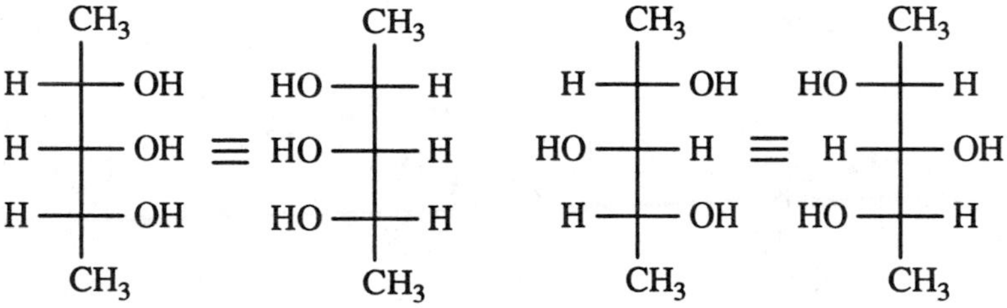

(b) 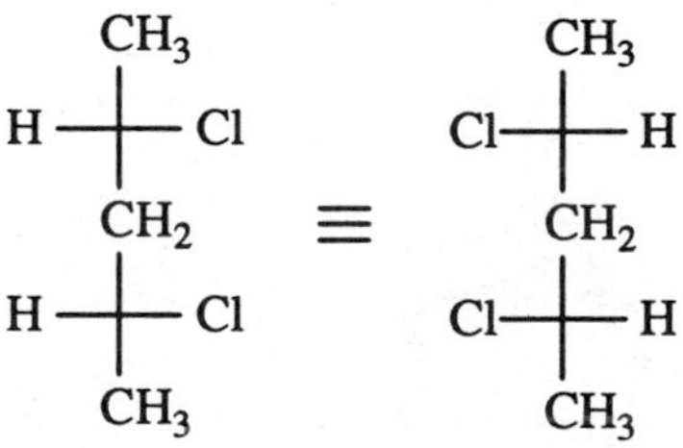

(c)

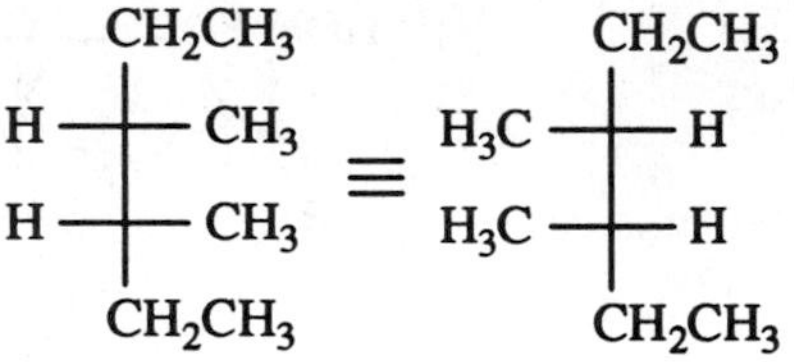

$$\begin{array}{c} CH_2CH_3 \\ H - C - CH_3 \\ H - C - CH_3 \\ CH_2CH_3 \end{array} \equiv \begin{array}{c} CH_2CH_3 \\ H_3C - C - H \\ H_3C - C - H \\ CH_2CH_3 \end{array}$$

(d)

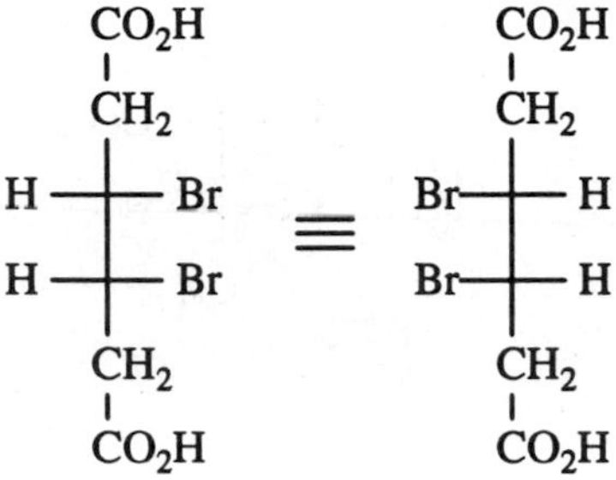

$$\begin{array}{c} CO_2H \\ CH_2 \\ H - C - Br \\ H - C - Br \\ CH_2 \\ CO_2H \end{array} \equiv \begin{array}{c} CO_2H \\ CH_2 \\ Br - C - H \\ Br - C - H \\ CH_2 \\ CO_2H \end{array}$$

4.27 (a)

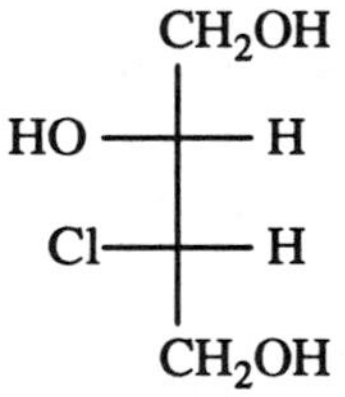

$$\begin{array}{c} CH_2OH \\ HO - C - H \\ Cl - C - H \\ CH_2OH \end{array}$$

enantiomer

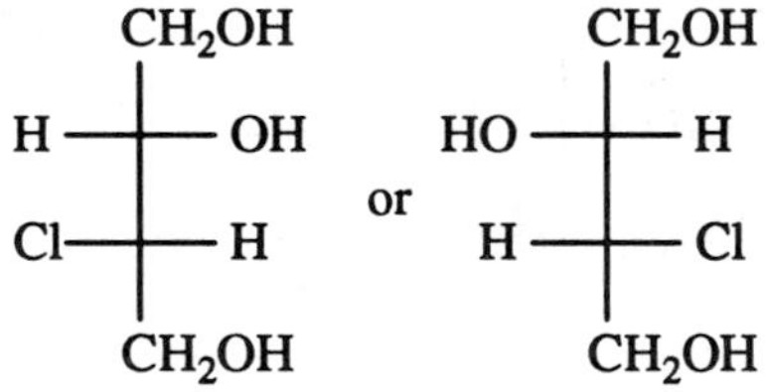

$$\begin{array}{c} CH_2OH \\ H - C - OH \\ Cl - C - H \\ CH_2OH \end{array} \quad \text{or} \quad \begin{array}{c} CH_2OH \\ HO - C - H \\ H - C - Cl \\ CH_2OH \end{array}$$

diastereomer

(b)

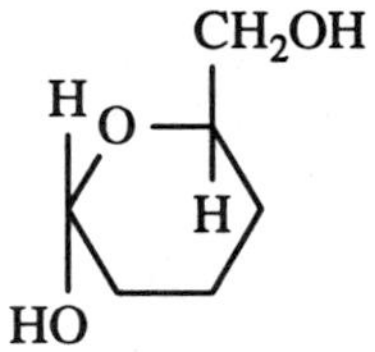

CH_2OH, H, O, H, HO

enantiomer

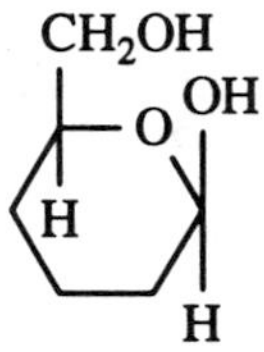

CH_2OH, OH, O, H, H

diastereomer

(c)

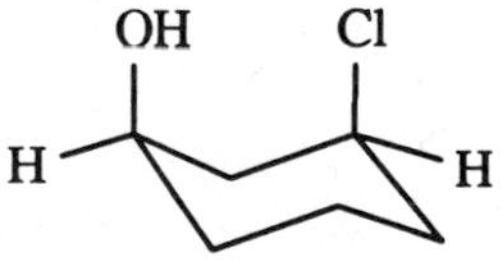

OH, Cl, H, H

enantiomer

Cl, H, H, OH

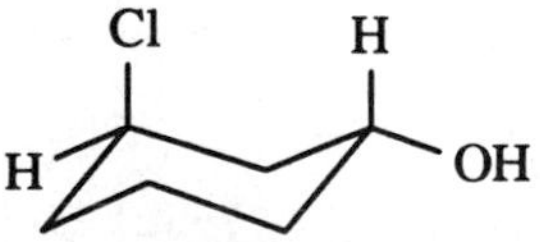

diastereomer

(d) 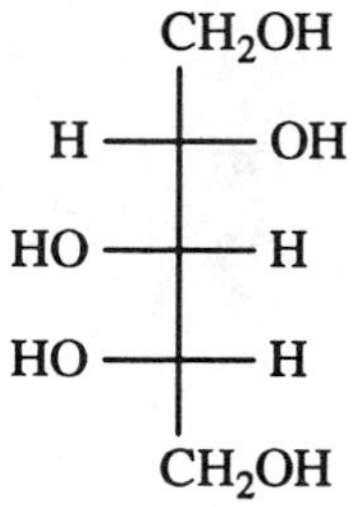

enantiomer

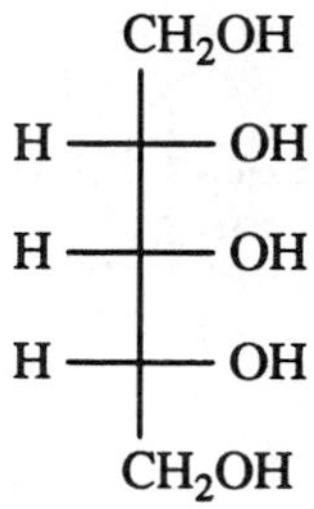 and 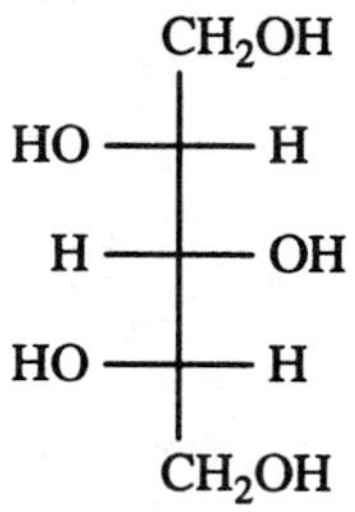

are diastereomers.
There are others.

4.28 (a) (1*S*,2*S*) (b) (1*S*,3*R*,4*R*) (c) (1*R*,2*S*)

4.29 (a) 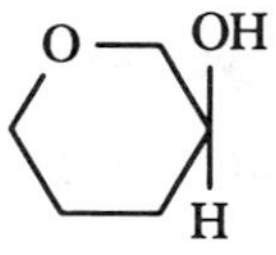 and 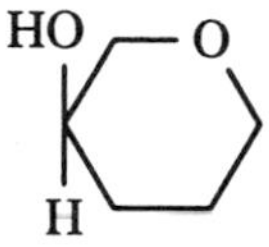 or 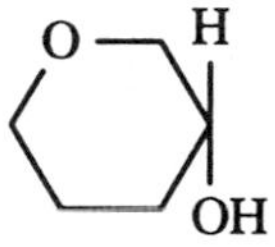

(b) 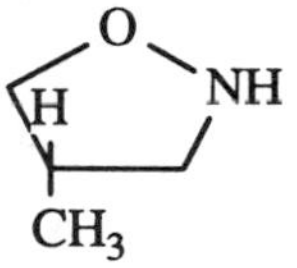 and 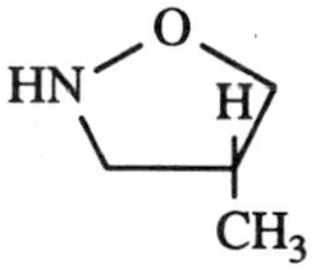 or 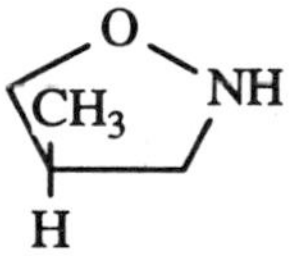

(c) 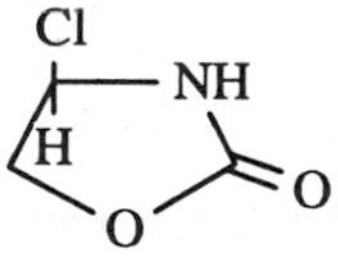 and 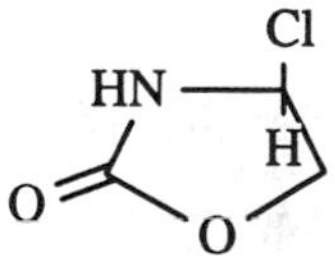 or 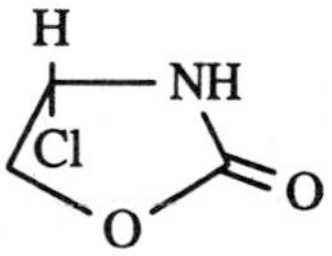

4.30 (a) 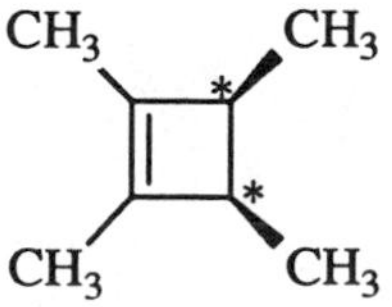 (b) and (c) none (d) 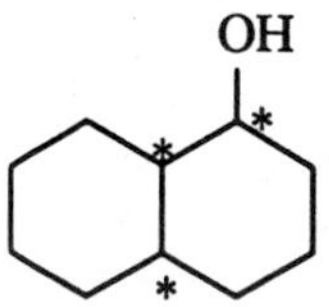

5

Alkyl Halides; Substitution and Elimination Reactions

Answers to Additional Problems

5.1 (a) $CH_2{=}CHCH_2Cl$ (b) $CH_2{=}CHBr$ (c) $C_6H_5CH_2Br$

(d) $CH_3CH_2CH_2CH_2I$ (e) $BrCH_2CH(CH_3)_2$ (f) cyclohexyl–Br

5.2 (a) *tert*-butyl bromide (b) isopropyl chloride

(c) methylene chloride (d) *sec*-butyl iodide

(e) benzyl chloride (f) allyl bromide

5.3 (a) $CH_3CH_2CH_2F$, 1-fluoropropane $CH_3CH(F)CH_3$, 2-fluoropropane

(b) $CH_3CH_2CH_2CH_2Br$, 1-bromobutane

$(CH_3)_2CHCH_2Br$, 1-bromo-2-methylpropane

$CH_3CH_2\overset{*}{C}H(Br)CH_3$, 2-bromobutane

$(CH_3)_3CBr$, 2-bromo-2-methylpropane

5.4 (a) Cl, H_3C / C=C \ H, C(H)(Br)(H_3C)

(b) CO_2H / H–C–Br / CH_2CH_3 or CO_2H / Br–C(H)–CH_2CH_3

(c) Cl, H, C=C, H, H, C=C, H, CH_3

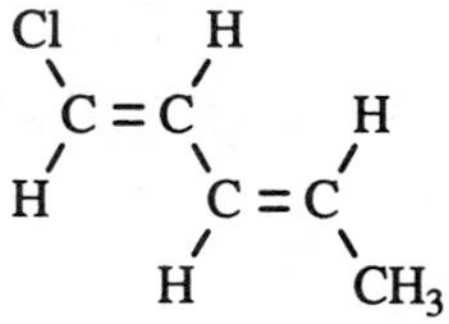

(d) CH_3; Br—H; H—Br; Cl—H; CH_2CH_3 or H_3C, C–C, Br, H, H, CH_2CH_3, C, Cl, H, Br

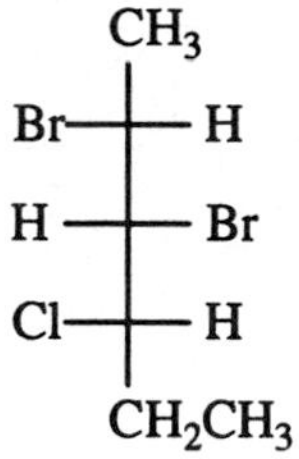

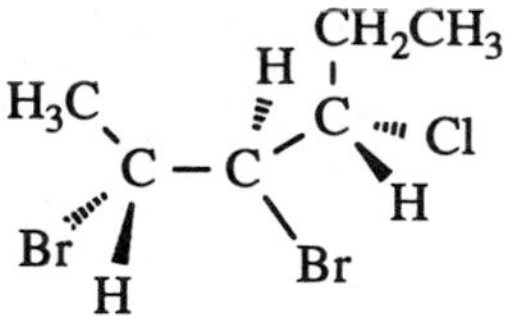

5.5 (a) Br (aryl); $CH_2$$CH_2Cl$ (alkyl)

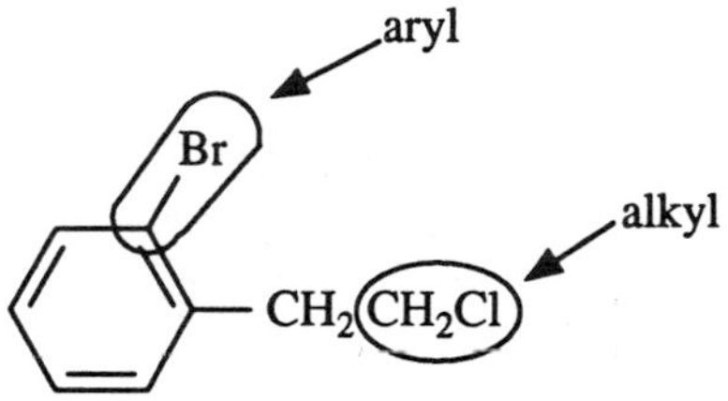

(b) aryl → Cl; alkyl → CH_2Cl; CH; aryl → Cl; Cl ← vinylic

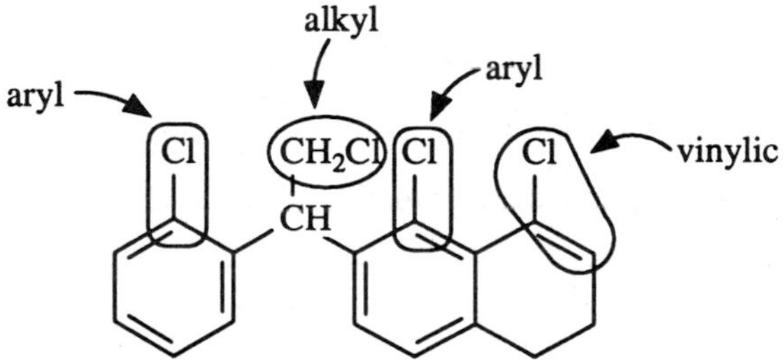

(c) vinylic: BrCH=CH CH=CHBr

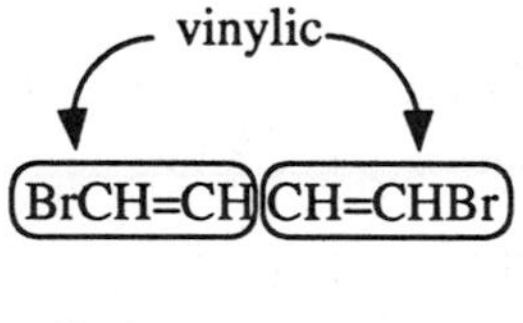

(d) alkyl (allylic) → Cl-CH_2; vinylic → Cl on C; C=CH–; aryl → Cl

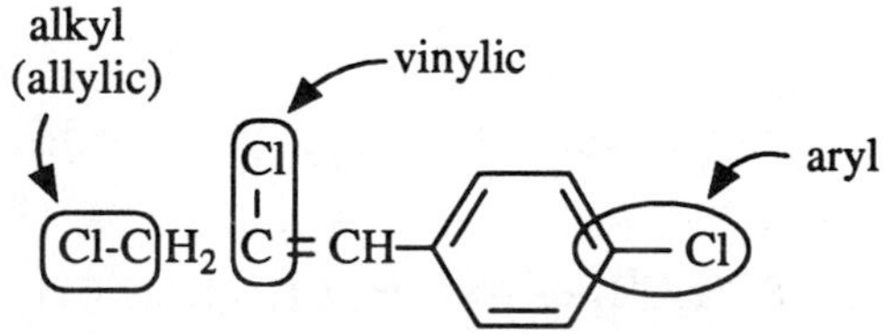

5.6 (a) 3° (b) 3° (c) 1°

5.7 (a)

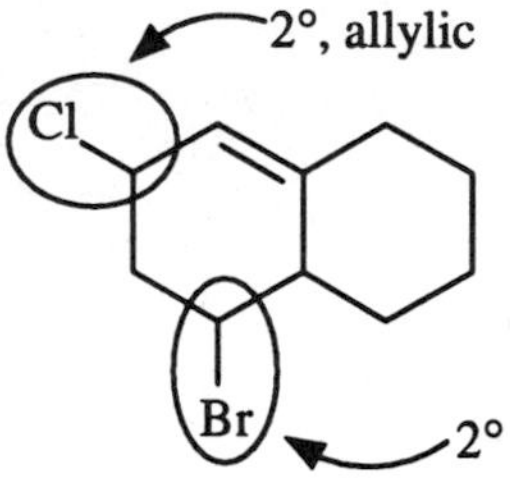

(b)

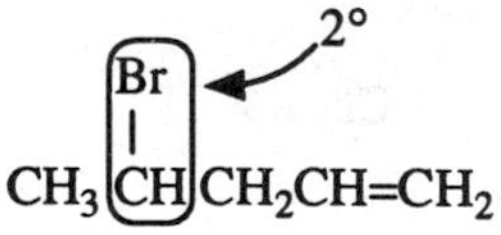

(c)

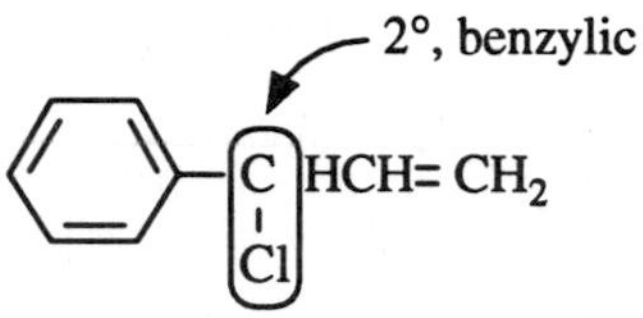

5.8 (a) The methoxide ion is a base in this reaction because it abstracts a proton from water.

(b) a nucleophile, because it displaces an iodide ion in a substitution reaction.

(c) a base, because it removes a proton from the bromopropane.

(d) a nucleophile, because it displaces a bromide ion in a substitution reaction.

5.9 (a) $CH_3C(Br)(CH_3)CH_3 + CH_3O^- \longrightarrow CH_2{=}C(CH_3)CH_3 + CH_3OH + Br^-$

(b) $C_6H_5CH_2Cl + CH_3S^- \longrightarrow C_6H_5CH_2SCH_3 + Cl^-$

5.10 (a) $CH_3(CH_2)_4Br + Na^+\ {}^-OH \longrightarrow CH_3(CH_2)_4OH + NaBr$

(b) $CH_3(CH_2)_4Br + Na^+\ {}^-SCH_2CH_3 \longrightarrow CH_3(CH_2)_4SCH_2CH_3 + NaBr$

(c) $CH_3(CH_2)_4Br + Na^+\ {}^-CN \longrightarrow CH_3(CH_2)_4CN + NaBr$

(d) $CH_3(CH_2)_4Br + Na^+\ I^- \longrightarrow CH_3(CH_2)_4I + NaBr$

(e) $CH_3(CH_2)_4Br + Na^+ \ ^-SCN \longrightarrow CH_3(CH_2)_4SCN + NaBr$

(f) $CH_3(CH_2)_4Br + HC{\equiv}C^- \ Na^+ \longrightarrow CH_3(CH_2)_4C{\equiv}CH + NaBr$

5.11 (a) $^-OC(=O)CH_3$

(b) ^-CN

(c) H_2O (not ^-OH, which would cause elimination)

(d) $^-SCH_3$

5.12 (a) [:N≡C····C(H)····:I:]$^-$ → C≡N: (cyclohexane, H) + $:\ddot{I}:^-$

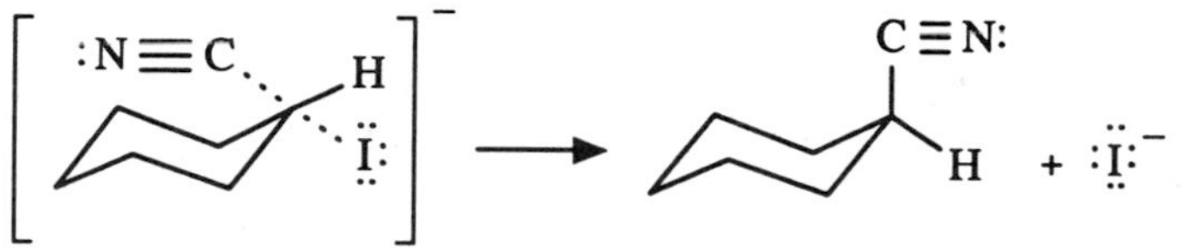

(b) [:ÖH ···· C (H) ···· :Cl: ; H, CH_3]$^-$ → :ÖH, H, H, CH_3 + $:\ddot{Cl}:^-$

(c) [H_3C, H, :Br, H, $\ddot{S}CH_3$]$^-$ →

H_3C, H, H, $\ddot{S}CH_3$ + $:\ddot{Br}:^-$

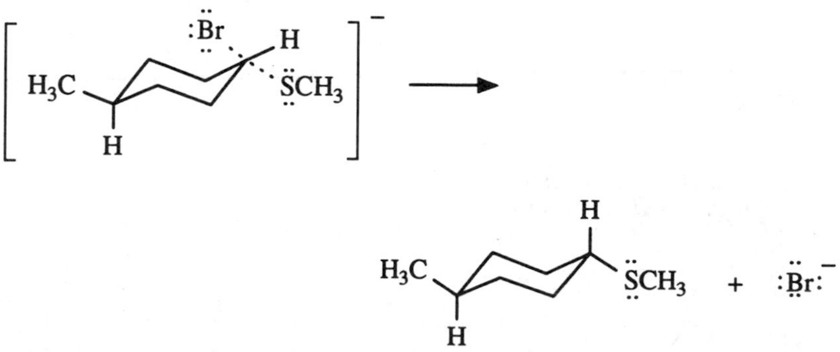

(d) [$CH_3CH_2\ddot{O}$ ···· C ···· $:\ddot{Cl}:$; H, H_3C, C_6H_5]$^-$ → $CH_3CH_2\ddot{O}$ – C ; H, C_6H_5, CH_3 + $:\ddot{Cl}:^-$

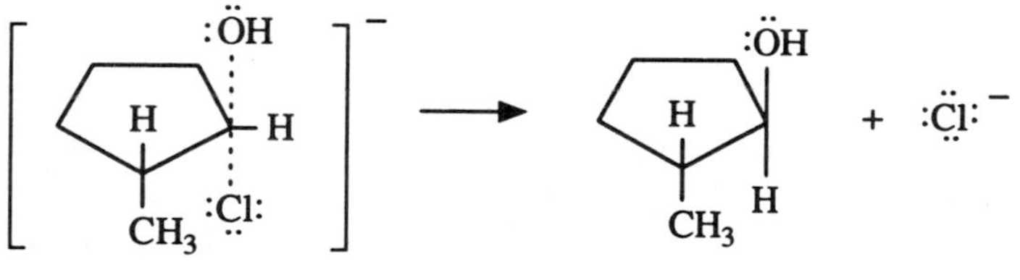

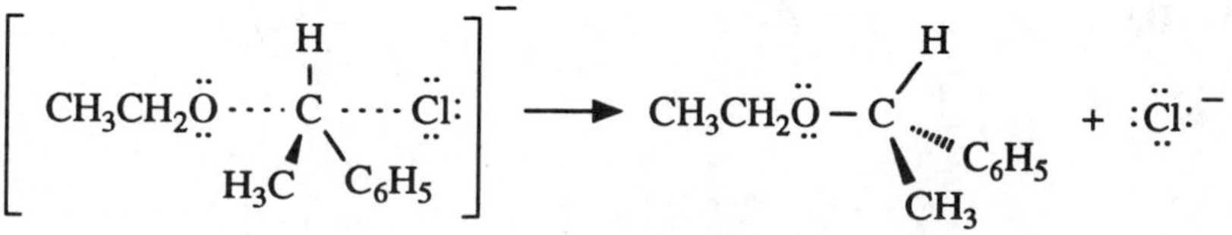

(e) 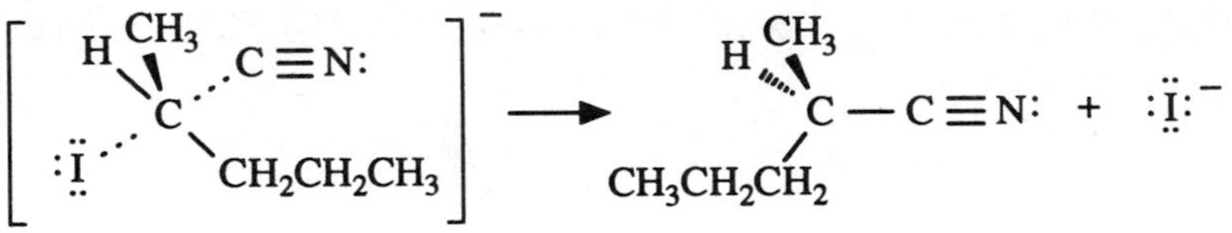 $[H(CH_3)C(\cdots C\equiv N{:})(CH_2CH_2CH_3)\cdots {:}\ddot{I}]^- \longrightarrow H(CH_3)C(CH_3CH_2CH_2)-C\equiv N{:} + {:}\ddot{I}{:}^-$

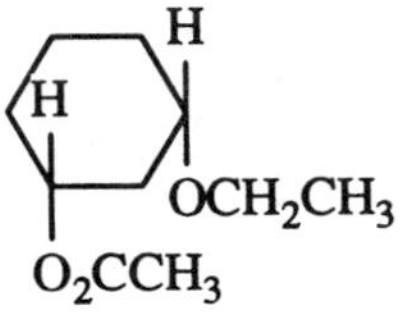

5.13 (a) 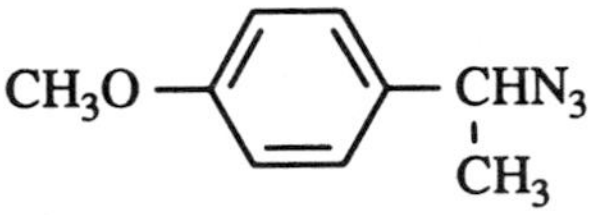 $CH_3O-C_6H_4-CHN_3$ (with CH_3 on CH)

(b) cyclohexane ring with H, H, OCH_2CH_3, O_2CCH_3

(c) CH_2SCH_3 on decahydroquinoline ring, N–C_6H_5

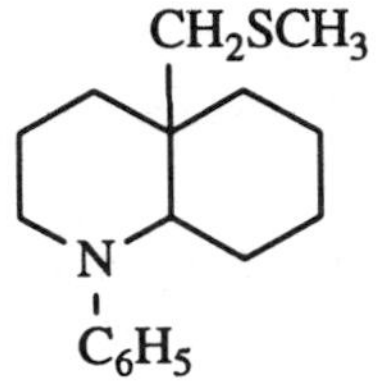

(d) $CH_3C{\equiv}CCH_3$

(e) $C_6H_5CHCO_2^-$ with OCH_3 on CH

5.14 (a) 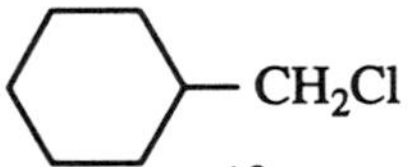 cyclohexyl–CH_2Cl

1°

(b) $(CH_3)_2CHCH_2CHClCH_3$

less hindered

(c) 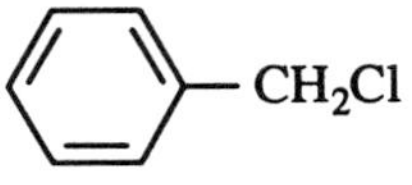 phenyl–CH_2Cl

The other compound is aryl and nonreactive.

5.15 (a) 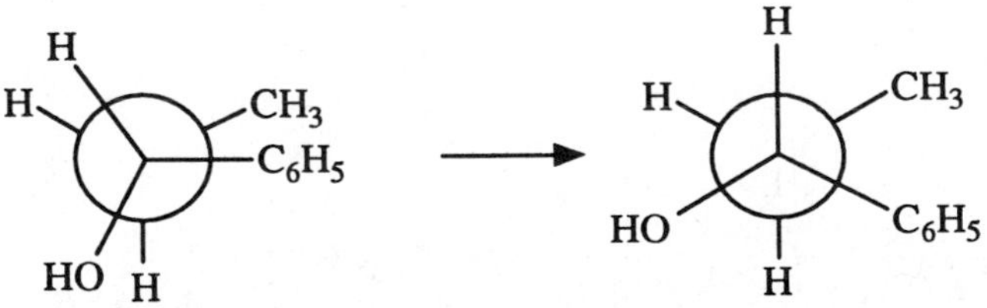 Newman projection (H, H, HO, H, CH_3, C_6H_5) $\longrightarrow$ Newman projection (H, H, HO, H, CH_3, C_6H_5)

(b) 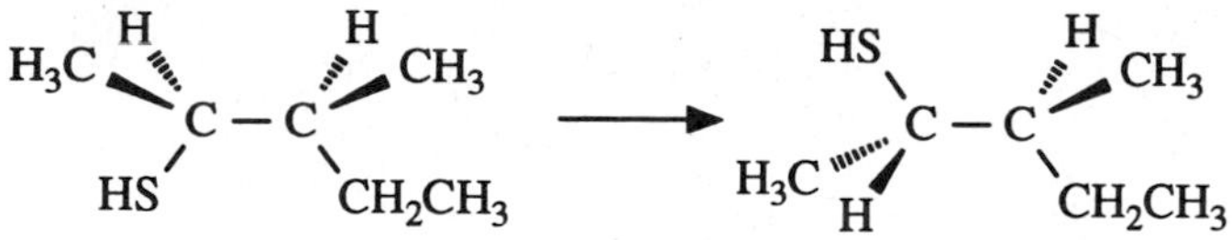 [stereochemical structures: H_3C, H, HS on C—C bearing H, CH_3, CH_2CH_3 → HS, H_3C, H on C—C bearing H, CH_3, CH_2CH_3]

(c) [decalin structure with H, H, OCH_3, CH_3O]

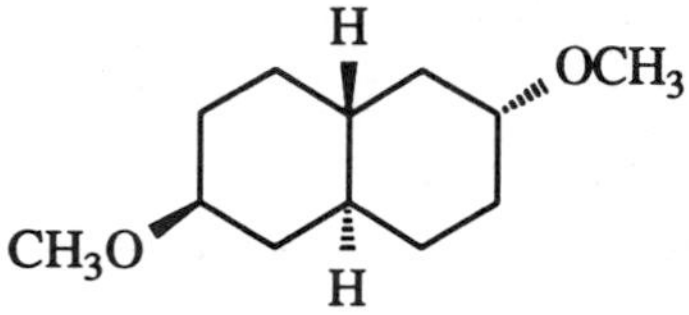

(d) 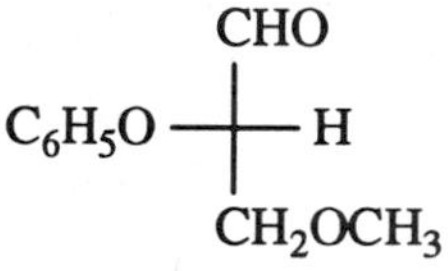 CHO; C_6H_5O — H; CH_2OCH_3

5.16 The I^- is lost in the first step of the reaction sequence, which is the rate-controlling step.

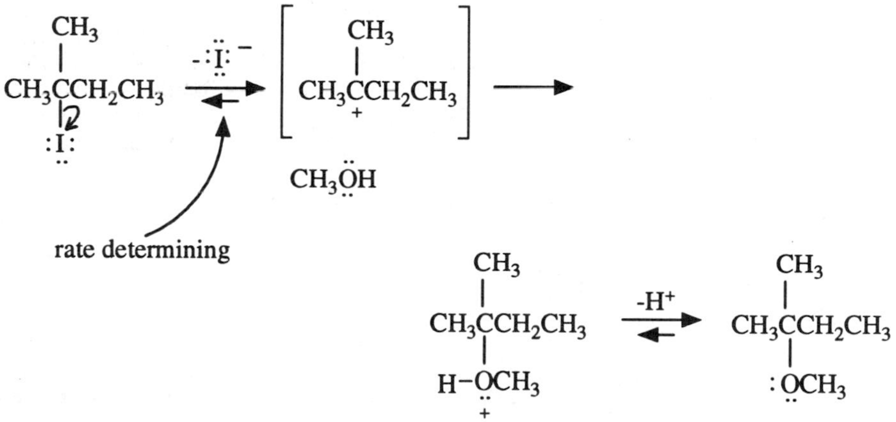

$CH_3CCH_2CH_3$ (with CH_3 and :I:) $\underset{}{\overset{-:I:^-}{\rightleftharpoons}}$ [$CH_3\overset{+}{C}CH_2CH_3$ (with CH_3)] ⟶ ($CH_3\ddot{O}H$)

rate determining

$CH_3CCH_2CH_3$ (with CH_3 and $H-\overset{+}{O}CH_3$) $\overset{-H^+}{\rightleftharpoons}$ $CH_3CCH_2CH_3$ (with CH_3 and $:\ddot{O}CH_3$)

5.17 In each case, the selection is based upon the structure that will yield the more stable carbocation.

(a) 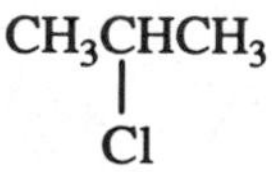 CH_3CHCH_3 (with Cl)

(b) $(CH_3)_3CBr$

(c) 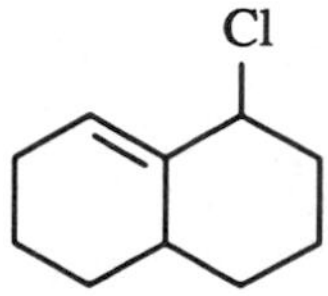 [bicyclic structure with Cl]

an allylic halide

5.18 (a) 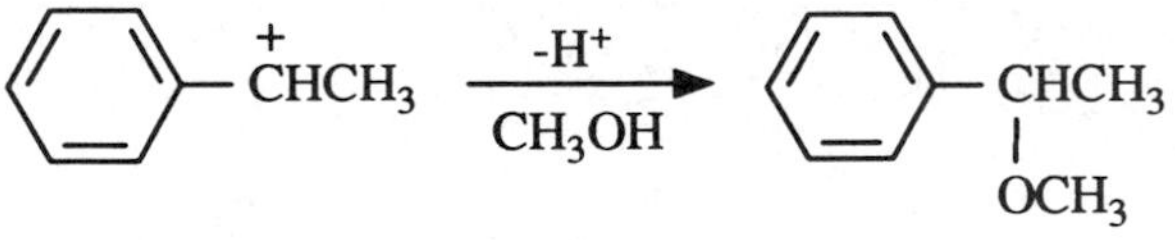 $C_6H_5\overset{+}{C}HCH_3 \xrightarrow[CH_3OH]{-H^+} C_6H_5CH(OCH_3)CH_3$

(b) 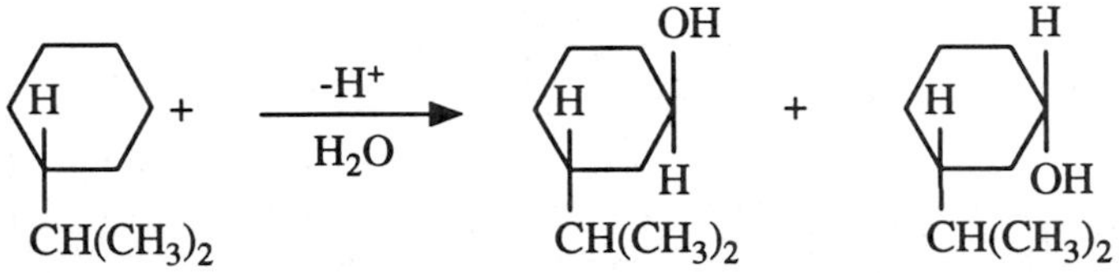 cation + $\xrightarrow[H_2O]{-H^+}$ product (OH, H) + product (H, OH), each bearing $CH(CH_3)_2$

5.19 (d) and (c) << (b) < (a): The order is based upon the relative order of carbocation stability of the carbocation that would be generated from each alkyl (or aryl) halide. Primary and aryl halides do not form cations under ordinary conditions.

5.20 The 4-methoxybenzyl cation would be more stable because the unshared valence electrons of the oxygen can be donated to help stabilize the positive charge (the last resonance structure shown). The resonance structures for the 4-nitrobenzyl cation show that this ion is less stabilized because (1) the nitro group does not help share the positive charge and (2) one resonance structure contains adjacent positive charges.

(a) 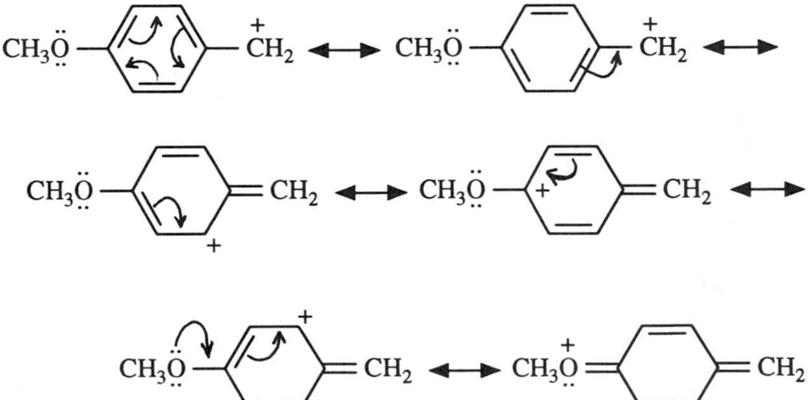 $CH_3\ddot{O}-C_6H_4-\overset{+}{C}H_2 \longleftrightarrow CH_3\ddot{O}-C_6H_4-\overset{+}{C}H_2 \longleftrightarrow$

$CH_3\ddot{O}-C_6H_4{}^{+}{=}CH_2 \longleftrightarrow CH_3\ddot{O}-C_6H_4{}^{+}{=}CH_2 \longleftrightarrow$

$CH_3\ddot{O}-C_6H_4{}^{+}{=}CH_2 \longleftrightarrow CH_3\overset{+}{O}{=}C_6H_4{=}CH_2$

(b) 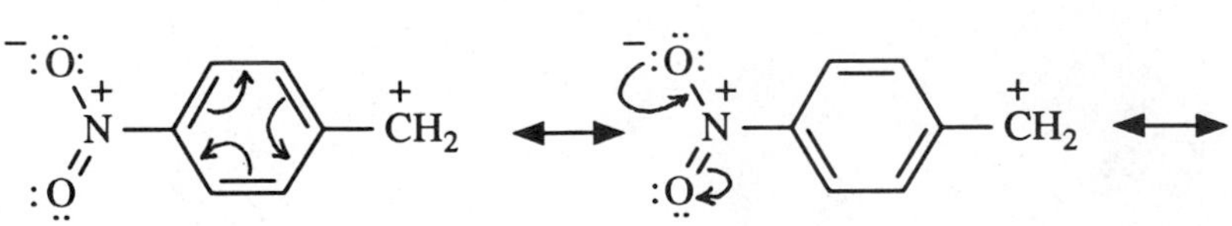 $O_2N^{+}-C_6H_4-\overset{+}{C}H_2 \longleftrightarrow O_2N^{+}-C_6H_4-\overset{+}{C}H_2 \longleftrightarrow$

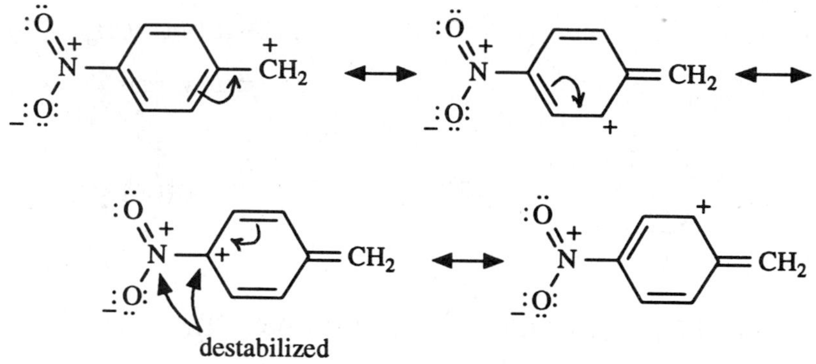

:Ö: N+ :Ö:⁻ — C₆H₄ — $\overset{+}{C}H_2$ ⟷ :Ö: N+ :Ö:⁻ — ring = CH_2 (+) ⟷

:Ö: N+ :Ö:⁻ — ring (+) = CH_2 ⟷ :Ö: N+ :Ö:⁻ — ring (+) = CH_2

destabilized

5.21 $CH_3CH_2C(CH_3)(Cl)CH_3 \xrightarrow{-Cl^-} [CH_3CH(H)-\overset{+}{C}(CH_3)CH_3] \longrightarrow CH_3CH{=}C(CH_3)_2 + H_3\overset{+}{O}$

HÖH

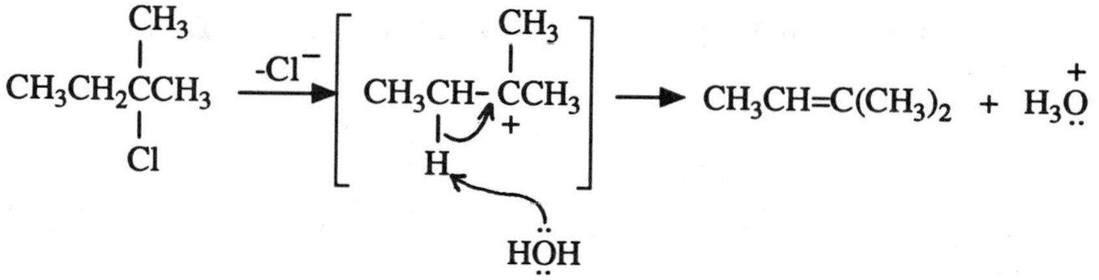

5.22 (a) 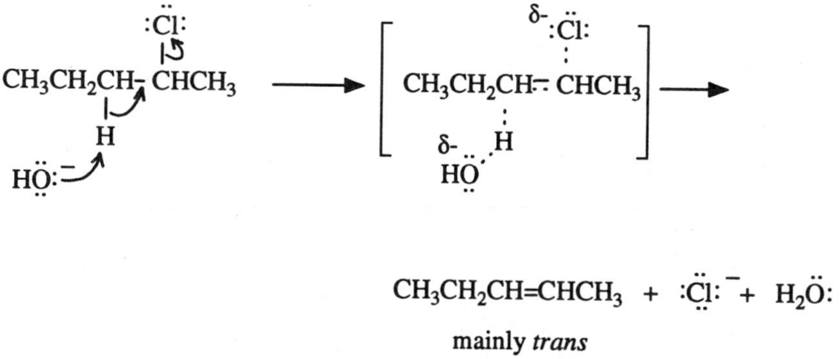 $CH_3CH_2CH(H)-CH(:\ddot{C}l:)CH_3$ + $H\ddot{O}:^-$ ⟶ $[CH_3CH_2CH\text{---}CHCH_3$, δ^- :Cl:, δ^- HÖ···H$]$ ⟶

$CH_3CH_2CH{=}CHCH_3 + :\ddot{C}l:^- + H_2\ddot{O}:$

mainly *trans*

(b) 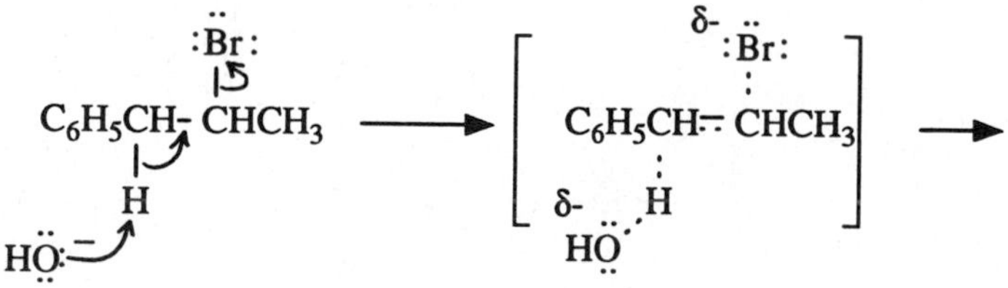 $C_6H_5CH(H)-CH(:\ddot{B}r:)CH_3$ + $H\ddot{O}:^-$ ⟶ $[C_6H_5CH\text{---}CHCH_3$, δ^- :Br:, δ^- HÖ···H$]$ ⟶

$C_6H_5CH{=}CHCH_3$ + $:\ddot{Br}:^-$ + $H_2\ddot{O}:$

mainly *trans*

5.23 (a) C_6H_5–$CH_2CH{=}CHCH_3$ + C_6H_5–$CH_2CH_2CH{=}CH_2$

most

(b) $CH_3CH_2CH{=}C(CH_3)_2$ + $CH_3CH_2CH_2C(CH_3){=}CH_3$

most

5.24 (b) < (a) < (c). The order of reactivity of alkyl halides in both E1 and E2 reactions is 3° > 2° > 1°.

5.25 (b) < (a) < (c) < (d): The ranking is based upon the amount of alkyl substitution on the double-bond carbons for compounds (b) and (a) and upon steric hindrance for compounds (a) and (c). Alkene (d) is the most stable because the double bond is conjugated the with benzene ring.

5.26 (a) [cyclohexene ring bearing H and CH_3 on one carbon, $(CH_3)_3C$ on a ring carbon, and H]

(b) $(CH_3)_3C$–(cyclohexene)–CH_3 + [cyclohexene ring bearing H and CH_3, $(CH_3)_3C$, and H]

most

5.27 (a) cyclopentyl–Br

(b) C_6H_5–$CH(Br)CH_2$–C_6H_5

(c) $(CH_3CH_2)_3CBr$

(d) [Structure: cyclohexane ring bearing H_3C, CH_3, CH_3, Br, CH_3, CH_3, CH_3, H substituents]

stereochemistry not specified

Br and H *trans*

Because of the steric hindrance in (d), we would also expect the following alkene product (probably as the major product):

[Structure: cyclohexane ring with H_3C, CH_3, H_3C, $=CH_2$, H_3C, CH_3 substituents]

5.28 (a) [cyclohexyl]$-CH_2OH$ + [cyclohexylidene]$=CH_2$

(b) $CH_3CH_2CH_2CH_2OC(CH_3)_3$ + $CH_2{=}CHCH_2CH_3$

(c) $CH_2{=}C(CH_3)_2$ + $CH_3C(CH_3)_2-SCH_3$ (little, if any)

(d) $C_6H_5CH(CN)CH_3$ + $C_6H_5CH{=}CH_2$

5.29 The *tert*-butoxide ion is a sterically hindered base that preferentially attacks the least sterically hindered hydrogen, thus producing more of the Hofmann product: $CH_3CH_2CH_2C(CH_3){=}CH_2$

5.30 (d) < (c) < (a) < (b)

5.31 (a) $CH_3CH_2OCH_3$; S_N2

(b) $(CH_3)_2C{=}CH_2$; E2

(c) $(CH_3)_2C{=}CH_2$; E2

(d) no reaction because the halogen compound is an aryl halide

(e) $CH_3CH(CN)CH_3$; S_N2

(f) $CH_3CH_2NH_2$; S_N2

(g) $(CH_3)_2C{=}CH_2$; E2

(h) 1-methylcyclohexanol (cyclohexane ring bearing CH_3 and OH) ; and S_N1 1-methylcyclohexene (ring with double bond)–CH_3 ; E1

(i) $CH_2{=}CHCH_2OH$; S_N1 because the solvent is polar

(j) $(CH_3)_2C{=}CHCH_3$; E2

(k) $C_6H_5C(CN)(CH_3)CH_2CH_3$; S_N1 and $C_6H_5C(CH_3){=}CHCH_3$; E1

$C_6H_5C(OCH_2CH_3)(CH_3)CH_2CH_3$ would also be observed (S_N1).

(l) $(CH_3)_2C{=}CHCH_3$; E2

5.32 (a) cyclopropyl–$CH_2CH_2OC_6H_5$ (S_N2)

(b) piperidinium ring with $+N$ bearing CH_3 and H, I^-

(c) 3-methylcyclopentene (ring with double bond)–CH_3 (E2) + cyclopentane ring bearing CH_3 and SC_6H_5, with H, H (S_N2)

major product

(d) $(CH_3)_3CCH{=}CHCl$ (E2)

(e) [3-methylcyclohexene structure with CH_3] (E2)

(f) $(CH_3)_2C(CH(CO_2CH_2CH_3)_2)CH{=}O$, drawn as:

$$\begin{array}{l} \quad\;\; O \\ \quad\;\; \| \\ (CH_3)_2CCH \\ \quad\;\; | \\ \quad CH(CO_2CH_2CH_3)_2 \end{array}$$

(S_N2)

5.33 (a) Treatment with I^- would give a larger percent of substitution because I^- is a better nucleophile (more polarizable) than Cl^-.

(b) Treatment with $^-SCH_3$ will give a larger percent of substitution. Sulfur is more polarizable than oxygen; therefore, $^-SCH_3$ is more nucleophilic than $^-OCH_3$. Also $^-OCH_3$ is a stronger base, a property that would increase the rate of elimination.

(c) Heating $CH_3CH_2CH_2CBr(CH_3)_2$ with ethanol would give a greater percent of substitution. The reaction leading to alkene would be slower than the elimination reaction of the other bromoalkane; consequently, substitution for $CH_3CH_2CH_2CBr(CH_3)_2$ would be more competitive.

(d) $CH_3CH_2CH_2Br$ because it would form the least stable alkene.

(e) with $NaOCH_3$ because it is less sterically hindered.

5.34 (a) $(CH_3)_2CHBr + C_6H_5O^- \longrightarrow C_6H_5OCH(CH_3)_2 + Br^-$

(b) $(CH_3)_2CHBr + CH_3O^- \longrightarrow (CH_3)_2CHOCH_3 + Br^-$

(c) $CH_3CH_2CH_2Cl + CH_3CH_2CH_2O^- \longrightarrow (CH_3CH_2CH_2)_2O + Cl^-$

(d) $CH_3I + OH^- \longrightarrow CH_3OH + I^-$

(e) $CH_3I + CN^- \longrightarrow CH_3CN + I^-$

(f) $CH_3CH_2CHBrCH_3 + KOH \xrightarrow[CH_3CH_2OH]{heat}$

$CH_3CH{=}CHCH_3$ + H_2O + KBr

(mainly *trans*)

(g) 2 $CH_3CH_2CH_2CH_2Br$ + Na_2S ⟶ $(CH_3CH_2CH_2CH_2)_2S$ + 2 NaBr

or $CH_3CH_2CH_2CH_2Br$ + $CH_3CH_2CH_2CH_2S^-$ ⟶

(h) 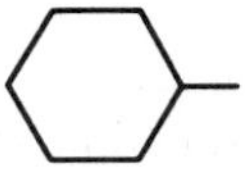 cyclohexyl–Br + KOH $\xrightarrow[CH_3CH_2OH]{heat}$

cyclohexene + H_2O + KBr

5.35 (a) $C_6H_5CH_2Br$ + $CH_3CO_2^-\ Na^+$ ⟶ product

(b) $C_6H_5CH_2Br$ + $Na^+\ I^-$ ⟶ product

(c) CH_3I + $C_6H_5NH_2$ ⟶ $C_6H_5\overset{+}{N}H_2CH_3\ I^-$ $\xrightarrow{OH^-}$ product

(d) 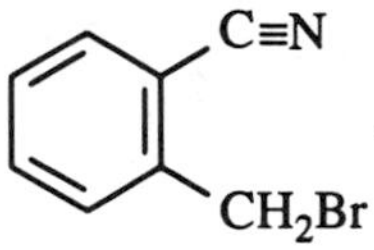 o-($C{\equiv}N$)$C_6H_4CH_2Br$ + $(CH_3)_2NH$ ⟶ o-($C{\equiv}N$)$C_6H_4CH_2\overset{+}{N}H(CH_3)_2$

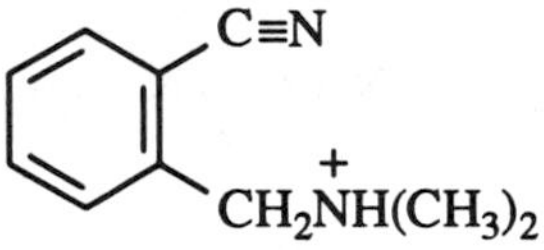

$\xrightarrow{OH^-}$ product

(e) $Br(CH_2)_4Br$ + Na_2S ⟶ $[BrCH_2CH_2CH_2CH_2S^-]$ ⟶ product

6

Free-Radical Reactions

Answers to Additional Problems

6.1 (a) $H:\ddot{C}\cdot$ with H above and below C ($\cdot CH_3$)

(b) $H:\ddot{C}:\dot{C}:\ddot{C}:H$ with H above and below each C ($CH_3\dot{C}HCH_3$)

(c) $H:\ddot{C}:\ddot{O}\cdot$ with H above and below C ($CH_3O\cdot$)

(d) cyclohexyl radical: $\dot{C}H$ in a six-membered ring of C atoms, each remaining C bearing two H

(e) $H:\ddot{C}:C:\ddot{O}\cdot$ with H above and below the first C and $\ddot{O}:$ double-bonded to the second C ($CH_3CO_2\cdot$)

(f) phenoxy radical: $C_6H_5\ddot{O}\cdot$

6.2 (a) $:\ddot{Br}:\ddot{Br}: \longrightarrow\ :\ddot{Br}\cdot\ +\ \cdot\ddot{Br}:$

(b) $H\ddot{O}:\ddot{O}H \longrightarrow H\ddot{O}\cdot\ +\ \cdot\ddot{O}H$

(c) $C_6H_5C(=\ddot{O}:)O:OC(=\ddot{O}:)C_6H_5 \longrightarrow C_6H_5C(=\ddot{O}:)O\cdot\ +\ \cdot OC(=\ddot{O}:)C_6H_5$

(d) $CH_3CH_2:H \longrightarrow CH_3\dot{C}H_2\ +\ H\cdot$

(e) $C_6H_5CH_2:H \longrightarrow C_6H_5\dot{C}H_2\ +\ H\cdot$

(f) $CH_2{=}CHCH_2:H \longrightarrow CH_2{=}CH\dot{C}H_2\ +\ H\cdot$

6.3 Initiation:

$$Cl{-}Cl \xrightarrow{h\nu} 2\,Cl\cdot$$

Propagation:

$$C_6H_5CH_2{-}H + Cl\cdot \longrightarrow C_6H_5\dot{C}H_2 + HCl$$

$$C_6H_5\dot{C}H_2 + Cl{-}Cl \longrightarrow C_6H_5CH_2Cl + Cl\cdot$$

Termination:

$$Cl\cdot \; \cdot CH_2C_6H_5 \longrightarrow C_6H_5CH_2Cl$$

$$Cl\cdot \; \cdot Cl \longrightarrow Cl{-}Cl$$

$$C_6H_5\dot{C}H_2 \; \cdot CH_2C_6H_5 \longrightarrow C_6H_5CH_2CH_2C_6H_5$$

6.4 Stereoisomers are not shown.

(a) $ClCH_2CH(CH_3)_2 + (CH_3)_3CCl$

(b) $ClCH_2C(CH_3)_2CH_2CH(CH_3)_2 + (CH_3)_3CCHClCH(CH_3)_2 +$

$(CH_3)_3CCH_2CCl(CH_3)_2 + (CH_3)_3CCH_2CH(CH_3)CH_2Cl$

(c) cyclohexyl–CH_2Cl + 1-chloro-1-methylcyclohexane + 2-chloro-1-methylcyclohexane + 3-chloro-1-methylcyclohexane + 4-chloro-1-methylcyclohexane (Cl–cyclohexane–CH_3)

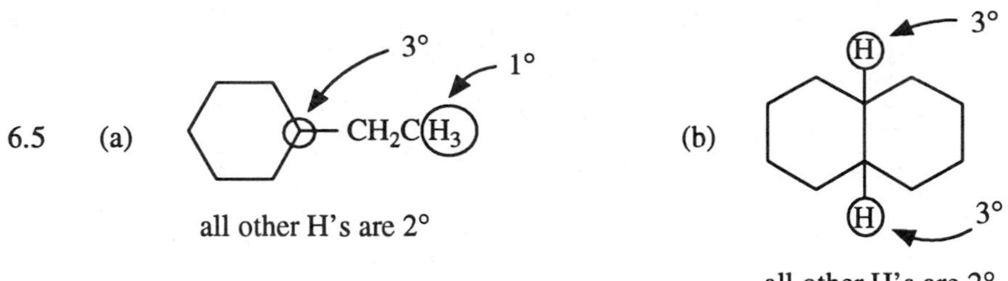

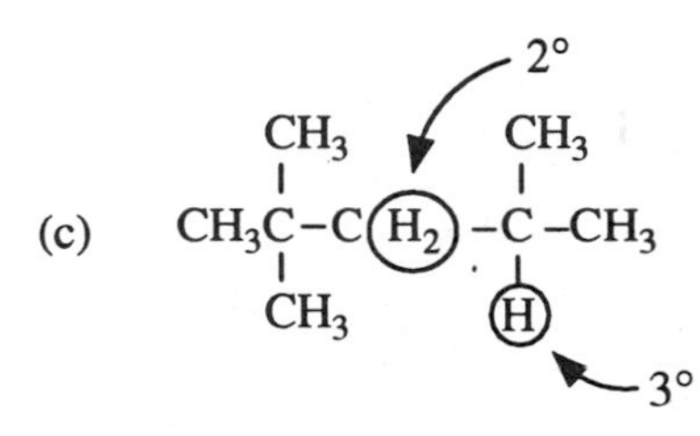

6.6 Compound (b) would react more rapidly with bromine in sunlight because it contains two tertiary hydrogens. Compound (a) has no tertiary hydrogens.

6.7 (a) 20 secondary hydrogens and 2 tertiary hydrogens; therefore, the ratio of secondary to tertiary hydrogens is 10 : 1.

The following ratios are 1° : 2° : 3° hydrogens.

(b) 15 : 2 : 1

(c) 6 : 12 : 2 or 3: 6: 1

(d) 6 : 4: 0 or 3 : 2 : 0

6.8 In both cases, hydrogen (2) would be extracted at a faster rate than hydrogen (1).

6.9 (a)

$$\mathrm{RS{-}H} + 89\ \mathrm{kcal/mol} \longrightarrow \mathrm{RS\cdot} + \mathrm{H\cdot}$$

$$\mathrm{\cdot CH_3} + \mathrm{H\cdot} \longrightarrow \mathrm{CH_4} + 104\ \mathrm{kcal/mol}$$

$$\mathrm{\cdot CH_3} + \mathrm{RSH} \longrightarrow \mathrm{CH_4} + \mathrm{RS\cdot} + 15\ \mathrm{kcal/mol}$$

The reaction is exothermic by 15 kcal/mol.

(b)

$$C_6H_5CH_2{-}H + 66\ \text{kcal/mol} \longrightarrow C_6H_5\dot{C}H_2 + H\bullet$$

$$\bullet CH_3 + H\bullet \longrightarrow CH_4 + 104\ \text{kcal/mol}$$

$$\bullet CH_3 + C_6H_5CH_3 \longrightarrow CH_4 + C_6H_5\dot{C}H_2 + 38\ \text{kcal/mol}$$

The reaction is exothermic by 38 kcal/mol.

6.10 $CH_3CH_2CH(CH_3)_2$

the number of 1° hydrogens (9) × relative rate of 1.0 = 9.0

the number of 2° hydrogens (2) × relative rate of 3.0 = 6.0

the number of 3° hydrogens (1) × relative rate of 4.5 = 4.5

sum 19.5

% 1° monochlorination: $\frac{9.0}{19.5} \times 100 = 46\%$

% 2° monochlorination: $\frac{6.0}{19.5} \times 100 = 31\%$

% 3° monochlorination: $\frac{4.5}{19.5} \times 100 = 23\%$

6.11 (a)  $\dot{C}H_2{-}CH{=}CH{-}CH_2CH(=O) \longleftrightarrow CH_2{=}CH{-}\dot{C}H{-}CH_2CH(=O)$

(b)

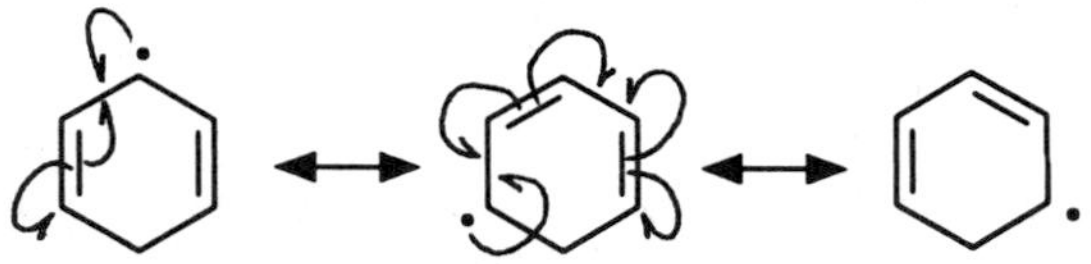

6.12 (a) 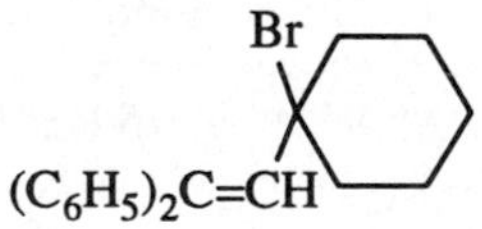 (racemic) $(C_6H_5)_2C{=}CH$, Br (b) 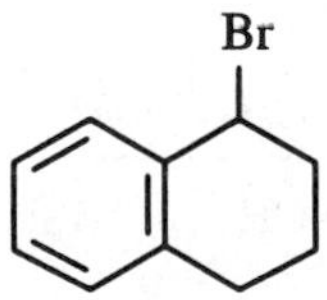 Br (racemic)

(c) *cis* and *trans* $CH_3CH_2CH_2CH(Br)CH{=}CHCH_3$ + $CH_3CH_2CH_2CH_2CH{=}CHCH_2Br$

6.13 (a) $BrCH_2\text{-}CH{=}CHCOH$ (with C=O on the carbonyl carbon)

(b) 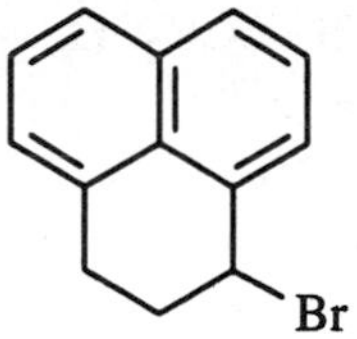 [structure: bromo-substituted dihydrophenalene, Br]

(c) 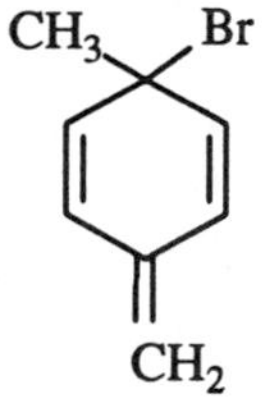 [structure: CH_3, Br on ring carbon; $=CH_2$]

(d) 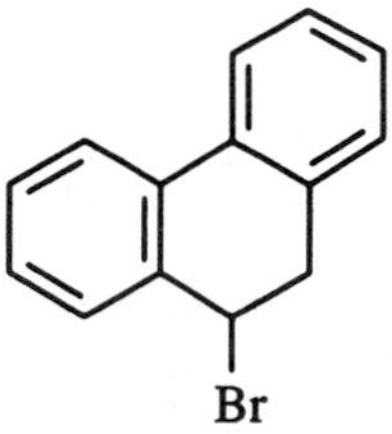 [structure: 9-bromo-9,10-dihydrophenanthrene, Br]

6.14 Initiation: $Br_2 \xrightarrow{h\nu} 2\ Br\cdot$

Propagation:

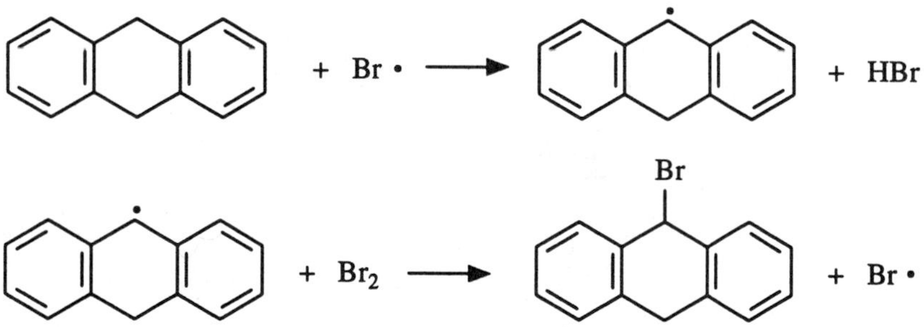

[9,10-dihydroanthracene] + Br $\cdot$ ⟶ [radical] + HBr

[radical] + Br_2 ⟶ [Br-substituted product] + Br $\cdot$

Termination: combination of any two radicals, such as:

$Br\cdot + Br\cdot \longrightarrow Br_2$

6.15 (a) $CH_3OCH_2CH_3 + O_2 \longrightarrow HOOCH_2OCH_2CH_3 + CH_3OCHCH_3$ (with OOH on CH)

(major)

(b) $CH_3CHCH(=O) + O_2 \longrightarrow CH_3CHC(=O)OOH$ (each with CH_3 on CH)

$CH_3CHCH(=O) + CH_3CHC(=O)OOH \longrightarrow 2\ CH_3CHC(=O)OH$ (each with CH_3 on CH)

(c) 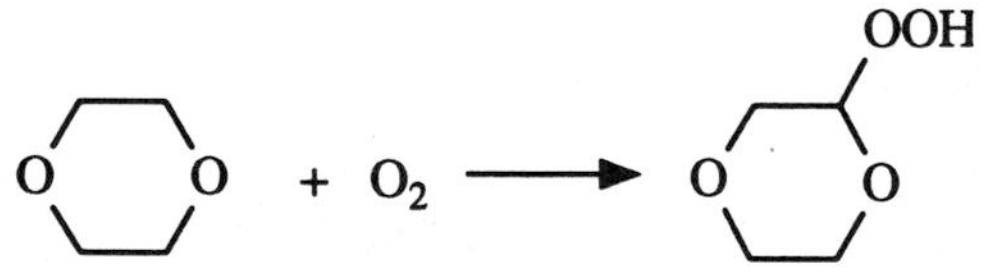 O-ring-O $+ O_2 \longrightarrow$ O-ring-O with OOH

6.16 (a) $x\,CH_2{=}CHBr \xrightarrow{\text{peroxide}} -\!(CH_2-CH)_x\!-$ with Br on CH

(b) $x\,CH_2{=}CHOCH_3 \xrightarrow{\text{peroxide}} -\!(CH_2-CH)_x\!-$ with OCH_3 on CH

(c) 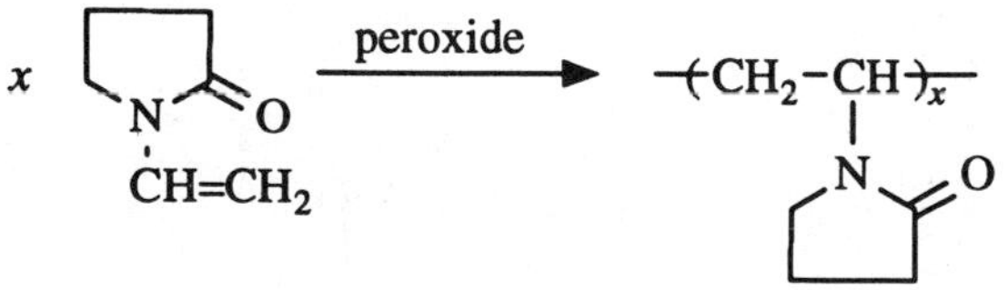 x N-ring=O with $CH{=}CH_2$ on N $\xrightarrow{\text{peroxide}} -\!(CH_2-CH)_x\!-$ with N-ring=O on CH

(d) $x\,(CH_3)_3Si\text{-}CH{=}CH_2 \xrightarrow{\text{peroxide}} -\!(CH_2-CH)_x\!-$ with $Si(CH_3)_3$ on CH

7

Alcohols

Answers to Additional Problems

7.1 (a) $CH_3C(CH_3)_2CH_2OH$ (1°)

(b) $CH_3CH(OH)-C(CH_3)_2-OH$ (OH on CH: 2°; OH on C: 3°)

(c) $HOCH_2-$cyclohexane$-C(CH_3)(OH)$ (CH_2OH: 1°; ring OH: 3°)

7.2 (a) propyl alcohol (b) allyl alcohol

(c) benzyl alcohol (d) isopropyl alcohol

(e) cyclobutyl alcohol (f) ethylene glycol

7.3 (a) (1) keto; (2) 4-hydroxy-2-butanone

(b) (1) carboxyl; (2) 3-hydroxybutanoic acid

(c) (1) hydroxyl; (2) 3,5-heptadiene-1,2-diol

(d) (1) aldehyde; (2) 2,3,4-trihydroxybutanal

7.4 (a) $HOCH_2CH=CHCH_2CH_2OH$ (left OH: 1° and allylic; right OH: 1°)

(b) $HOCH_2CH_2$–C_6H_4–$CH(OH)CH_3$

OH: 2° and benzylic; $HOCH_2$: 1°

(c) OH: 2°, benzylic, and allylic

(d) $HOCH_2$–C_6H_4–$CH{=}CHCH_2OH$

$CH{=}CHCH_2OH$: 1° and allylic; $HOCH_2$: 1° and benzylic

(e) OH: 3°; HO: 3° and benzylic

(f) $CH(OH)CH_3$ on naphthalene

OH: 2° and benzylic

7.5 (a) 1-Bromo-3,5-dimethylcyclohexane (CH_3, CH_3, Br)

(b) $CH_3(CH_2)_{16}CH{=}CHCH_2CH_2Br$

(c) 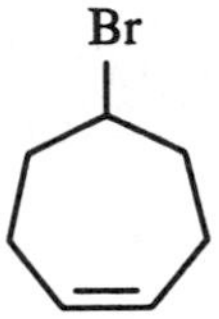 [structure: 4-bromocycloheptene]

Br

(d) 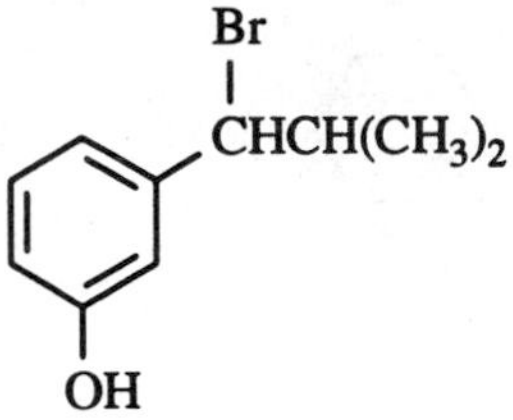 [structure]

Br

$CHCH(CH_3)_2$

OH

In each case, the halogen could be Cl^- or I^-. In (d), the -OH on the benzene ring is a phenol. It cannot be prepared by a simple S_N2 reaction. Therefore

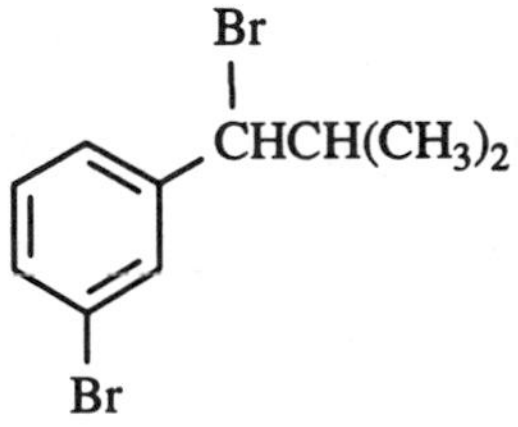

Br

$CHCH(CH_3)_2$

Br

is an incorrect answer.

7.6 (a) 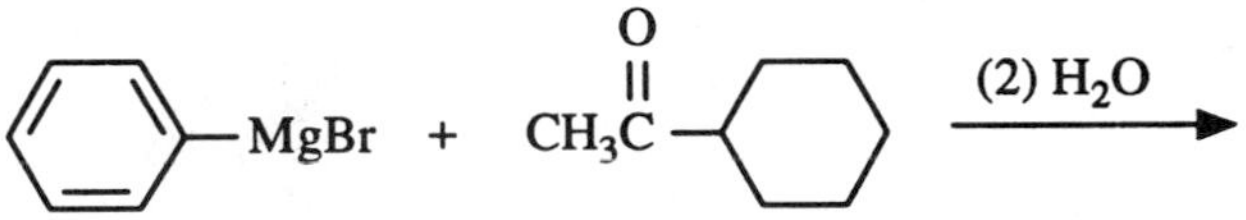 C_6H_5—MgBr + $CH_3C(=O)$—C_6H_{11} $\xrightarrow{(2)\ H_2O}$

CH_3MgBr + C_6H_5—C(=O)—C_6H_{11} $\xrightarrow{(2)\ H_2O}$

C_6H_{11}—MgBr + $CH_3C(=O)$—C_6H_5 $\xrightarrow{(2)\ H_2O}$

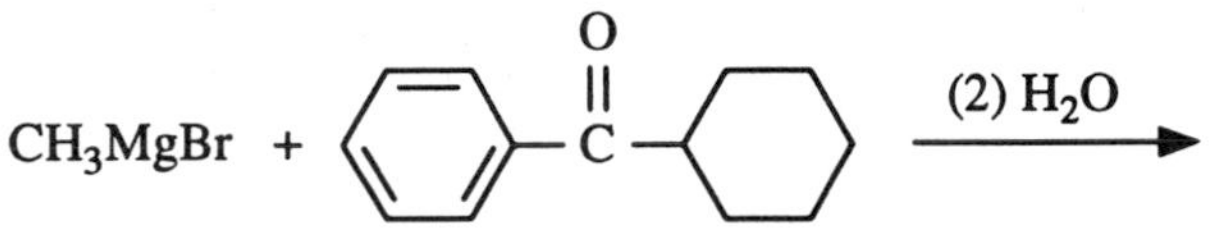

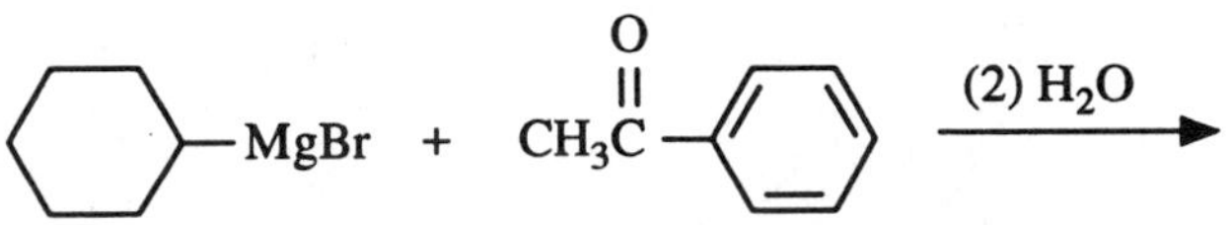

(b) 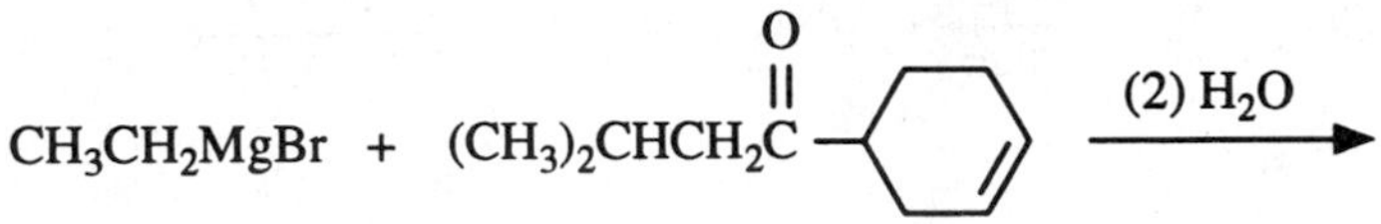 CH_3CH_2MgBr + $(CH_3)_2CHCH_2C(=O)$—(cyclohexenyl) $\xrightarrow{(2)\ H_2O}$

$(CH_3)_2CHCH_2MgBr$ + $CH_3CH_2C(=O)$—(cyclohexenyl) $\xrightarrow{(2)\ H_2O}$

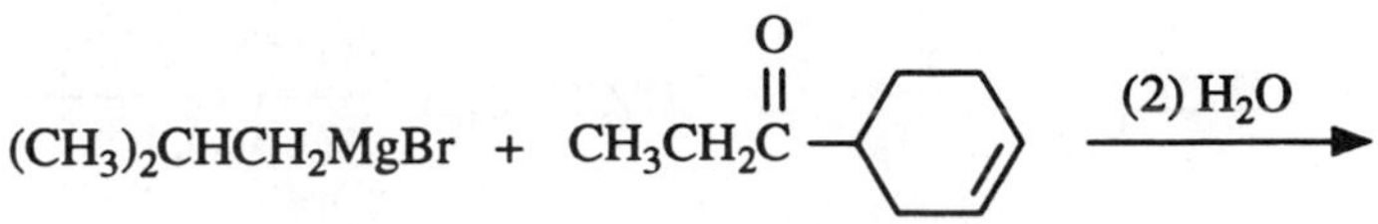

$$\text{(cyclohexenyl)-MgBr} + CH_3CH_2\overset{O}{\overset{\|}{C}}\text{-}CH_2CH(CH_3)_2 \xrightarrow{(2)\ H_2O}$$

7.7 (a) $CH_2{=}CHMgI$ (b) $CH_2{=}CH\overset{OMgI}{\overset{|}{C}}(CH_3)_2$

(c) $CH_2{=}CH_2 + CH_3OMgI$ (d) $CH_3CH_2CO_2H + Mg^{2+} + Br^-$

7.8 (a) $(CH_3)_2CHOH + CH_3CH_2OH$ (b) $(CH_3)_3COH + CH_3CH_2OH$

(c) $(CH_3)_3COH$ (d) $C_6H_5CH_2OH$

7.9 (a) and (b) $\text{(cyclohexyl)-}\overset{OH}{\overset{|}{C}}HCH_2CH_2CH_3$

7.10 (a) $CH_3\overset{OH}{\overset{|}{C}}H(CH_2CH_3)_2$ (b) 1-(CH=CHCH₃)-cyclohexan-1-ol: cyclohexane ring bearing OH and $CH{=}CHCH_3$

(c) $\text{(cyclohexyl)-}\overset{OH}{\overset{|}{C}}HCH_3$ (d) $C_6H_5CH_2OH$

7.11 (a) $(CH_3)_2CHBr \xrightarrow[\text{ether}]{Mg} (CH_3)_2CHMgBr \xrightarrow[(2)\ H_2O,\ H^+]{(1)\ HCHO}$

(b) $CH_3CH_2CH_2Br \xrightarrow[\text{ether}]{Mg} CH_3CH_2CH_2MgBr \xrightarrow[(2)\ H_2O,\ H^+]{(1)\ CH_3\overset{O}{\overset{\|}{C}}CH_2CH_3}$

(c) $$CH_3CH_2Br \xrightarrow[\text{ether}]{\text{Mg}} CH_3CH_2MgBr \xrightarrow[\text{(2) } H_2O,\ H^+]{\text{(1) } CH_3\overset{\overset{O}{\|}}{C}CH_3}$$

In (b) and (c), other combinations could be used.

7.12 (a) $$CH_3CH_2CH_2CH_2Br \xrightarrow[\text{ether}]{\text{Mg}} CH_3CH_2CH_2CH_2MgBr$$

$$\xrightarrow[\text{(2) } H_2O,\ H^+]{\text{(1) } CH_3CHO} CH_3CH_2CH_2CH_2\underset{\underset{OH}{|}}{C}HCH_3$$

In (a), you might have used $CH_3CH_2CH_2CH_2CHO$ + CH_3MgI

(b) $$(CH_3)_3CCl \xrightarrow[\text{ether}]{\text{Mg}} (CH_3)_3CMgCl \xrightarrow[\text{(2) } H_2O,\ H^+]{\text{(1) HCHO}} CH_3\overset{\overset{CH_3}{|}}{\underset{\underset{CH_3}{|}}{C}}CH_2OH$$

(c) $$C_6H_5Br \xrightarrow[\text{ether}]{\text{Mg}} C_6H_5MgBr \xrightarrow[\text{(2) } H_2O,\ H^+]{\text{(1) } C_6H_5CHO} (C_6H_5)_2CHOH$$

7.13 (a) S_N1 (b) S_N1 (c) S_N2

7.14 (a) $$CH_3CH_2CH_2CH_2\ddot{O}H \overset{H^+}{\rightleftharpoons} CH_3CH_2CH_2CH_2\overset{+}{O}H_2 \xrightarrow[S_N2]{:\ddot{Br}:^-}$$

$$\left[\overset{\delta-}{:\ddot{Br}} \cdots \underset{\underset{CH_2CH_2CH_3}{|}}{\overset{H\ \ H}{C}} \cdots \overset{\delta+}{OH_2} \right] \longrightarrow :\ddot{Br}CH_2CH_2CH_2CH_3 + H_2\ddot{O}:$$

(b)
$$CH_3CH_2\underset{CH_3}{\overset{CH_3}{C}}-\ddot{O}H \xrightleftharpoons{H^+} CH_3CH_2\underset{CH_3}{\overset{CH_3}{C}}-\overset{+}{\ddot{O}}H_2 \xrightarrow{S_N1} \left[CH_3CH_2\underset{CH_3}{\overset{CH_3}{C}}\cdots\ddot{O}H_2\right]^+$$

$$\longrightarrow CH_3CH_2\overset{+}{C}(CH_3)_2 + H_2\ddot{O}: \xrightarrow{:\ddot{Br}:^-} \left[\overset{\delta+}{CH_3CH_2\underset{CH_3}{\overset{CH_3}{C}}}\cdots\overset{\delta-}{:\ddot{Br}:}\right] \longrightarrow CH_3CH_2\underset{CH_3}{\overset{CH_3}{C}}-\ddot{Br}:$$

7.15 (a) < (c) < (b). The order of reactivity of alcohols with HBr is 3° > 2° > 1°.

7.16
$$(R)\text{-}CH_3CH_2\underset{:\ddot{O}H}{CH}CH_3 \xrightleftharpoons{H^+} (R)\text{-}CH_3CH_2\underset{\overset{+}{\ddot{O}}H_2}{CH}CH_3 \xrightleftharpoons{-H_2\ddot{O}:} CH_3CH_2\overset{+}{C}HCH_3$$

achiral

$$\xrightleftharpoons{H_2\ddot{O}:} (R)(S)\text{-}CH_3CH_2\underset{:\overset{+}{O}(H)-H}{CH}CH_3 \xrightleftharpoons{-H^+} (R)(S)\text{-}CH_3CH_2\underset{:\ddot{O}H}{CH}CH_3$$

7.17 (a) $(CH_3)_3CCH_2Cl$ + $(CH_3)_2\underset{Cl}{C}CH_2CH_3$

(b) $(C_6H_5)_3CCH_2Cl$ + $(C_6H_5)_2\underset{Cl}{C}CH_2C_6H_5$

(c) (cyclohexane ring: CH_3, CH_3 on one carbon; Br on adjacent carbon) + (cyclohexane ring: CH_3, Br on one carbon; CH_3 on adjacent carbon)

7.18 (a)

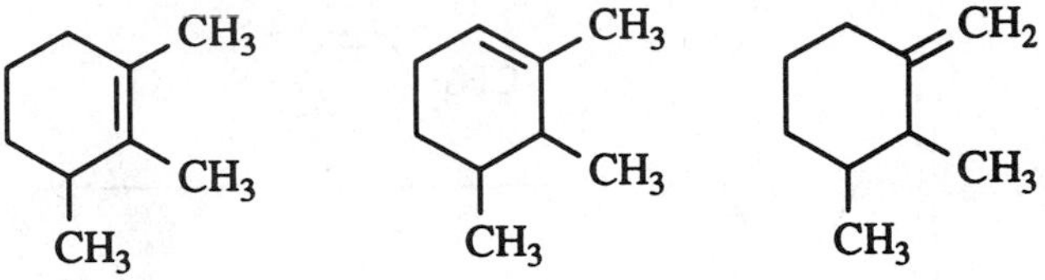

(1,2,3-trimethylcyclohexene, labels: CH_3, CH_3, CH_3) most

(2,3-dimethyl-1-methylcyclohexene isomer, labels: CH_3, CH_3, CH_3)

(methylenecyclohexane, labels: CH_2, CH_3, CH_3)

(b) $CH_3CH{=}CH$–C_6H_5

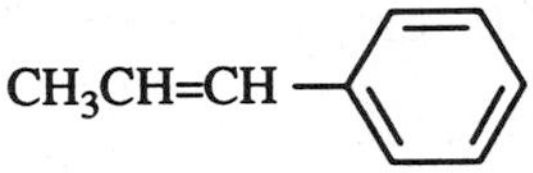

only alkene product

7.19 (a) $CH_3CH_2CH_2CH_2OS(=O)_2$–C_6H_4–CH_3

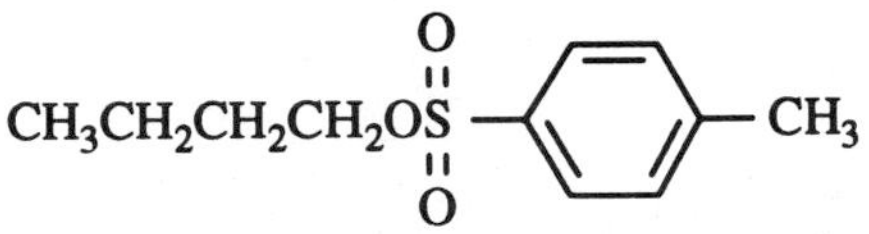

(b) CH_3CHONO_2 with CH_2CH_3 on the CH

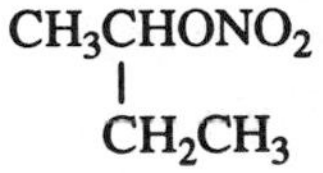

(c) $(CH_3)_3COSO_3H$

7.20 (1) $(CH_3)_2CHOH$ + CH_3–C_6H_4–$S(=O)_2$-Cl + pyridine ⟶

$(CH_3)_2CHOS(=O)_2$–C_6H_4–CH_3 + pyridinium (N–H, +) Cl^-

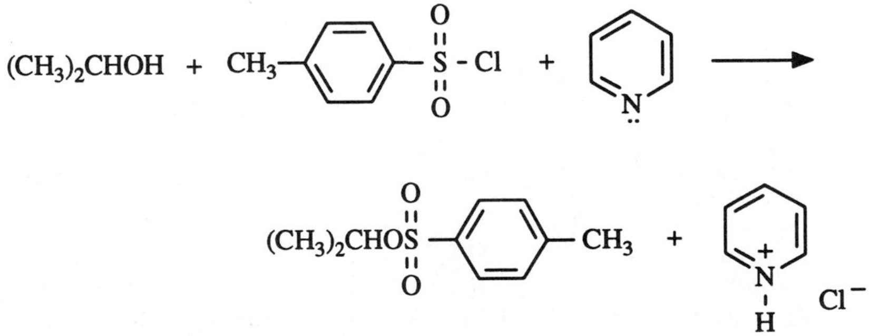

(2) $(CH_3)_2CH\ddot{O}SO_2$–C_6H_4–CH_3 $\xrightarrow{-\ CH_3C_6H_4SO_2\ddot{O}:^-}$

$[CH_3\overset{+}{C}HCH_3]$ $\xrightarrow{H_2\ddot{O}:}$ $CH_3CH(\overset{+}{O}H_2)CH_3$ $\xrightarrow{-H^+}$ $CH_3CH(:\ddot{O}H)CH_3$

7.21 (e) < [(d), (a)] < [(b), (f), (g)] < (c). Compounds (d) and (a) are at the same oxidation level, as are compounds (b), (f), and (g).

7.22 (a) $CH_3CH_2CH(OH)CH_3 + H_2CrO_4 \xrightarrow{heat}$ product

(b) $CH_3CH_2CH_2OH + H_2CrO_4 \xrightarrow{heat}$ product

(c) $CH_3CH_2CH_2CH_2OH + CrO_3 \cdot 2$ pyridine $\xrightarrow{25°}$ product

7.23 (a) next higher: $C_6H_5C(=O)OH$

next lower: $C_6H_5CH_2OH$

(b) next higher: $CH_3C{\equiv}CCH_3$

next lower: $CH_3CH_2CH_2CH_3$

(c) next higher: $CH_3CH_2CH(=O)$

next lower: $CH_3CH_2CH_3$

(d) next higher: 1,1,2-trichlorocyclohexane (Cl, Cl, Cl)

next lower: chlorocyclohexane (Cl)

7.24 (a) $C_6H_5CH_2Br \xrightarrow{OH^-} C_6H_5CH_2OH \xrightarrow{CrO_3 \cdot 2\ pyridine}$ product

(b) 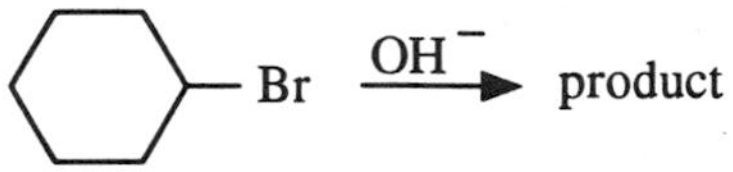 $C_6H_{11}Br$ (cyclohexyl–Br) $\xrightarrow{OH^-}$ product

(c) $CH_3CH_2CH_2CH(Br)CH_3 \xrightarrow{OH^-} CH_3CH_2CH_2CH(OH)CH_3 \xrightarrow{H_2CrO_4,\ heat}$ product

(d) $(CH_3)_2CHCH_2Br \xrightarrow{OH^-} (CH_3)_2CHCH_2OH \xrightarrow{H_2CrO_4,\ heat}$ product

7.25 (a) (*R*)-$CH_3CH_2CH(CH_3)OH$

(b) 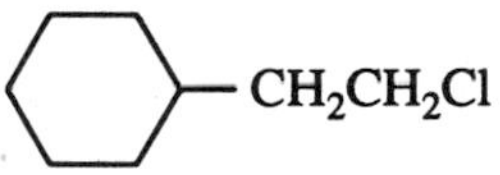 cyclohexyl–CH_2CH_2Cl

(c) 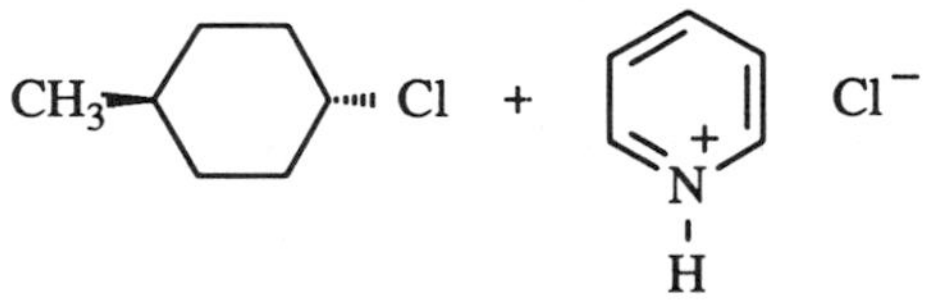 CH_3–cyclohexane–Cl + pyridinium ($C_5H_5N^+H$) Cl^-

(d) $(C_6H_5)_2C(CH_2CH_3)OH$

(e) 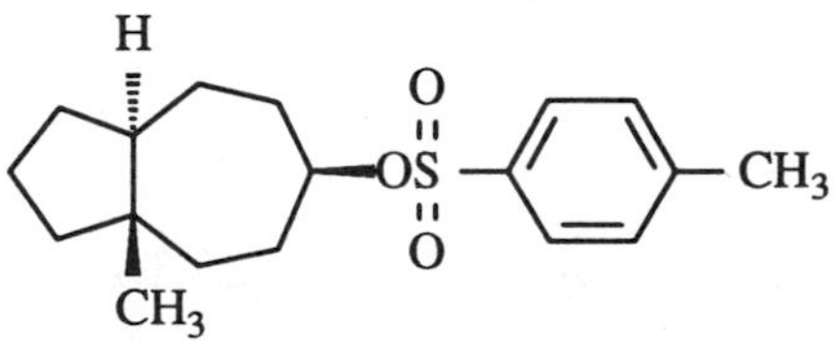 H, CH_3 bicyclic –OSO_2–C_6H_4–CH_3

(f) (*R*)-CH_3CHDCl

(g) $(CH_3)_2C(OH)CH_2CH(=O)$

(h) (*R*)-$CH_3CH(O^-Na^+)CH_2CH_3$

7.26 (a)

$$CH_3CH_2CH(\ddot{O}H)CH_2CH_2 \underset{}{\overset{H^+}{\rightleftharpoons}} CH_3CH_2CH(\overset{+}{\ddot{O}}H_2)CH_2CH_3 \xrightarrow{-H_2\ddot{O}:}$$

$$CH_3CH_2\overset{+}{C}HCH_2CH_3 \xrightarrow{:\ddot{Br}:^-} CH_3CH_2CH(:\ddot{Br}:)CH_2CH_3$$

(b)

$$(R)\text{-}CH_3CH_2CH(CH_3)\ddot{O}H + CH_3S(=O)(=\ddot{O}:)-\ddot{Cl}: \longrightarrow CH_3CH_2CH(CH_3)-\overset{+}{\ddot{O}}(H)-S(=O)(CH_3)(\ddot{O}:^-)-\ddot{Cl}: \xrightarrow{-HCl}$$

$$CH_3CH_2CH(CH_3)-\ddot{O}-SO_2CH_3 \xrightarrow[-CH_3SO_3^-]{^-:\ddot{O}H} (S)\text{-}H\ddot{O}-CH(CH_3)CH_2CH_3$$

7.27 (a) $SOCl_2$ + pyridine to avoid possible rearrangement.

(b) HI since no rearrangement will take place. If an S_N2 reaction is desired, then the alcohol should be converted to the tosylate and the tosylate treated with I^-.

(c) HCl because this alcohol or its chlorosulfite cannot undergo S_N2 reaction.

(d) $SOCl_2$ + pyridine because HCl is less reactive with 1° alcohols.

7.28 (a)

$$2\ CH_3CH_2CH(OH)CH_3 + 2Na \longrightarrow 2\ CH_3CH_2CH(O^-\ Na^+)CH_3 + H_2$$

(b)

$$CH_3CH_2CH(OH)CH_3 + NaOH \longrightarrow \text{no appreciable reaction}$$

(c)

$$CH_3CH_2CH(OH)CH_3 + HI \longrightarrow CH_3CH_2CH(I)CH_3 + H_2O$$

(d) $CH_3CH_2\underset{\displaystyle OH}{\underset{|}{C}}HCH_3 \xrightarrow{H_2SO_4,\ heat}$ *cis* and *trans*-$CH_3CH{=}CHCH_3$ (*most*) + $CH_3CH_2CH{=}CH_2$

(e) $CH_3CH_2\underset{\displaystyle OH}{\underset{|}{C}}HCH_3 \xrightarrow{KMnO_4,\ ^{-}OH} CH_3CH_2\overset{\displaystyle O}{\overset{\|}{C}}CH_3$

(f) $CH_3CH_2\underset{\displaystyle OH}{\underset{|}{C}}HCH_3 + SOCl_2 \longrightarrow CH_3CH_2\underset{\displaystyle Cl}{\underset{|}{C}}HCH_3 + SO_2 + HCl$

(g) $CH_3CH_2\underset{\displaystyle OH}{\underset{|}{C}}HCH_3 + NaH \longrightarrow CH_3CH_2\underset{\displaystyle O^-\ Na^+}{\underset{|}{C}}HCH_3 + H_2$

(h) $CH_3CH_2\underset{\displaystyle OH}{\underset{|}{C}}HCH_3 + NaCN \longrightarrow$ no reaction

(i) $CH_3CH_2\underset{\displaystyle OH}{\underset{|}{C}}HCH_3 + PCl_3 \longrightarrow CH_3CH_2\underset{\displaystyle Cl}{\underset{|}{C}}HCH_3 + HOPCl_2$

7.29 (a) $CH_3CH_2CH_2CH_2OH + H_2CrO_4 \xrightarrow{heat} CH_3CH_2CH_2CO_2H$

(b) $CH_3CH_2CH_2CH_2OH + SOCl_2 \xrightarrow{pyridine} CH_3CH_2CH_2CH_2Cl$

(c) $CH_3CH_2CH_2CH_2OH + HI \longrightarrow CH_3CH_2CH_2CH_2I$

(d) $CH_3CH_2CH_2CH_2OH \xrightarrow{CrO_3 \cdot 2\ pyridine} CH_3CH_2CH_2\overset{\displaystyle O}{\overset{\|}{C}}H$

$$\xrightarrow[(2)\ H_2O,\ H^+]{(1)\ CH_3MgBr} CH_3CH_2CH_2\underset{}{\overset{OH}{\overset{|}{C}}}HCH_3 \xrightarrow{H_2CrO_4,\ heat} CH_3CH_2CH_2\overset{O}{\overset{\|}{C}}CH_3$$

(e) $$CH_3CH_2CH_2CH_2OH \xrightarrow{CrO_3 \cdot 2\ pyridine} CH_3CH_2CH_2\overset{O}{\overset{\|}{C}}H$$

(f) $$CH_3CH_2CH_2CH_2OH \xrightarrow{HBr} CH_3CH_2CH_2CH_2Br \xrightarrow[ether]{Mg}$$

$$CH_3CH_2CH_2CH_2MgBr \xrightarrow[(2)\ H_2O,\ H^+]{(1)\ HCHO} CH_3CH_2CH_2CH_2CH_2OH$$

(g) $$CH_3CH_2CH_2CH_2OH \xrightarrow{HBr} CH_3CH_2CH_2CH_2Br \xrightarrow[ether]{Mg}$$

[from (f)]

$$CH_3CH_2CH_2CH_2MgBr \xrightarrow[(2)\ H_2O,\ H^+]{(1)\ CH_3CHO} CH_3CH_2CH_2CH_2\underset{OH}{\underset{|}{C}}HCH_3$$

(h) $$CH_3CH_2CH_2CH_2OH \xrightarrow{HBr} CH_3CH_2CH_2CH_2Br \xrightarrow[ether]{Mg}$$

$$CH_3CH_2CH_2CH_2MgBr \xrightarrow[(2)\ H_2O,\ H^+]{(1)\ CH_3CH_2CH_2CH_2CHO}$$

$$(CH_3CH_2CH_2CH_2)_2CHOH$$

(i) $$CH_3CH_2CH_2CH_2OH \xrightarrow[(2)\ CN^-]{(1)\ HBr} CH_3CH_2CH_2CH_2CN$$

7.30 (a) $CH_3C(=O)CH_3 \xrightarrow[(2)\ H_2O,\ H^+]{(1)\ CH_3MgBr} (CH_3)_3COH$

(b) $CH_3C(=O)CH_3 \xrightarrow[(2)\ H_2O,\ H^+]{(1)\ CH_3CH_2MgBr} (CH_3)_2C(OH)CH_2CH_3$

(c) $CH_3C(=O)CH_3 \xrightarrow[(2)\ H_2O,\ H^+]{(1)\ NaBH_4} (CH_3)_2CHOH \xrightarrow{HBr} (CH_3)_2CHBr$

$\xrightarrow[ether]{Mg} (CH_3)_2CHMgBr \xrightarrow{D_2O} (CH_3)_2CHD$

8

Ethers, Epoxides, and Sulfides

Answers to Additional Problems

8.1 (a) benzyl phenyl ether

(b) diallyl ether or di-2-propenyl ether

(c) 1-ethoxy-2-methoxybenzene

(d) 2-isopropoxyethanol

(e) *sec*-butyl isopropyl ether

(f) (*S*)-4-chloro-4-methyl-1-hexene oxide or (*S*)-(2-chloro-2-methylbutyl)oxirane

8.2 (a) CH_3OCH_3

(b) 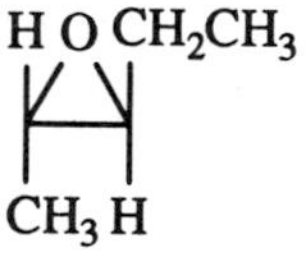 H O CH_2CH_3 / CH_3 H

(c) 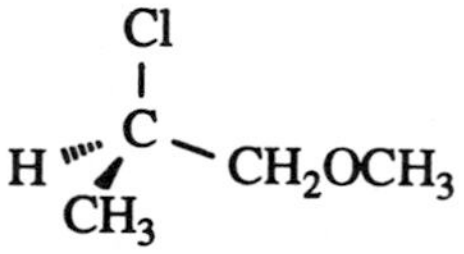 Cl / H C CH_2OCH_3 / CH_3

(d) 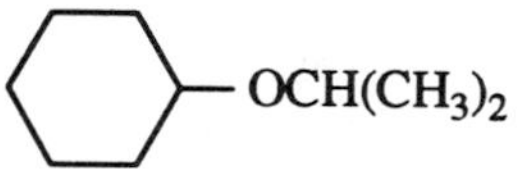 $OCH(CH_3)_2$

(e) $CH_3OCH_2CH_2OH$

(f) 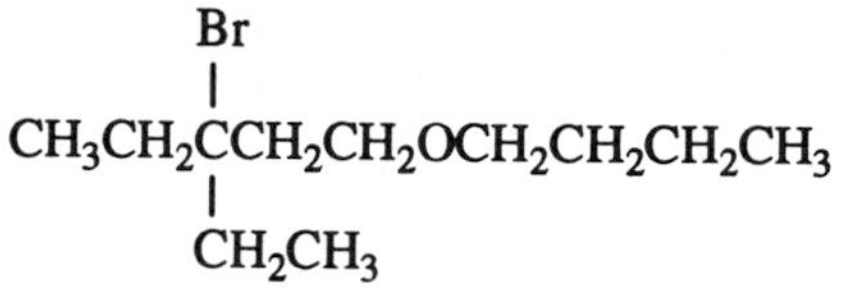 $CH_3CH_2C(Br)(CH_2CH_3)CH_2CH_2OCH_2CH_2CH_2CH_3$

(g) 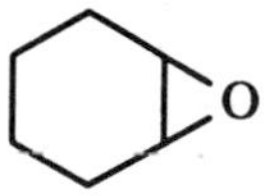 O

(h)  $CH_3CH_2CH_2CH\text{-}CH_2$ (epoxide, O bridging CH and CH_2)

(i) 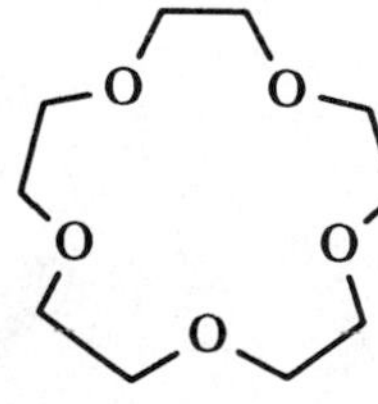 O O O O O

(j) 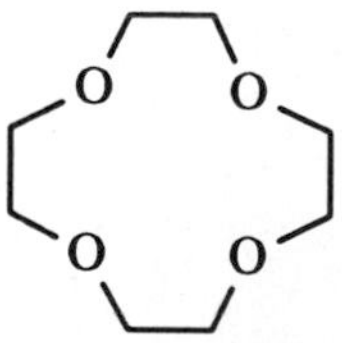 O O O O

8.3 (a) $CH_2{=}CH_2$ (b) $CH_3CH_2OCH_2CH_3$ (c) CH_3CH_2OH

8.4 (a) no reaction (b) no reaction (c) no reaction

(d) $(CH_3)_2CHCH_2Br$ + $CH_3CH_2CH_2Br$

(e) no reaction (f) no reaction

8.5 (a) $C_6H_5-CH_2Br + {}^{-}OCH(CH_3)_2 \longrightarrow C_6H_5-CH_2OCH(CH_3)_2$

(b) $CH_3CH_2CH_2CH_2Br + {}^{-}OC(CH_3)_3 \longrightarrow CH_3CH_2CH_2CH_2OC(CH_3)_3$

(c) $CH_3CH_2CH_2CH_2Br + {}^{-}O-C_6H_5 \longrightarrow$

$CH_3CH_2CH_2CH_2O-C_6H_5$

8.6 $C_6H_5-OCH_2-C_6H_5 \xrightarrow{H^+} C_6H_5-\overset{+}{O}(H)-CH_2-C_6H_5$ (I^- attacks CH_2)

$\longrightarrow C_6H_5-OH + ICH_2-C_6H_5$

8.7 (a) CH_2-CH_2 (epoxide, bridged by O) $\xrightarrow{HI} HOCH_2CH_2I \xrightarrow{HI} ICH_2CH_2I$

(b) CH_2-CH_2 (epoxide, bridged by O) $+ (CH_3)_2NH \longrightarrow HOCH_2CH_2N(CH_3)_2$

(c) CH_2-CH_2 (epoxide, bridged by O) $+ C_6H_5MgBr \longrightarrow BrMgOCH_2CH_2C_6H_5$

(d) CH_2-CH_2 (epoxide, bridged by O) + $CH_3CH_2OH \xrightarrow{H^+} HOCH_2CH_2OCH_2CH_3$

(e) CH_2-CH_2 (epoxide, bridged by O) + $H_2O \xrightarrow{NaOH} HOCH_2CH_2OH$

(f) CH_2-CH_2 (epoxide, bridged by O) + $CH_3CH_2OH \xrightarrow{Na^+\ ^-OCH_2CH_3} HOCH_2CH_2OCH_2CH_3$

8.8 (a) CH_3CH-CH_2 (epoxide, bridged by O) $\xrightarrow{HI} CH_3CH(I)-CH_2(OH) \xrightarrow{HI} CH_3CHICH_2I$

(b) CH_3CH-CH_2 (epoxide, bridged by O) + $(CH_3)_2NH \longrightarrow CH_3CH(OH)-CH_2N(CH_3)_2$

(c) CH_3CH-CH_2 (epoxide, bridged by O) + $C_6H_5MgBr \longrightarrow CH_3CH(OMgBr)-CH_2C_6H_5$

(d) CH_3CH-CH_2 (epoxide, bridged by O) + $CH_3CH_2OH \xrightarrow{H^+}$

$CH_3CH(OCH_2CH_3)-CH_2OH$ + $CH_3CH(OH)-CH_2OCH_2CH_3$

(e) CH_3CH-CH_2 (epoxide, bridged by O) + $H_2O \xrightarrow{NaOH} CH_3CH(OH)-CH_2OH$

(f) CH_3CH-CH_2 (epoxide, bridged by O) + $CH_3CH_2OH \xrightarrow{Na^+\ ^-OCH_2CH_3} CH_3CH(OH)-CH_2OCH_2CH_3$

8.9 (structures: 1-methylcyclopentene oxide, protonated by H^+, attacked by $CH_3CH_2\ddot{O}H$; loss of H^+ gives the trans product with OH, CH_3, OCH_2CH_3)

H⁺ ⇌ → ; $-H^+$ ⇌

$HOCH_2CH_3$ (+) ; $:OCH_2CH_3$; $\ddot{O}H$; CH_3

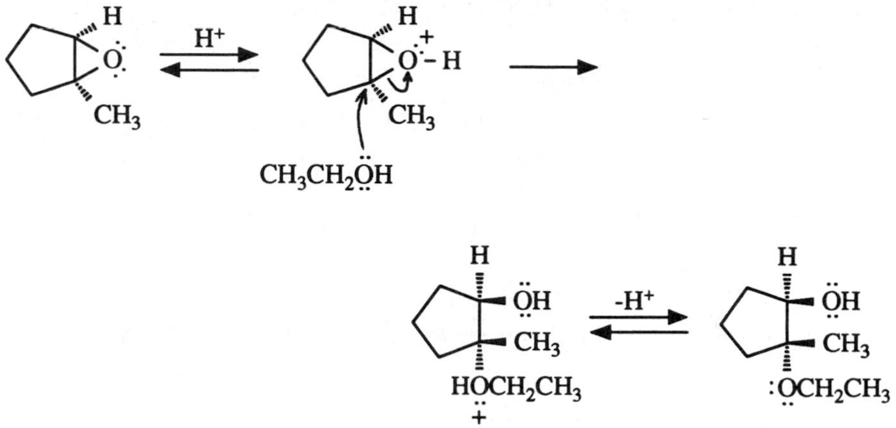

8.10 (a)  cyclohexyl–S — S–cyclohexyl (b) 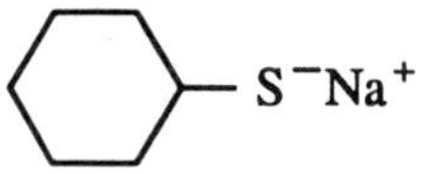 cyclohexyl–S^-Na^+

(c) 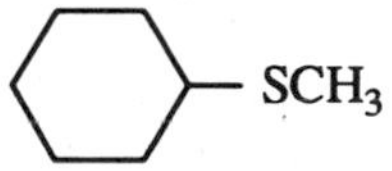 cyclohexyl–SCH_3 (d) 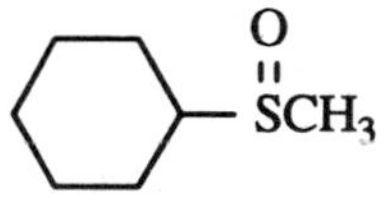 cyclohexyl–S(=O)CH_3

8.11 excess CH_3CH_2Br + Na_2S ⟶ $(CH_3CH_2)_2S$ + 2 NaBr

8.12 (a) $CH_3CH_2CH_2CH_2O^-\ K^+$ (b) $(CH_3)_2CHOCH_2CH_2CH_2CH_3$

(c) $(CH_3)_2C(Br)OCH_2CH_2CH_2CH_3$ (d) $CH_2{=}C(CH_3)OCH_2CH_2CH_2CH_3$

8.13 (a) 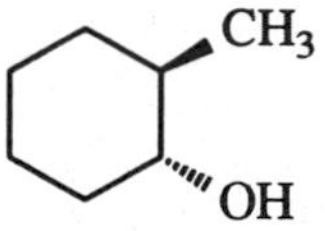 trans-2-methylcyclohexanol (CH_3, OH) (b) 2 CH_3I + o-(iodomethyl)phenol (OH, CH_2I)

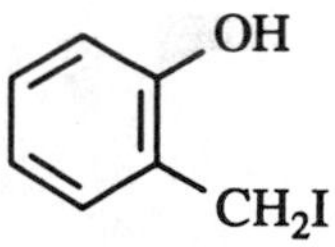

(c) $C_6H_5CH(OH)CH_2C_6H_5$ (d)  piperidine-NCH_2CH_2OH

(e) $HOCH_2CH_2Br$ + $HOCH_2CH_2OH$

(f) $C_6H_5CH(OH)-CH(OCH_3)-C_6H_4-OCH_3$ + $C_6H_5CH(OCH_3)-CH(OH)-C_6H_4-OCH_3$

(g) no reaction

8.14 (a) $CH_3CH_2CH_2CH_2OH \xrightarrow{K} CH_3CH_2CH_2CH_2O^-K^+$

$CH_3CH_2CH_2CH_2OH \xrightarrow{HI} CH_3CH_2CH_2CH_2I$

$CH_3CH_2CH_2CH_2O^-K^+ + CH_3CH_2CH_2CH_2I \longrightarrow (CH_3CH_2CH_2CH_2)_2O$

(b) $CH_3CH_2CH_2CH_2I$ from (a) $\xrightarrow{HS^-}$ $CH_3CH_2CH_2CH_2SH$

(c) $CH_3CH_2CH_2CH_2SH$ from (b) $\xrightarrow{^-OH}$ $CH_3CH_2CH_2CH_2S^-$

$\xrightarrow{CH_3CH_2CH_2CH_2I}$ $(CH_3CH_2CH_2CH_2)_2S$

9

Spectroscopy I

Answers to Additional Problems

9.1 (a) 2.78 μm (b) 6.90 μm (c) 9.52 μm

9.2 (a) 3030 cm^{-1} (b) 1385 cm^{-1} (c) 1818 cm^{-1}

9.3 (a) < (d) < (b) < (c)

9.4 (a) The -OH band of the alcohol at around 3300 cm^{-1} (3.0 μm) would distinguish that compound from the aldehyde and the ketone. The aldehyde and ketone would show strong carbonyl absorption at about 1700-1750 cm^{-1} (5.7-5.9 μm). The C-H absorption of the CHO group at around 2900 cm^{-1} (3.45 μm) would distinguish the aldehyde from the ketone.

(b) The characteristic hydrogen-bonded O-H group of the carboxylic acid at about 3300 cm^{-1} (3.0 μm) would distinguish acetic acid from the other two compounds. The ester could be distinguished from the ketone by a strong band around 1110-1300 cm^{-1} (7.7-9.0 μm) arising from C-O absorption.

(c) The secondary amine could be distinguished from the other two compounds by the presence of a single band around 3300 cm^{-1} (3.0 μm). The tertiary amine could be distinguished from the hydrocarbon by C-N absorption around 900-1300 cm^{-1} (8-11 μm).

9.5 The pairs of compounds in (b), (c), and (d) could not be distinguished by infrared spectroscopy.

(b) Both compounds contain the same functional groups, OH and phenyl.

(c) Both the internal double bond and the internal triple bond are symmetrical and, thus, infrared inactive.

(d) Both compounds contain only N-H groups as functional groups. The compounds could not be distinguished if the spectrum of only one of the two compounds is available; however, if both spectra were available, they could be distinguished by the relative intensities of the N-H and C-H absorption.

9.6 (a) the terminal C≡C band at about 2220 cm^{-1} (4.5 μm)

(b) the carbonyl band at about 1700 cm^{-1} (5.9 μm)

(c) the O-H band at about 3300 cm^{-1} (3.0 μm)

(d) the NH_2 double peak at about 3300 cm^{-1} (3.0 μm)

9.7 (a) The NH_2 band of propylamine at around 3300 cm^{-1} (3.0 μm) would be the most useful band for distinguishing the two compounds.

(b) the carbonyl band of $CH_3C{\equiv}C\overset{\overset{O}{\|}}{C}H$ at around 1690 cm^{-1} (5.9 μm)

(c) the carbonyl band of $C_6H_5\overset{\overset{O}{\|}}{C}Cl$ at about 1780 cm^{-1} (5.6 μm)

9.8 The resonance structures show that the C-O bond of phenol has substantial double-bond character. Therefore, we would expect that the stretching vibration would require more energy and that the infrared absorption would be shifted to higher, more energetic frequencies (shorter wavelength). Indeed, the range for phenol C-O absorption is 1140-1230 cm^{-1} (8.13-8.77 μm), while that for an alcohol is about 1075 cm^{-1} (9.3 μm).

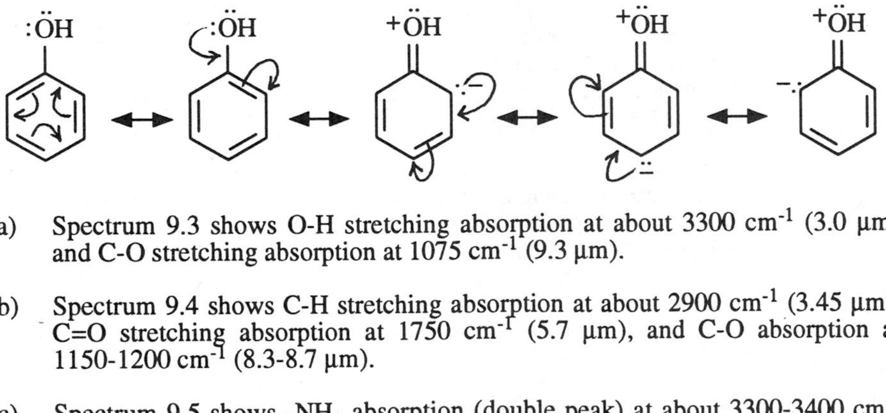

:ÖH ⟷ :ÖH ⟷ +ÖH ⟷ +ÖH ⟷ +ÖH

9.9 (a) Spectrum 9.3 shows O-H stretching absorption at about 3300 cm^{-1} (3.0 μm) and C-O stretching absorption at 1075 cm^{-1} (9.3 μm).

(b) Spectrum 9.4 shows C-H stretching absorption at about 2900 cm^{-1} (3.45 μm), C=O stretching absorption at 1750 cm^{-1} (5.7 μm), and C-O absorption at 1150-1200 cm^{-1} (8.3-8.7 μm).

(c) Spectrum 9.5 shows $-NH_2$ absorption (double peak) at about 3300-3400 cm^{-1} (2.9-3.0 μm). The C-N absorption is more difficult to assign but absorption bands do appear at 900 cm^{-1} (11.0 μm) and at 1060 cm^{-1} (9.4 μm).

(d) Spectrum 9.1 shows typical carboxyl O-H absorption at 3000-3500 cm^{-1} (2.9-3.3 μm), and carbonyl absorption at 1710 cm^{-1} (5.8 μm).

(e) Spectrum 9.2 shows aldehyde C-H absorption at 2720 cm^{-1} (3.7 μm) and carbonyl absorption at 1730 cm^{-1} (5.8 μm). Note the lack of C-O absorption in the 900-1300 cm^{-1} (7.7-11.0 μm) range.

9.10 (a) RNH_2, spectrum 9.10 (b) ArOH, spectrum 9.6

(c) ROH, spectrum 9.9 (d) RNHR′, spectrum 9.8

(e) RCO_2R', no match (f) RCHO, spectrum 9.7

9.11 compound (c), $CH_3CH(CH_3)CH_2C(=O)OH$

$$\begin{array}{c} O \\ \| \\ CH_3\underset{\substack{|\\ CH_3}}{C}HCH_2COH \end{array}$$

Reasons: The carboxyl O-H absorption at 3000-3500 cm^{-1} (2.9-3.3 μm) is evident, as is the carbonyl absorption at 1700 cm^{-1} (5.9 μm). The spectrum does not show aryl C-C absorption around 1600 cm^{-1} (6.25 μm). Therefore, compound (c) is selected.

9.12 compound (c), $CH_2{=}CHCH_2CH_2\overset{O}{\overset{\|}{C}}CH_3$

Reasons: The spectrum shows carbonyl absorption, which rules out compound (a). The choice among compounds (b), (c), and (d) is based upon the absence of C≡C absorption at 2000-2500 cm^{-1} (4-5 μm), the presence of a C=C-H band at 2900 cm^{-1} (3.45 μm), and a C=C band at 1640 cm^{-1} (6.1 μm).

9.13 (a) $CH_3CH_2CH(CH_2CH_3)CH_2CH_3$ — $\underline{CH_3}CH_2CHC(H_2)CH_3$, with branch C(H₂) and CH_3 (circled: H_2, H_2; underlined: CH_3)

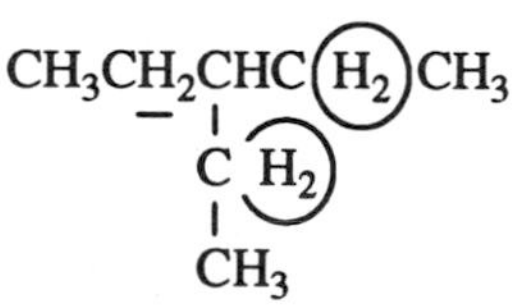

(b) benzene ring with substituents: OCH_3 (top), H (upper left, underlined), (H) (upper right, circled), H (lower left), H (lower right), H (bottom)

(c) $CH_2{=}CH{-}CH{=}CH{-}CH{=}CH_2$ drawn with H atoms: H, (H) circled, $\underline{H}$, H

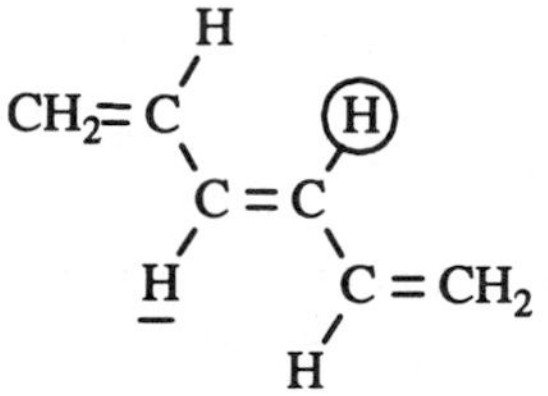

(d) $\underline{CH_3}\text{-}O\overset{O}{\overset{\|}{C}}\text{-}\underset{\substack{|\\ OCH_3}}{CH}\text{-}\overset{O}{\overset{\|}{C}}OC(H_3)$ (circled: H_3)

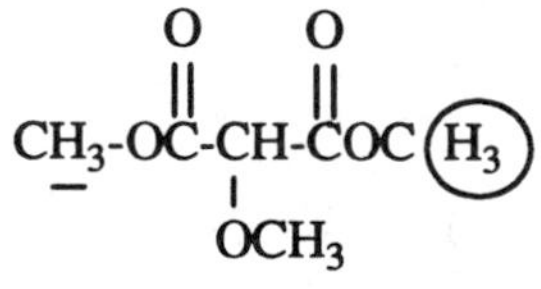

9.14 (a) $CH_3CH_2\underset{}{\overset{\substack{OH\\ |}}{C}}HCH_3$ — labels: a (CH_3), b (CH_2), c (CH), d (CH_3), e (OH)

(b) $CH_3\overset{\substack{CH_3\\ |}}{C}HCH_2CH_2OH$ — labels: a (CH_3 groups), b (CH), c (CH_2), d (CH_2), e (OH)

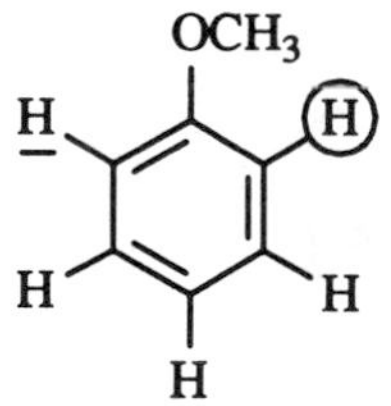

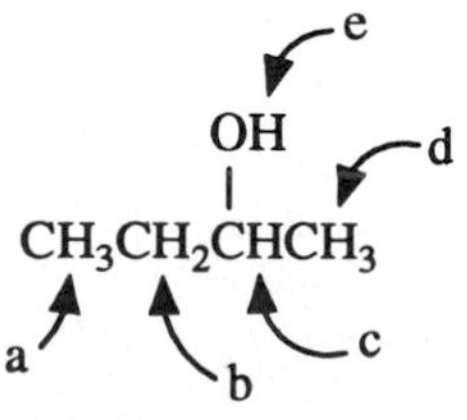

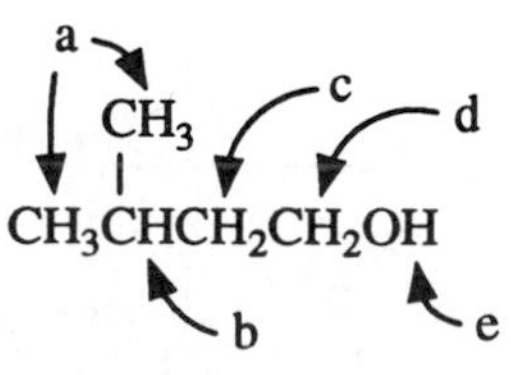

(c)

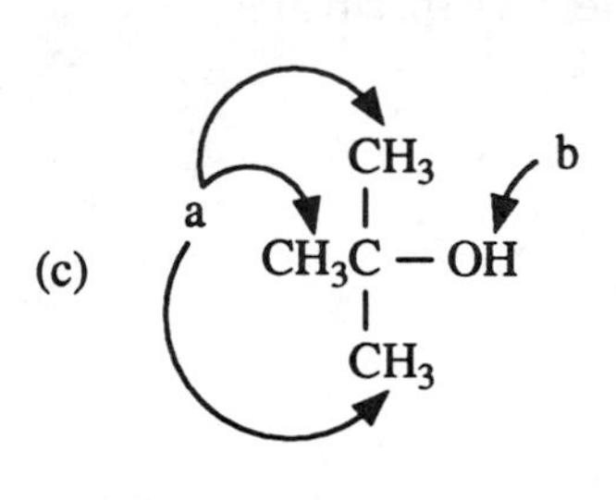

(d)

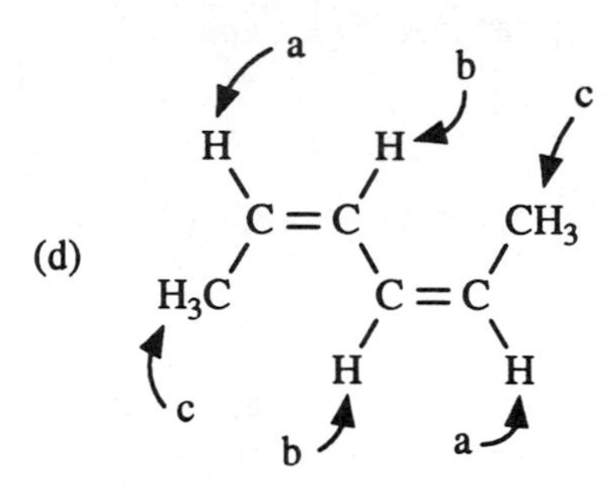

(e) c, b, a, d, c; CH_3, CH_3

(f) c, CH_3, b, a, d, f, e, OH

(g) a b c b a

$(CH_3)_2CH$ — — $CH(CH_3)_2$

c

(h) b c

CH_3O — — Br

a b c

(i) a b c d c b a

$CH_3CH_2CH_2CH_2CH_2CH_2CH_3$

9.15 A concentrated solution of the alcohol contains a greater number of hydrogen-bonded protons; therefore, more protons are deshielded by two adjacent oxygens:

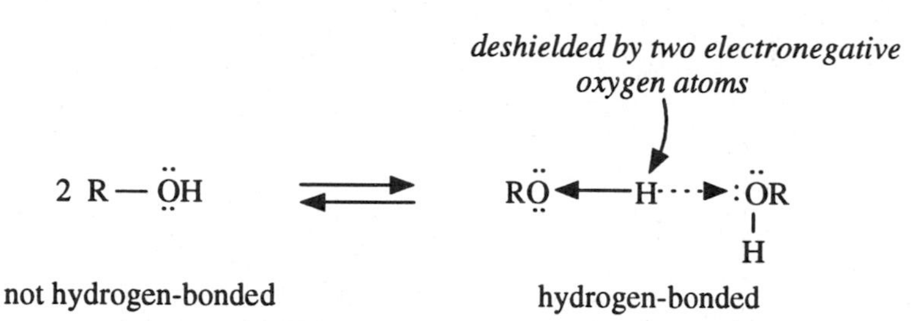

As the concentration of ROH increases, the equilibrium shifts to the right.

9.16 (a) (2)

(b) 1

9.17 (f) < (d) < (e) < (a) < (b) < (c)

9.18 (d) < (b) < (a) < (c)

9.19 $\frac{62.5}{20.8} = 3.0$ $\frac{20.8}{20.8} = 1$

Therefore, the proton ration is 3:1.

Compound (a) has a proton ration of 6:4, or 3:2 and is not compatible. Compound (b) has a proton ratio of 6:2, and is thus compatible with the spectrum. The proton ratio is 3:1. Compound (c) would exhibit only a singlet and, therefore, does not fit the data.

9.20 (a) 9 : 2 : 1 (b) 6 : 2 : 2 : 2 or 3 : 1 : 1 : 1

9.21 (a) triplet (3 peaks) (b) sextet (6 peaks)

(c) septet (7 peaks) (d) quartet (4 peaks)

(e) singlet (1 peak) (f) doublet (2 peaks)

9.22 (a)

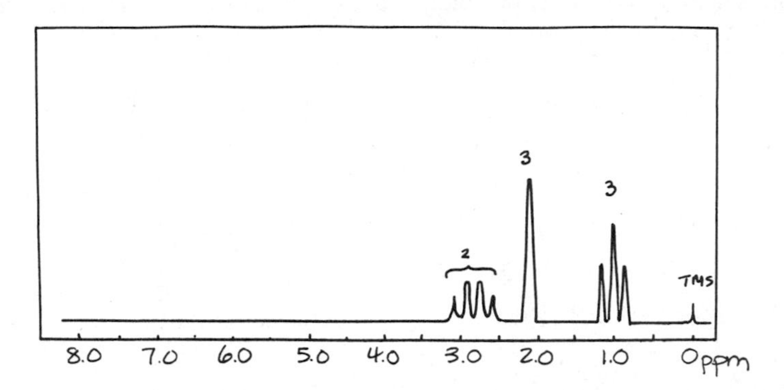
3
3
2
TMS
8.0
7.0
6.0
5.0
4.0
3.0
2.0
1.0
0 ppm

(b)

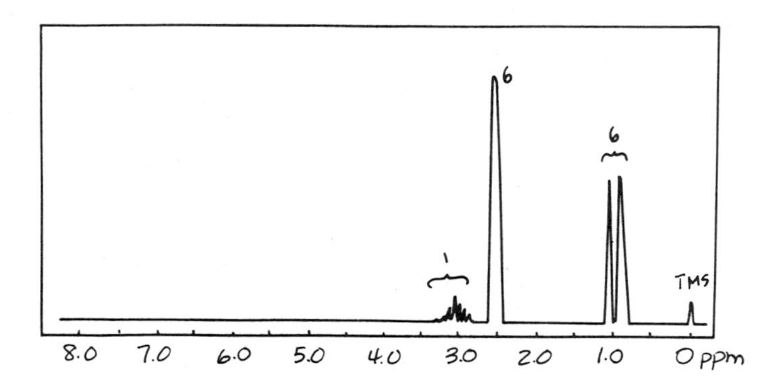
6
6
1
TMS
8.0
7.0
6.0
5.0
4.0
3.0
2.0
1.0
0 ppm

(c)

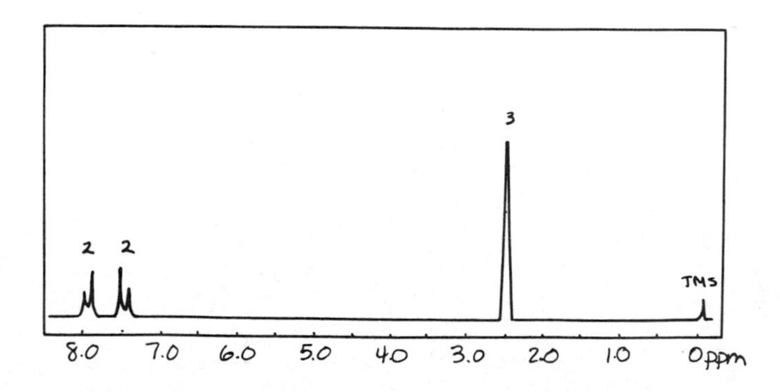
3
2 2
TMS
8.0
7.0
6.0
5.0
4.0
3.0
2.0
1.0
0 ppm

9.23 (a) Look for the multiplicity of the offset absorption of the aldehyde proton. In $C_6H_5C\underline{H}O$, the absorption would be a singlet because there are no neighboring protons. The offset absorption of $C_6H_5CH_2C\underline{H}O$, however, would be a triplet because there are two neighboring protons. $C_6H_5C\underline{H}_2CHO$ would also show a doublet in the 2-3 ppm region.

(b) If the aromatic absorption shows a pair of doublets, then the compound is *p*-$HOC_6H_4CH_3$. If, however, the pattern is a multiplet, then the compound is the *m* isomer.

(c) If the spectrum consists of a single singlet, then the structure is $(CH_3)_2C{=}C(CH_3)_2$. However, if a multiplet is observed, the structure is $(CH_3CH_2)_2C{=}CH_2$.

(d) If the spectrum consists of a doublet and a small multiplet (septet), the compound is $(CH_3)_2CHBr$. However, if the pattern consists of many small peaks, the compound is $CH_3CH_2CH_2Br$.

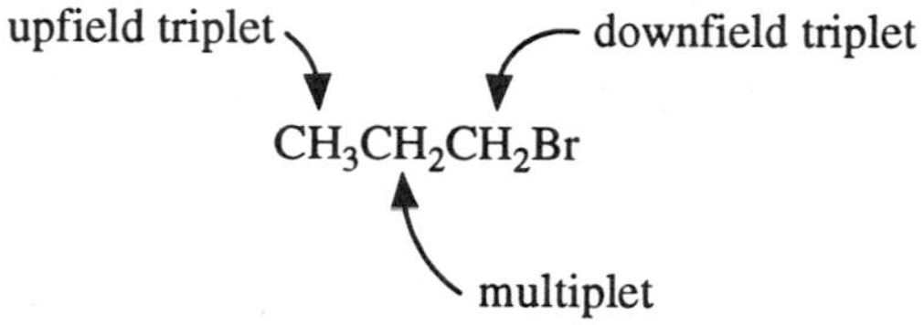

upfield triplet ↘ ↙ downfield triplet

$CH_3CH_2CH_2Br$

↖ multiplet

9.24 compound (c), $(CH_3)_3CNH_2$

Reasons: There are only two compounds in the list that have a proton ratio of 9 : 2 - compounds (b) and (c). These two compounds can be distinguished by the chemical shift of the protons with area 2. Thc -$C\underline{H}_2$-Cl protons should be deshielded and downfield, while the -$N\underline{H}_2$ protons should be found farther upfield. Because the absorption is upfield (δ 1.1), the spectrum is that of compound (c).

9.25 compound (b), $CH_3O-C_6H_4-OCH_3$

Reasons: Only compounds (b) and (d) have proton ratios of 6 : 4, or 3 : 2. Compound (b) was selected on the basis of the chemical shift of the absorption band at about δ 3.7. Had the spectrum been that of compound (d), the chemical shift of that absorption should have been farther upfield.

9.26

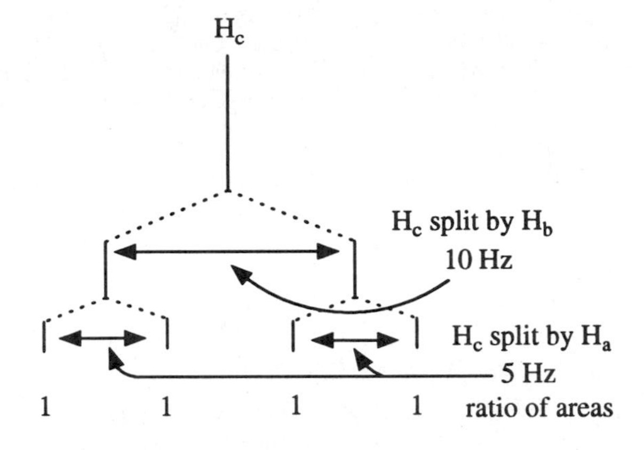

9.27 compound (a), $CH_3CHClCHCl_2$

Reasons: Compounds (b) and (c) contain only two types of equivalent carbons, not three. Compound (d) would show one singlet, one triplet, and one quartet in the off-resonance decoupled spectrum. Only compound (a) fits the splitting patterns:

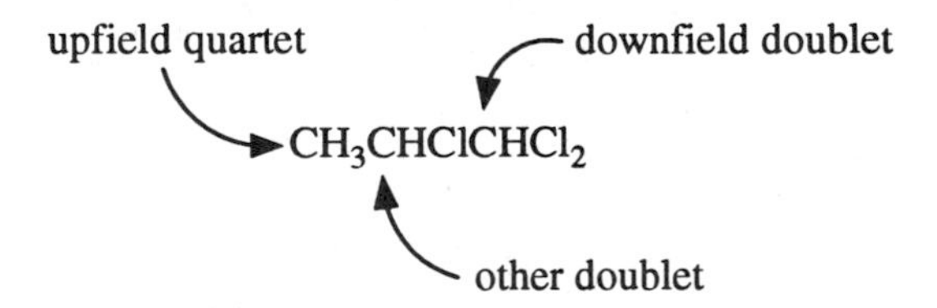

9.28 compound (c), $(CH_3)_2C{=}O$

Reasons: The compound contains two types of carbon atoms, one bonded to three hydrogens and the other bonded to zero hydrogens. The ^{13}C-H off-resonance decoupled spectrum shows a quartet, indicating that one of the carbons is bonded to three hydrogens. This spectrum also shows a singlet, indicating that the other carbon is bonded to zero hydrogens.

9.29 compound (c), $(CH_3CH_2CH_2)_2O$

Reasons: Only compounds (b) and (d) contain three nonequivalent carbons. These two compounds can be distinguished by the chemical shifts of the three peaks in the proton decoupled spectrum. Compound (b) would show two downfield peaks ($\underline{C}H_3O\underline{C}H_2CH_3$) and one upfield peak ($CH_3OCH_2\underline{C}H_3$). Compound (c), on the other hand, would show two upfield peaks ($\underline{C}H_3\underline{C}H_2CH_2)_2O$ and one downfield peak $(CH_3CH_2\underline{C}H_2)_2O$.

The off-resonance decoupled spectrum is not easily interpreted because of overlapping peaks from the two triplets.

9.30 compound (b), $(CH_3)_2CHCH_2CH{=}CH_2$

Reasons: Compound (c) can be eliminated because it contains only two sets of equivalent carbon atoms, not five. Compound (a) can be eliminated because the spectra show no band in the δ 200-220 region (C=O group). Compound (b) is chosen over (d) because of the downfield doublet (at about δ 140) and triplet (at about δ 120), denoting the two carbons of the $-CH{=}CH_2$ grouping. Compound (d) would be expected to show two doublets in this region.

9.31 The structure is

$$C_6H_5CH_2\underset{\displaystyle C_6H_5}{\underset{|}{C}}HOH\ .$$

Reasons: The infrared spectrum shows -OH. The ^1H NMR spectrum shows the following absorptions:

10 aryl protons: two C_6H_5-

triplet (area 1): $\diagdown\!\!\diagup C\underline{H}CH_2-$

doublet (area 2): $\diagdown\!\!\diagup CHC\underline{H}_2-$

singlet (area 1): $-O\underline{H}$

The triplet at δ 4.8 is downfield suggesting that this CH is bonded to O.

9.32 The compound is

$$CH_3CH{=}CH\overset{\displaystyle O}{\overset{\|}{C}}H\ .$$

Reasons: The infrared spectrum shows aldehyde CH and C=O bands. The ^1H NMR spectrum shows an offset doublet, suggesting $-\overset{|}{C}HC\underline{H}O$. The upfield doublet must be $C\underline{H}_3\overset{|}{C}H-$. The multiplets far downfield (6-7 ppm) suggest $-C\underline{H}{=}C\underline{H}-$.

9.33 The compound is $ClCH_2CH_2CH_2OH$.

Reasons: The infrared spectrum shows absorption for an OH group. The ^1H NMR spectrum shows a singlet at about δ 4.4, which can be assigned to this O$\underline{H}$. The quintet at δ 1.9 must come from $-CH_2C\underline{H}_2CH_2-$. The downfield multiplet must arise from two CH_2 groups of $-C\underline{H}_2CH_2C\underline{H}_2-$.

9.34 The compound is $C_6H_5CH_2CH_2OH$.

Reasons: The infrared spectrum shows OH, and the ^{1}H NMR spectrum shows a singlet at about δ 7.2 (area 5), a typical phenyl (C_6H_5-) peak. The remainder of the molecular formula, C_2H_4, can be assigned as follows:

triplet at δ 3.7 (area 2): $-CH_2C\underline{H_2}OH$
triplet at δ 2.6 (area 2): $-C\underline{H_2}CH_2OH$.

10

Alkenes and Alkynes

Answers to Additional Problems

10.1 (a) iodoethene

(b) cyclopropylethene

(c) (*E*)-2-chloro-2,4-pentadienal

(d) (*E*)-3,4-dimethyl-3-decene

(e) 1,2-pentadiene

(f) (*Z*)-1-bromo-2-methyl-2-butene

(g) (*E*)-1-bromo-2-mcthyl-2-butene

(h) (*Z*)-butenedioic acid

10.2 (a) 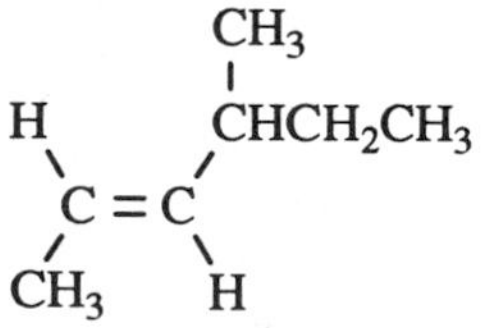

(b) 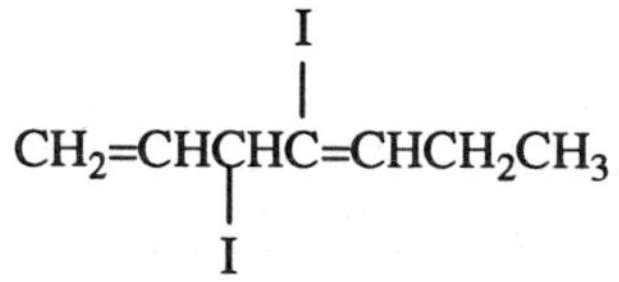

(c) 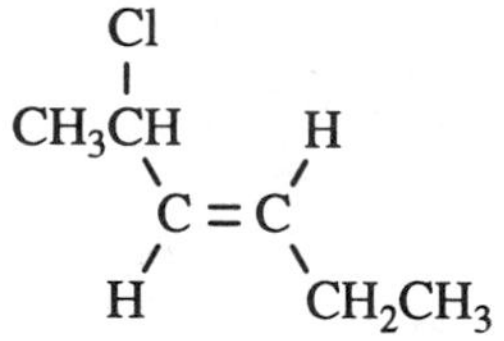

(d) 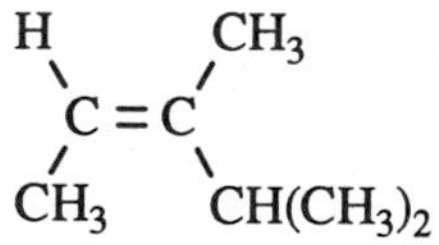

(e) 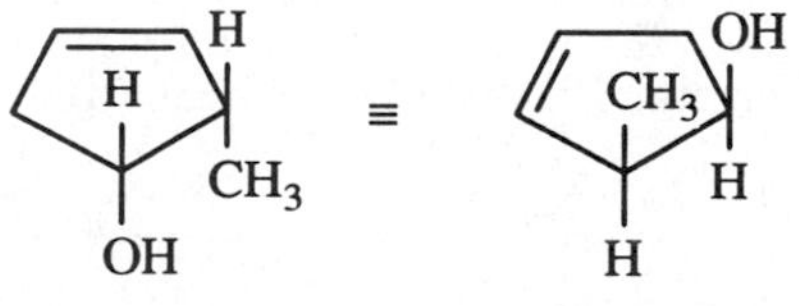

(f) 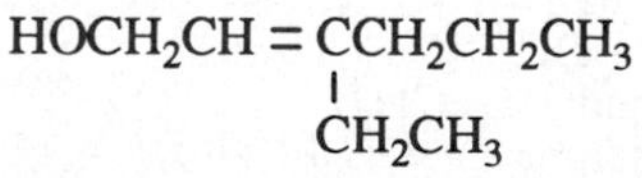

10.3 (a) 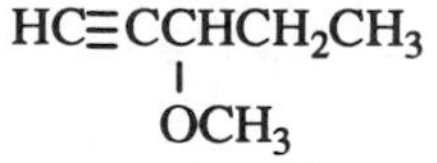 $HC{\equiv}CCHCH_2CH_3$ (with OCH_3 on the CH carbon)

(b) $BrCH_2C{\equiv}CCH_3$

(c) 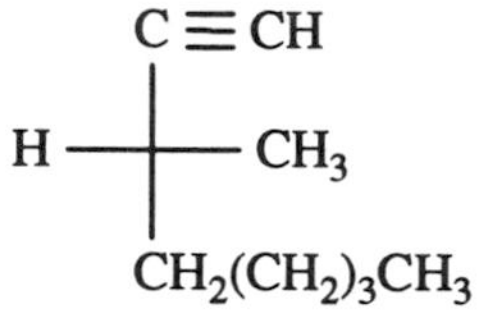 Fischer projection: top $C{\equiv}CH$; left H; right CH_3; bottom $CH_2(CH_2)_3CH_3$

or

$C{\equiv}CH$ on C, with H (hashed), H_3C (wedge), and $(CH_2)_4CH_3$

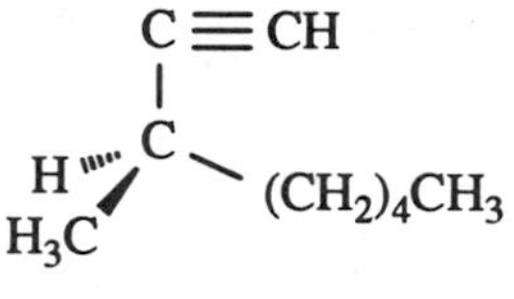

(d) 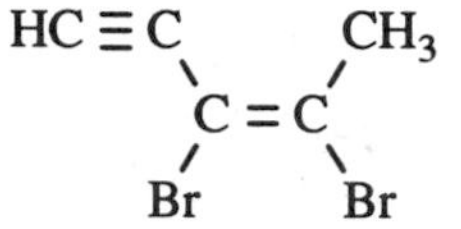 $HC{\equiv}C$ and CH_3 on top, $C{=}C$, Br and Br on bottom

$$\begin{matrix} HC{\equiv}C & & CH_3 \\ & C{=}C & \\ Br & & Br \end{matrix}$$

10.4 Compound (a) is (*Z*) and compounds (b), (c), and (d) are (*E*). The priority assignments are as follows:

(a) The bromo and propyl substituents are of higher priority.

(b) The methyl and chloro substituents are of higher priority.

(c) The fluoro and $CH_2C(CH_3)_3$ substituents are of higher priority.

(d) The methyl keto and the methyl ester groups are higher priority.

10.5 (a) *cis* (b) *trans* (c) The 1,2 double bond is *trans*

The terminal double bond in (c) cannot have *cis* and *trans* forms.

10.6 (a) a carboxyl group (b) no

(c) yes (d) $(CH_3)_2C{=}CHCO_2H$

10.7 The spectrum shows integration for 12 protons. This fact eliminates compounds (b) and (c) since they each contain 14 protons. The spectrum for compound (a) should show a singlet with area of nine. This is not observed in the spectrum of the unknown. Therefore, the spectrum must be that of compound (d).

Assigning the signals:

the signal at about δ 1.0 (area 3): $-C\underline{H_3}$
the multiplet between δ 1 and 2 (area 4): $-C\underline{H_2}C\underline{H_2}-$
the triplet at about δ 3.6 (area 2): $-OC\underline{H_2}CH_2CH_2CH_3$
the signals at about δ 4 (areas 1 and 1): $=C\underline{H_2}$
the four peaks between δ 6 and 7 (area 1): $CH_2{=}C\underline{H}O-$

10.8 (a) $C_6H_5CH_2Br$ (b) $C_6H_5C(Br)(CH_3)_2$ or $C_6H_5CH(CH_2Br)CH_3$

(c) $CH_3CH_2CH_2CH_2CH_2CH(Br)\text{-}CH_2Br$

10.9 (a) $HC{\equiv}CH \xrightarrow{NaNH_2} HC{\equiv}C{:}^{-}\,Na^{+} \xrightarrow[S_N2]{CH_3CH_2CH_2Br} CH_3CH_2CH_2C{\equiv}CH$

(b) $CH_3(CH_2)_3CH{=}CH_2 \xrightarrow{Br_2} CH_3(CH_2)_3CH(Br){-}CH_2Br \xrightarrow[(2)\ H_2O,\ H^+]{(1)\ NaNH_2} CH_3(CH_2)_3C{\equiv}CH$

10.10 (a) and (b) $CH_3CH_2CH(Cl)CH_3$ (racemic)

(c) $CH_2{=}CHCl$ and CH_3CHCl_2 (d) $(CH_3)_2CHCl$

(e) Cl, H, CH_3, CH_3 + CH_3, H, Cl, CH_3 (both racemic)

10.11 (a) $\xrightarrow{Br\cdot} CH_3CH_2CH_2\dot{C}HCH_2Br \xrightarrow[-Br\cdot]{HBr} CH_3CH_2CH_2CH_2CH_2Br$

(b) $\xrightarrow{H^+} C_6H_5\overset{+}{C}{=}CHC_6H_5 \xrightarrow{Cl^-} C_6H_5C(Cl){=}CHC_6H_5 \xrightarrow{H^+}$

$C_6H_5\overset{+}{C}(Cl)CH_2C_6H_5 \xrightarrow{Cl^-} C_6H_5C(Cl)_2CH_2C_6H_5$

(c) $\xrightarrow{H^+}$ $CH_3\overset{+}{C}HCH_2CH_2CH_2CH_3$ + $CH_3CH_2\overset{+}{C}HCH_2CH_2CH_3$ $\xrightarrow{Br^-}$

racemic $CH_3\underset{}{\overset{Br}{C}}HCH_2CH_2CH_2CH_3$ + $CH_3CH_2\underset{Br}{C}HCH_2CH_2CH_3$

10.12 (a) $CH_3CH_2\overset{OH}{C}HCH_3$ (b) $(CH_3)_2\overset{OH}{C}CH(CH_3)_2$

(c) 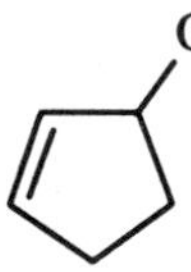 OH (by way of the allylic cation)

10.13 (c) < (a) < (b) < (d)

10.14 (a) 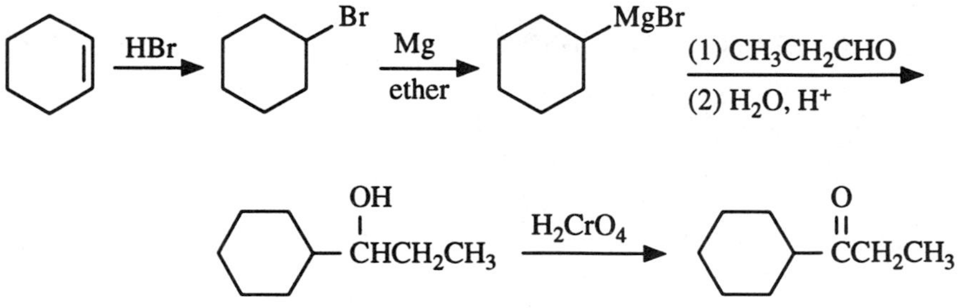 $\xrightarrow{HBr}$ Br $\xrightarrow[\text{ether}]{Mg}$ MgBr $\xrightarrow[(2)\ H_2O,\ H^+]{(1)\ CH_3CH_2CHO}$

$-\overset{OH}{C}HCH_2CH_3$ $\xrightarrow{H_2CrO_4}$ $-\overset{O}{\overset{||}{C}}CH_2CH_3$

(b) 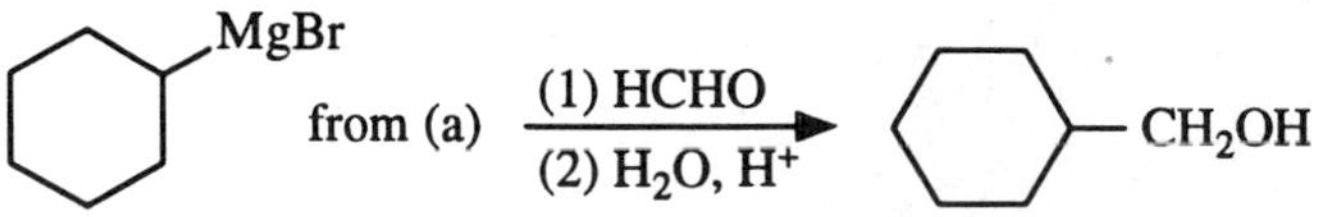 MgBr from (a) $\xrightarrow[(2)\ H_2O,\ H^+]{(1)\ HCHO}$ $-CH_2OH$

(c) $\xrightarrow{C_6H_5CO_3H}$ O $\xrightarrow[(2)\ H_2O,\ H^+]{(1)\ CH_3MgI}$ OH CH_3

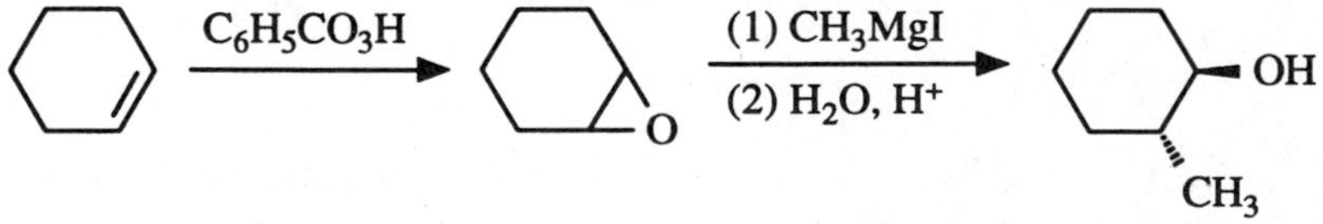

(d) 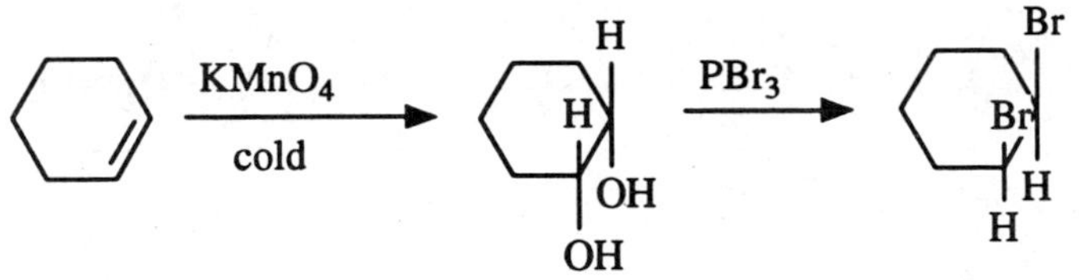 KMnO$_4$, cold → (H, H, OH, OH) → PBr$_3$ → (Br, Br, H, H)

10.15 All products are achiral or racemic.

(a) 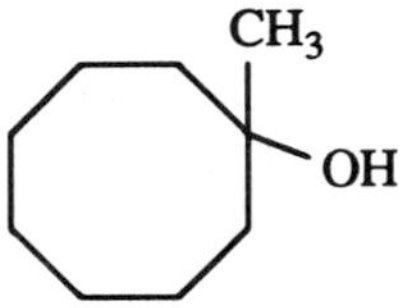 CH_3, OH

(b) 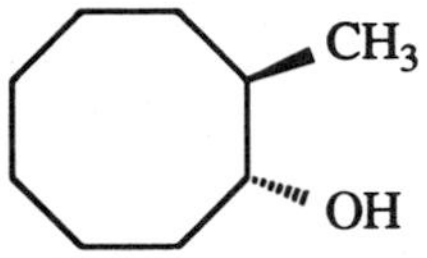 CH_3, OH

(c) 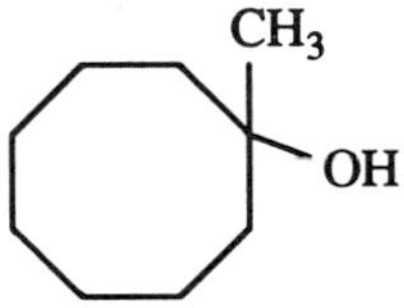 CH_3, OH

(d) 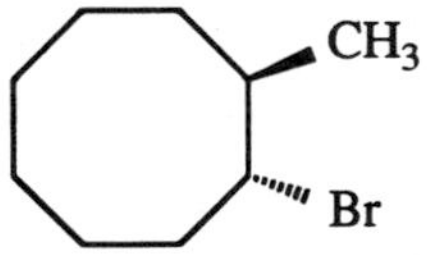 CH_3, Br

(e) 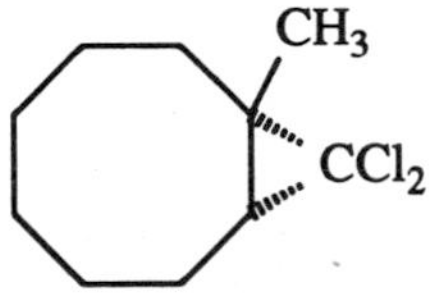 CH_3, CCl_2

(f) 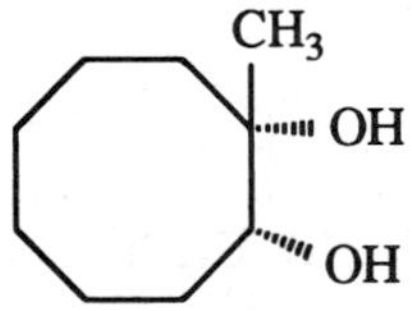 CH_3, OH, OH

(g) 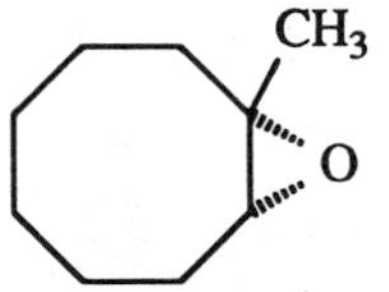 CH_3, O

(h) 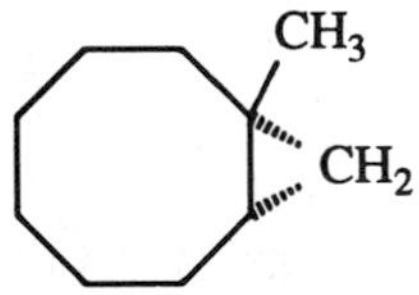 CH_3, CH_2

10.16 (a) 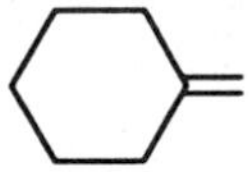 $=CHCH_3$ (more substituted) (b) *trans*

(c) (*E*) (d) $CH_2{=}CHCO_2CH_2CH_3$ (conjugated)

10.17 (a) Methylcyclopentane (cyclopentane ring with CH_3)

(b) $C_6H_5CH_2CH_2C_6H_5$

(c) Tricyclic structure with H (wedge) and H (wedge)

10.18 (a) $C_6H_5\overset{O}{\overset{\|}{C}}OH + CO_2$

(b) $HO\overset{O}{\overset{\|}{C}}(CH_2)_5\overset{O}{\overset{\|}{C}}OH$

(c) $CH_3\overset{O}{\overset{\|}{C}}OH + CH_3\overset{O}{\overset{\|}{C}}CH_2CH_2CH_3$

(d) $HO\overset{O}{\overset{\|}{C}}\underset{C_6H_5}{\underset{|}{C}}HCH_2CH_2\underset{C_6H_5}{\underset{|}{C}}H\overset{O}{\overset{\|}{C}}OH$ (*meso*)

10.19 (a) $HO\overset{O}{\overset{\|}{C}}CH_2CH_2\underset{CH_3}{\underset{|}{C}}HCH_2\overset{O}{\overset{\|}{C}}OH$

(b) $HO\overset{O}{\overset{\|}{C}}-\overset{O}{\overset{\|}{C}}(CH_2)_4\overset{O}{\overset{\|}{C}}CH_2CH_2\overset{O}{\overset{\|}{C}}OH$

(c) $H\overset{O}{\overset{\|}{C}}CH_2\overset{O}{\overset{\|}{C}}H$ + 2-formylcyclohexanone ($H\overset{O}{\overset{\|}{C}}$ on cyclohexane ring with ring C=O)

10.20 (a) $(CH_3CH_2)_2C{=}C(CH_2CH_3)_2 \xrightarrow{O_3} (CH_3CH_2)_2C$ (ozonide ring: O–O / O) $C(CH_2CH_3)_3$

$$\xrightarrow[H^+]{H_2O_2} 2\ CH_3CH_2\overset{O}{\overset{\|}{C}}CH_2CH_3$$

(b) $HOCH_2C{\equiv}CCH_2OH \xrightarrow[Pd]{1\ H_2}$ (H)(HOCH$_2$)C=C(H)(CH$_2$OH)

(c) $CH_3C{\equiv}CCH_3 + H_2O \xrightarrow{Hg^{2+}} \left[CH_3\overset{OH}{\overset{|}{C}}{=}CHCH_2 \right] \longrightarrow CH_3\overset{O}{\overset{\|}{C}}CH_2CH_3$

an enol

10.21 (a) cycloheptane bearing $O\overset{O}{\overset{\|}{C}}CH_3$ and CH_2Cl on the same carbon

(b) $CH_3(CH_2)_4\overset{CH_3}{\overset{|}{\underset{OH}{\underset{|}{C}}}}CH_2HgO\overset{O}{\overset{\|}{C}}CH_3$

(racemic)

(c) $CH_3(CH_2)_4\overset{CH_3}{\overset{|}{\underset{HO}{\underset{|}{C}}}}CH_3$

(d) bicyclic structure with CH$_3$ and OH (major) + bicyclic structure with OH and CH$_3$

(major)

10.22 (a) (1) (CH$_3$CH$_2$)(H)C=C(H)(CH$_3$) $\xrightarrow[(2)\ H_2O_2,\ H^+]{(1)\ O_3}$ $CH_3CH_2CO_2H + CH_3CO_2H$

(2) $CH_3CH_2(H)C=C(CH_3)(H)$ ⟶ same products

(b) (1) CH_3CH_2, H, HO — C - C — H, CH_3, OH + HO, OH — CH_3CH_2, H — C - C — H, CH_3

(2) CH_3CH_2, H, HO — C - C — CH_3, H, OH + HO, OH — CH_3CH_2, H — C - C — CH_3, H

(c) (1) CH_3CH_2, H — C — C (O) — H, CH_3 + CH_3CH_2, H — C — C (O) — H, CH_3 (*trans*, racemic)

(2) CH_3CH_2, H — C — C (O) — CH_3, H + CH_3CH_2, H — C — C (O) — CH_3, H (*cis*, racemic)

(d) (1) and (2) $CH_3CH_2CH_2CH_2CH_3$

(e) (1) CH_3CH_2, H, Br — C - C — Br, H, CH_3 + Br, CH_3CH_2, H — C · C — H, CH_3, Br

(2) CH_3CH_2, H, Br — C - C — Br, CH_3, H + Br, CH_3CH_2, H — C · C — CH_3, H, Br

(f) (1) and (2) $CH_3CH_2CO_2H$ + CH_3CO_2H

10.23 (a)

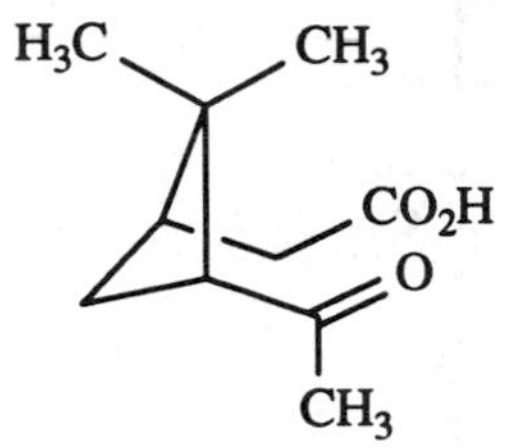

H_3C CH_3

CO_2H

O

CH_3

(b)

H_3C CH_3

O

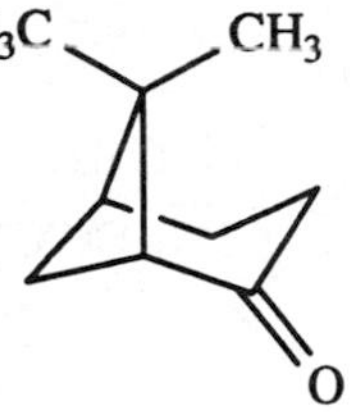

$+ \ CO_2$
(from HCO_2H)

10.23 (a)

(b) $CH_3CH_2CH{=}CH_2$

(c) HBr, peroxide

(d) H_2O, H^+

(e) (1) BH_3; (2) H_2O_2, OH^-

(f) $Br_2 \ + \ H_2O$

(g) $HC{\equiv}CH_2CH_3$

10.25 (a) HBr, peroxide or (1) BH_3; (2) Br_2

(b) Cl_2, CH_3OH

(c) Br_2

(d) (1) BH_3; (2) H_2O_2, OH^-

(e) (1) Br_2; (2) $NaNH_2$ followed by H_2O; (3) 2 HBr

10.26 (a) $CH_3C{\equiv}CCH_2CH_2CH_3 \ + \ Li \xrightarrow{NH_3}$ *trans*-$CH_3CH{=}CHCH_2CH_2CH_3$

(b) $CH_3C{\equiv}CCH_2CH_2CH_3 \ + \ 1\ H_2 \xrightarrow{\text{poisoned Pd catalyst}}$

cis-$CH_3CH{=}CHCH_2CH_2CH_3$

(c) $(CH_3)_3COH \xrightarrow[\text{warm}]{H_2SO_4} CH_2{=}C(CH_3)CH_3 \xrightarrow[Pt]{H_2} CH_3CH(CH_3)CH_3$

11

Aromaticity and Benzene; Electrophilic Aromatic Substitution

Answers to Additional Problems

11.1 (a) 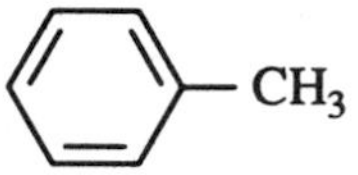 [benzene ring]—CH_3 (b) 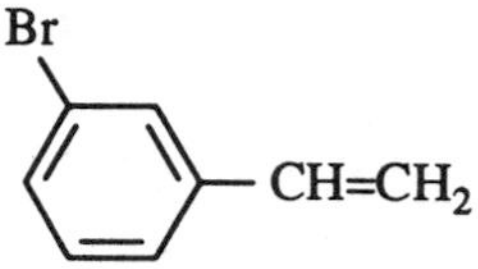 [benzene ring with Br]—$CH{=}CH_2$

(c) 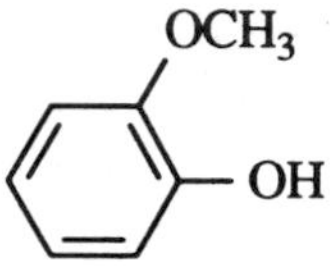 [benzene ring with OCH_3]—OH   (d) 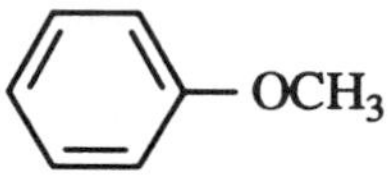 [benzene ring]—OCH_3

(e) 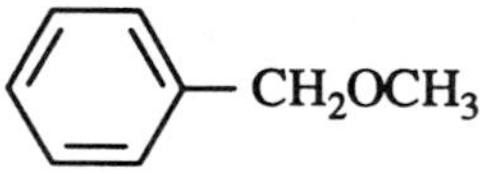 [benzene ring]—CH_2OCH_3 (f)  O_2N—[benzene ring]—NH_2

11.2 (a) *o*-methylbenzaldehyde

(b) *m*-methylaniline; *m*-aminotoluene

(c) *p*-chlorobenzoic acid

(d) 3-bromo-5-chlorotoluene

11.3 Compound (a) $C_6H_5CH_2CHOH$ (with C_6H_5 on the CHOH carbon)

The infrared spectrum shows a broad band in the -OH/NH region (around 3300 cm^{-1}, 3.0 μm). Of the structures given, only compounds (a) and (b) are alcohols. The sum of the proton integration areas in the NMR spectrum is 14, and the total number of aryl protons is 10. Of the two alcohols (a) and (b), only compound (a) has 10 aryl protons and a total of 14 protons. Compound (b) has five aryl protons and a total of 12 protons; therefore, it does not fit the data.

assignment of NMR signals:

10 aryl protons: 2 C_6H_5-
triplet at δ 4.8 (area 1): $-CH_2CHOH$
doublet at δ 3.0 (area 2): $-CH_2CHOH$
singlet at δ 2.1 (area 1): -OH

11.4 Compound (c). The NMR spectrum shows a pair of triplets upfield and a pair of doublets in the aromatic region. Of the structures given, only (c) and (d) would have a pair of triplets in their NMR spectra. Only compound (d) is *para*-substituted and thus would also have a pair of doublets in the aromatic region. Note that the signals for -O<u>H</u> and -N<u>H</u>$_2$ are superimposed in this NMR spectrum to give what appears to be a singlet with area 3.

11.5 Because of the pair of doublets in the aromatic region of the spectrum, we assume that the compound is probably a *p*-disubstituted benzene. The singlet (area 3) at about δ 2.2 falls in the region expected for a CH_3 group on benzene. The singlet at about δ 3.8 must arise from a CH_3 group bonded to an electronegative atom (oxygen). Therefore, the structure is:

CH_3—C$_6$H$_4$—OCH_3

11.6 (a) 18 pi electrons; 4n+2 where n = 4; aromatic

(b) Not aromatic as drawn but shown to be [tropylium oxide structure, ring +, O−], which has 6 pi electrons and is aromatic.

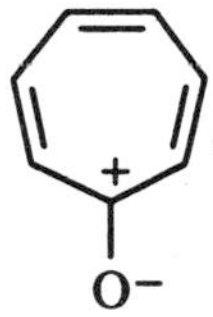

(c) not aromatic. [bicyclic structure; arrow: *sp*^{3}, not *sp*^{2}]

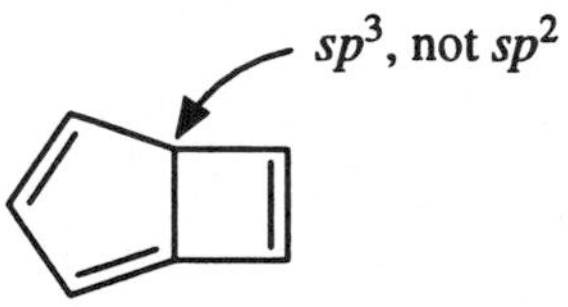

(d) 14 pi electrons; 4n + 2 where n = 3; aromatic. Does show aromaticity by NMR but not by chemical stability.

11.7 $Cl_2 + FeCl_3 \rightleftharpoons Cl^+ + FeCl_4^-$

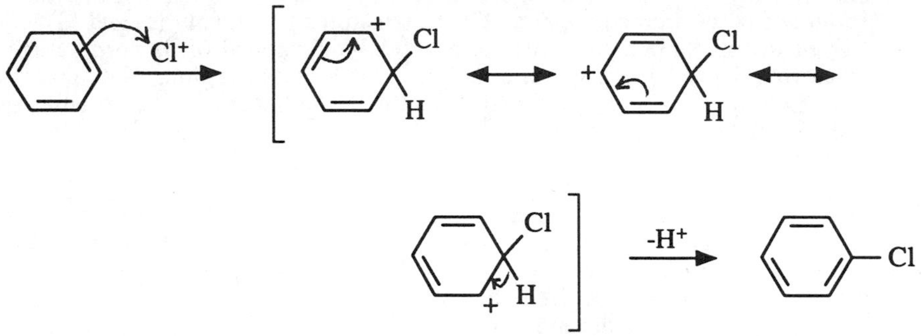

Cl^+ → [Cl, H, + ↔ + Cl, H ↔ Cl, H, +] $\xrightarrow{-H^+}$ Cl

11.8 (a), (b), and (d) have *m*-director; (c) has a *o,p*-director

11.9 (e) < (d) < (c) < (f) < (b) < (a)

11.10 (a) + CH_3CH_2CHBr (with CH_3) $\xrightarrow{AlCl_3}$ $CHCH_2CH_3$ (with CH_3)

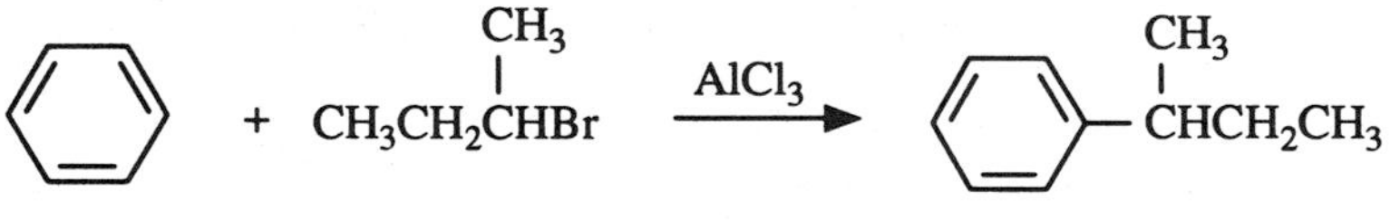

(b) CO_2H + Cl_2 $\xrightarrow{FeCl_3}$ Cl, CO_2H

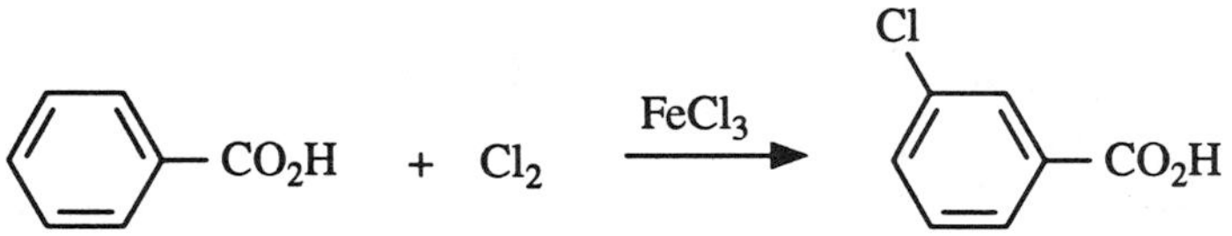

(c) Br + CH_3CCl (with O) $\xrightarrow{AlCl_3}$ CH_3C (with O) — Br

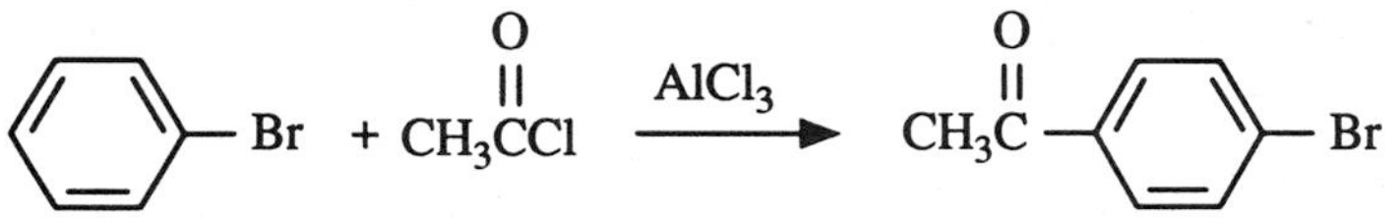

11.11 (a) O_2N, H, +, OH ↔ O_2N, H, +, $\ddot{O}H$ ↔

O_2N, H, $\overset{+}{O}H$ ↔ O_2N, H, +, $\ddot{O}H$

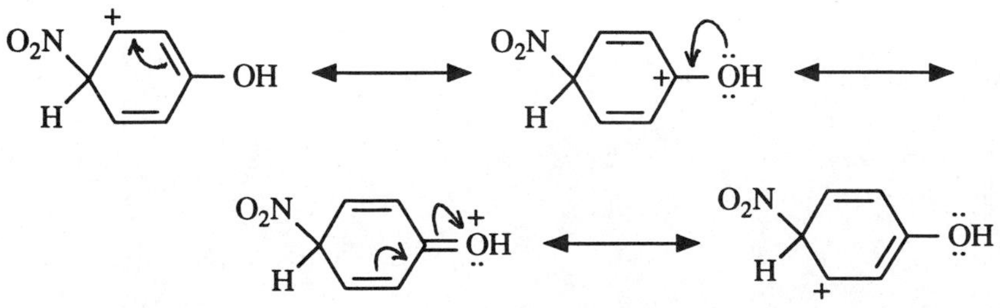

(b) 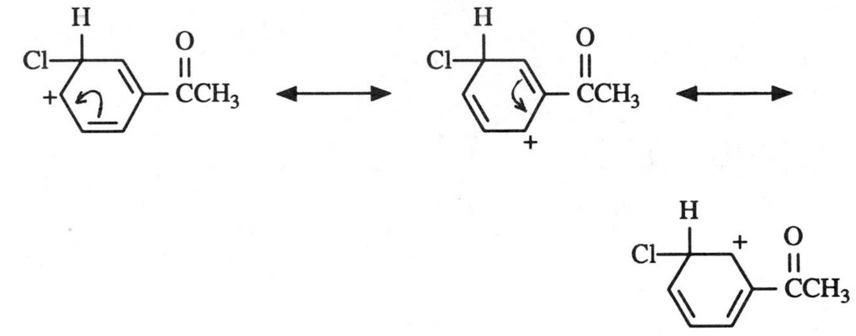 [resonance structures of the chloro-substituted carbocation intermediate with $\overset{O}{\overset{\|}{C}}CH_3$ substituent; Cl, H labels]

11.12 (a) 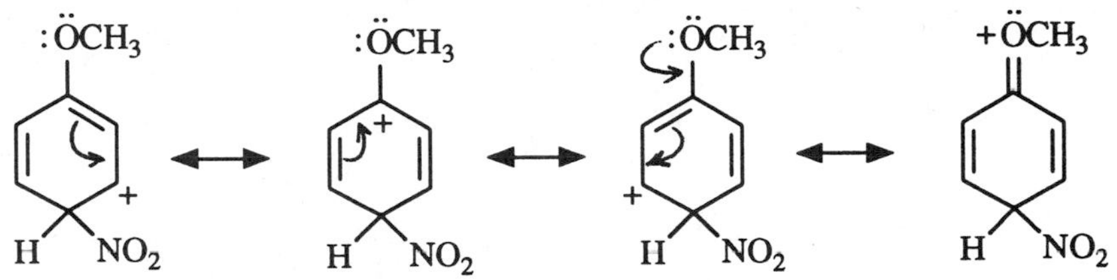 [resonance structures of the intermediate with $:\ddot{O}CH_3$, H and NO_2 substituents, ending with $+\ddot{O}CH_3$]

(b) 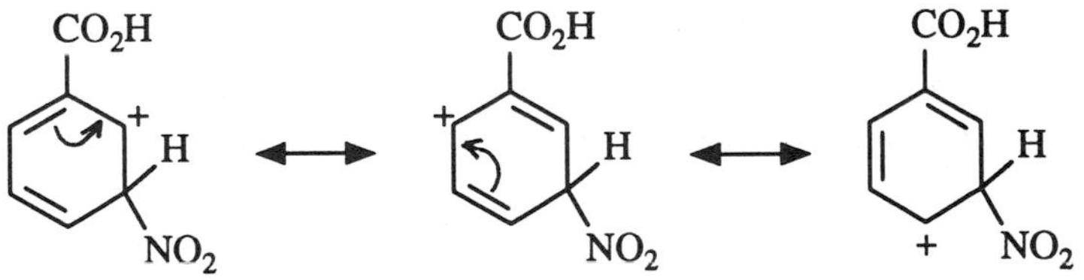 [resonance structures of the intermediate with CO_2H, H and NO_2 substituents]

11.13 (a) 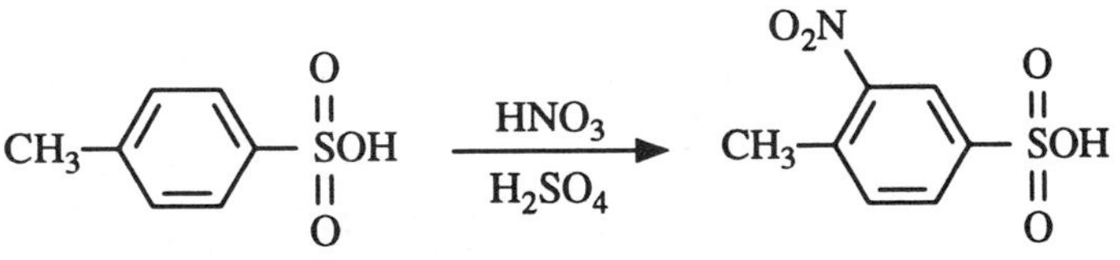 $CH_3-C_6H_4-SO_2OH \xrightarrow[H_2SO_4]{HNO_3}$ (O_2N-substituted) $CH_3-C_6H_3(NO_2)-SO_2OH$

(b) $O_2N-C_6H_4-OH \xrightarrow[H_2SO_4]{HNO_3} O_2N-C_6H_3(NO_2)-OH$

(c) $Br-C_6H_4-CH_3 \xrightarrow[H_2SO_4]{HNO_3} Br-C_6H_3(NO_2)-CH_3$

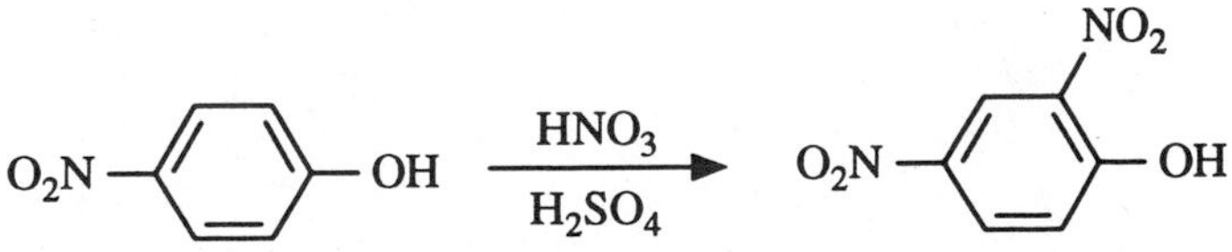

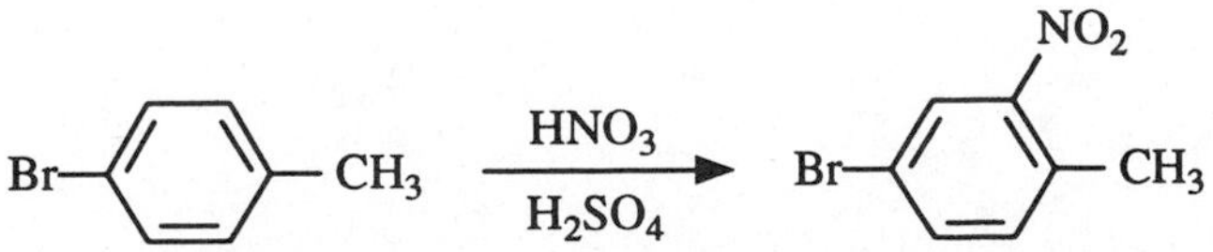

11.14 (a) Toluene ($C_6H_5CH_3$) $\xrightarrow[H_2SO_4]{HNO_3}$ p-nitrotoluene (CH_3, NO_2) $\xrightarrow[heat]{CrO_3}$ p-nitrobenzoic acid ($\overset{O}{\overset{\|}{C}}OH$, NO_2) $\xrightarrow[2)\ buffer\ pH]{1)\ Fe/HCl}$ product

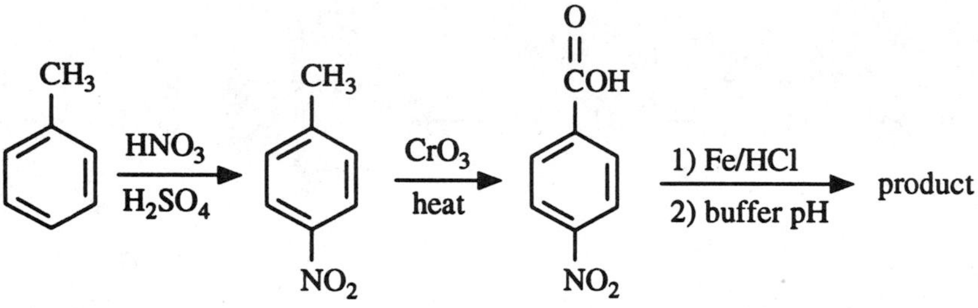

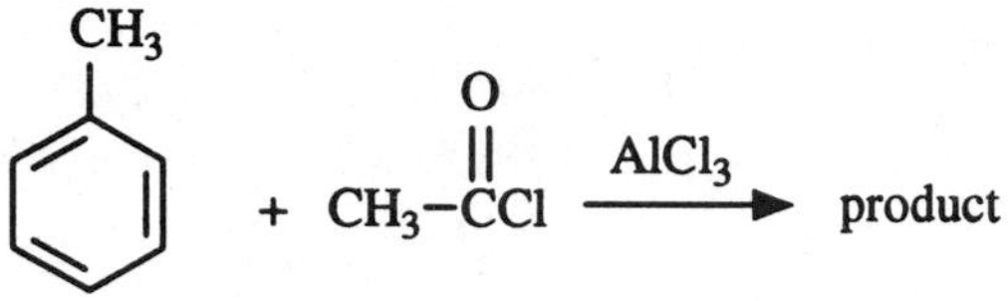

(b) Toluene (CH_3) + $CH_3-\overset{O}{\overset{\|}{C}}Cl$ $\xrightarrow{AlCl_3}$ product

(c) Benzene $\xrightarrow[SO_3]{H_2SO_4}$ benzenesulfonic acid (SO_3H) $\xrightarrow[AlCl_3]{CH_3Cl}$ product

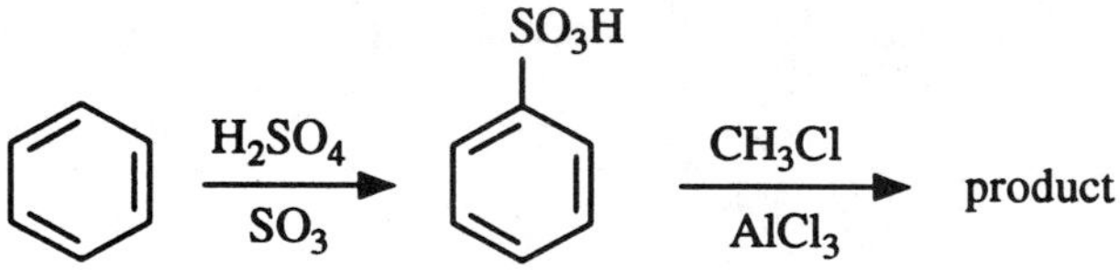

12

Substituted Benzenes

Answers to Additional Problems

12.1 (a) 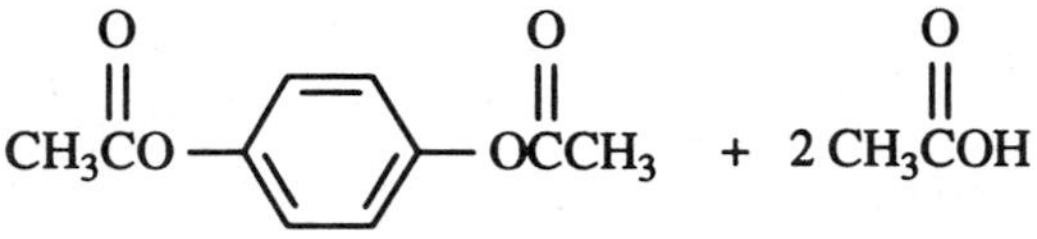 $CH_3\overset{O}{\overset{\|}{C}}O-C_6H_4-O\overset{O}{\overset{\|}{C}}CH_3 \;+\; 2\,CH_3\overset{O}{\overset{\|}{C}}OH$

(b) 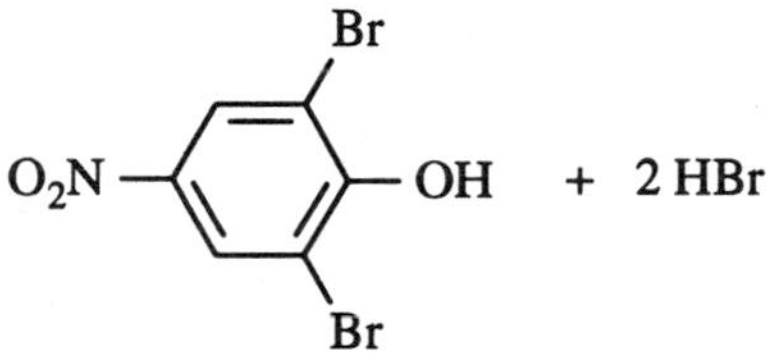 $O_2N-C_6H_2(Br)_2-OH \;+\; 2\,HBr$

(c) $CH_3CH_2O\overset{O}{\overset{\|}{C}}-C_6H_2(Cl)_2-OH \;+\; 2\,HCl$

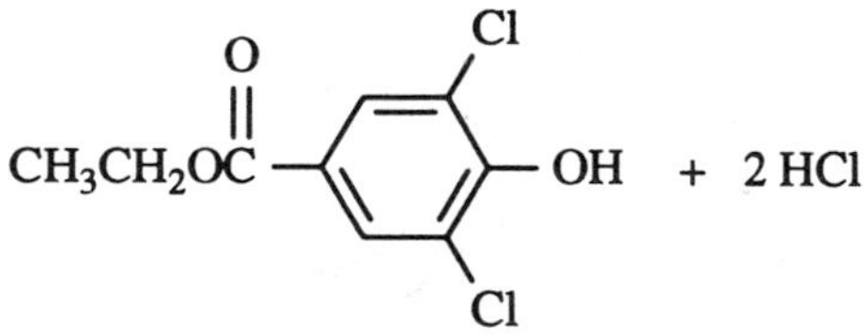

(d) 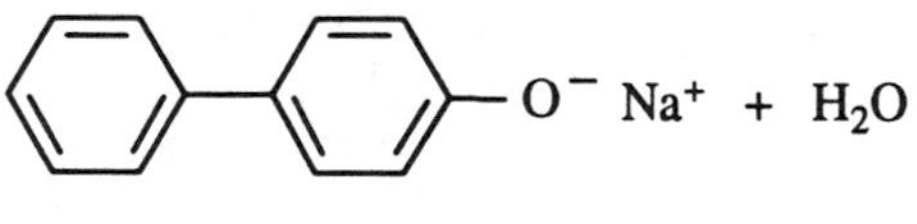 $C_6H_5-C_6H_4-O^-\,Na^+ \;+\; H_2O$

(e) 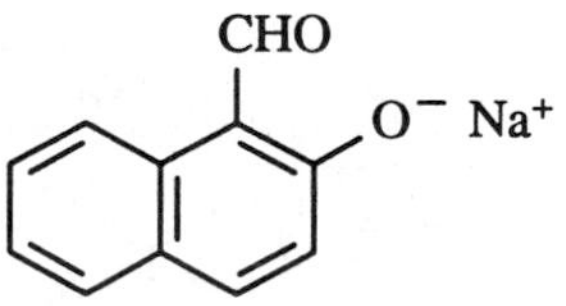 CHO, $O^-\,Na^+$

12.2 (a)

Benzene $\xrightarrow[AlCl_3]{CH_3CCl\ (O)}$ acetophenone ($C_6H_5CCH_3$, C=O) $\xrightarrow[H_2SO_4]{HNO_3}$ m-nitroacetophenone (CCH_3 (C=O), NO_2) $\xrightarrow[2)\ NaOH]{1)\ Fe/HCl}$

m-aminoacetophenone (CCH_3 (C=O), NH_2) $\xrightarrow[HCl]{NaNO_2}$ (CCH_3 (C=O), $N_2^+\ Cl^-$) $\xrightarrow[heat]{H_2O}$ product

(b)

Benzene $\xrightarrow[H_2SO_4]{HNO_3}$ $C_6H_5NO_2$ $\xrightarrow[2)\ NaOH]{1)\ Fe/HCl}$ $C_6H_5NH_2$ $\xrightarrow[excess]{Cl_2}$

2,4,6-trichloroaniline (NH_2; Cl, Cl, Cl) $\xrightarrow[HCl]{NaNO_2}$ ($N_2^+\ Cl^-$; Cl, Cl, Cl) $\xrightarrow{H_3PO_2}$ product

(c)

Benzene $\xrightarrow[AlCl_3]{CH_3Cl}$ $C_6H_5CH_3$ $\xrightarrow[H_2SO_4]{HNO_3}$ p-nitrotoluene (CH_3, NO_2) $\xrightarrow[2)\ NaOH]{1)\ Fe/HCl}$

p-toluidine (CH_3, NH_2) $\xrightarrow[HCl]{NaNO_2}$ (CH_3, $N_2^+\ Cl^-$)

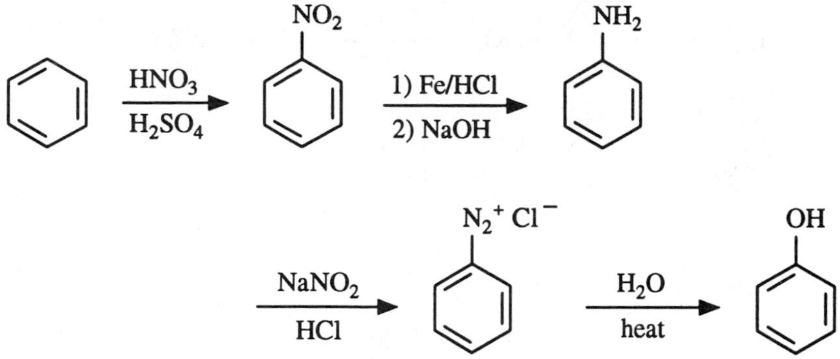

NO_2 NH_2

$\xrightarrow[H_2SO_4]{HNO_3}$ $\xrightarrow[2)\ NaOH]{1)\ Fe/HCl}$

$N_2^+\ Cl^-$ OH

$\xrightarrow[HCl]{NaNO_2}$ $\xrightarrow[heat]{H_2O}$

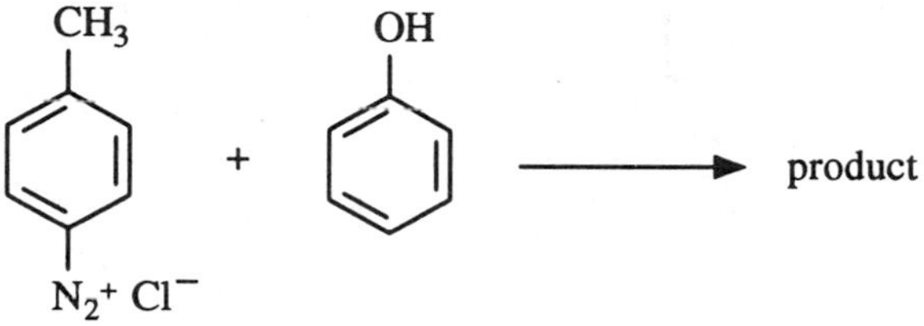

CH_3 OH

\+ product

$N_2^+\ Cl^-$

12.3 (a) OH + HO_3S — $N_2^+Cl^-$

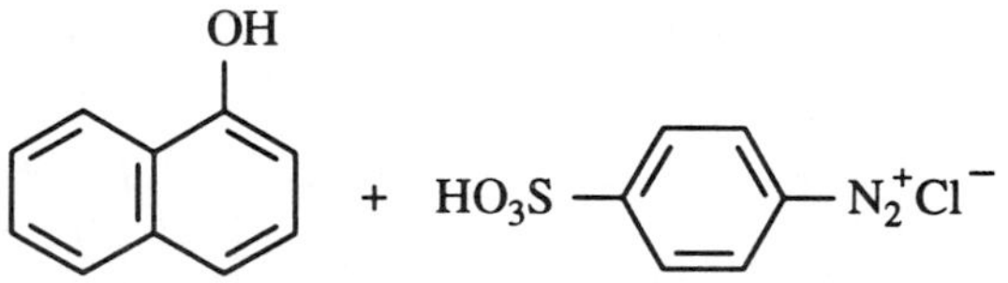

(b) OH + HO_3S — $N_2^+Cl^-$

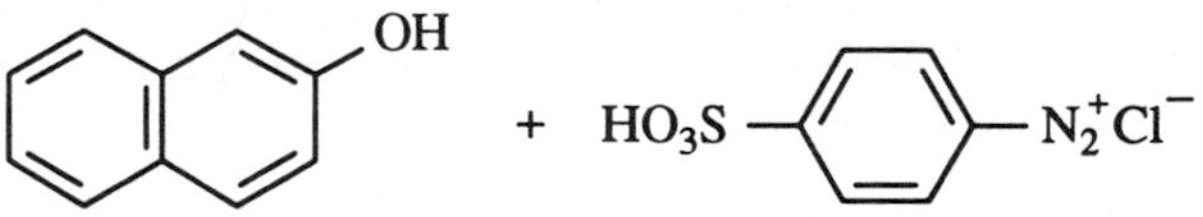

(c) 2 NH_2 OH NaO_3S SO_3Na + $Cl^-\ \overset{+}{N_2}$ — $N_2^+Cl^-$

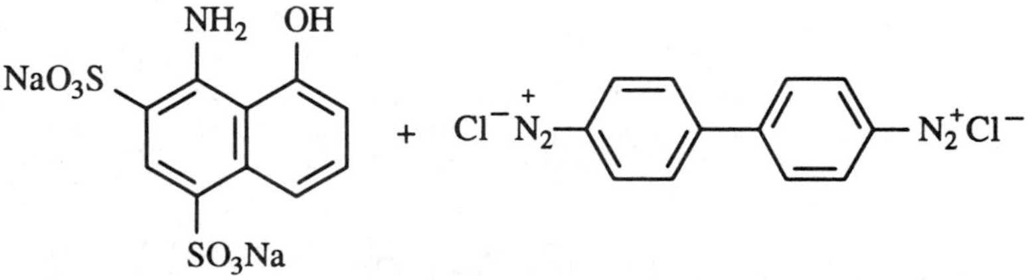

12.4 (a) Cl–C$_6$H$_4$–NO_2 $\xrightarrow[\text{heat}]{\text{aqueous NaOH}}$ product

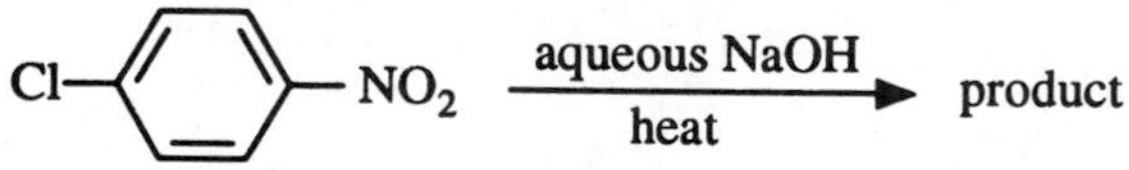

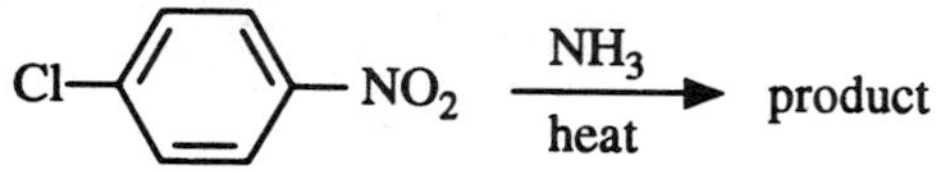

(b) Cl–C$_6$H$_4$–NO_2 $\xrightarrow[\text{heat}]{NH_3}$ product

(c) Cl–C$_6$H$_4$–NO_2 $\xrightarrow[CH_3OH \text{ heat}]{Na^+ \ ^-OCH_3}$ product

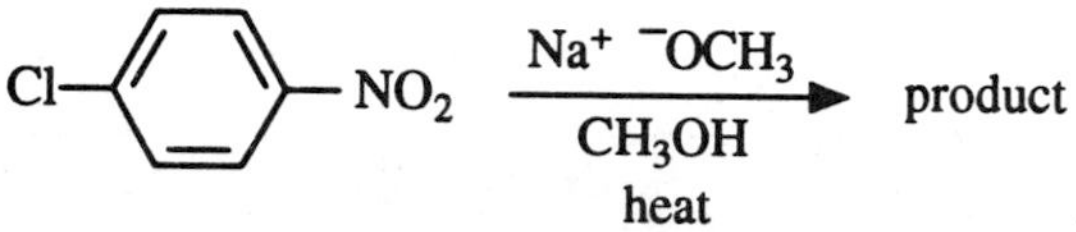

12.5 (a)

NO_2, Cl (ortho) $\xrightarrow{\text{piperidine (N–H)}}$ NO_2, N-piperidinyl (ortho)

O_2N–C$_6$H$_4$–Cl $\xrightarrow{\text{piperidine (N–H)}}$ O_2N–C$_6$H$_4$–N(piperidinyl)

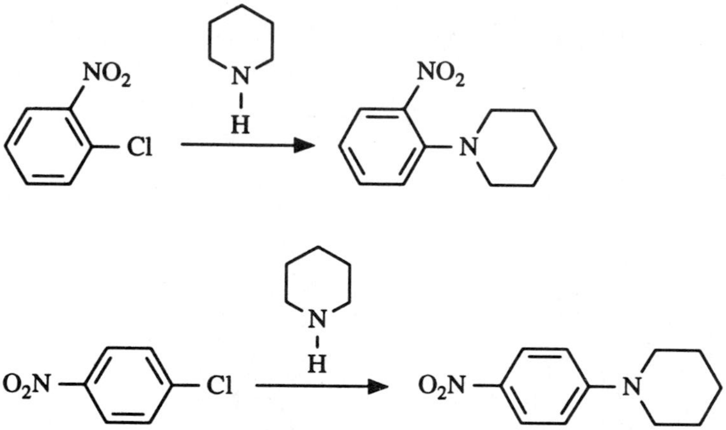

(b) The intermediate for *o* or *p* substitution is resonance stabilized, with added stabilization provided by the nitro group. The intermediate for *m* substitution has no added stabilization.

resonance structures for p substitution:

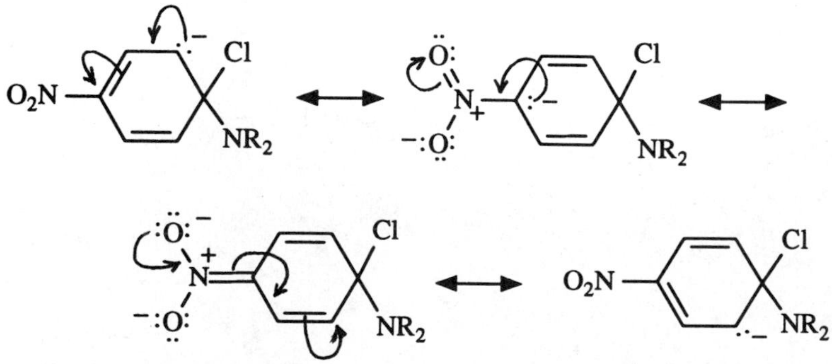

O_2N, Cl, NR_2 ⟷ :Ö:, N+, –:Ö:, Cl, NR_2 ⟷

:Ö:–, N+, –:Ö:, Cl, NR_2 ⟷ O_2N, Cl, NR_2

resonance structures for m substitution:

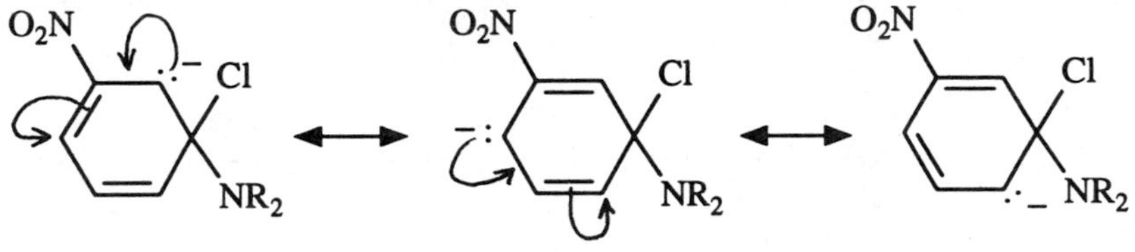

O_2N ... Cl, NR_2 $\longleftrightarrow$ O_2N ... Cl, NR_2 $\longleftrightarrow$ O_2N ... Cl, NR_2

12.6 (a) O_2N–C_6H_3(NO_2)–NH_2

(b) O_2N–C_6H_4–SS–C_6H_4–NO_2

(c) O_2N–C_6H_2(NO_2)(NO_2)–NHNH$_2$

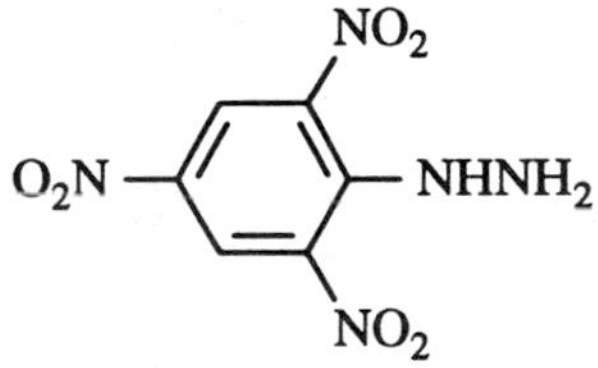

(d) $CH_3\overset{O}{\overset{\|}{C}}NH$–$C_6H_4$–$SO_2$–$C_6H_4$–$NO_2$

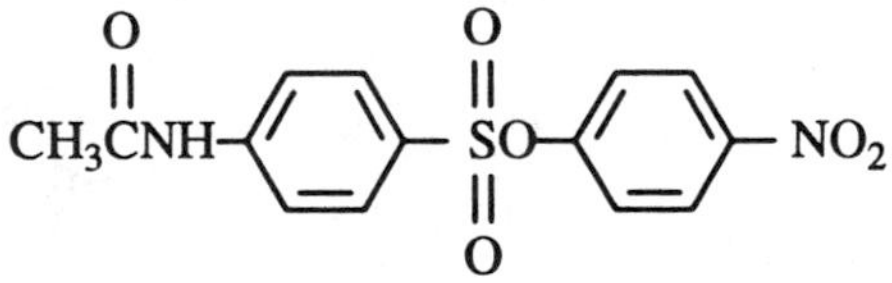

(e) O_2N, NH_2, O_2N, NH_2

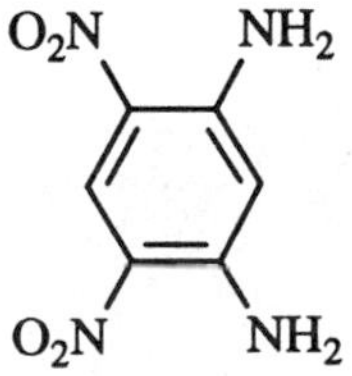

12.7 (a) C_6H_6 $\xrightarrow[H_2SO_4]{HNO_3}$ C_6H_5–NO_2 $\xrightarrow[FeBr_3]{Br_2}$ product

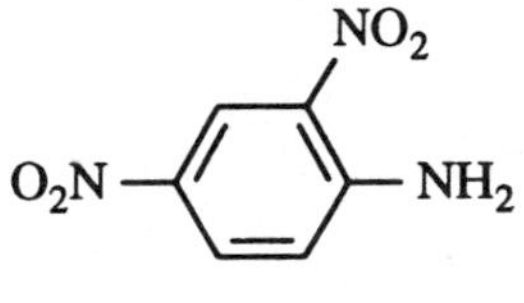

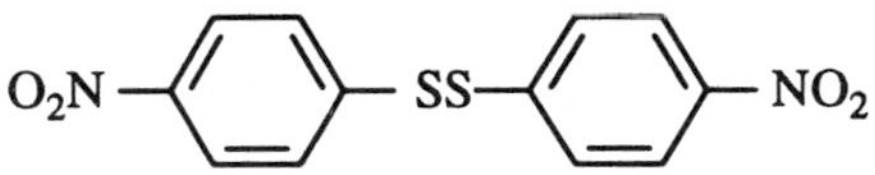

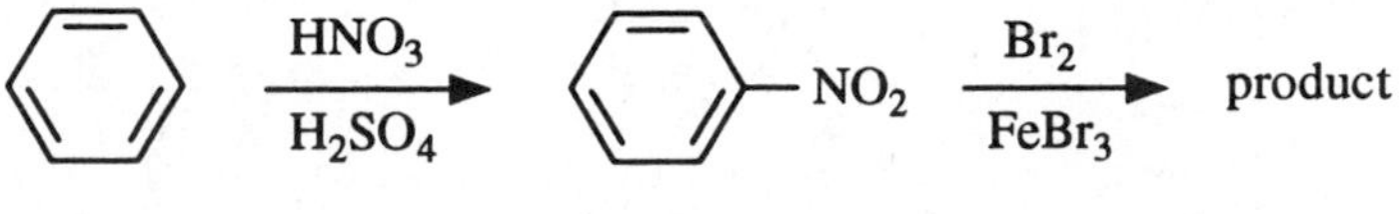

(b) 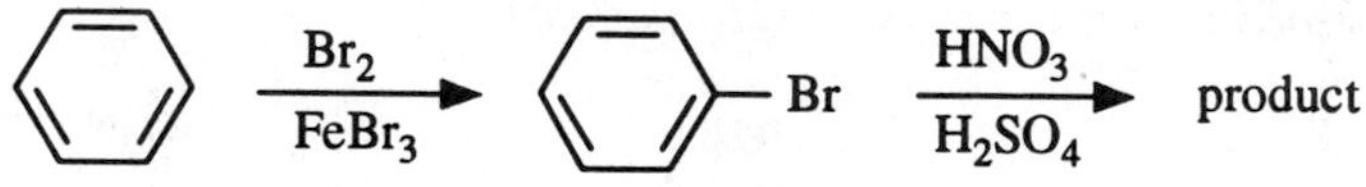 benzene $\xrightarrow[FeBr_3]{Br_2}$ C_6H_5—Br $\xrightarrow[H_2SO_4]{HNO_3}$ product

(c) benzene $\xrightarrow[SO_3]{H_2SO_4}$ C_6H_5—SO_3H $\xrightarrow[AlCl_3]{CH_3I}$ product

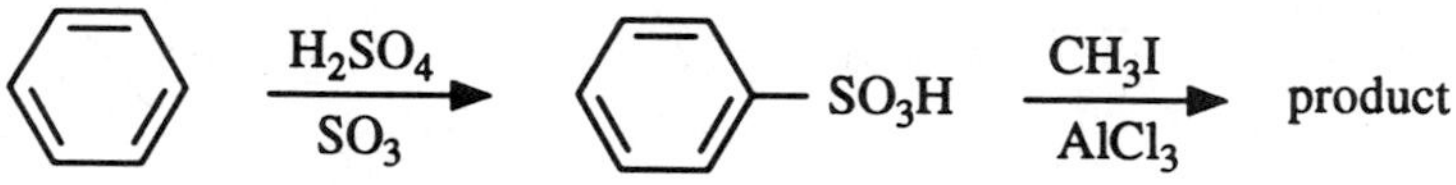

(d) 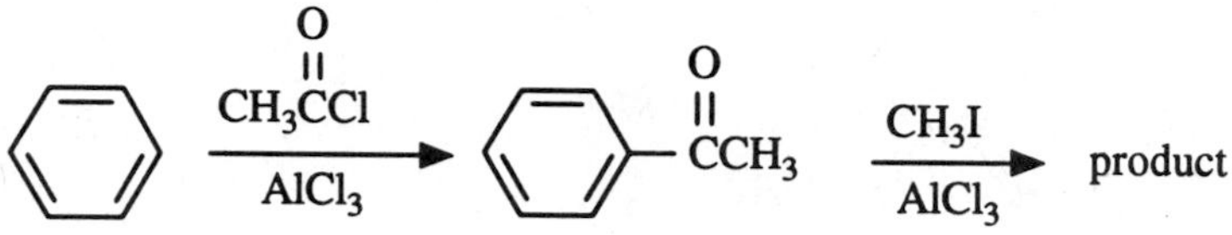 benzene $\xrightarrow[AlCl_3]{CH_3\overset{O}{\overset{\|}{C}}Cl}$ C_6H_5—$\overset{O}{\overset{\|}{C}}CH_3$ $\xrightarrow[AlCl_3]{CH_3I}$ product

12.8 C_6H_5—$\overset{O}{\overset{\|}{C}}O$—$C_6H_5$ $\xrightarrow[FeBr_3]{Br_2}$ C_6H_5—$\overset{O}{\overset{\|}{C}}O$—$C_6H_4$—Br

deactivated (ring on the C=O side) — activated (ring on the O side)

+ the *o* isomer

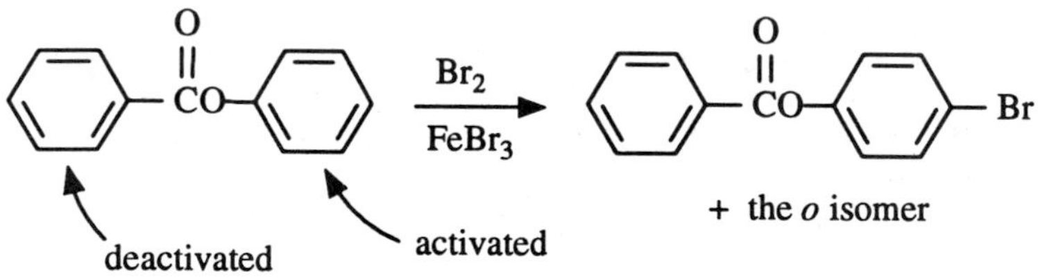

12.9 (a) 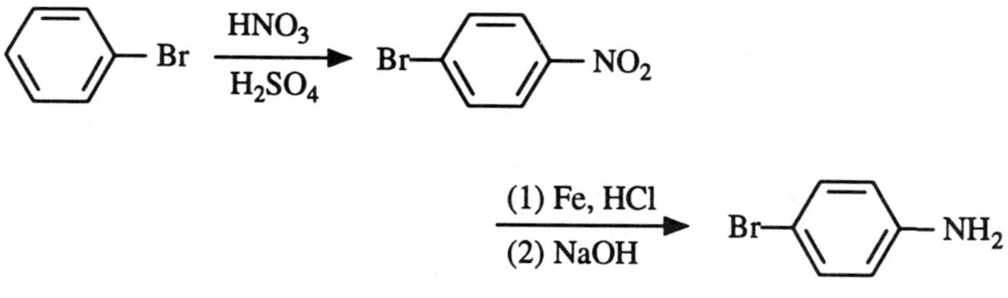 C_6H_5—Br $\xrightarrow[H_2SO_4]{HNO_3}$ Br—C_6H_4—NO_2

$\xrightarrow[(2)\ NaOH]{(1)\ Fe,\ HCl}$ Br—C_6H_4—NH_2

Bromination of $C_6H_5NH_2$ yields primarily polybrominated products.

(b) 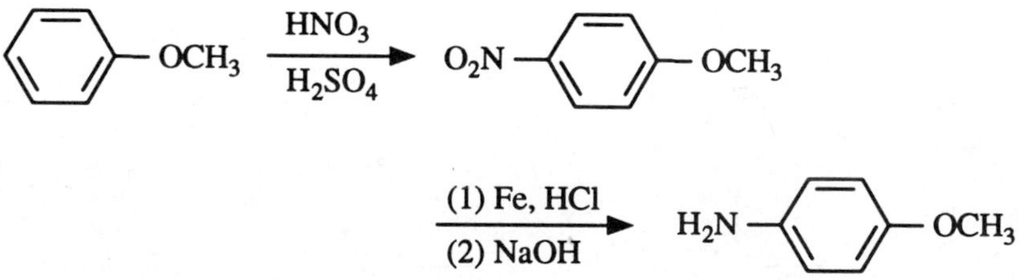 C_6H_5—OCH_3 $\xrightarrow[H_2SO_4]{HNO_3}$ O_2N—C_6H_4—OCH_3

$\xrightarrow[(2)\ NaOH]{(1)\ Fe,\ HCl}$ H_2N—C_6H_4—OCH_3

(c) $C_6H_5OCCH_3$ (with C=O) $\xrightarrow[H_2SO_4]{HNO_3}$ $O_2N-C_6H_4-OCCH_3$ (with C=O)

$\xrightarrow[(2)\ NaOH]{(1)\ Fe,\ HCl}$ $H_2N-C_6H_4-OCCH_3$ (with C=O)

(d) $CH_3O-C_6H_4-NH_2$ (from b) $\xrightarrow[HCl,\ cold]{NaNO_2}$

$CH_3O-C_6H_4-N_2^+Cl^-$ $\xrightarrow[KCN]{CuCN}$ $CH_3O-C_6H_4-CN$

12.10 (a) C_6H_6 $\xrightarrow[AlCl_3]{CH_3I}$ $C_6H_5-CH_3$ $\xrightarrow[h\nu]{Cl_2}$ $C_6H_5-CH_2Cl$

(b) $2\ C_6H_6 + ClCH_2CH_2Cl \xrightarrow{AlCl_3} C_6H_5CH_2CH_2C_6H_5 \xrightarrow{NBS}$

$C_6H_5CH_2CH(Br)C_6H_5 \xrightarrow[heat]{(CH_3)_3CO^-K^+} C_6H_5CH{=}CHC_6H_5 \xrightarrow[^-OH]{cold\ KMnO_4}$

$C_6H_5CH(OH)-CH(OH)C_6H_5$

(c) C_6H_6 $\xrightarrow[AlCl_3]{CH_3I}$ $C_6H_5-CH_3$ $\xrightarrow[FeBr_3]{Br_2}$ $Br-C_6H_4-CH_3$

$\xrightarrow{\text{separate from } o \text{ isomer}}$ $\xrightarrow[heat]{KMnO_4,\ H^+}$ $Br-C_6H_4-COH$ (with C=O)

13

Aldehydes and Ketones

Answers to Additional Problems

13.1 (a) $CH_3CH_2C(=O)CH_2CH_3$

(b) $CH_3CH(Cl)CH(=O)$

(c) $ClCH_2CH_2CH_2CH(=O)$

(d) $C_6H_5CH_2C(=O)CH_2C_6H_5$

(e) $ClCH_2C(=O)CH_2Cl$

(f) $Cl_2CHC(=O)CH_3$

13.2 (a) 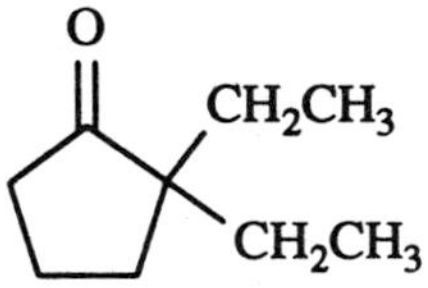 2,2-diethylcyclopentanone (cyclopentane ring with C=O and adjacent carbon bearing CH_2CH_3, CH_2CH_3)

(b) H, CH_3 on one carbon; H, $C(=O)CH_3$ on the other: $CH_3CH=CHC(=O)CH_3$ (C=C, cis H's)

(c) cycloheptane-1,4-dione (seven-membered ring with two C=O)

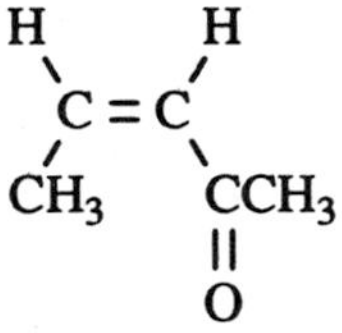

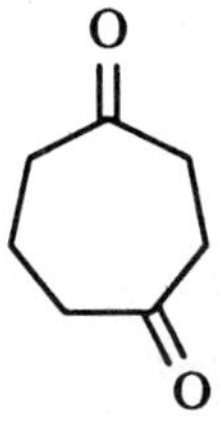

(d) $CH_3CH(OH)CH_2CH_2CH(=O)$

(e) 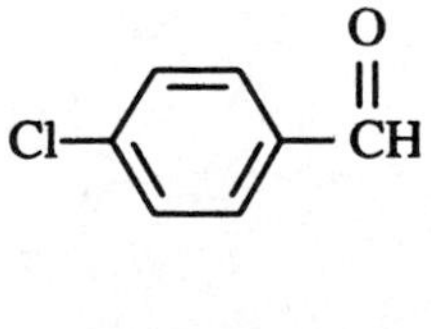 $Cl-C_6H_4-CH(=O)$ (para)

(f) $CH_2=CHCH(=O)$

13.3 (a) $$CH_3C\equiv CCH_3 \xrightarrow{H_2O,\ Hg^{2+}} CH_3\overset{\overset{\large O}{||}}{C}CH_2CH_3 \quad \text{(by way of enol)}$$

(b) $$CH_3CH_2Br \xrightarrow[\text{ether}]{Mg} CH_3CH_2MgBr \xrightarrow[(2)\ H_2O,\ H^+]{(1)\ CH_3CHO} CH_3\overset{\overset{\large OH}{|}}{C}HCH_2CH_3$$

$$\xrightarrow[\text{heat}]{H_2CrO_4} CH_3\overset{\overset{\large O}{||}}{C}CH_2CH_3$$

(c) $$CH_3\overset{\overset{\large OH}{|}}{C}HCH_2CH_3 \xrightarrow[\text{heat}]{H_2CrO_4} CH_3\overset{\overset{\large O}{||}}{C}CH_2CH_3$$

(d) $$CH_2{=}CHCH_2CH_3 \xrightarrow{H_2O,\ H^+} CH_3\overset{\overset{\large OH}{|}}{C}HCH_2CH_3 \xrightarrow[\text{heat}]{H_2CrO_4} CH_3\overset{\overset{\large O}{||}}{C}CH_2CH_3$$

13.4 (a) $$C_6H_5\overset{\overset{\large CO_2H}{|}}{C}HCH_2\overset{\overset{\large O}{||}}{C}C_6H_5$$

(b) (2-chlorophenyl)$-\overset{\overset{\large O}{||}}{C}C_6H_5$ + Cl$-$(4-phenylene)$-\overset{\overset{\large O}{||}}{C}C_6H_5$

13.5 (a) $$C_6H_5CH_3 \xrightarrow{NBS} C_6H_5CH_2Br \xrightarrow{OH^-} C_6H_5CH_2OH$$

$$\xrightarrow[25^\circ]{CrO_3 \cdot 2\ \text{pyridine}} C_6H_5CHO$$

(b) $CH_3CH{=}CH_2 \xrightarrow[(2)\ H_2O_2,\ OH^-]{(1)\ BH_3} CH_3CH_2CH_2OH$

$\xrightarrow[25°]{CrO_3 \cdot 2\ pyridine} CH_3CH_2\overset{O}{\overset{\|}{C}}H$

(c) $CH_3CH{=}CH_2 \xrightarrow{H_2O,\ H^+} CH_3\overset{OH}{\overset{|}{C}}HCH_3 \xrightarrow[heat]{H_2CrO_4} CH_3\overset{O}{\overset{\|}{C}}CH_3$

(d) $C_6H_6 \xrightarrow[AlCl_3]{CH_3\overset{O}{\overset{\|}{C}}Cl} C_6H_5\overset{O}{\overset{\|}{C}}CH_3 \xrightarrow[H_2SO_4]{HNO_3}$ (benzene ring bearing $\overset{O}{\overset{\|}{C}}CH_3$ and NO_2 meta)

13.6 (a) $C_6H_5\overset{O}{\overset{\|}{C}}CH_2CH_3$

Note the typical ethyl pattern: upfield triplet (area 3) and dowfield quartet (area 2).

(b) CH_3–(para-phenylene)–$\overset{O}{\overset{\|}{C}}CH_3$

Note that the relative areas must be doubled to be in accord with the molecular formula.

(c) $CH_3\underset{Cl}{\underset{|}{C}}H\overset{O}{\overset{\|}{C}}CH_3$

13.7 (a) $CCl_3\overset{O}{\overset{\|}{C}}H$ (b) $CH_3OCH_2\overset{O}{\overset{\|}{C}}H$

(c) $NCCH_2\overset{O}{\overset{\|}{C}}CH_3$ (d) $(CH_3)_2N-C_6H_4-CHO$

In each case the compound selected contains the more electron-withdrawing substituent bonded to the carbonyl carbon.

13.8 (d) < (b) < (c) < (a)

13.9 (a) 1-hydroxycyclohexanecarbonitrile (cyclohexane ring bearing OH and CN on the same carbon) (b) $O_2N-C_6H_4-\overset{OH}{\overset{|}{C}}HCN$

(c) $\overset{OH}{\overset{|}{C}}H_2CN$ (d) Fischer projections: CN / HO—H / H—OH / CH_2OH + CN / H—OH / H—OH / CH_2OH

13.10 (a) $CH_3CH_2CH_2\overset{OH}{\overset{|}{C}}HOH \overset{H^+}{\rightleftharpoons} CH_3CH_2CH_2\overset{O}{\overset{\|}{C}}H + H_2O$

(b) 1-methyl... 2-methylcyclopentane-1,1-diol (ring with OH, OH, CH_3) $\overset{H^+}{\rightleftharpoons}$ 2-methylcyclopentanone (ring with O, CH_3)

(c) 2,6-dihydroxytetrahydropyran (HO, O, OH) $\overset{H^+}{\rightleftharpoons}$ $O=CH-CH_2-CH_2-CH_2-CH(OH)OH$

$\overset{H^+}{\rightleftharpoons} O=CH-CH_2-CH_2-CH_2-CH=O + H_2O$

13.11 (a) $\rightleftharpoons$ $(CH_3)_2CHCH_2CH(OH)OCH_2CH_3$ $\xrightleftharpoons{CH_3CH_2OH,\ H^+}$

$(CH_3)_2CHCH_2CH(OCH_2CH_3)_2$ + H_2O

(b) $\rightleftharpoons$ cyclohexyl–CH(OH)OCH₂CH=CH₂ $\xrightleftharpoons{CH_2{=}CHCH_2OH,\ H^+}$

cyclohexyl–$CH(OCH_2CH{=}CH_2)_2$ + H_2O

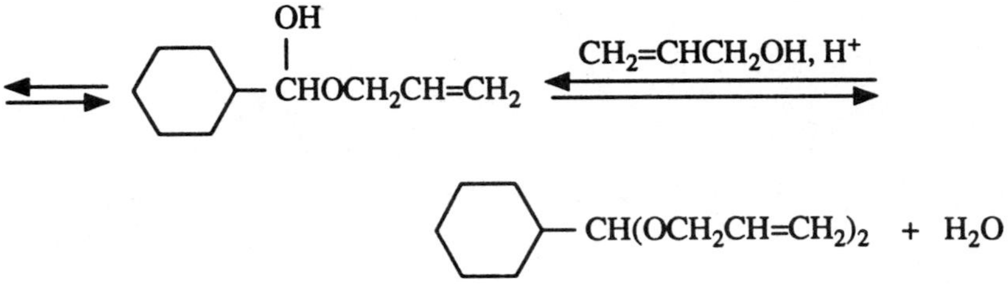

13.12 (a) 2-methoxytetrahydropyran (OCH_3 on ring carbon adjacent to ring O)

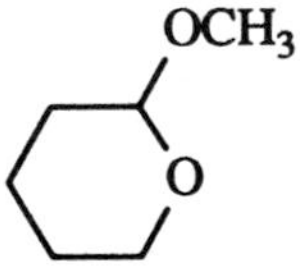

(b) CH_3CH in a ring with O–CH_2 / O–CH_2 (2-methyl-1,3-dioxolane)

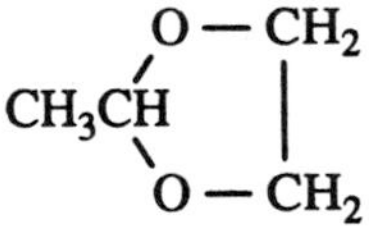

(c) cyclohexane spiro-fused to a 1,3-dioxolane, bearing CH_2CH_2CN and CH_3 on one adjacent carbon and CH_3 on the other adjacent carbon

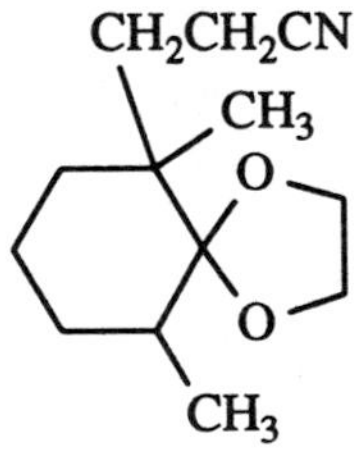

13.13 (a) $\xrightleftharpoons{H_2O,\ H^+}$ 3-methylcyclohexane with OH and $OCH_2CH_2CH_2OH$ on C-1 $\xrightleftharpoons{H_2O,\ H^+}$

3-methylcyclohexanone (=O) + $HOCH_2CH_2CH_2OH$

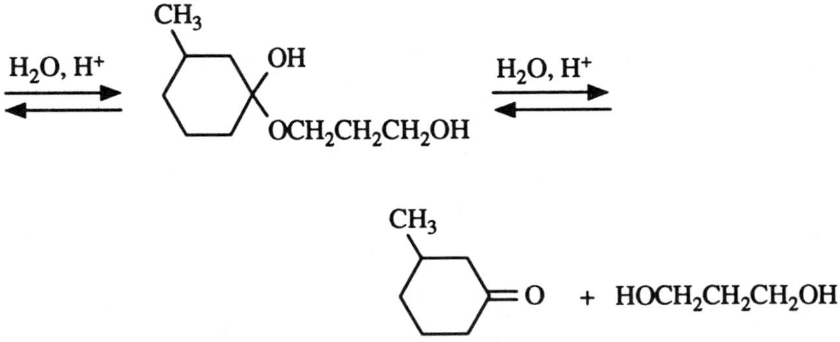

(b) $\xrightleftharpoons{H_2O,\ H^+}$ bicyclic alkene with bridge carbon bearing HO and OCH_3 + CH_3OH $\xrightleftharpoons{H_2O,\ H^+}$

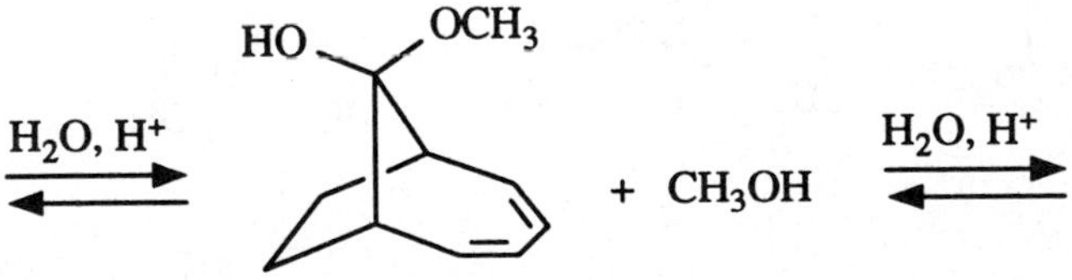

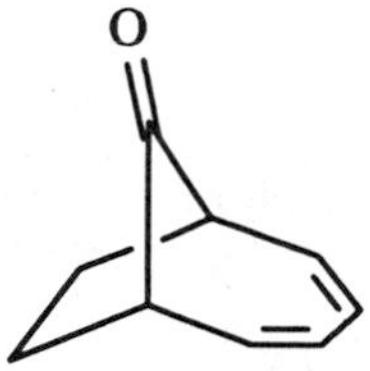

O

+ CH_3OH

(c) 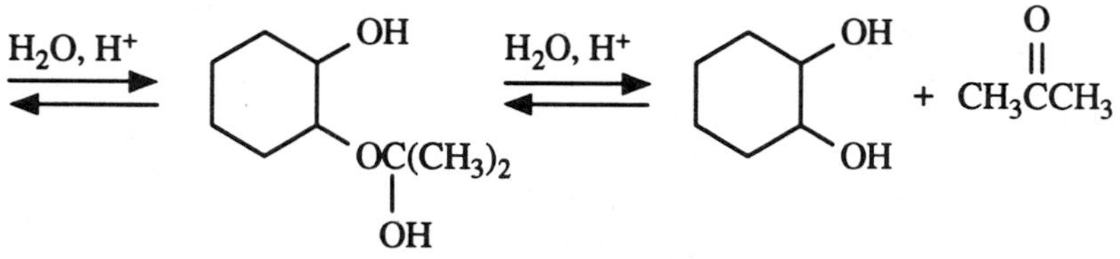 $\xrightleftharpoons{H_2O,\ H^+}$ OH, $OC(CH_3)_2$, OH $\xrightleftharpoons{H_2O,\ H^+}$ OH, OH + $CH_3\overset{O}{\overset{\|}{C}}CH_3$

13.14 (a) 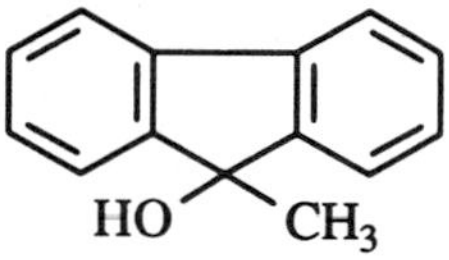 HO, CH_3

(b) (2*R*,4*R*)- and (2*S*,4*R*)- $C_6H_5\underset{CH_3}{\underset{|}{C}}HCH_2\overset{OH}{\overset{|}{C}}HCH_3$

(c) 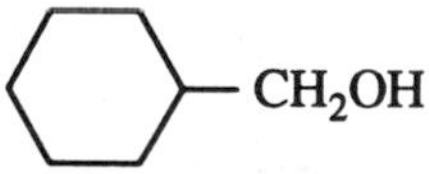 CH_2OH

(d) 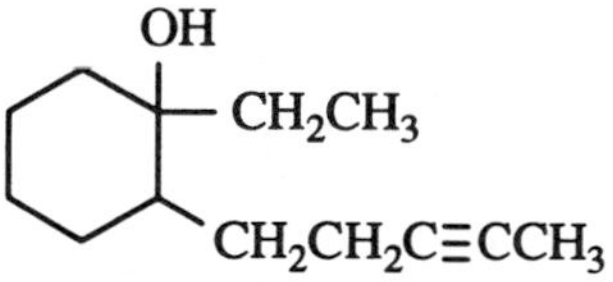 OH, CH_2CH_3, $CH_2CH_2C{\equiv}CCH_3$

13.15 (a) $CH_3CH_2CH_2MgBr \xrightarrow[(2)\ H_2O,\ H^+]{(1)\ CH_3CH_2CH_2\overset{O}{\overset{\|}{C}}CH_2CH_2CH_3}$

(b) $C_6H_5MgBr \xrightarrow[(2)\ H_2O,\ H^+]{(1)\ CH_3CH_2\overset{O}{\overset{\|}{C}}H}$ or $CH_3CH_2MgBr \xrightarrow[(2)\ H_2O,\ H^+]{(1)\ C_6H_5\overset{O}{\overset{\|}{C}}H}$

(c) $CH_3CH_2CH_2CH_2MgBr \xrightarrow[(2)\ H_2O,\ H^+]{(1)\ H\overset{O}{\overset{\|}{C}}H}$

(d) $C_6H_5MgBr \xrightarrow[(2)\ H_2O,\ H^+]{(1)\ CH_3C(=O)CH_2CH_3}$ or

$CH_3MgBr \xrightarrow[(2)\ H_2O,\ H^+]{(1)\ C_6H_5C(=O)CH_2CH_3}$ or

$CH_3CH_2MgBr \xrightarrow[(2)\ H_2O,\ H^+]{(1)\ C_6H_5C(=O)CH_3}$

13.16 (a) $ClCH_2CH_2CH{=}O \xrightarrow[H^+]{HOCH_2CH_2OH} ClCH_2CH_2CH(OCH_2CH_2O) \xrightarrow[ether]{Mg}$

$ClMgCH_2CH_2CH(OCH_2CH_2O) \xrightarrow[(2)\ H_2O,\ H^+]{(1)\ CH_3CHO} CH_3CH(OH)CH_2CH_2CH{=}O$

(b) $BrCH_2CH_2CH_2CH{=}O \xrightarrow[H^+]{HOCH_2CH_2OH} BrCH_2CH_2CH_2CH(OCH_2CH_2O) \xrightarrow[ether]{Mg}$

$BrMg(CH_2)_3CH(OCH_2CH_2O) \xrightarrow[(2)\ H_2O,\ H^+]{(1)\ \text{cyclohexanone}}$ 1-hydroxycyclohexyl with OH and $CH_2CH_2CH_2CHO$

13.17 (a) cycloheptane-1,2-dione bis(phenylhydrazone): ring with $=NNHC_6H_5$ and $=NNHC_6H_5$

(b) $CH_3O-C_6H_4-C(=NNHCONH_2)CH_2CH_3$

(c) 4,4-dimethylcyclohexenyl–N(morpholine) + H_2O

13.18 (a)

$CH_3CHO \overset{H^+}{\rightleftharpoons} CH_3CH{=}\overset{+}{O}H \overset{CH_3OH}{\rightleftharpoons} CH_3CH(OH)\overset{+}{O}(H)CH_3 \overset{-H^+}{\rightleftharpoons} CH_3CH(OH)OCH_3 \overset{H^+}{\rightleftharpoons}$

$CH_3CH(\overset{+}{O}H_2)OCH_3 \overset{-H_2O}{\rightleftharpoons} CH_3CH{=}\overset{+}{O}CH_3 \overset{CH_3OH}{\rightleftharpoons} CH_3CH(\overset{+}{O}(H)CH_3)OCH_3 \overset{-H^+}{\rightleftharpoons} CH_3CH(OCH_3)OCH_3$

(b)

$CH_3CHO \overset{CH_3NH_2}{\rightleftharpoons} CH_3CH(O^-)\overset{+}{N}H_2CH_3 \rightleftharpoons CH_3CH(OH)NHCH_3 \overset{H^+}{\rightleftharpoons} CH_3CH(\overset{+}{O}H_2)NHCH_3$

$\overset{-H_2O}{\rightleftharpoons} CH_3CH{=}\overset{+}{N}(H)CH_3 \overset{-H^+}{\rightleftharpoons} CH_3CH{=}NCH_3$

(c)

$$CH_3CH{=}\ddot{O}: \xrightleftharpoons{(CH_3)_2\ddot{N}H} CH_3CH(\ddot{O}:^-)(\overset{+}{N}H(CH_3)_2) \rightleftharpoons CH_3CH(:\ddot{O}H)(:N(CH_3)_2) \xrightleftharpoons{H^+} CH_3CH(\overset{+}{\ddot{O}}H_2)(:N(CH_3)_2)$$

$$\xrightleftharpoons{-H_2\ddot{O}:} CH_2(H){-}CH{=}\overset{+}{N}(CH_3)_2 \xrightleftharpoons{-H^+} CH_2{=}CH{-}:N(CH_3)_2$$

13.19 (a) $C_6H_5CHO + CH_2{=}P(C_6H_5)_3 \longrightarrow$

(b) $C_6H_5CHO \xrightarrow[(2)\ H_2O,\ H^+]{(1)\ NaBH_4}$

(c) $C_6H_5CHO + NH_2NHC({=}O)NH_2 \xrightleftharpoons{H^+}$

13.20 (a) $[(C_6H_5)_3P{=}CHCH_3 \longleftrightarrow (C_6H_5)_3\overset{+}{P}{-}\overset{-}{\ddot{C}}HCH_3$

(b) $CH_3CH_2CH{=}CHCH_2OCH_3$

(c) $C_6H_5CH{=}CHCH{=}CH_2$

13.21 (a) cyclohexanone (=O) $+ C_6H_5CH_2CH_2CH{=}P(C_6H_5)_3 \longrightarrow$

(b) cycloheptanone (=O) $+ (CH_3)_2CHCH{=}P(C_6H_5)_3 \longrightarrow$

13.22 (a) $CH_3CH_2Br \xrightarrow[\text{ether}]{\text{Mg}} CH_3CH_2MgBr \xrightarrow[(2)\ H_2O,\ H^+]{(1)\ CH_3CH_2C(=O)CH_2CH_3}$

$CH_3CH_2C(OH)(CH_2CH_3)_2$

(b) 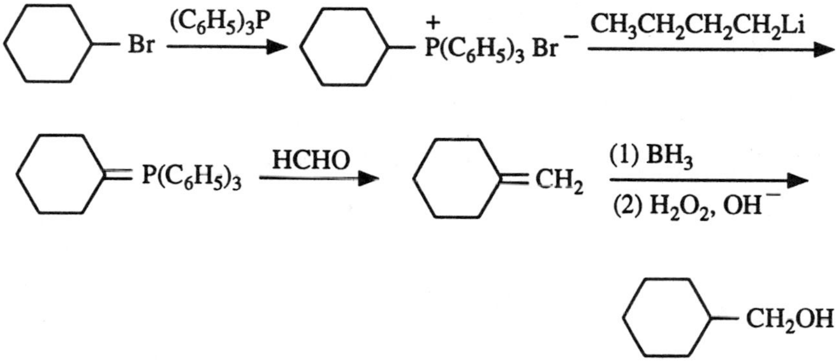 $C_6H_{11}Br \xrightarrow{(C_6H_5)_3P} C_6H_{11}\overset{+}{P}(C_6H_5)_3\ Br^- \xrightarrow{CH_3CH_2CH_2CH_2Li}$

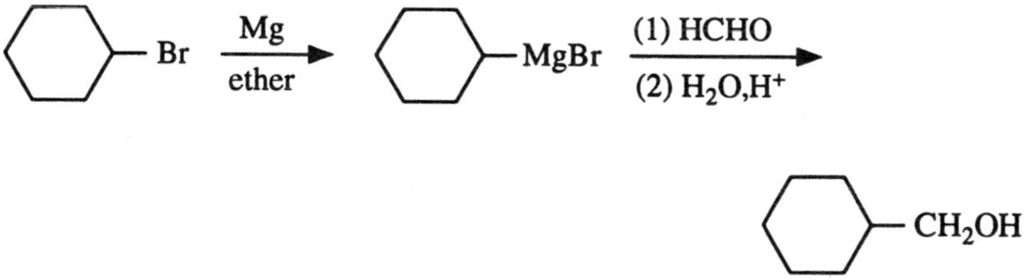

$C_6H_{10}{=}P(C_6H_5)_3 \xrightarrow{HCHO} C_6H_{10}{=}CH_2 \xrightarrow[(2)\ H_2O_2,\ OH^-]{(1)\ BH_3}$

$C_6H_{11}CH_2OH$

(c) $C_6H_{11}Br \xrightarrow[\text{ether}]{\text{Mg}} C_6H_{11}MgBr \xrightarrow[(2)\ H_2O, H^+]{(1)\ HCHO}$

$C_6H_{11}CH_2OH$

13.23 (a) 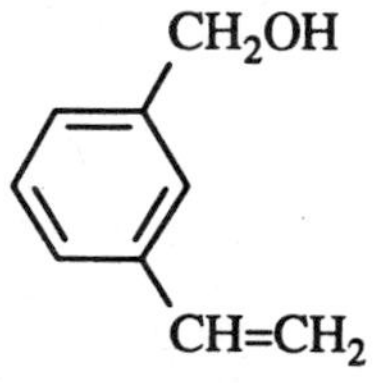 benzene ring with CH_2OH and $CH{=}CH_2$ (para) (b) 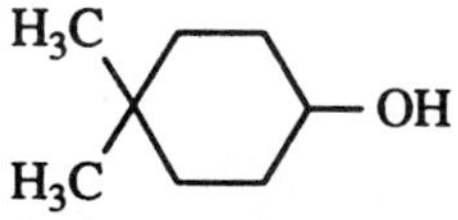 4,4-dimethylcyclohexanol (H_3C, H_3C, OH) (c) 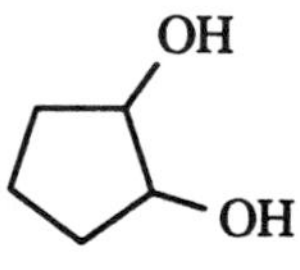 cyclopentane-1,2-diol (OH, OH)

13.24 (a) $CH_3CH_2CH_2CH_2OH$ (b) $CH_3CH_2CH_2C_6H_5$

(c) $C_6H_5CH_2CH_2CH_3$ (d) $C_6H_5CH(NHCH_3)CH_2CH_3$

13.25 (a) 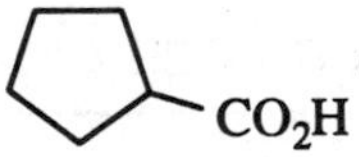 Cyclopentane–CO_2H or its salt (b) 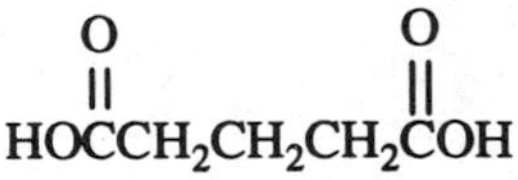 $HO\overset{O}{\overset{\|}{C}}CH_2CH_2CH_2\overset{O}{\overset{\|}{C}}OH$

(c) $C_6H_5CH{=}\underset{CH_2CH_3}{\underset{|}{C}}\overset{O}{\overset{\|}{C}}OH$ (d) 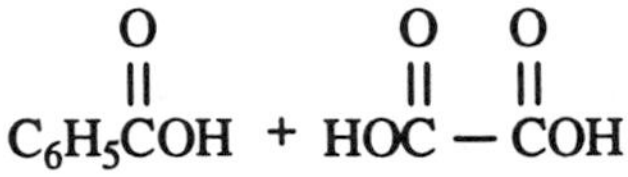 $C_6H_5\overset{O}{\overset{\|}{C}}OH$ + $HO\overset{O}{\overset{\|}{C}}-\overset{O}{\overset{\|}{C}}OH$

13.26 (a)  $N{\equiv}CCH_2\overset{O}{\overset{\|}{C}}OCH_3 \rightleftharpoons N{\equiv}CCH{=}\overset{OH}{\overset{|}{C}}OCH_3$

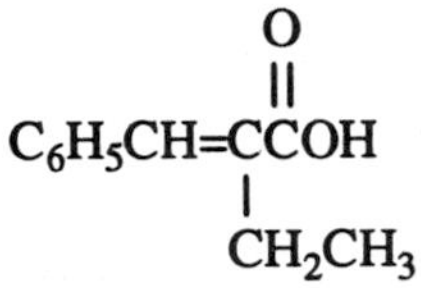

(b) 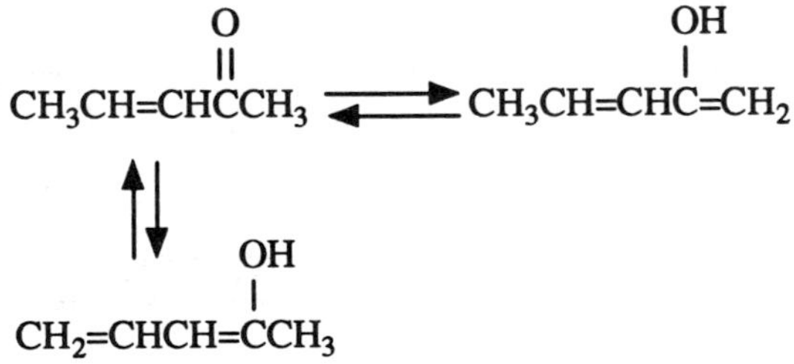 $CH_3CH{=}CH\overset{O}{\overset{\|}{C}}CH_3 \rightleftharpoons CH_3CH{=}CH\overset{OH}{\overset{|}{C}}{=}CH_2$

$\updownarrows$

$CH_2{=}CHCH{=}\overset{OH}{\overset{|}{C}}CH_3$

13.27 (a) 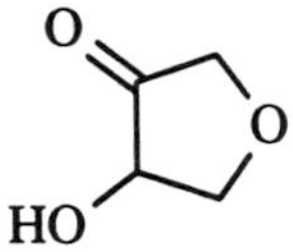 HO, O (b) 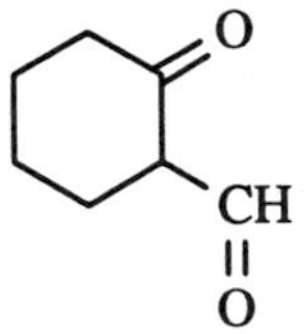 CH, O (c) 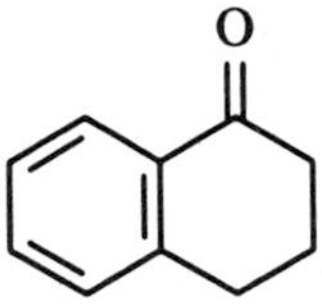 O

13.28 (a) 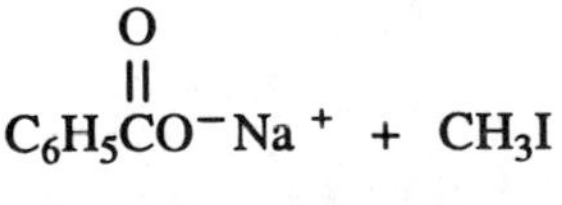 $C_6H_5\overset{O}{\overset{\|}{C}}O^-Na^+$ + CH_3I (b) Br, Br, O (c) O, CH_3, Br

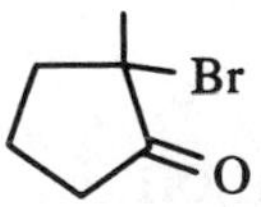

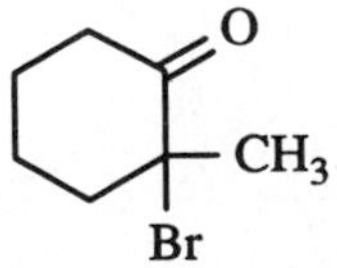

13.29 (a) $$C_6H_5\overset{O}{\overset{\|}{C}}CH_3 \xrightarrow[(2)\ H_2O,\ H^+]{(1)\ C_6H_5Li} (C_6H_5)_2\overset{OH}{\overset{|}{C}}CH_3$$

(b) $$C_6H_5\overset{O}{\overset{\|}{C}}CH_3 \xrightarrow[(2)\ H_2O,\ H^+]{(1)\ NaBH_4} C_6H_5\overset{OH}{\overset{|}{C}}HCH_3$$

(c) $$C_6H_5\overset{O}{\overset{\|}{C}}CH_3 + CH_3CH_2CH{=}P(C_6H_5)_3 \longrightarrow$$

$$C_6H_5\underset{CH_3}{\underset{|}{C}}{=}CHCH_2CH_3 \quad + (C_6H_5)_3PO$$

(d) $$C_6H_5\overset{O}{\overset{\|}{C}}CH_3 \xrightarrow[(2)\ H_2O,\ H^+]{(1)\ NaOI} C_6H_5CO_2H + CHI_3$$

(e) $$C_6H_5\overset{O}{\overset{\|}{C}}CH_3 \xrightarrow[\text{heat, pressure}]{H_2,\ Ni} C_6H_5\overset{OH}{\overset{|}{C}}HCH_3 \text{ or } C_6H_{11}\overset{OH}{\overset{|}{C}}HCH_3 \text{ (cyclohexyl)}$$

depending on conditions

(f) $$C_6H_5\overset{O}{\overset{\|}{C}}CH_3 \underset{}{\overset{NH_2OH,\ H^+}{\rightleftharpoons}} C_6H_5\underset{CH_3}{\underset{|}{C}}{=}NOH$$

(g) $$C_6H_5\overset{O}{\overset{\|}{C}}CH_3 \overset{\text{excess } CH_3OH,\ H^+}{\rightleftharpoons} C_6H_5\underset{CH_3}{\underset{|}{\overset{OCH_3}{\overset{|}{C}}}}OCH_3 + H_2O$$

(h) $C_6H_5C(=O)CH_3 \xrightarrow[\text{Ni}]{NH_3\ +\ H_2} C_6H_5CH(CH_3)NH_2$

(i) $C_6H_5C(=O)CH_3 \xrightarrow[H^+]{Br_2} C_6H_5C(=O)CH_2Br$

(j) $C_6H_5C(=O)CH_3 \xrightarrow{\text{Zn(Hg), HCl}} C_6H_5CH_2CH_3$

(k) $C_6H_5C(=O)CH_3 + NH_2NHC(=O)NH_2 \underset{}{\overset{H^+}{\rightleftharpoons}} C_6H_5C(CH_3){=}NNHC(=O)NH_2$

(l) $C_6H_5C(=O)CH_3$ + morpholine (N–H) $\overset{H^+}{\rightleftharpoons}$ $C_6H_5C(=CH_2)$ - N-morpholine

(m) $C_6H_5C(=O)CH_3 + HNO_3 \xrightarrow{H_2SO_4}$ m-O_2N-C_6H_4-$C(=O)CH_3$

(n) $C_6H_5C(=O)CH_3 \xrightarrow[(2)\ CH_3CH_2O^-]{(1)\ NH_2NH_2} C_6H_5CH_2CH_3$

(o) $C_6H_5C(=O)CH_3 \xrightarrow[(2)\ H_2O,\ H^+]{(1)\ CH_3MgBr} C_6H_5C(OH)(CH_3)CH_3$

(p) $C_6H_5C(=O)CH_3 \xrightarrow[(2)\ H_2,\ Ni]{(1)\ HCN,\ NaCN} C_6H_5C(OH)(CH_3)CH_2NH_2$

$$\begin{array}{ccc} \overset{\displaystyle O}{\|} & & \overset{\displaystyle OH}{|} \\ C_6H_5CCH_3 & \xrightarrow[\text{(2) } H_2\text{, Ni}]{\text{(1) HCN, NaCN}} & C_6H_5CCH_2NH_2 \\ & & \underset{\displaystyle CH_3}{|} \end{array}$$

(q)

$$\begin{array}{ccc} \overset{\displaystyle O}{\|} & & \overset{\displaystyle OH}{|} \\ C_6H_5CCH_3 & \xrightarrow[\text{(2) } H_2O\text{, } H^+]{\text{(1) } LiAlH_4} & C_6H_5CHCH_3 \end{array}$$

(r)

$$\begin{array}{ccc} \overset{\displaystyle O}{\|} & & \\ C_6H_5CCH_3 & \xrightarrow[\text{(2) } H_2\text{, Ni}]{\text{(1) } CH_3CH_2NH_2\text{, } H^+} & C_6H_5CHNHCH_2CH_3 \\ & & \underset{\displaystyle CH_3}{|} \end{array}$$

14

Carboxylic Acids

Answers to Additional Problems

14.1 (a) acetyl (b) acetoxy (c) benzoyl

14.2 (a) $C_6H_5CH(CH_3)CO_2H$ (b) $(CH_3)_2CHCO_2H$

(c) $CH_3CH_2CO_2H$ (d) o-ethoxybenzoic acid: benzene ring bearing CO_2H and OCH_2CH_3 on adjacent carbons

14.3 (a) $(CH_3)_3CCH_2Br \xrightarrow[\text{ether}]{\text{Mg}} (CH_3)_3CCH_2MgBr \xrightarrow[\text{(2) } H_2O,\ H^+]{\text{(1) } CO_2}$ product

(b) $HC(=O)$–C_6H_4–Br (para) $\xrightarrow[H^+,\ -H_2O]{HOCH_2CH_2OH}$ 2-(4-bromophenyl)-1,3-dioxolane $\xrightarrow[\text{(3) } H_2O,\ H^+]{\text{(1) Mg, ether; (2) } CO_2}$ product

(c) cyclopentyl–Br $\xrightarrow[\text{(2) } CO_2;\ \text{(3) } H_2O,\ H^+]{\text{(1) Mg, ether}}$ product

14.4 (a) $C_6H_5CH_2Br \xrightarrow{CN^-} C_6H_5CH_2CN \xrightarrow[\text{heat}]{H_2O,\ H^+} C_6H_5CH_2CO_2H$

$C_6H_5CH_2Br \xrightarrow[\text{ether}]{\text{Mg}} C_6H_5CH_2MgBr \xrightarrow[\text{(2) } H_2O,\ H^+]{\text{(1) } CO_2} C_6H_5CH_2CO_2H$

(b) Cyclohexyl–Br $\xrightarrow{CN^-}$ Cyclohexyl–CN $\xrightarrow[\text{heat}]{H_2O,\ H^+}$ Cyclohexyl–CO_2H

Cyclohexyl–Br $\xrightarrow[\text{ether}]{\text{Mg}}$ Cyclohexyl–MgBr $\xrightarrow[\text{(2) } H_2O,\ H^+]{\text{(1) } CO_2}$ Cyclohexyl–CO_2H

14.5 (a) $C_6H_5CH{=}O \xrightarrow[\text{or other oxidizing agent}]{H_2CrO_4,\ KMnO_4} C_6H_5CO_2H$

(b) $C_6H_5CH_3 \xrightarrow[\text{heat}]{H_2CrO_4 \text{ or } KMnO_4} C_6H_5CO_2H$

(c) $C_6H_5CO_2CH_3 + H_2O \xrightarrow[\text{heat}]{H^+} C_6H_5CO_2H$

(d) $C_6H_5CN + H_2O \xrightarrow[\text{heat}]{H^+} C_6H_5CO_2H$

(e) *trans*-$C_6H_5CH{=}CHC_6H_5 \xrightarrow[\text{(2) } H^+]{\text{(1) hot aqueous } KMnO_4} C_6H_5CO_2H$

14.6 (a) A Grignard synthesis would be superior to the nitrile synthesis. As its first step, the nitrile synthesis would have an S_N2 reaction of a hindered secondary halide. This reaction would probably give a poor yield of product.

$CH_3CH_2CH(Br)CH(CH_3)_2 \xrightarrow[\text{ether}]{\text{Mg}} CH_3CH_2CH(MgBr)CH(CH_3)_2$

$\xrightarrow[\text{(2) } H_2O,\ H^+]{\text{(1) } CO_2} CH_3CH_2CH(CO_2H)CH(CH_3)_2$

(b) A nitrile synthesis would be superior to the Grignard synthesis because the starting alkyl halide contains an -OH group, which would destroy a Grignard reagent.

$$\underset{\displaystyle OH}{CH_3CH_2\overset{|}{C}HCH_2CH_2Cl} \xrightarrow{CN^-} \underset{\displaystyle OH}{CH_3CH_2\overset{|}{C}HCH_2CH_2CN}$$

$$\xrightarrow[\text{heat}]{H_2O,\ H^+} \underset{\displaystyle OH}{CH_3CH_2\overset{|}{C}HCH_2CH_2CO_2H}$$

The γ-hydroxy acid would probably spontaneously form a lactone, or cyclic ester.

$$\underset{\displaystyle OH}{CH_3CH_2\overset{|}{C}HCH_2CH_2CO_2H} \xrightarrow{-H_2O} CH_3CH_2\text{-(γ-lactone ring: 5-membered ring containing O, with C=O)}$$

(c) A Grignard synthesis would be superior to the nitrile synthesis because aryl halides do not undergo S_N2 reactions.

$$C_6H_5Br \xrightarrow[\text{ether}]{Mg} C_6H_5MgBr \xrightarrow[(2)\ H_2O,\ H^+]{(1)\ CO_2} C_6H_5CO_2H$$

14.7 (a) $CH_3(CH_2)_6CH_2OH \xrightarrow{HBr} CH_3(CH_2)_6CH_2Br \xrightarrow[\text{ether}]{Mg} CH_3(CH_2)_6CH_2MgBr$

$$\xrightarrow[(2)\ H_2O,\ H^+]{(1)\ \overset{O}{\widehat{CH_2CH_2}}} CH_3(CH_2)_6CH_2CH_2CH_2OH \xrightarrow[\text{heat}]{H_2CrO_4} CH_3(CH_2)_8CO_2H$$

(b) $CH_3(CH_2)_6CH_2OH \xrightarrow{HBr} CH_3(CH_2)_6CH_2Br \xrightarrow[\text{ether}]{Mg} CH_3(CH_2)_6CH_2MgBr$

$$\xrightarrow[(2)\ H_2O,\ H^+]{(1)\ CO_2} CH_3(CH_2)_7CO_2H$$

$$CH_3(CH_2)_6CH_2OH \xrightarrow{HBr} CH_3(CH_2)_6CH_2Br \xrightarrow{CN^-} CH_3(CH_2)_6CH_2CN$$

$$\xrightarrow[\text{heat}]{H_2O,\ H^+} CH_3(CH_2)_6CH_2CO_2H$$

(c) $CH_3(CH_2)_6CH_2OH \xrightarrow[\text{heat}]{H_2CrO_4} CH_3(CH_2)_6CO_2H$

(d) $CH_3(CH_2)_6CH_2OH \xrightarrow[\text{heat}]{H_2SO_4} CH_3(CH_2)_5CH{=}CH_2$

$\xrightarrow[\text{(2) } H^+]{\text{(1) hot aqueous } KMnO_4} CH_3(CH_2)_5CO_2H$

14.8 (a) methylenecyclopentane ($=CH_2$) $\xrightarrow[\text{or HBr, peroxide}]{\text{(1) } BH_3\text{; (2) } Br_2}$ cyclopentyl–CH_2Br

$\xrightarrow[\text{(3) } H_2O, H^+]{\text{(1) Mg, ether; (2) } CO_2}$ product

(b) methylenecyclopentane ($=CH_2$) $\xrightarrow[\text{(2) } H_2O_2, OH^-]{\text{(1) } BH_3}$ cyclopentyl–CH_2OH $\xrightarrow[\text{heat}]{H_2CrO_4}$ product

(c) $CH_3CH_2CH(=O) \xrightarrow[CN^-]{HCN} CH_3CH_2CH(OH)CN \xrightarrow[\text{heat}]{H_2O, H^+} CH_3CH_2CH(OH)C(=O)OH$

(d) $CH_3CH_2CH{=}CHCH_2CH_3 \xrightarrow[\text{(2) } H^+]{\text{(1) hot aqueous } KMnO_4} CH_3CH_2CO_2H$

14.9 (a) $C_6H_5SO_3^-$ (b) $C_6H_5O^-$ (c) $CH_3CH_2CO_2^-$

14.10 $pK_a = -\log(K_a)$

(a) $-\log(1.28 \times 10^{-5}) = 5 - \log(1.28) = 5 - 0.11 = 4.89$

(b) 4.28

14.11 Let x = concentration of RCO_2^- and H^+ in *M*

Using the mass law equation,

$$1.51 \times 10^{-5} = \frac{x^2}{0.0300}$$

$$\begin{aligned} x^2 &= 0.0300 \times (1.51 \times 10^{-5}) \\ &= 0.0453 \times 10^{-5} \\ &= 45.3 \times 10^{-8} \\ x &= 6.73 \times 10^{-4} \\ [H^+] &= 6.73 \times 10^{-4}\ M \end{aligned}$$

14.12 $RCO_2H \rightleftharpoons RCO_2^- + H^+$

If the H^+ concentration is 0.00184 *M*, the RCO_2^- concentration is also 0.00184 *M* and the undissociated RCO_2H concentration must be (0.200-0.00184) *M*.

$$K_a = \frac{(0.00184)(0.00184)}{(0.200 - 0.00184)} = \frac{3.39 \times 10^{-6}}{0.198}$$

$$K_a = 1.71 \times 10^{-5}$$

If 0.100 mol of NaOH were added to the solution, it would react with 0.100 mol of the acid generating 0.100 mol of RCO_2^-.

$$\begin{aligned} K_a &= 1.71 \times 10^{-5} \\ [RCO_2H] &= 0.100 \\ [RCO_2^-] &= (0.200 - 0.100) = 0.100 \\ [H^+] &= ? \\ 1.71 \times 10^{-5} &= \frac{(0.100)(x)}{(0.100)} \\ x &= 1.71 \times 10^{-5} \end{aligned}$$

14.13 (a) CH_3SH because sulfur, being a larger atom than oxygen, is better able to carry a negative charge.

(b) CH_3CH_2OH because oxygen is more electronegative than is nitrogen.

(c) $CH_3CH(Br)COOH$ because Br has a greater inductive electron-withdrawl effect than does I.

(d) C_6H_5–OH because its anion can be resonance stabilized.

14.14 (f) < (c) < (e) < (b) < (d) < (a)

14.15

(a) Cl-	(b) F-	(c) CH_3O-
(d) C_6H_5-	(e) CH_2=CH-	(f) Br-

14.16

$$H-C(=O)-OH \overset{H^+}{\rightleftharpoons} H-C(=\overset{+}{O}H)-OH \overset{CH_3CH_2OH}{\rightleftharpoons} H-C(OH)(OH)-\overset{+}{O}(H)-CH_2CH_3 \overset{-H^+}{\rightleftharpoons}$$

$$H-C(OH)(OH)-O-CH_2CH_3 \overset{H^+}{\rightleftharpoons} H-C(OH)(\overset{+}{O}H_2)-OCH_2CH_3 \overset{-H_2O}{\rightleftharpoons} H-C(=\overset{+}{O}H)-OCH_2CH_3 \overset{-H^+}{\rightleftharpoons} H-C(=O)-OCH_2CH_3$$

14.17 (c) < (b) < (a) based on steric hindrance near the carbonyl carbon atom.

14.18

$$CH_3C(=O\cdots H-O)-CH_2-C(=O) \xrightarrow{heat} \left[CH_3C(=O)-CH_2-C(=O)-O-H \right] \longrightarrow$$

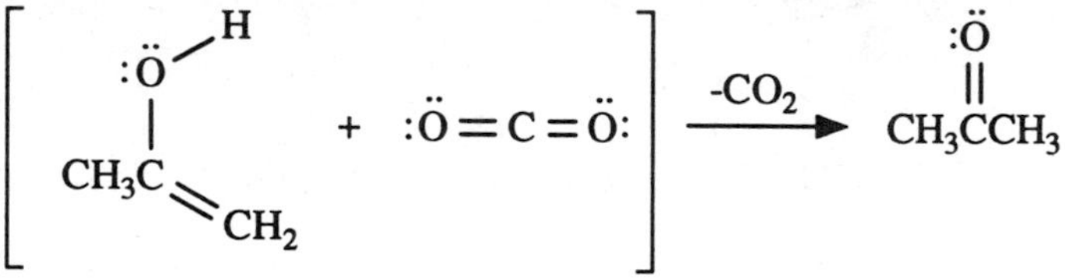

$$\left[\; CH_3C(\ddot{O}H){=}CH_2 \; + \; :\ddot{O}{=}C{=}\ddot{O}: \; \right] \xrightarrow{-CO_2} CH_3\overset{\overset{\large :\ddot{O}}{\|}}{C}CH_3$$

14.19 (a) 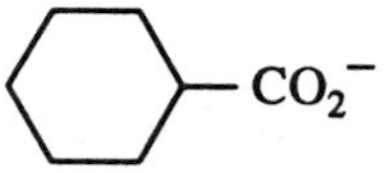 $C_6H_{11}{-}CO_2^-$ (cyclohexyl) $+ \; H_2O \; + \; CO_2$

(b) no reaction

(c) $CH_3CH_2CH_2CO_2^- \; + \; H_2O$

(d) $C_6H_5O^- \; + \; H_2O$

(e) no reaction

(f) 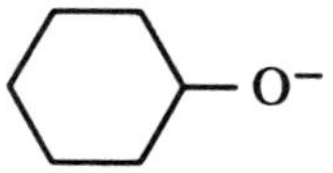 $C_6H_{11}{-}O^-$ (cyclohexyl) $+ \; NH_3$

(g) $CH_3CO_2^- \; + \; CH_3OH$

(h) $C_6H_5O^- \; + \; C_6H_{11}{-}OH$ (cyclohexyl)

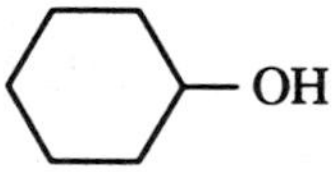

(i) no reaction

(j) $C_6H_5OH \; + \; CH_3CH_2CO_2^-$

14.20 (a) $C_6H_5CH_2OH$

(b) $(CH_3)_3CCH_2CO_2CH_3 \; + \; H_2O$

(c) 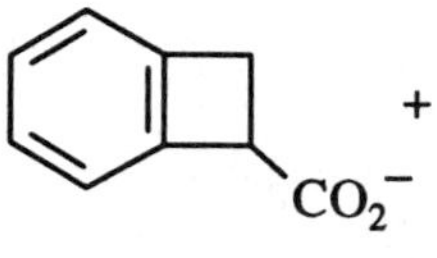 benzocyclobutene-CO_2^- $+ \; C_6H_6 \; + \; Mg^{2+} \; + \; Br^-$

(d) 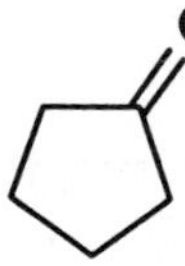 cyclopentanone (O) $+ \; CO_2$

15

Derivatives of Carboxylic Acids

Answers to Additional Problems

15.1 (a) *p-tert*-butylbenzoyl chloride

(b) isobutyl benzoate

(c) diethyl propanedioate or diethyl malonate

(d) isopropyl 2,3,3-trimethylbutanoate

(e) 3-butenoic anhydride

(f) 2-bromopentanoyl bromide

(g) 3-pentenenitrile

(h) hexanediamide

15.2 (a) ethanoic butanoic anhydride

(b) 3-butenoyl chloride

(c) cyclohexyl ethanoate

(d) methyl cyclohexanecarboxylate

(e) 3,3-dimethylbutanenitrile

(f) *N*-methyl-2-methylpropanamide

15.3 (a) $BrCH_2CH_2CH_2\overset{\overset{O}{\|}}{C}OCH_3$ (b) $CH_3O\overset{\overset{O}{\|}}{C}CH_2Cl$

(c) $ClCH_2O\overset{\overset{O}{\|}}{C}CH_3$ (d) $ClCH_2CH_2CN$

(e) $CH_3\overset{\overset{O}{\|}}{C}O\overset{\overset{O}{\|}}{C}H$ (f) $H\overset{\overset{O}{\|}}{C}N(CH_3)_2$

15.4 (a) $CH_3CH_2CH_2\overset{O}{\overset{\|}{C}}O\overset{O}{\overset{\|}{C}}CH_2CH_2CH_3$ (b) $CH_3CH_2CH_2CH_2\overset{O}{\overset{\|}{C}}F$

(c) $C_6H_5\overset{O}{\overset{\|}{C}}NHCH_2CH_3$ (d) $(CH_3)_2CHCH_2CN$

(e) $CH_3\underset{Cl}{\underset{|}{C}}H\overset{O}{\overset{\|}{C}}OCH_3$ (f) 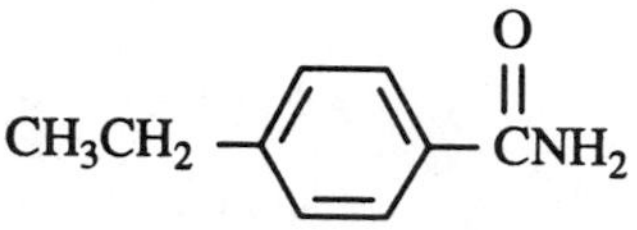 $CH_3CH_2-C_6H_4-\overset{O}{\overset{\|}{C}}NH_2$

(g) 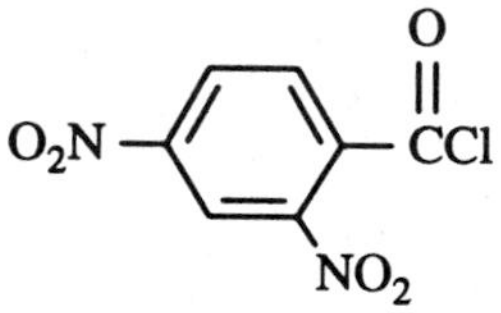 $O_2N-C_6H_3(NO_2)-\overset{O}{\overset{\|}{C}}Cl$

15.5 (a) $CH_3CH_2\overset{O}{\overset{\|}{C}}NH_2$ (b) $(CH_3)_2\overset{Br}{\overset{|}{C}}-\overset{O}{\overset{\|}{C}}OCH_2CH_3$

(c) $(C_6H_5)_2\underset{CN}{\underset{|}{C}}CH_2CH_3$

15.6 (a) $(CH_3)_2CH\underset{Br}{\underset{|}{C}}HCO_2H$ (b) $(CH_3)_2CHCH_2O\overset{O}{\overset{\|}{C}}CH_3$

(c) $(CH_3)_3C\overset{O}{\overset{\|}{C}}CH_3$ (d) 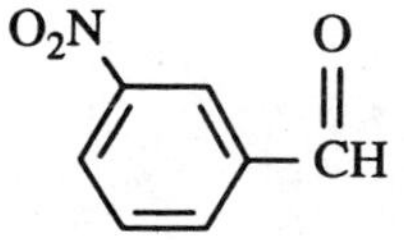 $O_2N-C_6H_4-\overset{O}{\overset{\|}{C}}H$

(e) 3,5-dimethylbenzophenone: $(CH_3)_2C_6H_3-\overset{O}{\overset{\|}{C}}-C_6H_5$ (methyl groups labelled H_3C)

(f) anthrone (9(10H)-anthracenone): tricyclic ring with C=O

(g) $CH_3\overset{O}{\overset{\|}{C}}CH(\overset{O}{\overset{\|}{C}}CH_3)_2$

15.7 (1) $CH_3)_2CHCH_2CH_2\overset{O}{\overset{\|}{C}}Cl + C_6H_6 \xrightarrow{AlCl_3}$

(2) $(CH_3)_2CHCH_2CH_2\overset{O}{\overset{\|}{C}}Cl + (C_6H_5)_2CuLi \longrightarrow (CH_3)_2CHCH_2CH_2\overset{O}{\overset{\|}{C}}C_6H_5$

(3) $[(CH_3)_2CHCH_2CH_2]_2CuLi + C_6H_5\overset{O}{\overset{\|}{C}}Cl \longrightarrow$

(All three routes give $(CH_3)_2CHCH_2CH_2\overset{O}{\overset{\|}{C}}C_6H_5$.)

15.8 (a) $CH_3CH_2CH_2\overset{O}{\overset{\|}{C}}OH + CH_3\overset{O}{\overset{\|}{C}}O\overset{O}{\overset{\|}{C}}CH_3 \xrightarrow{heat}$

or $CH_3CH_2CH_2\overset{O}{\overset{\|}{C}}Cl + CH_3CH_2CH_2CO_2^-\ Na^+ \longrightarrow$

(b) $O_2N-C_6H_4-CO_2H + SOCl_2 \longrightarrow$

(c) $CH_3\overset{O}{\overset{\|}{C}}Cl$ + excess $(CH_3CH_2)_2NH \longrightarrow$

(Acetic anhydride or an acetate ester could have been used in place of the acetyl chloride.)

15.9 (a) $CH_3\overset{O}{\overset{\|}{C}}OH \xrightarrow[(2)\ H_2O,\ H^+]{(1)\ LiAlH_4} CH_3CH_2OH \xrightarrow{HBr} CH_3CH_2Br \xrightarrow{CN^-} CH_3CH_2CN$

(b) $CH_3\overset{O}{\overset{\|}{C}}OH \xrightarrow{SOCl_2} CH_3\overset{O}{\overset{\|}{C}}Cl \xrightarrow[AlCl_3]{C_6H_6} CH_3\overset{O}{\overset{\|}{C}}C_6H_5$

(c) $CH_3\overset{O}{\overset{\|}{C}}OH \xrightarrow{SOCl_2} CH_3\overset{O}{\overset{\|}{C}}Cl \xrightarrow{(CH_3)_2NH} CH_3\overset{O}{\overset{\|}{C}}N(CH_3)_2$

(d) $CH_3\overset{O}{\overset{\|}{C}}OH + NaOH \longrightarrow CH_3\overset{O}{\overset{\|}{C}}O^-Na^+$

$CH_3\overset{O}{\overset{\|}{C}}OH \xrightarrow{SOCl_2} CH_3\overset{O}{\overset{\|}{C}}Cl$

(both products combine) $\longrightarrow CH_3\overset{O}{\overset{\|}{C}}O\overset{O}{\overset{\|}{C}}CH_3$

(e) $CH_3\overset{O}{\overset{\|}{C}}OH \xrightarrow[heat]{CH_3CH_2OH,\ H^+} CH_3\overset{O}{\overset{\|}{C}}OCH_2CH_3$

(f) $CH_3\overset{O}{\overset{\|}{C}}OH + Cl_2 \xrightarrow{PCl_3} ClCH_2\overset{O}{\overset{\|}{C}}OH$

15.10 (a) [phthalic anhydride structure: benzene ring fused to five-membered ring with two C=O and ring O]

(b) $O_2N-C_6H_4-\overset{O}{\overset{\|}{C}}O\overset{O}{\overset{\|}{C}}-C_6H_4-NO_2$

(c) [ring structure: O, O, O]

15.11 (a) $CH_3\overset{\overset{O}{\|}}{C}OH \xrightarrow{SOCl_2} CH_3\overset{\overset{O}{\|}}{C}Cl \xrightarrow{H\overset{\overset{O}{\|}}{C}ONa} CH_3\overset{\overset{O}{\|}}{C}O\overset{\overset{O}{\|}}{C}H$

(b) $(H)(CO_2H)C{=}C(H)(CO_2H)$ + $CH_3\overset{\overset{O}{\|}}{C}O\overset{\overset{O}{\|}}{C}CH_3 \longrightarrow$ [maleic anhydride ring: O, O, O] + $CH_3\overset{\overset{O}{\|}}{C}OH$

(c) $(HO\overset{\overset{O}{\|}}{C}CH_2)_2C(CH_3)_2$ + $CH_3\overset{\overset{O}{\|}}{C}O\overset{\overset{O}{\|}}{C}CH_3 \longrightarrow$ [ring: O, O, O, CH_3, CH_3] + $CH_3\overset{\overset{O}{\|}}{C}OH$

15.12 (a) $C_6H_5CH_2O\overset{\overset{O}{\|}}{C}CH_3 + CH_3\overset{\overset{O}{\|}}{C}OH$

(b) $C_6H_5NH\overset{\overset{O}{\|}}{C}CH_3 + CH_3\overset{\overset{O}{\|}}{C}O^{-}\ \overset{+}{N}H_3C_6H_5$

(c) $C_6H_5\overset{\overset{O}{\|}}{C}O\overset{\overset{O}{\|}}{C}C_6H_5 + C_6H_5\overset{\overset{O}{\|}}{C}O\overset{\overset{O}{\|}}{C}CH_3 + CH_3\overset{\overset{O}{\|}}{C}OH$

(d) (R)-glyceraldehyde: CHO–$C(H)(HO)$–CH_2OH $\xrightarrow{CH_3\overset{O}{\overset{\|}{C}}O\overset{O}{\overset{\|}{C}}CH_3}$ CHO–$C(H)(CH_3\overset{\|}{\underset{O}{C}}O)$–$CH_2O\overset{O}{\overset{\|}{C}}CH_3$

(R) (R)

(e) $C_6H_5O\overset{O}{\overset{\|}{C}}CH_3 + Na^+ \, {}^-O\overset{O}{\overset{\|}{C}}CH_3$

(f) $(CH_3)_2N\overset{O}{\overset{\|}{C}}CH_3 + (CH_3)_2\overset{+}{N}H_2 \, {}^-O\overset{O}{\overset{\|}{C}}CH_3$

15.13 (a) CH_3CH_2O–C_6H_4–$NH\overset{O}{\overset{\|}{C}}CH_3$ + CH_3CH_2O–C_6H_4–$\overset{+}{N}H_3 \, {}^-O\overset{O}{\overset{\|}{C}}CH_3$

(b) $HO_2CCH_2CH_2CH_2CO_2H$

(c) $Na^+ \, {}^-O_2CCH_2CH_2CO_2^- \, Na^+$

(d) C_6H_4 (ortho) bearing $\overset{O}{\overset{\|}{C}}NHCH_2CH_3$ and $CO_2^- \; H_3\overset{+}{N}CH_2CH_3$

15.14 (a) $CH_3\overset{:\ddot{O}}{\overset{\|}{C}}\text{–}Cl + H_2\ddot{N}C_6H_5 \longrightarrow CH_3\overset{:\ddot{O}:^-}{\overset{|}{C}}\text{–}\ddot{C}l: \; (\underset{+}{H_2NC_6H_5}) \xrightarrow{-:\ddot{C}l:}$

$CH_3\overset{\ddot{O}:}{\overset{\|}{C}}\overset{+}{N}HC_6H_5$ (N–H) $\xrightarrow{C_6H_5\ddot{N}H_2}$ $CH_3\overset{\ddot{O}:}{\overset{\|}{C}}\ddot{N}HC_6H_5 + C_6H_5\overset{+}{N}H_3$

(b) $CH_3C(=O)-OC(=O)CH_3 + H_2NC_6H_5 \longrightarrow CH_3C(-O^-)(-OC(=O)CH_3)(-\overset{+}{N}H_2C_6H_5) \xrightarrow{-\,^-OC(=O)CH_3}$

$CH_3C(=O)\overset{+}{N}H(H)C_6H_5 \xrightarrow{CH_3C(=O)O^-} CH_3C(=O)NHC_6H_5 + CH_3C(=O)OH$

15.15 (a) phthalic anhydride $\xrightarrow{CH_3CH_2OH,\ H^+}$ $C_6H_4(CO_2CH_2CH_3)(CO_2H)$

$\xrightarrow{SOCl_2}$ $C_6H_4(CO_2CH_2CH_3)(CCl(=O))$ $\xrightarrow{NH_3}$ product

(b) $C_6H_5OH \xrightarrow{HNO_3} O_2N-C_6H_4-OH \xrightarrow[o\text{ isomer}]{\text{separate from}}$

$\xrightarrow[\text{pyridine}]{CH_3C(=O)Cl \text{ or } (CH_3C(=O))_2O} O_2N-C_6H_4-OC(=O)CH_3$

(c) $HO_2C(CH_2)_3CO_2H \xrightarrow[\text{warm}]{(CH_3\overset{O}{\overset{\|}{C}})_2O}$ (cyclic glutaric anhydride: six-membered ring O=C–O–C=O) $\xrightarrow{CH_3OH,\ H^+}$

$CH_3O\overset{O}{\overset{\|}{C}}(CH_2)_3CO_2H \xrightarrow[(2)\ (CH_3)_2NH]{(1)\ SOCl_2}$ product

15.16 (a) $CH_3CH_2CH_2CO_2H$ + excess $CH_3OH \xrightarrow{H^+,\ \text{heat}}$ product

(b) $CH_3CH_2CH_2CO_2H \xrightarrow[PBr_3]{Br_2} CH_3CH_2\underset{Br}{\underset{|}{C}}H\overset{O}{\overset{\|}{C}}Br \xrightarrow{CH_3OH}$ product

(c) $CH_3CH_2CH_2CO_2H \xrightarrow[(2)\ H_2O,\ H^+]{(1)\ LiAlH_4} CH_3CH_2CH_2CH_2OH \xrightarrow{HBr}$

$\xrightarrow[\text{ether}]{Mg} CH_3CH_2CH_2CH_2MgBr \xrightarrow[(2)\ H_2O,\ H^+]{(1)\ CO_2}$

$CH_3(CH_2)_3CO_2H \xrightarrow[\text{heat}]{CH_3CH_2OH,\ H^+}$ product

In (c) the bromoalkane could have been treated with NaCN.

(d) (1) $CH_3CH_2CH_2CO_2H + CH_3CH_2CH_2CH_2OH \xrightarrow[\text{heat}]{H^+}$ product

(2) $CH_3CH_2CH_2CO_2H \xrightarrow{SOCl_2} CH_3CH_2CH_2\overset{O}{\overset{\|}{C}}Cl$

$\xrightarrow[\text{pyridine}]{CH_3CH_2CH_2CH_2OH}$ product

$$\text{(3) } CH_3CH_2CH_2\overset{\overset{\displaystyle O}{\|}}{C}OH \xrightarrow{CH_3OH,\ H^+} CH_3CH_2CH_2\overset{\overset{\displaystyle O}{\|}}{C}OCH_3 \xrightarrow{CH_3CH_2CH_2CH_2OH,\ H^+} \text{product}$$

In method (3), CH_3OH must be removed from the reaction mixture to drive the reaction to completion. You may have also converted butanoic acid to an acid chloride followed by reaction with butanol.

15.17 (a) [decalin ring bearing OH and a carbon with CH_2OH and CH_3] + CH_3OH

(b) $C_6H_5\overset{\overset{\displaystyle O}{\|}}{C}O^-\ Na^+ + O_2N{-}C_6H_4{-}O^-\ Na^+$

(c) $(CH_3)_3CCH_2\overset{\overset{\displaystyle O}{\|}}{C}O{-}C_6H_{11}$ + CH_3CH_2OH

(d) $CH_3CH_2CH_2CH_2OH$ + $C_6H_{11}{-}OH$

(e) $CH_2{=}CHCH_2\overset{\overset{\displaystyle OH}{|}}{C}(C_6H_5)_2$ + C_6H_5OH

15.18 The 1,3-*cis* substituents of (a) can be positioned on the same side of the ring and can react with each other. When the substituents are 1,3-*trans*, as in (b), the two groups cannot be positioned on the same side of the ring and, therefore, are too far apart to form a lactone.

(a) 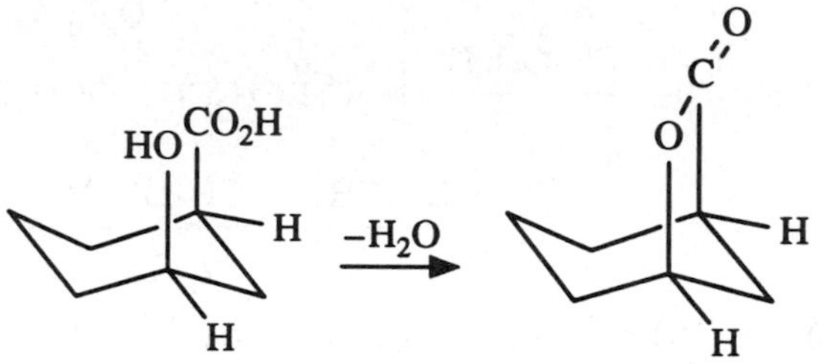 HO, CO_2H ... H $\xrightarrow{-H_2O}$ ... H

a, a — *a lactone*

(b) 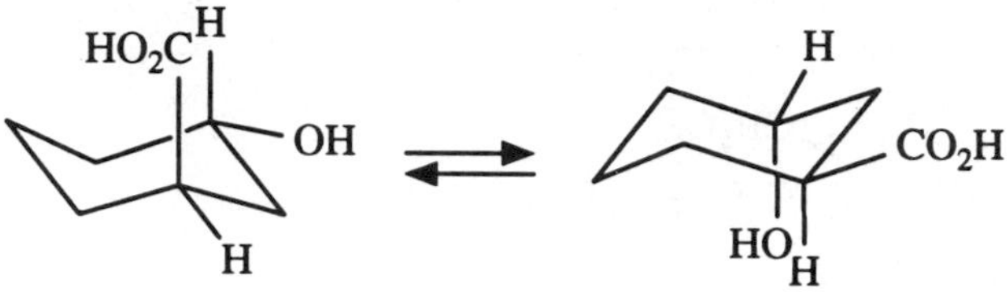 HO_2C, H, OH $\rightleftharpoons$ H, CO_2H, HO, H

a, e — *e, a*

15.19 (a) $LiAlH_4$ followed by H_2O, H^+

(b) Na, CH_3CH_2OH

(c) H_2, catalyst, heat, and pressure

Note $NaBH_4$ is not listed because it does not reduce esters.

15.20 (a)

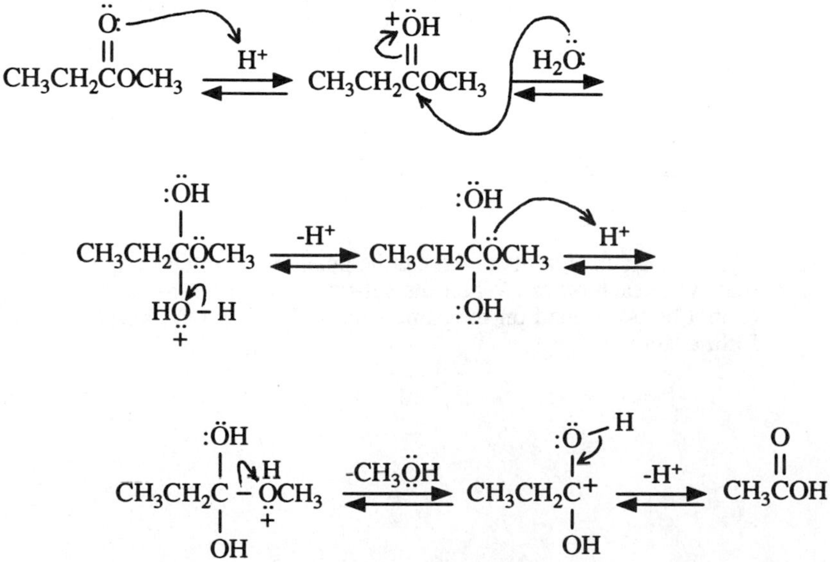

$$CH_3CH_2\overset{\overset{\ddot{O}:}{\|}}{C}OCH_3 \xrightleftharpoons{H^+} CH_3CH_2\overset{\overset{+\ddot{O}H}{\|}}{C}OCH_3 \xrightleftharpoons{H_2\ddot{O}:}$$

$$CH_3CH_2\underset{\underset{HO^+-H}{|}}{\overset{\overset{:\ddot{O}H}{|}}{C}}\ddot{O}CH_3 \xrightleftharpoons{-H^+} CH_3CH_2\underset{\underset{:\ddot{O}H}{|}}{\overset{\overset{:\ddot{O}H}{|}}{C}}\ddot{O}CH_3 \xrightleftharpoons{H^+}$$

$$CH_3CH_2\underset{\underset{OH}{|}}{\overset{\overset{:\ddot{O}H}{|}}{C}}-\overset{+}{O}(H)CH_3 \xrightleftharpoons{-CH_3\ddot{O}H} CH_3CH_2\underset{\underset{OH}{|}}{\overset{\overset{:\ddot{O}-H}{|}}{C^+}} \xrightleftharpoons{-H^+} CH_3\overset{\overset{O}{\|}}{C}OH$$

(b) $C_6H_5COCH_3$ (C=Ö:) + $^{-}:\ddot{O}H$ ⇌ $C_6H_5C(:\ddot{O}:^{-})(:\ddot{O}H)-\ddot{O}CH_3$ $\xrightarrow{-:\ddot{O}CH_3}$

$C_6H_5C(=O)\ddot{O}-H$ $\xrightarrow[-H_2O]{^{-}:\ddot{O}H}$ $C_6H_5C(=O)\ddot{O}:^{-}$

$$CH_3O^- + H_2O \longrightarrow CH_3OH + {}^{-}OH$$

15.21 (a) $C_6H_5C(CH_3)_2OH + CH_3OH$ (b) $C_6H_5C(OH)(CH_3)C_6H_5$

(c) $HOCH_2CH_2CH_2C(OH)(CH_3)_2$ (d) $(CH_3)_3COH + 2\ CH_3OH$

(e) $C_6H_5CH(OH)CH_3$ (f) $C_6H_5C(CH_3)_2OH$

15.22 (a) H_3C–(five-membered lactone ring, O in ring, C=O)

(b) $CH_3C(=O)CH_2CH_2C(=O)OCH_3$ $\xrightarrow[(2)\ H_2O,\ H^+]{(1)\ NaBH_4}$ $CH_3CH(OH)CH_2CH_2C(=\ddot{O}:)OCH_3$

$\overset{H^+}{\rightleftharpoons}$ $CH_3CH(H\ddot{O}:)CH_2CH_2C(=\overset{+}{\ddot{O}}H)OCH_3$ ⇌ H_3C–(ring with $\overset{+}{\ddot{O}}$–H, C bearing $\ddot{O}H$ and OCH_3) $\overset{-H^+}{\rightleftharpoons}$

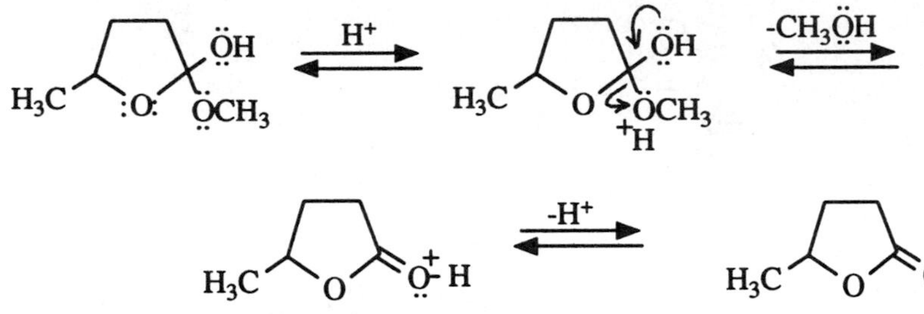

$$\text{H}_3\text{C}\text{-(2-methoxy-2-hydroxy-5-methyltetrahydrofuran)} \underset{}{\overset{H^+}{\rightleftharpoons}} \text{(protonated }\ddot{O}CH_3\text{)} \underset{}{\overset{-CH_3\ddot{O}H}{\rightleftharpoons}}$$

$$\text{H}_3\text{C}\text{-(lactone, }=\overset{+}{O}\text{-H)} \underset{}{\overset{-H^+}{\rightleftharpoons}} \text{H}_3\text{C}\text{-(lactone, }=\ddot{O}\text{)}$$

(c) No. Lithium aluminum hydride would reduce both the keto group and the ester group yielding:

$$CH_3\underset{|}{\overset{OH}{C}}HCH_2CH_2\overset{OH}{\underset{|}{C}}H_2 + CH_3OH$$

15.23 (a) $CH_3CH_2CH_2CH_2\overset{O}{\overset{\|}{C}}Cl + 2\ CH_3NH_2 \longrightarrow$

$$CH_3(CH_2)_3\overset{O}{\overset{\|}{C}}NHCH_3 + CH_3\overset{+}{N}H_3\ Cl^-$$

(b) $CH_3(CH_2)_3\overset{O}{\overset{\|}{C}}O\overset{O}{\overset{\|}{C}}(CH_2)_3CH_3 + 2\ CH_3NH_2 \longrightarrow$

$$CH_3(CH_2)_3\overset{O}{\overset{\|}{C}}NHCH_3 + CH_3(CH_2)_3\overset{O}{\overset{\|}{C}}O^-\ CH_3\overset{+}{N}H_3$$

(c) $CH_3(CH_2)_3\overset{O}{\overset{\|}{C}}OCH_3 + CH_3NH_2 \longrightarrow CH_3(CH_2)_3\overset{O}{\overset{\|}{C}}NHCH_3 + CH_3OH$

15.24 (a) 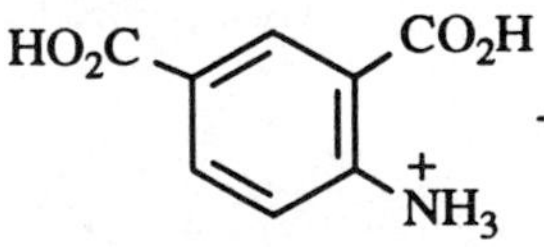 HO_2C–(benzene ring)–CO_2H, $\overset{+}{N}H_3$ $+ CH_3CO_2H$ (b) 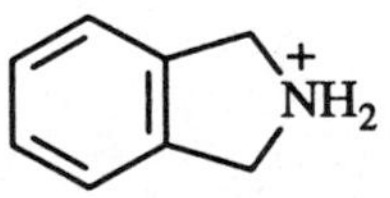 (isoindoline, $\overset{+}{N}H_2$)

(c) $C_6H_5CO_2^- + (CH_3)_2NH$

15.25 (a) $C_6H_5C(=O)NHCH_3 + H_2O + H^+ \longrightarrow C_6H_5CO_2H + CH_3\overset{+}{N}H_3$

(b) $C_6H_5C(=O)NHCH_3 + OH^- \longrightarrow C_6H_5CO_2^- + CH_3NH_2$

(c) $C_6H_5C(=O)NHCH_3 \xrightarrow[(2)\ H_2O,\ H^+]{(1)\ LiAlH_4} C_6H_5CH_2\overset{+}{N}H_2CH_3$

(d) $C_6H_5C(=O)NHCH_3 + CH_3OH + H^+ \longrightarrow C_6H_5CO_2CH_3 + CH_3\overset{+}{N}H_3$

15.26 (a) $(CH_3)_2CHC(=O)CH_3 \xrightarrow[(2)\ H_2O,\ H^+]{(1)\ LiAlH_4} (CH_3)_2CHCH(OH)CH_3$

(b) $(CH_3)_2CHCH_2CH(=O) \xrightarrow[(2)\ H_2O,\ H^+]{(1)\ LiAlH_4} (CH_3)_2CHCH_2CH_2OH$

(c) $C_6H_5CO_2H \xrightarrow[(2)\ H_2O,\ H^+]{(1)\ LiAlH_4} C_6H_5CH_2OH$

(d) $C_6H_5C(=O)Cl \xrightarrow[(2)\ H_2O,\ H^+]{(1)\ LiAlH_4} C_6H_5CH_2OH$

(e) $C_6H_5CO_2CH_3 \xrightarrow[(2)\ H_2O,\ H^+]{(1)\ LiAlH_4} C_6H_5CH_2OH + CH_3OH$

(f) $C_6H_5C(=O)NH_2 \xrightarrow[\text{(2) } H_2O,\ H^+ \text{ (3) neutralize}]{\text{(1) } LiAlH_4} C_6H_5CH_2NH_2$

15.27 (a) m-$CH_3C_6H_4CN$ (benzene ring with CN and CH_3 meta) (b) $NCCH_2CH_2CN$ (c) $CH_3C(=O)-C_6H_4-CN$ (para)

15.28 (a) $CH_3CH_2CH_2OH \xrightarrow{HBr} CH_3CH_2CH_2Br \xrightarrow{^-CN} CH_3CH_2CH_2CN$

(b) $C_6H_5CH_2OH \xrightarrow{HBr} C_6H_5CH_2Br \xrightarrow{^-CN} C_6H_5CH_2CN$

(c) $C_6H_6 \xrightarrow[H_2SO_4]{HNO_3} C_6H_5NO_2 \xrightarrow[\text{(2) neutralize}]{\text{(1) Fe, HCl}} C_6H_5NH_2$

$\xrightarrow[0°]{NaNO_2,\ HCl} C_6H_5\overset{+}{N}_2\ Cl^- \xrightarrow[KCN]{CuCN} C_6H_5CN$

15.29 (a) $C_6H_5CN + 2\ H_2O + H^+ \xrightarrow{100°} C_6H_5CO_2H + NH_4^+$

(b) $C_6H_5CN + 2\ H_2 \xrightarrow{\text{Raney Ni}} C_6H_5CH_2NH_2$

(c) $C_6H_5CN \xrightarrow[\text{(2) } H_2O,\ H^+ \text{ (3) neutralize}]{\text{(1) } LiAlH_4} C_6H_5CH_2NH_2$

(d) $C_6H_5CN + H_2O \xrightarrow[40°]{HCl} C_6H_5C(=O)NH_2$

15.30 (a) (1) $C_6H_{11}Br \xrightarrow{^-CN} C_6H_{11}CN \xrightarrow[heat]{H_2O,\ H^+}$ product

(cyclohexyl bromide → cyclohexyl cyanide)

(2) $C_6H_{11}Br \xrightarrow[ether]{Mg} C_6H_{11}MgBr \xrightarrow[(2)\ H_2O,\ H^+]{(1)\ CO_2}$ product

(b) (1) $CH_3CH_2CH_2Br \xrightarrow{^-CN} CH_3CH_2CH_2CN \xrightarrow[heat]{H_2O,\ H^+}$ product

(2) $CH_3CH_2CH_2Br \xrightarrow[ether]{Mg} CH_3CH_2CH_2MgBr \xrightarrow[(2)\ H_2O,\ H^+]{(1)\ CO_2}$ product

15.31 (a) $CH_3C(=O)OCH_2CH_3 + CH_3CO_2H$ (b) $CH_3C(=O)OCH_2CH_3 + HCl$

(c) $CH_3C(=O)OCH_2CH_3 + CH_3NH_2$ (d) $C_6H_5C(OH)(CH_2CH_3)_2$

The product in (d) could undergo dehydration to $C_6H_5C(CH_2CH_3){=}CHCH_3$

(e) $C_6H_5C(=O)CH_2CH_3$ (f) $HO_2CCH_2CH_2CO_2CH_2CH_3$

(g) $HOCH_2CH_2CH(CH_3)CH_2OH$ (h) $HOCH_2CH_2CH(CH_3)C(OH)(CH_3)_2$

15.32 (a)

$$CH_3CH_2CH{=}CH_2 \xrightarrow[\text{(2) } H_2O_2,\ ^{-}OH]{\text{(1) } BH_3} CH_3CH_2CH_2CH_2OH \xrightarrow[\text{heat}]{H_2CrO_4}$$

$$CH_3CH_2CH_2\overset{O}{\overset{\|}{C}}OH \xrightarrow{SOCl_2} CH_3CH_2CH_2\overset{O}{\overset{\|}{C}}Cl \xrightarrow{NH_3} \text{product}$$

(b)

$$C_6H_6 \xrightarrow[FeBr_3]{Br_2} C_6H_5Br \xrightarrow[\text{ether}]{Mg} C_6H_5MgBr \xrightarrow[\text{(2) } H_2O,\ H^+]{\text{(1) } CO_2}$$

$$C_6H_5CO_2H \xrightarrow{SOCl_2} C_6H_5\overset{O}{\overset{\|}{C}}Cl \xrightarrow{(CH_3)_2NH} \text{product}$$

(c)

$$CH_3CO_2H \xrightarrow{SOCl_2} CH_3\overset{O}{\overset{\|}{C}}Cl \xrightarrow{NH_3} CH_3\overset{O}{\overset{\|}{C}}NH_2$$

$$\xrightarrow[\text{(3) neutralize}]{\text{(1) } LiAlH_4 \quad \text{(2) } H_2O,\ H^+} \text{product}$$

16

Conjugate Additions

Answers to Additional Problems

16.1 (a) $CH_3C(C_6H_5)(I)CH{=}CH_2 + CH_2{=}C(C_6H_5)CH(I)CH_3 + CH_3C(C_6H_5){=}CHCH_2I + ICH_2C(C_6H_5){=}CHCH_3$

(b) $CH_3{-}C_6H_4{-}CH(Br)CH_2CH{=}CH{-}C_6H_4{-}CH_3$ +

$CH_3{-}C_6H_4{-}CH_2CH(Br)CH{=}CH{-}C_6H_4{-}CH_3$ +

$CH_3{-}C_6H_4{-}CH_2CH{=}CHCH(Br){-}C_6H_4{-}CH_3$

(c) $BrCH_2C(CH_3)(Br){-}C(CH_3){=}CH_2 + BrCH_2C(CH_3){=}C(CH_3)CH_2Br$

16.2 (a) (1) trans-4,5-dibromocyclohexene (Br, Br) (2) 3-bromocyclohexene (Br)

(b) (1) 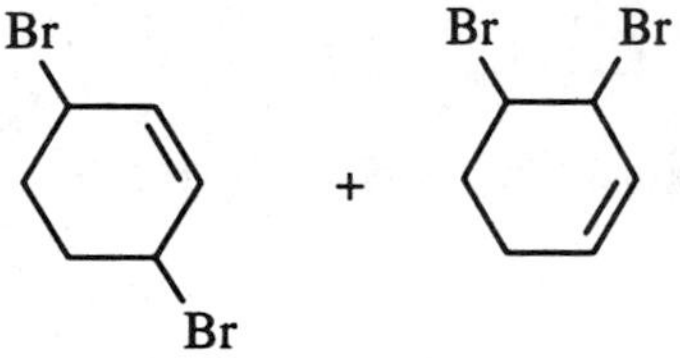 [structure: 3,6-dibromocyclohexene] + [structure: 3,4-dibromocyclohexene] (*cis* and *trans*)

(2) 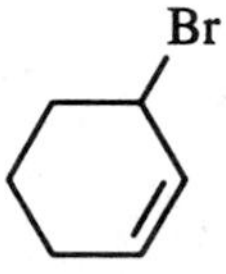 [structure: 3-bromocyclohexene] The 1,4-addition product is identical to the 1,2-addition product

(c) (1) 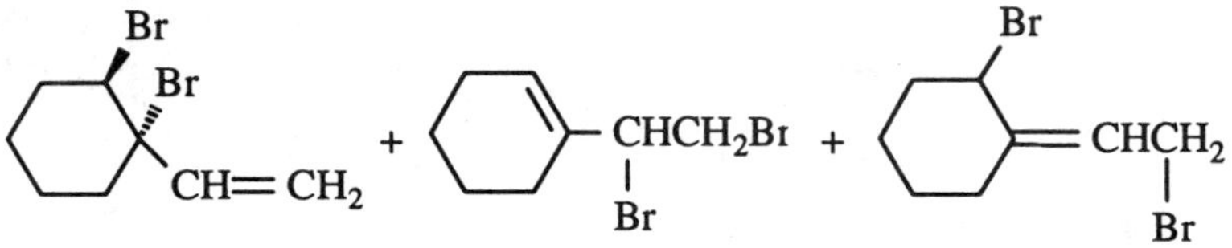 [structure: 1,2-dibromo-1-vinylcyclohexane, Br, Br, CH=CH₂] + [structure: cyclohexenyl–CHCH₂Br with Br] + [structure: 2-bromocyclohexylidene=CHCH₂Br]

(2) [structure: 1-bromo-1-vinylcyclohexane, Br, CH=CH₂] + [structure: cyclohexenyl–CHCH₃ with Br] +

[structure: 2-bromocyclohexylidene=CHCH₃] + [structure: cyclohexylidene=CHCH₂Br]

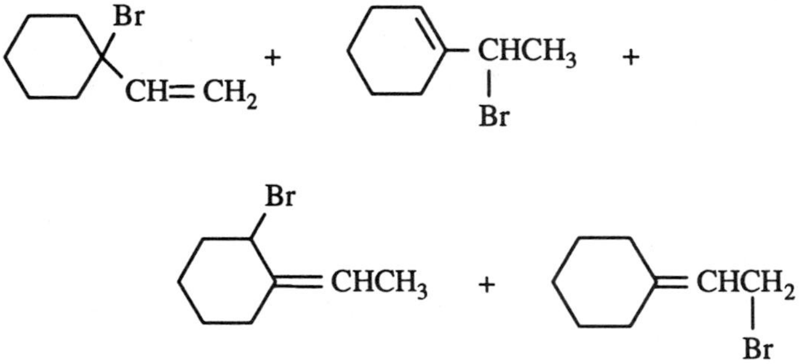

16.3 (a) $C_6H_5CH{=}CHC(=O)CH_3 + H_2O \xrightarrow{H^+} C_6H_5CH(OH)CH_2C(=O)CH_3$

(b) $C_6H_5CH{=}CHC(=O)CH_3 + CH_3MgI \longrightarrow C_6H_5CH{=}CHC(OMgI)(CH_3)CH_3$

(c) $C_6H_5CH{=}CHC(=O)CH_3 + HBr \longrightarrow C_6H_5CH(Br)CH_2C(=O)CH_3$

(d) $C_6H_5CH{=}CHC(=O)CH_3 \xrightarrow{CH_3NH_2} C_6H_5CH(NHCH_3)CH_2C(=O)CH_3$

$$\underset{}{\overset{H^+,\ CH_3NH_2}{\rightleftharpoons}} C_6H_5CH(NHCH_3)CH_2C(=NCH_3)CH_3$$

(e) $C_6H_5CH{=}CHC(=O)CH_3 \xrightarrow[HCN]{CN^-} C_6H_5CH(CN)CH_2C(=O)CH_3$

$$\overset{HCN,\ CN^-}{\rightleftharpoons} C_6H_5CH(CN)CH_2C(OH)(CN)CH_3$$

16.4 The reaction as shown is a 1,3-addition; not a 1,4-addition. The correct equation is:

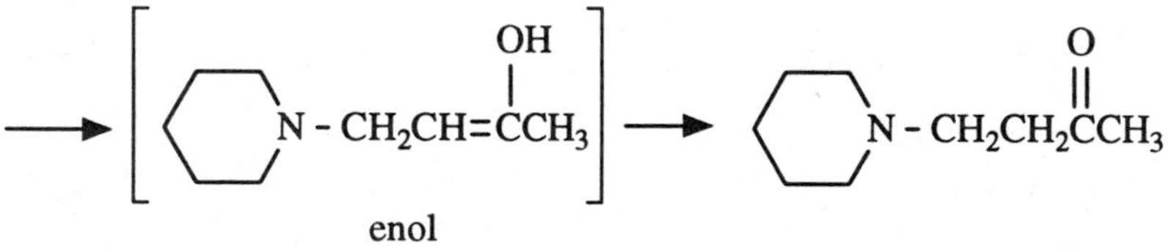

$$\longrightarrow \left[C_5H_{10}N{-}CH_2CH{=}C(OH)CH_3 \right] \longrightarrow C_5H_{10}N{-}CH_2CH_2C(=O)CH_3$$

enol

16.5 (a) 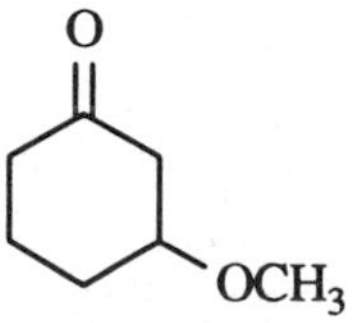 3-methoxycyclohexanone (ring with C=O and OCH_3)

(b) $C_6H_5C(=O)CH_2{-}CH_2N(CH_3)_2$

(c) 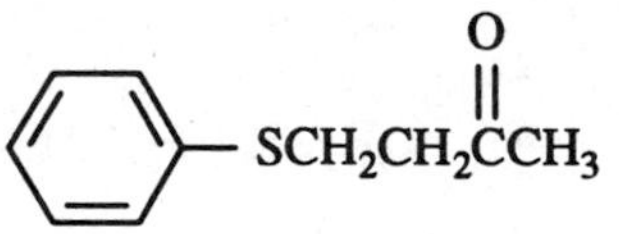 $SCH_2CH_2\overset{O}{\overset{\|}{C}}CH_3$ (d) 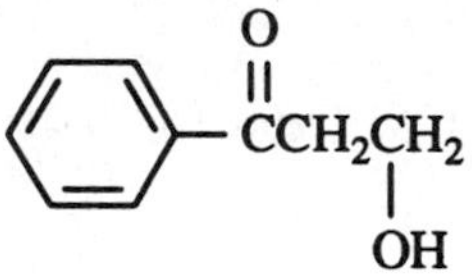 $\overset{O}{\overset{\|}{C}}CH_2CH_2$ OH

16.6 CH–CH$_2$–C(=O) NHNH$_2$ → N–N, H

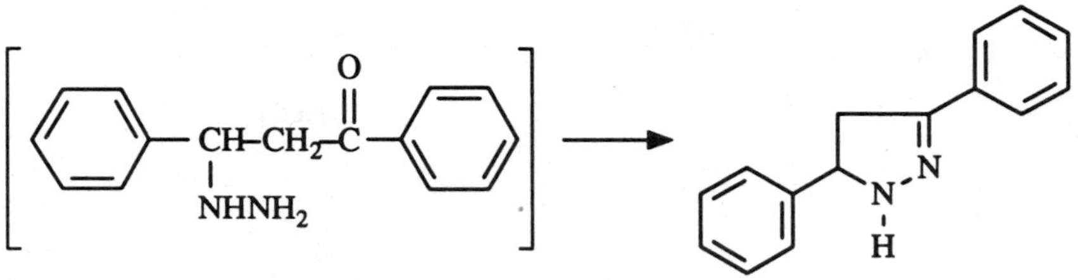

16.7 (a) 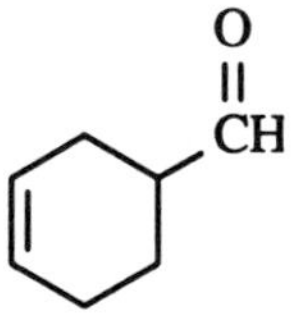 O, CH (b) 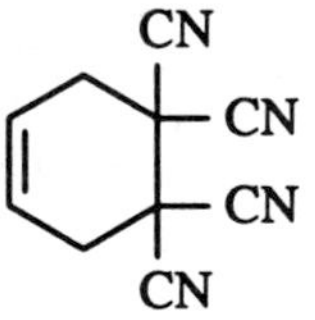 CN, CN, CN, CN

(c) 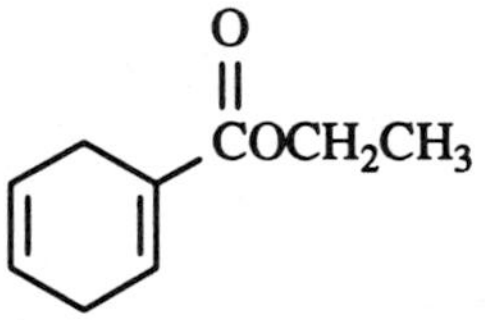 O, $COCH_2CH_3$ (d) 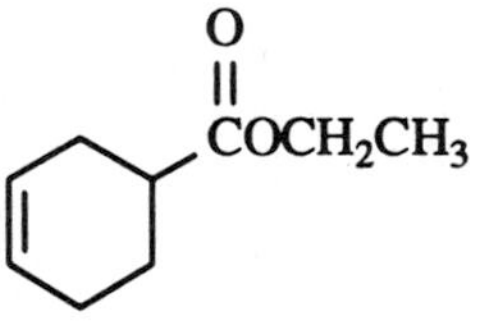 O, $COCH_2CH_3$

(e) 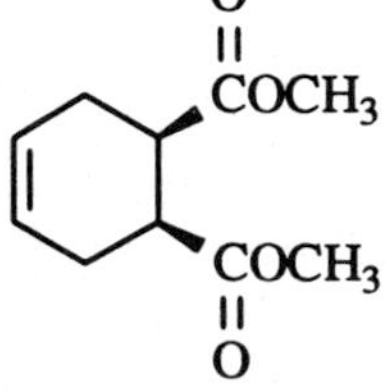 O, $COCH_3$, $COCH_3$, O (f) 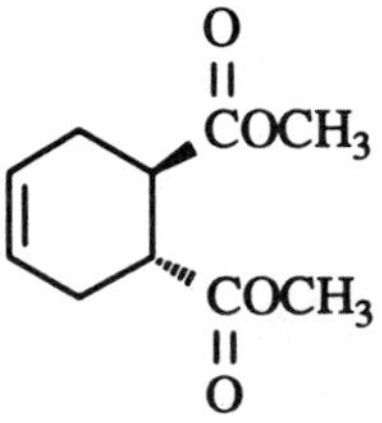 O, $COCH_3$, $COCH_3$, O

16.8 (a) 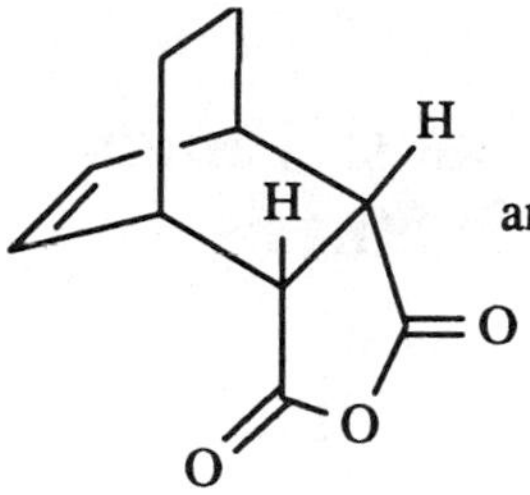 and/or

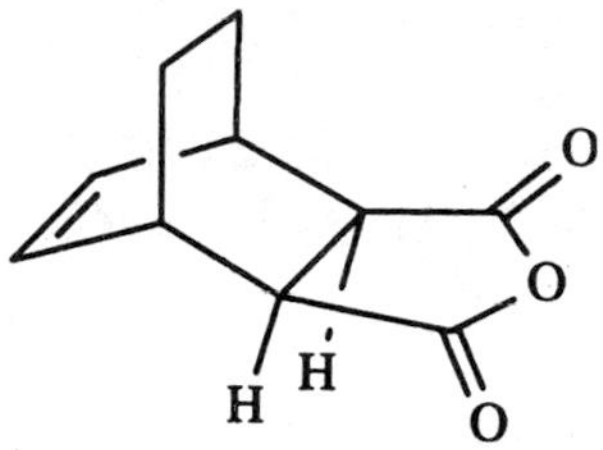

(b) 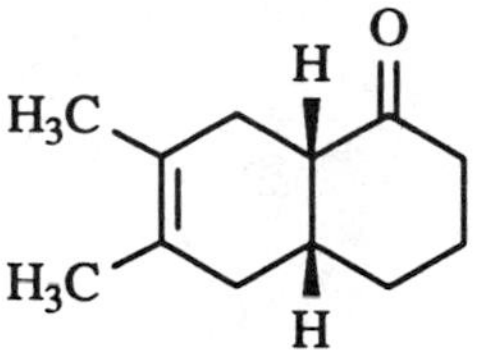 H_3C, H_3C, H, H, O

16.9 (a) 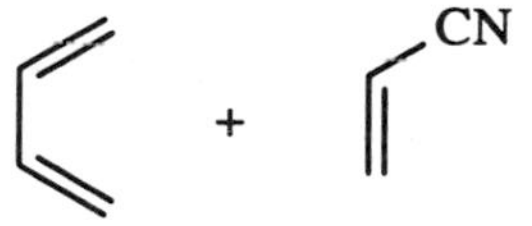 + CN (b) 2

(c) 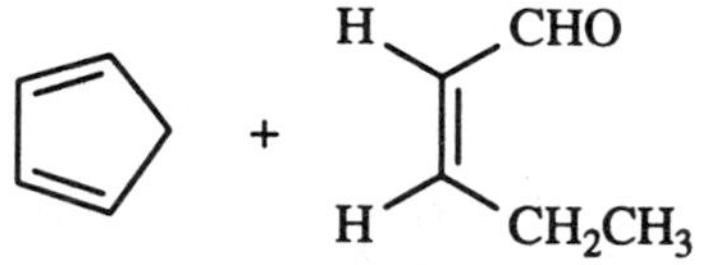 + H, CHO, H, CH_2CH_3

(d) 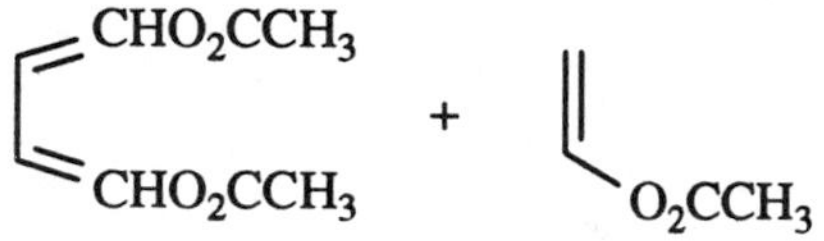 CHO_2CCH_3, CHO_2CCH_3 + O_2CCH_3

(e) 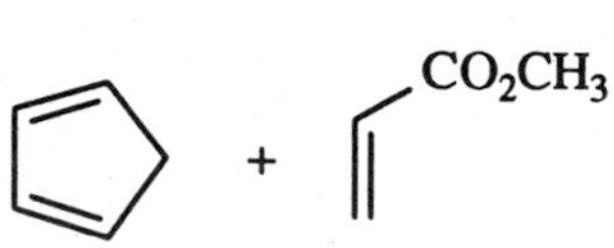 + CO_2CH_3 (f) 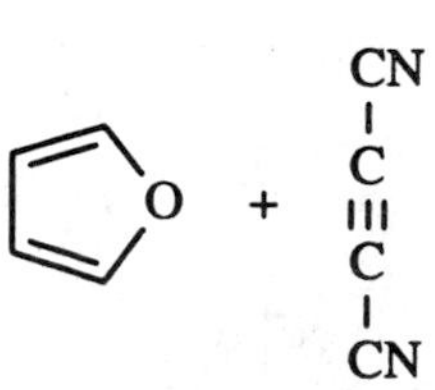 O + CN–C≡C–CN

17

Enolates and Carbanions: Building Blocks for Organic Synthesis

Answers to Additional Problems

17.1 $(CH_3)_2C(CO_2C_2H_5)_2 \xrightarrow[H_2O]{H^+} 2\ CH_3CH_2OH + (CH_3)_2C(COOH)_2$ (cyclic transition state: C=O···H–O)

$\xrightarrow[-CO_2]{heat}$ $(CH_3)_2C{=}C(OH)\text{–}OH$ $\longrightarrow$ $(CH_3)_2CHCOH$ (with C=O)

an "enol"

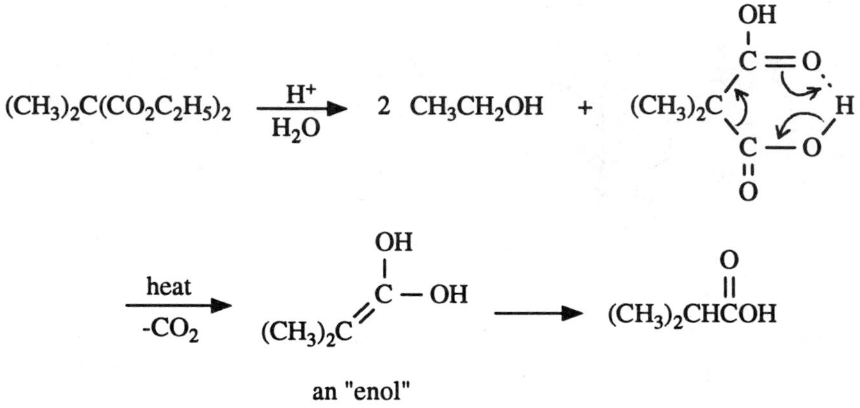

17.2 (a) (1) 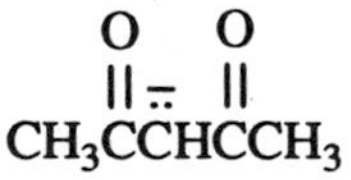 $CH_3C(=O)\ddot{C}^{-}HC(=O)CH_3$

(2) 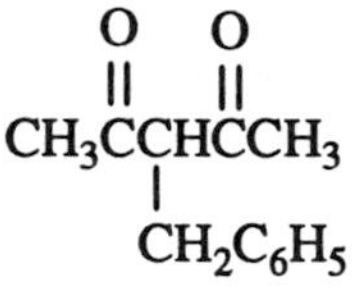 $CH_3C(=O)CH(CH_2C_6H_5)C(=O)CH_3$

(3) same as (2); no reaction

(b) (1) $^{-}:CH(CO_2C_2H_5)_2$

(2) 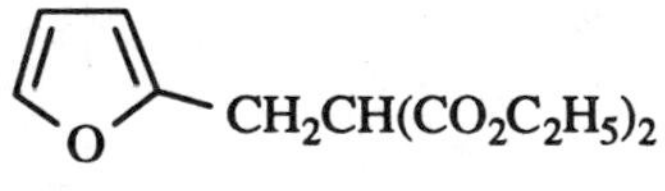 2-furyl–$CH_2CH(CO_2C_2H_5)_2$

(3) 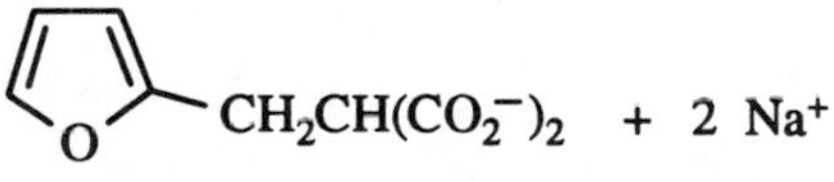 2-furyl–$CH_2CH(CO_2^-)_2$ + 2 Na^+

(4) 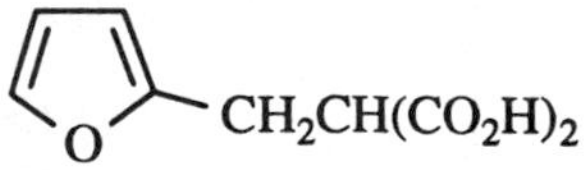 [furan-2-yl] $CH_2CH(CO_2H)_2$

(5)  [furan-2-yl] $CH_2CH_2CO_2H$ + CO_2

(c) (1) $C_2H_5O_2C\ddot{C}^{-}HCN$

(2) 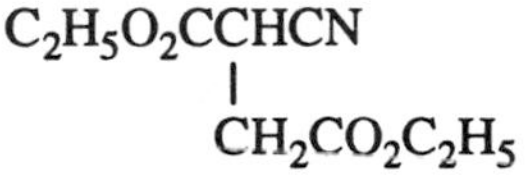 $C_2H_5O_2CCHCN$ with $CH_2CO_2C_2H_5$ on the central C

(3) 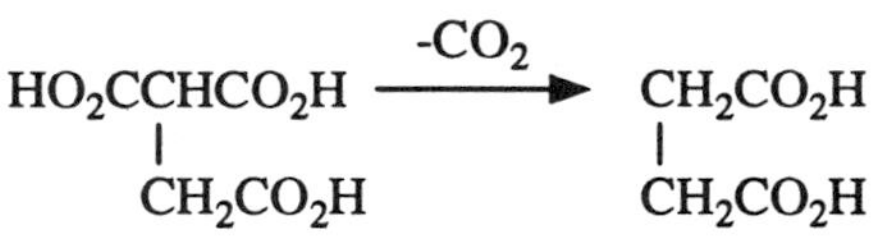 HO_2CCHCO_2H with CH_2CO_2H on the central C $\xrightarrow{-CO_2}$ CH_2CO_2H–CH_2CO_2H

(d) (1) 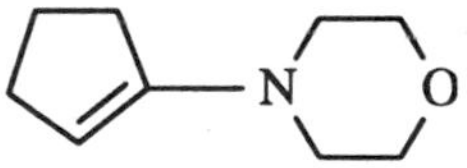 [1-morpholinocyclopentene]

(2) 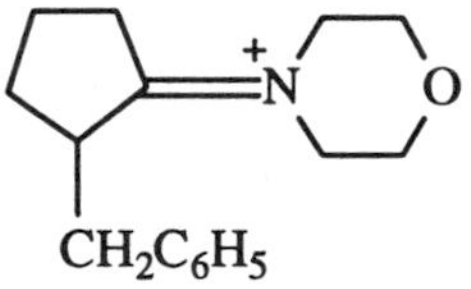 [2-benzylcyclopentylidene morpholinium ion, $\overset{+}{N}$] $CH_2C_6H_5$

(3) 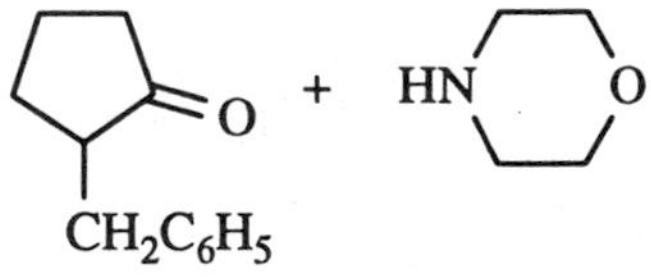 [2-benzylcyclopentanone, $CH_2C_6H_5$] + HN[morpholine]O

17.3 Compounds (b), (c), (d), (e), (f), and (g), but not (a) or (h). Compound (a) is a tertiary alkyl halide and compound (h) is a vinylic halide, neither of which will undergo an S_N2 reaction. Note that compound (g) would technically act as an acylating agent, not as an alkylating agent.

17.4 (a) CH_3CHO $\xrightarrow[H^+]{\text{morpholine (HN O)}}$ $CH_2{=}CH{-}N(morpholine)$ $\xrightarrow{CH_2{=}CHCH_2Br}$

$CH_2{=}CHCH_2CH_2CH{=}\overset{+}{N}(morpholine)$ $\xrightarrow{H_2O,\ H^+}$ product

(b) cyclohexanone $\xrightarrow[H^+]{\text{morpholine (HN O)}}$ 1-morpholinocyclohexene $\xrightarrow{BrCH_2C(O)CH_3}$

2-($CH_2C(O)CH_3$)cyclohexylidene-$\overset{+}{N}$(morpholine) $\xrightarrow[H^+]{H_2O}$ product

(c) cycloheptanone $\xrightarrow{\text{morpholine (HN O)}}$ 1-morpholinocycloheptene $\xrightarrow{CH_3C(O)Cl}$

2-($C(O)CH_3$)cycloheptylidene-$\overset{+}{N}$(morpholine) $\xrightarrow{H_2O,\ H^+}$ product

17.5 (a) $(CH_3)_2CHCH_2CHO$ $\xrightarrow[-H_2O]{OH^-}$ $(CH_3)_2CH\ddot{C}^{-}HCHO$ $\xrightarrow{(CH_3)_2CHCH_2CHO}$

$$(CH_3)_2CHCH_2CH(O^-)-CH(CH(CH_3)_2)CHO \xrightarrow[-OH^-]{H_2O} (CH_3)_2CHCH_2CH(OH)CH(CH(CH_3)_2)CHO$$

(b) $$CH_3-C_6H_4-CH_2CHO \xrightarrow[-H_2O]{OH^-} CH_3-C_6H_4-\ddot{C}^-HCHO$$

$$\xrightarrow{CH_3-C_6H_4-CH_2CHO} CH_3-C_6H_4-CH_2C(O^-)H-CH(C_6H_4-CH_3)CHO$$

$$\xrightarrow[-OH^-]{H_2O} CH_3-C_6H_4-CH_2CH(OH)CH(C_6H_4-CH_3)CHO$$

17.6 (a) $C_6H_5-CH(OH)-C_6H_{10}(CHO)$ (b) $C_6H_5-C(=O)CH=CH-C_4H_3O$ (2-furyl)

(c) $$\left[CH_3C(=O)CH_2CH_2CH_2CH_2\ddot{C}^-HCHO\right] \rightarrow \text{1-methyl-1-}O^-\text{-2-CHO-cyclohexane} \xrightarrow[-OH^-]{H_2O} \text{1-methyl-1-HO-2-CHO-cyclohexane}$$

17.7 (a)

$$O_2N-C_6H_4-\overset{O}{\overset{\|}{C}}H + CH_3\overset{O}{\overset{\|}{C}}H \xrightarrow{NaOH}$$

$$O_2N-C_6H_4-\overset{OH}{\overset{|}{C}}HCH_2\overset{O}{\overset{\|}{C}}H \xrightarrow{-H_2O} O_2N-C_6H_4-CH{=}CH\overset{O}{\overset{\|}{C}}H$$

(b)

$$O_2N-C_6H_4-\overset{O}{\overset{\|}{C}}H + CH_2(CO_2C_2H_5)_2 \xrightarrow{Na^+\ ^-OCH_2CH_3}$$

$$O_2N-C_6H_4-\overset{OH}{\overset{|}{C}}HCH(CO_2C_2H_5)_2 \xrightarrow{-H_2O}$$

$$O_2N-C_6H_4-CH{=}C(CO_2C_2H_5)$$

(c)

$$O_2N-C_6H_4-\overset{O}{\overset{\|}{C}}H + C_6H_5\overset{O}{\overset{\|}{C}}CH_3 \xrightarrow{NaOH}$$

$$O_2N-C_6H_4-\overset{OH}{\overset{|}{C}}HCH_2\overset{O}{\overset{\|}{C}}C_6H_5 \xrightarrow{-H_2O} O_2N-C_6H_4-CH{=}CH\overset{O}{\overset{\|}{C}}C_6H_5$$

17.8 (a) $CH_3CH_2CH_2CH{=}C(CO_2C_2H_5)_2$

(b) $C_6H_5CH{=}CHCH{=}CHCO_2H$ (after decarboxylation)

(c) (2-furyl)$-CH{=}C(CN)_2$

(d) 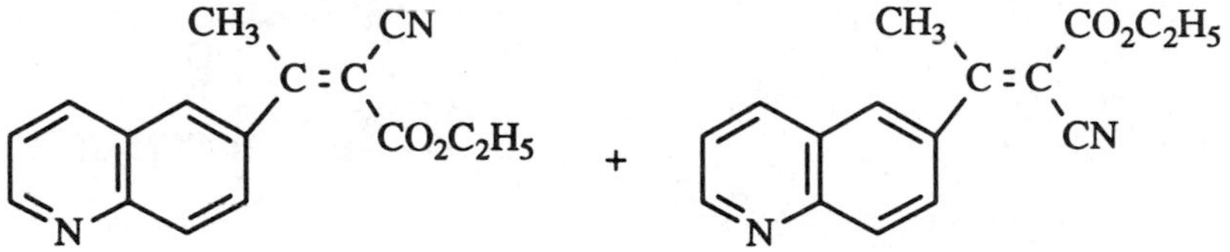 [Structures: 6-substituted quinoline with $C(CH_3)=C(CN)(CO_2C_2H_5)$ + 6-substituted quinoline with $C(CH_3)=C(CO_2C_2H_5)(CN)$ (the two geometric isomers)]

17.9 Compounds (a), (c), and (d) because each contains an alpha hydrogen. Compound (b) does not contain an alpha hydrogen.

17.10 (a)

$$CH_3CH_2\overset{O}{\overset{\|}{C}}OCH_3 \xrightarrow[(2)\ H_2O,\ H^+]{(1)\ CH_3O^-} CH_3CH_2\overset{O}{\overset{\|}{C}}\underset{CH_3}{\underset{|}{C}H}\overset{O}{\overset{\|}{C}}OCH_3$$

(b) no reaction

(c)

$$CH_3\overset{O}{\overset{\|}{C}}CH_2\overset{O}{\overset{\|}{C}}OCH_3 \xrightarrow[(2)\ H_2O,\ H^+]{(1)\ CH_3O^-} CH_3\overset{O}{\overset{\|}{C}}CH_2\overset{O}{\overset{\|}{C}}\underset{O=CCH_3}{\underset{|}{C}H}\overset{O}{\overset{\|}{C}}OCH_3$$

(d)

$$CH_3O\overset{O}{\overset{\|}{C}}CH_2CH_2CH_2\overset{O}{\overset{\|}{C}}OCH_3 \xrightarrow[(2)\ H_2O,\ H^+]{(1)\ CH_3O^-}$$

$$CH_3O\overset{O}{\overset{\|}{C}}CH_2CH_2CH_2\overset{O}{\overset{\|}{C}}\underset{CO_2CH_3}{\underset{|}{C}H}CH_2CH_2\overset{O}{\overset{\|}{C}}OCH_3$$

17.11 (a) The proton at position marked (1) is more acidic than the protons on the methyl group, marked (2). However, the anion at (1) can ring-close only to a four-membered ring, while the anion at position 2 can close to a six-membered ring. Because the two anions are in equilibrium, the more stable six-membered ring accumulates in solution.

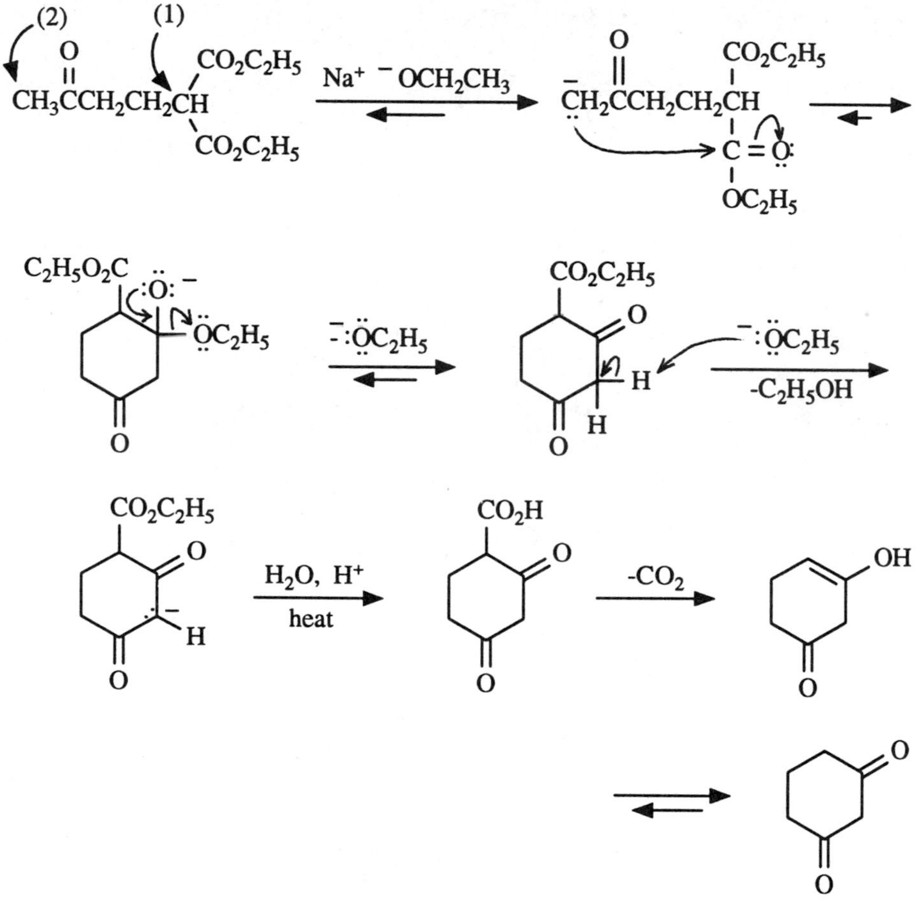

(2) (1)

$$CH_3\overset{O}{\overset{\|}{C}}CH_2CH_2CH(CO_2C_2H_5)_2 \xrightarrow{Na^+ \ ^-OCH_2CH_3} \overset{-}{C}H_2\overset{O}{\overset{\|}{C}}CH_2CH_2CH(CO_2C_2H_5)C(=O)OC_2H_5 \rightleftharpoons$$

$C_2H_5O_2C$ $:\ddot{O}:^-$ $\ddot{O}C_2H_5$ $\xrightarrow{^-:\ddot{O}C_2H_5}$ $CO_2C_2H_5$ O H H O $\xrightarrow[-C_2H_5OH]{^-:\ddot{O}C_2H_5}$

$CO_2C_2H_5$ O H O $\xrightarrow[heat]{H_2O,\ H^+}$ CO_2H O O $\xrightarrow{-CO_2}$ OH O

$\rightleftharpoons$ O O

(b) $$\overset{O}{\overset{\|}{C}}CH_2CH_2CH_2CH_2\overset{O}{\overset{\|}{C}}OC_2H_5 \xrightarrow{Na^+ \ ^-OC_2H_5}$$

$$\overset{O}{\overset{\|}{C}}\overset{-}{C}HCH_2CH_2CH_2\overset{\ddot{O}:}{\overset{\|}{C}}OC_2H_5 \longrightarrow$$

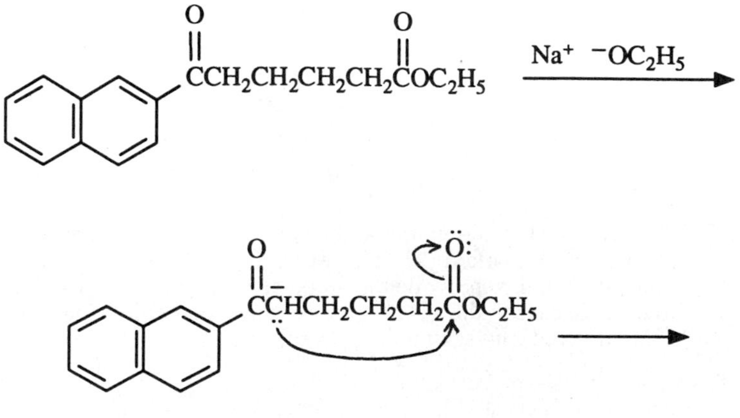

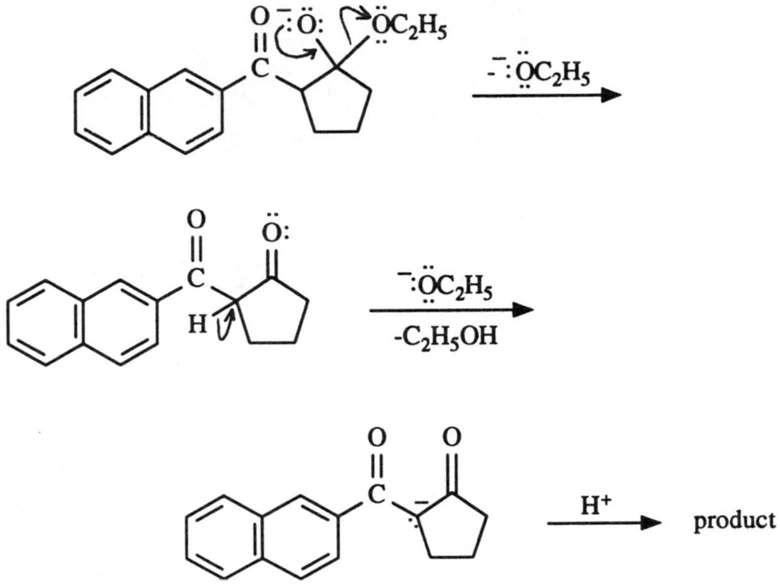

$$\xrightarrow{^{-}:\ddot{O}C_2H_5}$$

$$\xrightarrow[-C_2H_5OH]{^{-}:\ddot{O}C_2H_5}$$

$\xrightarrow{H^+}$ product

17.12 (a) $\longrightarrow$ $CH(CO_2C_2H_5)_2$ $\rightleftharpoons$ $^{-}:C(CO_2C_2H_5)_2$ $\xrightarrow[heat]{H_2O,\ H^+}$

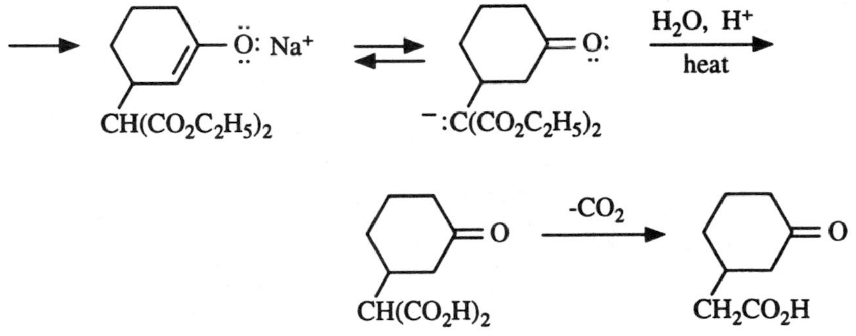

$CH(CO_2H)_2$ $\xrightarrow{-CO_2}$ CH_2CO_2H

(b) $\longrightarrow$ $Na^+\ ^{-}:\ddot{O}$ $CH(CO_2C_2H_5)_2$ $\rightleftharpoons$ $^{-}:C(CO_2C_2H_5)_2$

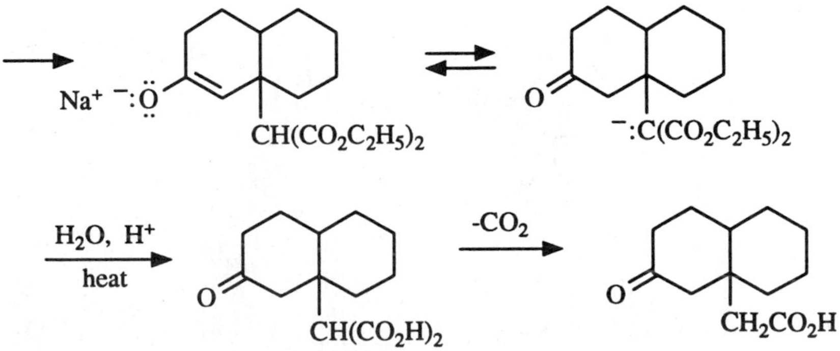

$\xrightarrow[heat]{H_2O,\ H^+}$ $CH(CO_2H)_2$ $\xrightarrow{-CO_2}$ CH_2CO_2H

(c) 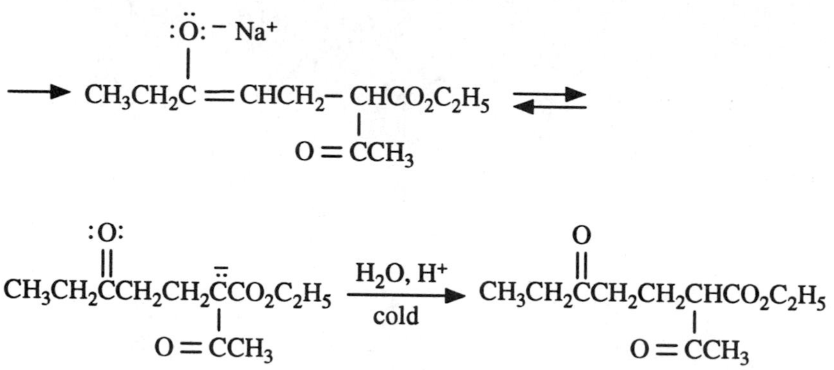 $\longrightarrow$ $CH_3CH_2C(\ddot{O}:^- Na^+){=}CHCH_2{-}CH(C(=O)CH_3)CO_2C_2H_5 \rightleftharpoons$

$CH_3CH_2C(=\!:\!O\!:)CH_2CH_2\ddot{C}^-(C(=O)CH_3)CO_2C_2H_5 \xrightarrow[\text{cold}]{H_2O,\ H^+} CH_3CH_2C(=O)CH_2CH_2CH(C(=O)CH_3)CO_2C_2H_5$

17.13 (a) 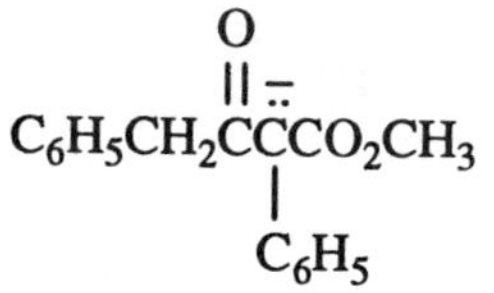 $C_6H_5CH_2C(=O)\ddot{C}^-(C_6H_5)CO_2CH_3$

(b) $CH_3CH_2CH(OH)CH(CH_3)CH(=O)$

(c) [iminium structure: 1-methyl-3,4-dihydronaphthalen-2(1H)-ylidene piperidinium, N^+, CH_3]

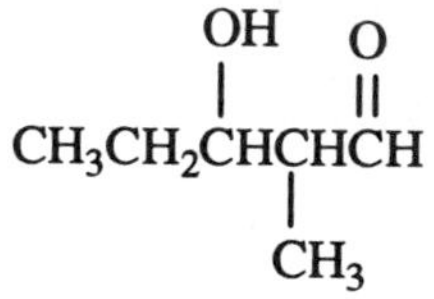

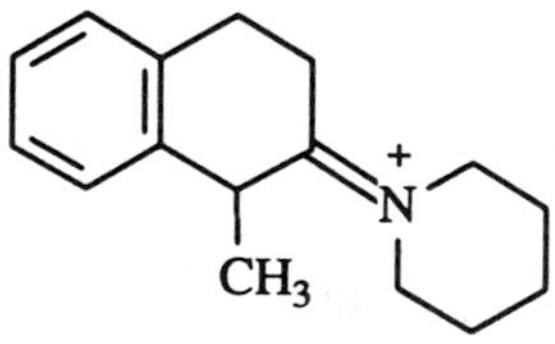

(d) 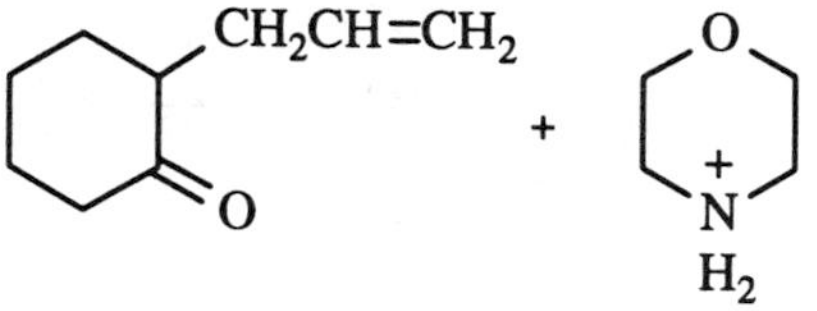 [2-($CH_2CH{=}CH_2$)cyclohexanone] + [morpholinium, O, $\overset{+}{N}H_2$]

(e) 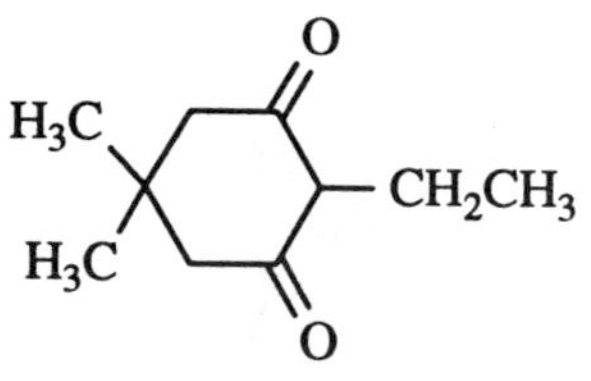 [5,5-dimethyl (H_3C, H_3C) 2-ethyl (CH_2CH_3) cyclohexane-1,3-dione, O, O]

17.14 (a) 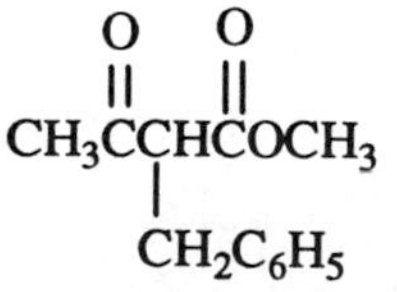 $CH_3C(=O)CH(CH_2C_6H_5)C(=O)OCH_3$

(b) 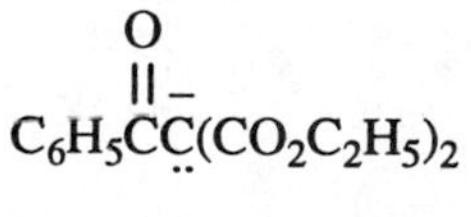 $C_6H_5C(=O)\ddot{C}^-(CO_2C_2H_5)_2$

(c) [bicyclo[2.2.2]octan-2-one bearing $CH_2CH(CO_2C_2H_5)_2$ at C-3]

$CH_2CH(CO_2C_2H_5)_2$

(d) $$C_6H_5\overset{OH}{\overset{|}{C}}HCH_2NO_2 \xrightarrow{-H_2O} C_6H_5CH=CHNO_2$$

17.15 (a) $$2\ CH_3CH_2CH_2CH_2\overset{O}{\overset{\|}{C}}H \xrightarrow[\text{(aldol)}]{OH^-} \text{product}$$

(b) $$CH_3CH_2CH_2CH_2\overset{O}{\overset{\|}{C}}H \xrightarrow{\text{HN(morpholine)O}} CH_3CH_2CH_2CH=CH-N(\text{morpholine})O$$

$$\xrightarrow[\text{(2) } H_2O,\ H^+]{\text{(1) } C_6H_5CH_2Br} \text{product}$$

(c) $$CH_3CH_2CH_2CH_2\overset{O}{\overset{\|}{C}}H \xrightarrow[\text{or } KMnO_4]{H_2CrO_4} CH_3CH_2CH_2CH_2\overset{O}{\overset{\|}{C}}OH$$

$$\xrightarrow[\text{heat}]{CH_3CH_2OH,\ H^+} \text{product}$$

(d) $$2\ CH_3CH_2CH_2CH_2\overset{O}{\overset{\|}{C}}OC_2H_5 \text{ from (c)} \xrightarrow[\text{(2) } H_2O,\ H^+ \text{ (cold)}]{\text{(1) } Na^+\ {}^-OC_2H_5 \text{ (ester condensation)}} \text{product}$$

18

Amines

Answers to Additional Problems

18.1 (a) 1° (b) 2° (c) quaternary

(d) 1° (e) both 1°

18.2 Compound (b) could be resolved because it has a quaternary nitrogen bonded to four different groups.

18.3 (a) CH_3O_2C—C_6H_4—NH_2 (b) $C_6H_5CH_2NHCH_2CH_3$

18.4 (a) (1) $LiAlH_4$; (2) H_2O, H^+; (3) NaOH

(b) NH_3 (c) Br_2, NaOH

(d) CH_3NH_2, H_2, Pt (e) (1) $LiAlH_4$; (2) H_2O, H^+; (3) NaOH

18.5 (a) $\overset{+}{N}(CH_3)_2$ I^-

(b) CN, $CH_2\overset{+}{N}H(CH_3)_2$ Br^- or CN, $CH_2N(CH_3)_2$ + $(CH_3)_2\overset{+}{N}H_2$ Br^-

18.6 (a) Hofmann rearrangement. The required starting alkyl halide for the Gabriel synthesis would be a 3° alkyl halide, which would not undergo the S_N2 reaction that is the first step.

(b) Gabriel synthesis. The starting material would be a benzylic halide, which undergoes S_N2 reactions readily.

18.7 (a) $(CH_3)_2CH$–(cyclohexane)=O $\xrightarrow[\text{Pt}]{NH_3,\ H_2}$ product

(b) $CH_3CH_2\overset{O}{\overset{\|}{C}}H \xrightarrow[CN^-]{HCN} CH_3CH_2\underset{OH}{\underset{|}{C}}HCN \xrightarrow[(2)\ H_2O,\ H^+;\ (3)\ NaOH]{(1)\ LiAlH_4}$ product

(c) $CH_3(CH_2)_3\overset{O}{\overset{\|}{C}}H \xrightarrow[\text{Pt}]{(CH_3)_2NH,\ H_2}$ product

18.8 (a) $CH_3CH_2NH_3^+$ (b) $H_2\overset{+}{N}$ (ring) O (c) $CH_3\underset{^+NH_3}{\underset{|}{C}}HCO_2H$

18.9

Amine	K_a	pK_a	K_b	pK_b
$C_4H_9NH_2$	1.69×10^{-11}	10.77	5.89×10^{-4}	3.23
$C_5H_{11}NH_2$	2.34×10^{-11}	10.63	4.27×10^{-4}	3.37
$(CH_3CH_2)_2NH$	3.24×10^{-11}	10.49	3.09×10^{-4}	3.51
$C_5H_{10}NH$	7.53×10^{-12}	11.12	1.32×10^{-3}	2.88

18.10 Compound (d), the 2° amine, is the most basic compound in the list. The unshared electrons of the nitrogen of compound (a) and compound (c) are delocalized and are not as available as those of compound (d) for bonding with a proton. Compound (d) is more basic than compound (b) because the inductive electron-releasing effect of two alkyl groups is greater than that of one group.

18.11 Compound (c), the amide, is the least basic of the group because the unshared valence electrons of its nitrogen are delocalized by the oxygen of the carbonyl group. These electrons are thus not available for bonding with a proton.

18.12 (a) $CH_3\overset{O}{\overset{\|}{C}}\overset{+}{N}H_3$ is more acidic because its conjugate base, an amide, is far less basic than an amine.

(b) $C_6H_5\overset{+}{N}H_3$ because its conjugate base is resonance stabilized and is thus a weaker base.

(c) $(CH_3CH_2)_2\overset{+}{N}H_2$ because its conjugate base is a weaker base than piperidine. Piperidine is the stronger base because there is less steric hindrance around the nitrogen atom.

(d) The *p*-nitroanilinium ion because the strongly electron-withdrawing nitrogen of the nitro group destabilizes the conjugate acid. Furthermore, the nitro group stabilizes the conjugate base by resonance.

18.13 (a) no reaction

(b) $(CH_3)_3CCH_2N(CH_3)_2$ + $CH_2{=}CH_2$

(c) N-CH_3 (piperidine ring) + $CH_2{=}CH_2$

18.14 (a) $C_6H_5N_2^+$ Cl^-

(b) $\overset{+}{N}$-H (pyridinium ring) Cl^-

(c) (cyclohexyl)$\overset{+}{N}HCH_2CH_3$

(d) $H_2NCH_2CH_2CH_2CH_2NH_2$

(e) $(CH_3CH_2)_2NNO$

(f) (phthalimide ring, O, O) $NCH_2CO_2C_2H_5$

(g) $C_6H_5CH{=}NCH_2CH_3$

(h) $CH_3CH_2NH\overset{O}{\overset{\|}{C}}CH_3$ + $CH_3\overset{O}{\overset{\|}{C}}O^-\ \overset{+}{N}H_3CH_2CH_3$

19

Polycyclic and Heterocyclic Aromatic Compounds

Answers to Additional Problems

19.1 (a) 22

(b) 20. This number of pi electrons does not fit the 4n + 2 rule. Not all of the rings are aromatic. The two "double bonds" have substantial double-bond character; that is, they can undergo addition reactions.

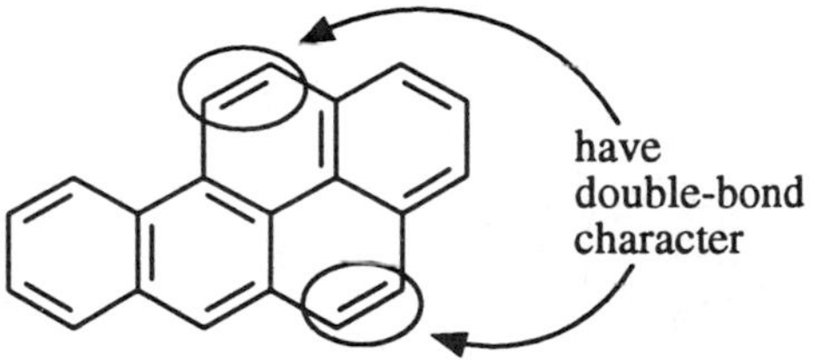

have double-bond character

19.2 (a) NO_2

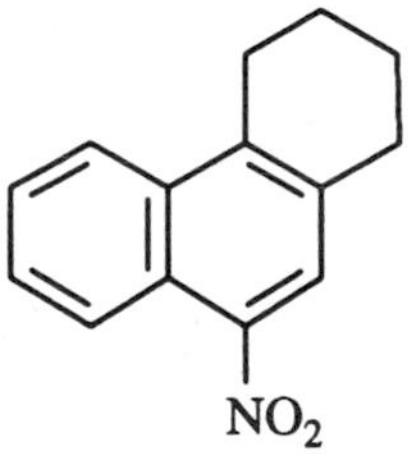

(b) Br, OCH_3

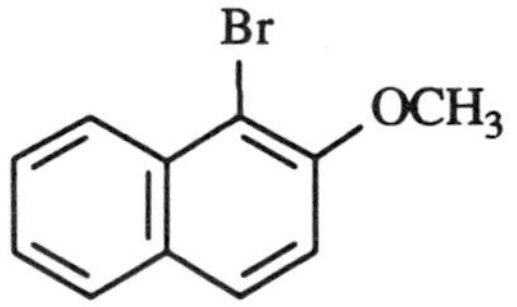

(c) CO_2H —[O]→ CO_2H, CO_2H, CO_2H —$-H_2O$→ CO_2H, O, O, O

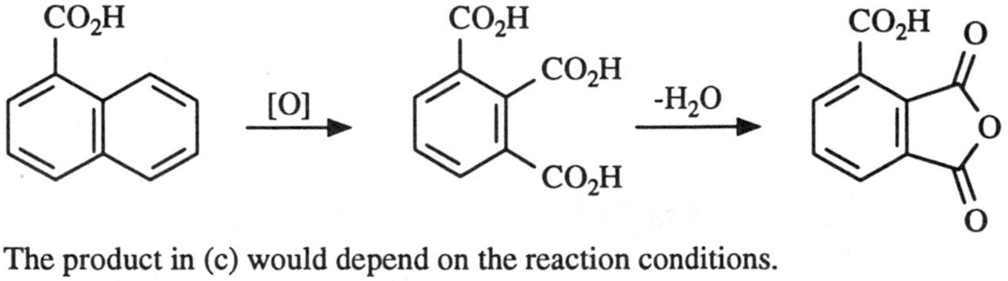

The product in (c) would depend on the reaction conditions.

(d) 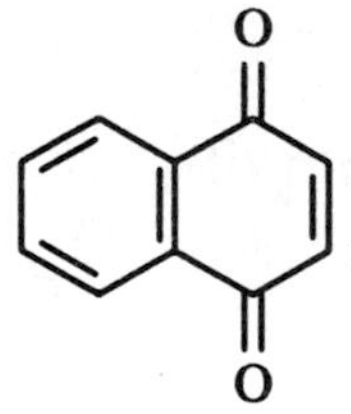 [structure: 1,4-naphthoquinone] The quinone could be oxidized further to *o*-phthalic acid or its anhydride.

(e) 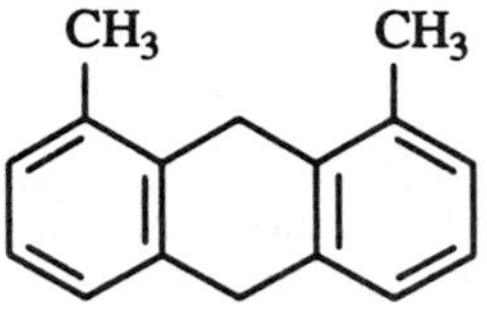 [structure: 1,8-dimethyl-9,10-dihydroanthracene, CH_3, CH_3]

(f) 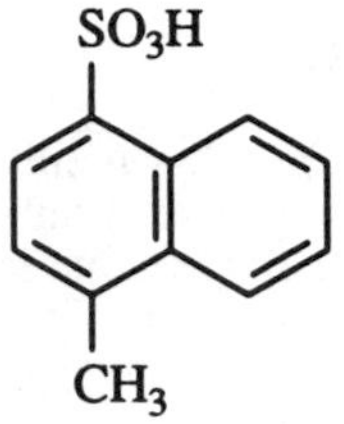 [structure: 4-methyl-1-naphthalenesulfonic acid, SO_3H, CH_3]

19.3 (a) 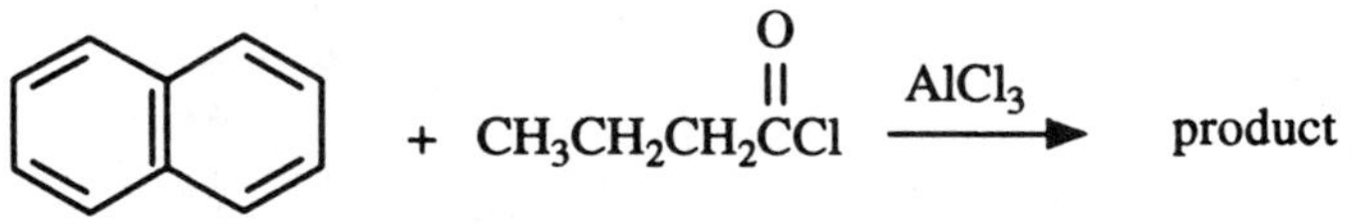 naphthalene + $CH_3CH_2CH_2\overset{O}{\overset{\|}{C}}Cl \xrightarrow{AlCl_3}$ product

(b) naphthalene $\xrightarrow[H_2SO_4]{HNO_3}$ 1-nitronaphthalene (NO_2) $\xrightarrow[\text{(2) neutralize}]{\text{(1) Fe, HCl}}$ product

(c) naphthalene $\xrightarrow[AlCl_3]{C_6H_5\overset{O}{\overset{\|}{C}}Cl}$ 1-naphthyl ketone ($\overset{O}{\overset{\|}{C}}C_6H_5$) $\xrightarrow[\text{(2) } H_2O,\ H^+]{\text{(1) } C_6H_5MgBr}$ product

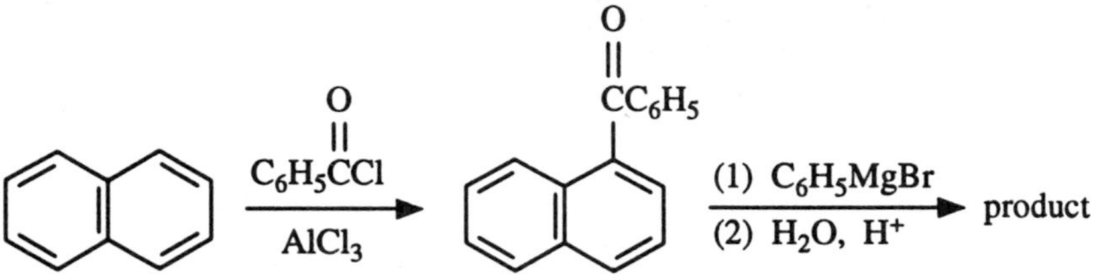

(d) 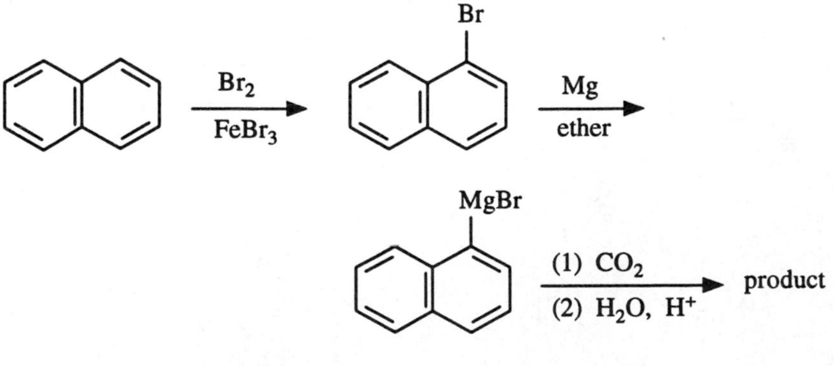 naphthalene $\xrightarrow[FeBr_3]{Br_2}$ 1-bromonaphthalene (Br) $\xrightarrow[\text{ether}]{Mg}$

1-naphthylmagnesium bromide (MgBr) $\xrightarrow[\text{(2) } H_2O,\ H^+]{\text{(1) } CO_2}$ product

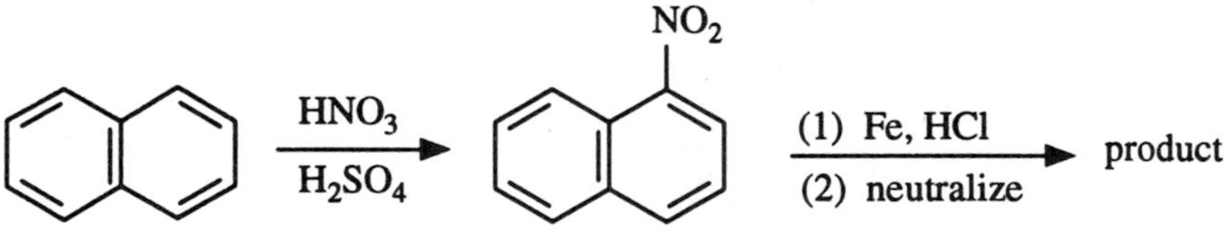

19.4 (a) 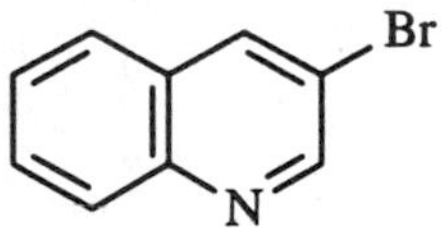 (b)

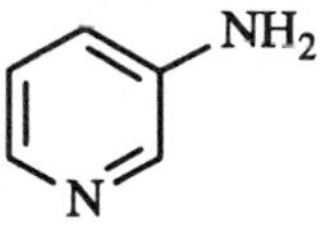

(c) 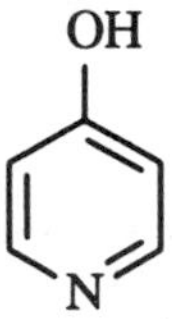 (d)

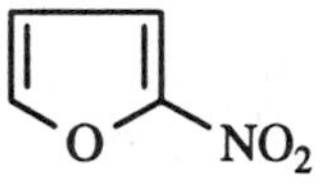

19.5 (a) 3-nitropyrrole (b) 3-methylisoquinoline

(c) 2-iodopyrimidine (d) 2-methylthiazole

(e) 5-isopropylquinoline (f) 4-methyl-2-furansulfonic acid

19.6 The first, noncharged resonance structure is the major contributor:

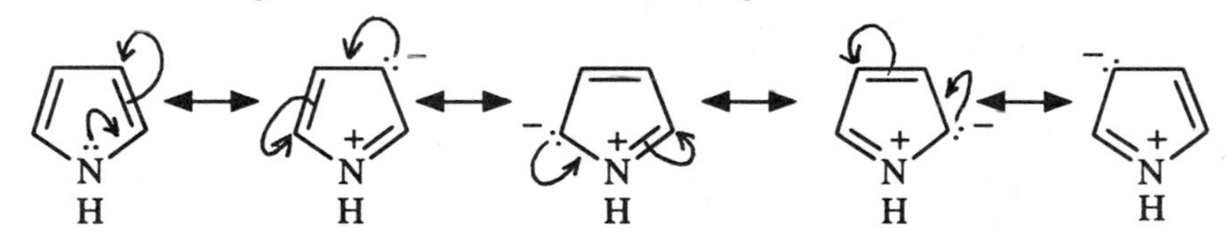

19.7 Compounds (a) and (c) contain activated aromatic rings and can undergo coupling reaction with benzenediazonium chloride. Compounds (b) and (d) are deactivated toward electrophilic substitution and would not undergo diazonium coupling.

(a) 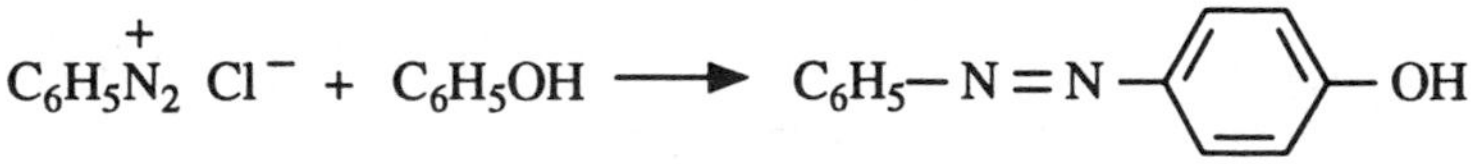 $C_6H_5\overset{+}{N}_2\ Cl^- + C_6H_5OH \longrightarrow C_6H_5{-}N{=}N{-}C_6H_4{-}OH$

(b) 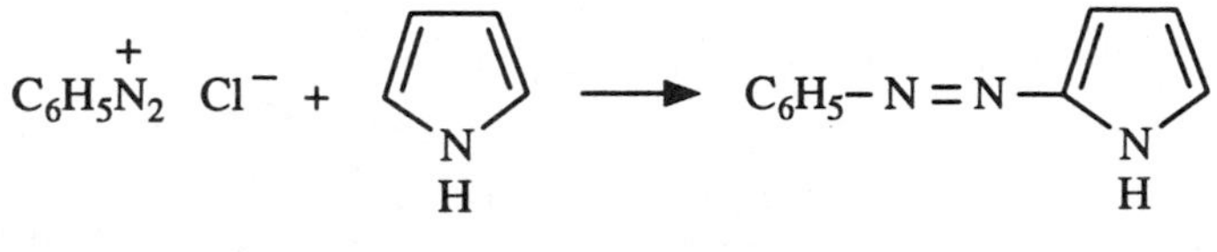 $C_6H_5\overset{+}{N}_2\ Cl^- +$ pyrrole $\longrightarrow C_6H_5{-}N{=}N{-}$(2-pyrrolyl)

20

Natural Products: Studies in Organic Synthesis

Answers to Additional Problems

20.1 (b), (c), and (d). Compound (a) has only 7 carbons. The naturally occuring terpenes have 5, 10, 15, etc. carbons; that is, they have units of isoprene.

20.2 (a) 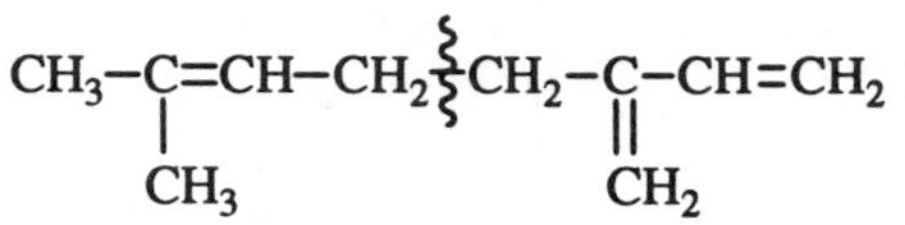 $CH_3-\underset{\underset{CH_3}{|}}{C}=CH-CH_2 \wr CH_2-\underset{\underset{CH_2}{||}}{C}-CH=CH_2$

(b) 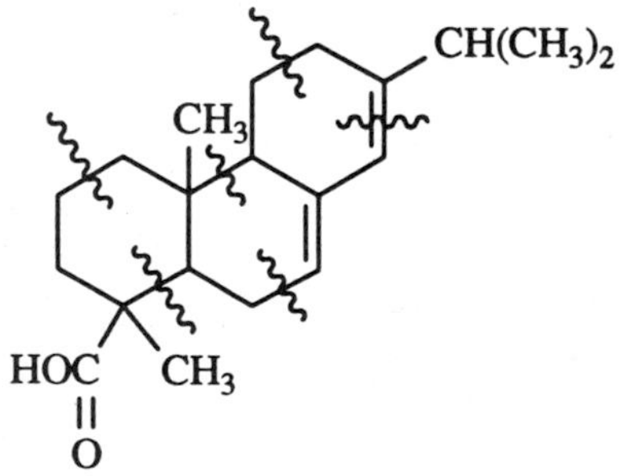 CH(CH3)2, CH3, HOC(=O), CH3

20.3 (a) C_{10}, monoterpene (b) C_{15}, sesquiterpene

(C) C_{30}, triterpene (d) C_{15}, sesquiterpene

20.4 (a) and (d) are alkaloids because they have a nitrogen atom.

(c) and (d) are terpenes.

20.5 $$\text{enantiomeric excess} = \frac{\text{excess of desired enantiomer}}{\text{total yield}} \times 100\%$$

$$90\% = \frac{x}{70\%} \times 100$$

x = 63% of the enantiomer in excess

Therefore, there is (100% - 63%) or 37% of the other enantiomer

20.6 They all contain a basic nitrogen atom. Consequently, they all form salts when treated with a mineral acid such as sulfuric acid.

20.7 (a)

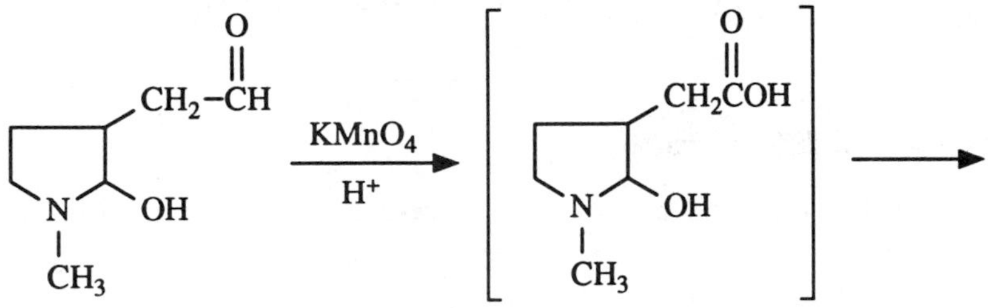

$CH_2-\overset{O}{\overset{||}{C}}H$ OH N CH_3 $\xrightarrow[H^+]{KMnO_4}$ [$CH_2\overset{O}{\overset{||}{C}}OH$ OH N CH_3] $\longrightarrow$

(b) HO CH_2OH N + H_2SO_4 $\xrightarrow{H_2O}$

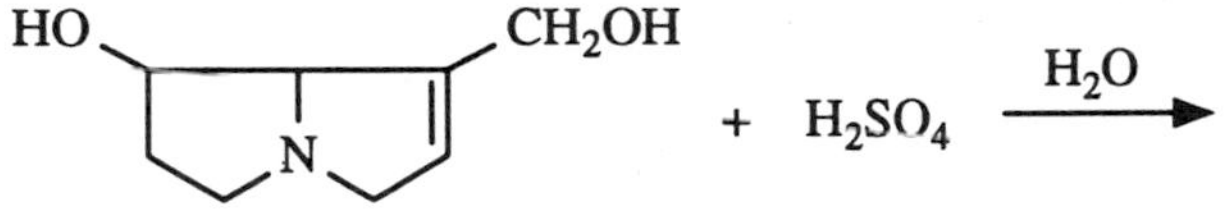

(c) CH_3 CH_3 O + $HOCH_2CH_2OH$ $\xrightarrow{H^+}$

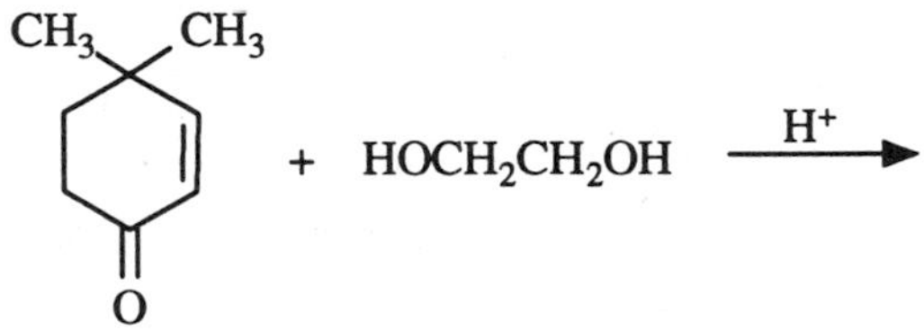

(d) =O + CH_2OCH_3 N H NH_2 $\longrightarrow$

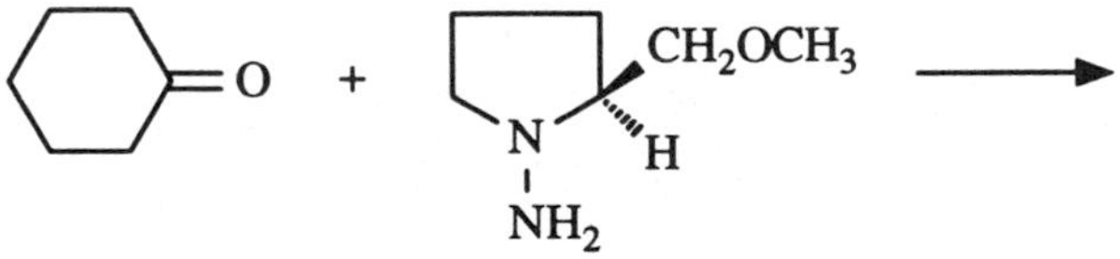

21

Pericyclic Reactions

Answers to Additional Problems

21.1 (a) cycloaddition (b) sigmatropic (c) cycloaddition

21.2 (a)

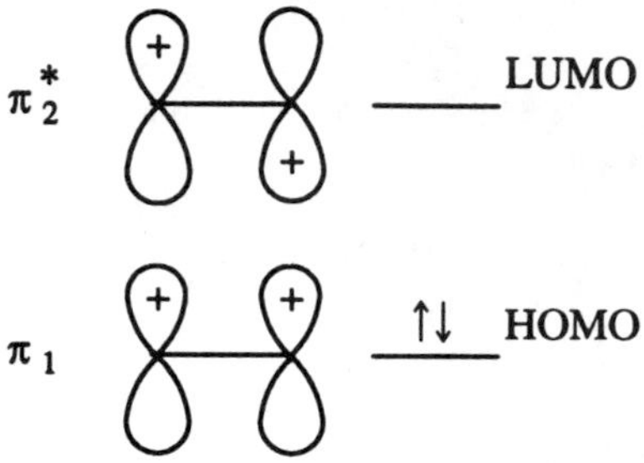

π_2^* LUMO

π_1 ↑↓ HOMO

(b) and (c)

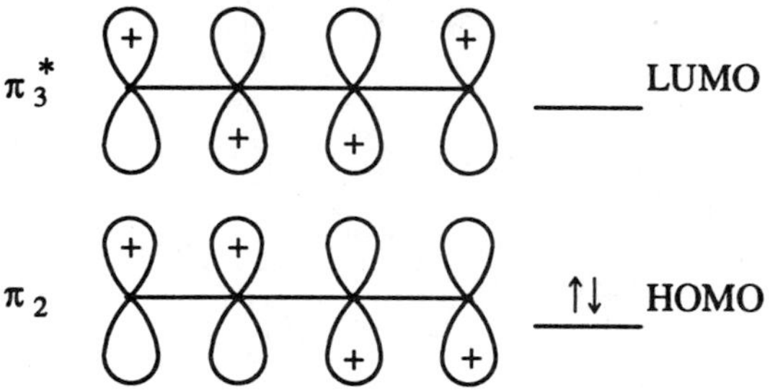

π_3^* LUMO

π_2 ↑↓ HOMO

21.3 (a) four pi electrons in three *p* orbitals:

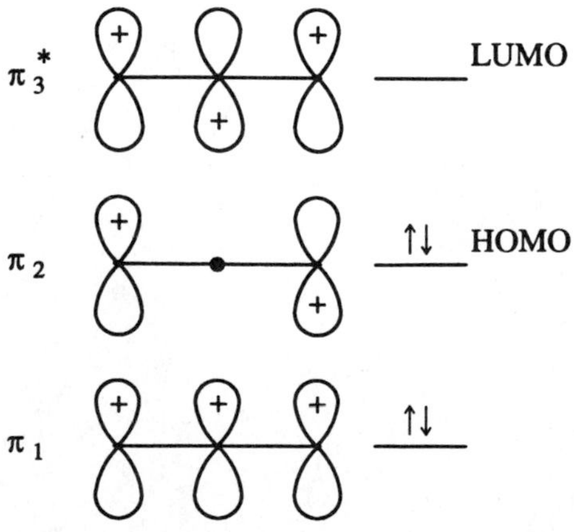

π_3^* LUMO

π_2 ↑↓ HOMO

π_1 ↑↓

(b) four pi electrons in four *p* orbitals:

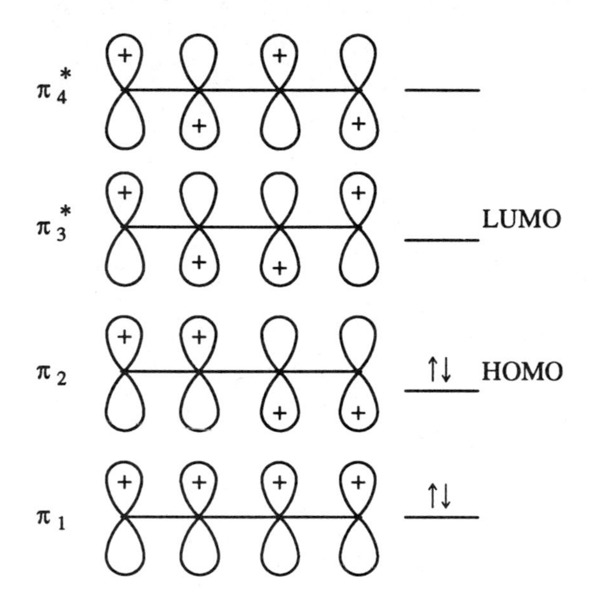

(c) two pi electrons in three *p* orbitals:

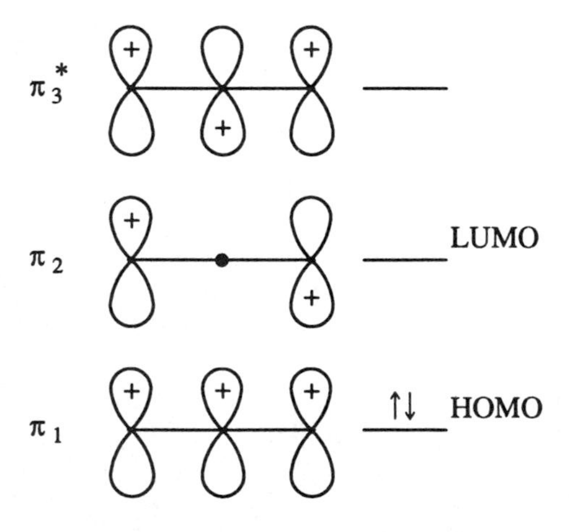

21.4 (a) + heat (Diels-Alder)

(b) 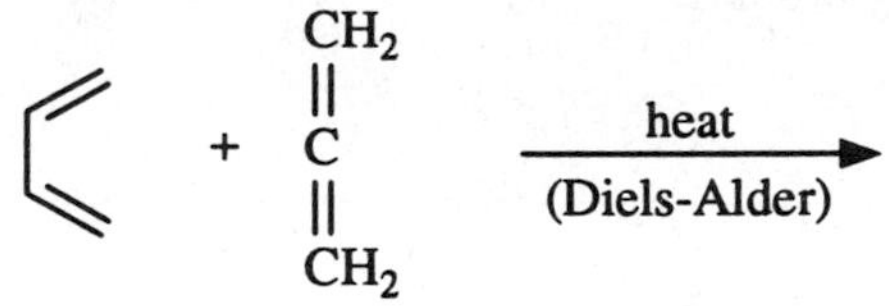 + $CH_2=C=CH_2$ $\xrightarrow[\text{(Diels-Alder)}]{\text{heat}}$

(c) + $CH_2=C(CH_3)_2$ $\xrightarrow{h\nu}$

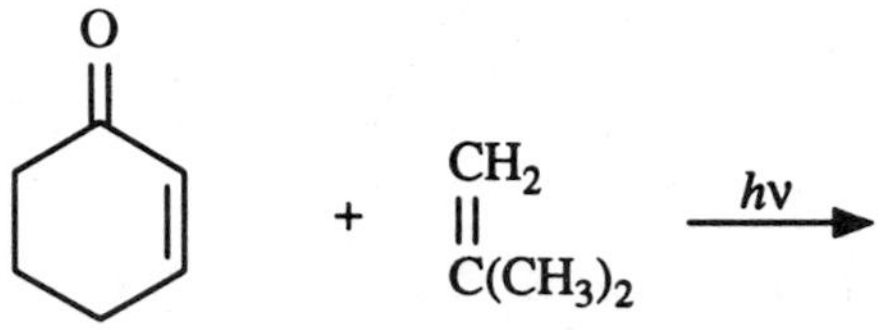

(d) 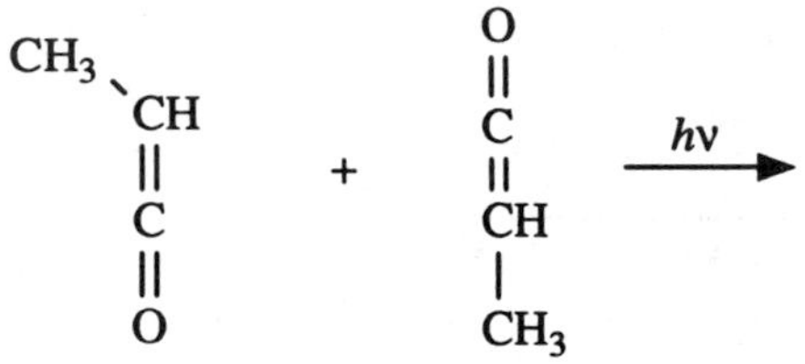 $CH_3CH=C=O$ + $O=C=CHCH_3$ $\xrightarrow{h\nu}$

21.5 (a) 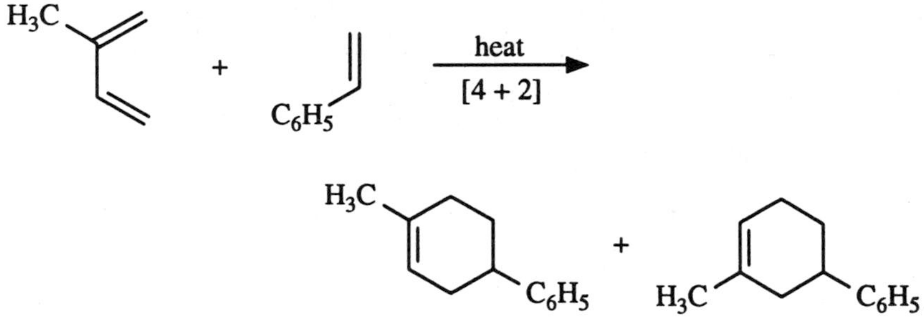 H_3C + C_6H_5 $\xrightarrow[\text{[4 + 2]}]{\text{heat}}$

H_3C C_6H_5 + H_3C C_6H_5

(b) 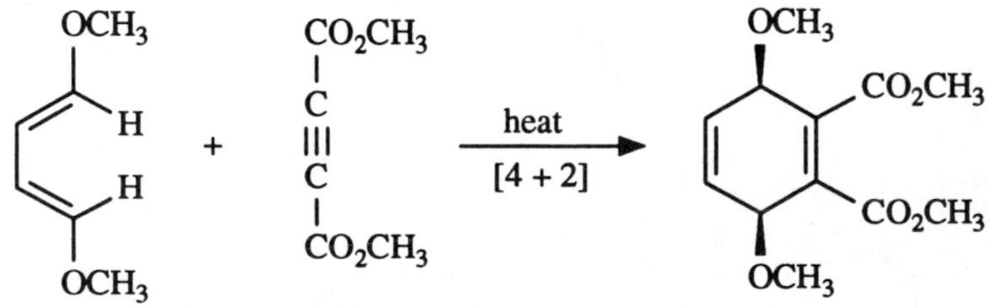 OCH_3 H H OCH_3 + CO_2CH_3–C≡C–CO_2CH_3 $\xrightarrow[\text{[4 + 2]}]{\text{heat}}$ OCH_3 CO_2CH_3 CO_2CH_3 OCH_3

(c) [reaction scheme: allyl 2-methoxyphenyl ether → heat, [3 + 3] sigmatropic rearrangement → 6-allyl-2-methoxycyclohexa-2,4-dienone → tautomerization → 2-allyl-6-methoxyphenol ($CH_2CH{=}CH_2$, OH, OCH_3)]

heat

[3 + 3] sigmatropic rearrangement

tautomerization

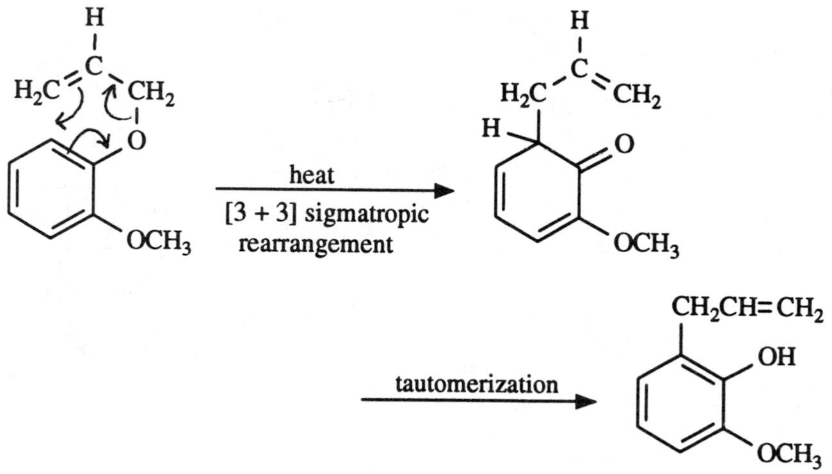

The last step in (c) is labeled "tautomerization instead of "[1,3] sigmatropic rearrangement" because tautomerization is a stepwise ionic reaction and not a concerted pericyclic reaction.

(d) [reaction scheme: trans-$C_6H_5CH{=}CHC_6H_5$ + $H_3C(H)C{=}C(H)CO_2CH_3$ → $h\nu$, [2 + 2] → two cyclobutane stereoisomers bearing C_6H_5, C_6H_5, CO_2CH_3, CH_3 and H substituents]

$h\nu$

[2 + 2]

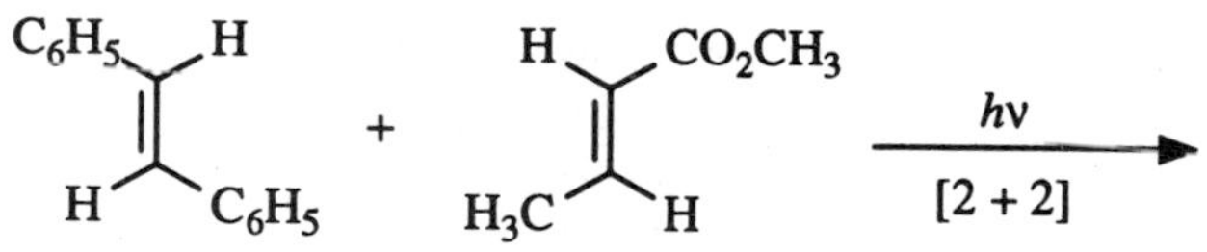

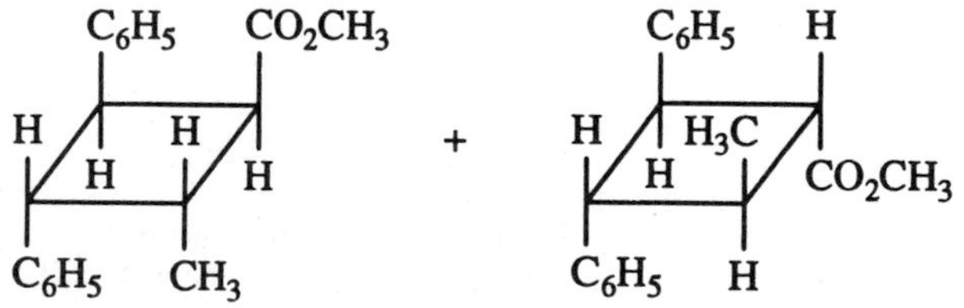

To determine the stereochemistry in (d), view the molecules as they approach each other before the reaction occurs. For example,

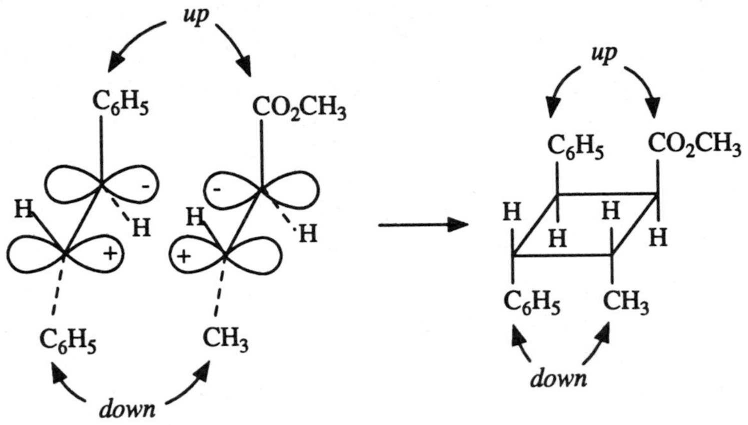

up

C_6H_5 CO_2CH_3

H H - + H H - + H

C_6H_5 CH_3

down

up

C_6H_5 CO_2CH_3

H H H H

C_6H_5 CH_3

down

H

(e)

$h\nu$
[4*n* + 2] photochemical
conrotatory

H

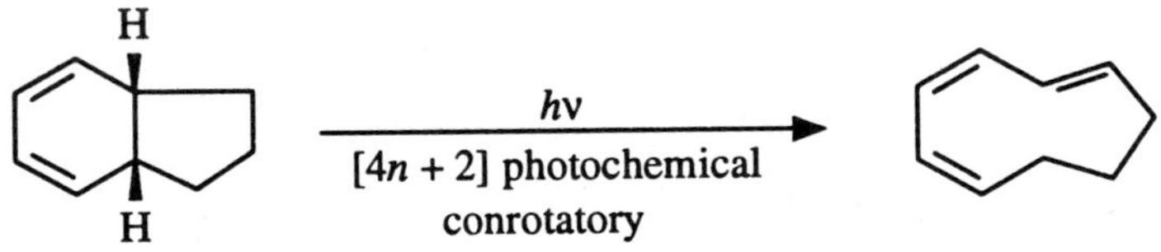

This reaction is very similar to the reverse reaction of the cyclization of 1,3,5-cyclohexatriene.

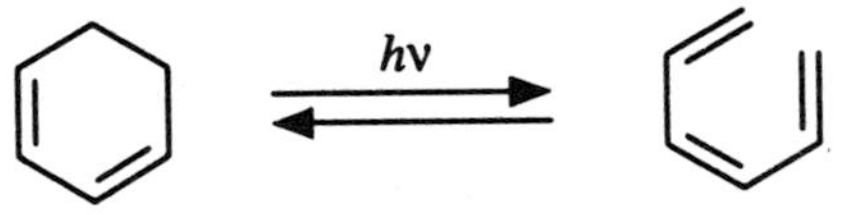

$h\nu$

To determine the direction of the movement of the groups in electrocyclic reactions, approach the reaction in steps (even though the actual reaction is concerted). Do not worry about the initial appearance of the product because you can redraw the structure stereochemically correct.

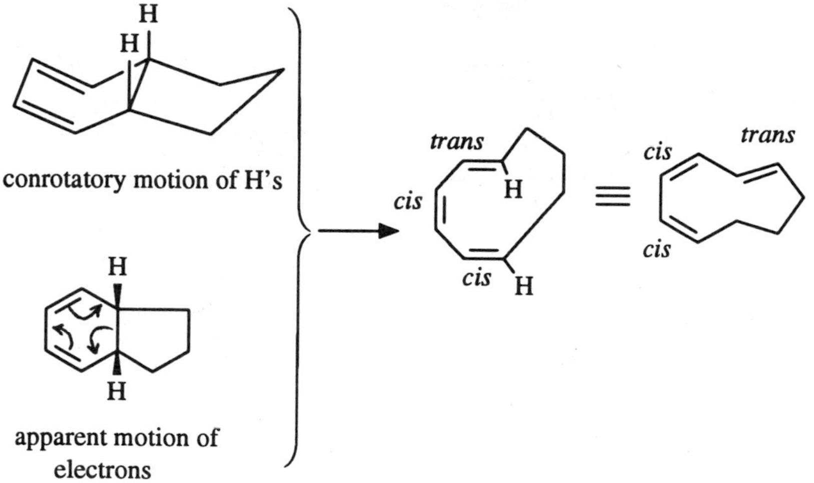

H
H

conrotatory motion of H's

H
H

apparent motion of electrons

trans
cis
H
cis H

≡

trans
cis
cis

(f)

heat
disrotatory

H
H

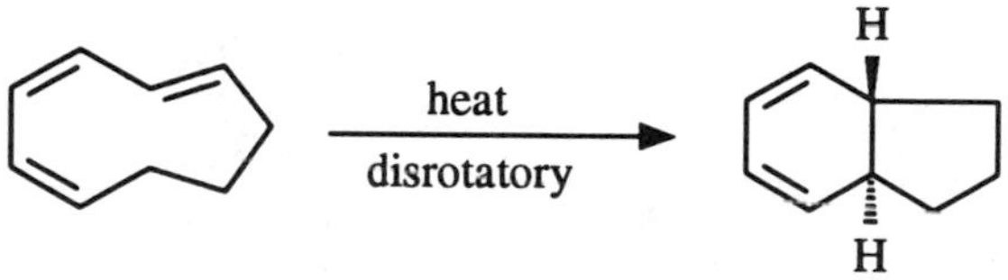

Compare this reaction with that in part (e). To return to a *cis* ring juncture would require light and conrotatory motion. Therefore, the ring juncture H atoms must be *trans* when heat is used.

21.6 (a) conrotatory (b) disrotatory

21.7 (a) *Step 1:*

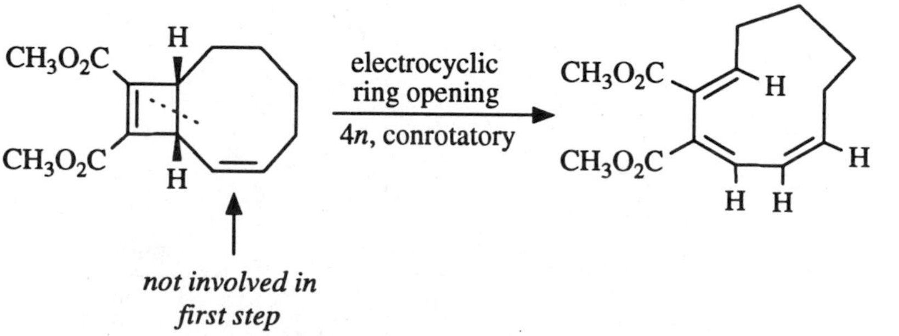

CH_3O_2C
CH_3O_2C
H
H

electrocyclic ring opening
4*n*, conrotatory

CH_3O_2C
CH_3O_2C
H
H
H H

not involved in first step

(A ten-membered ring can accommodate a *trans* double bond with H "inside" the ring.)

Step 2: Redraw the structure.

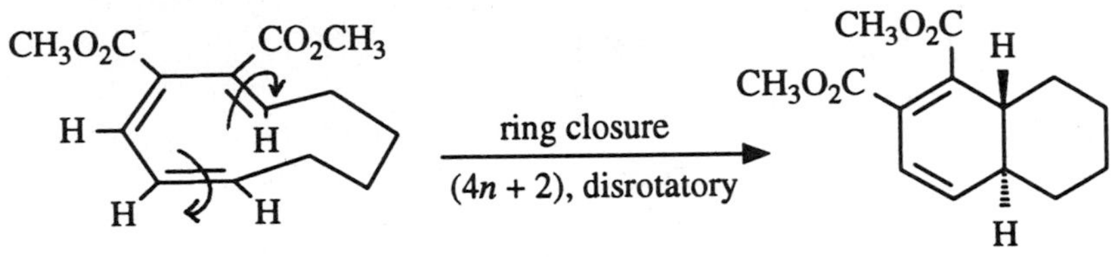

CH_3O_2C CO_2CH_3 H H H H

ring closure

(4*n* + 2), disrotatory

CH_3O_2C CH_3O_2C H H

(b)

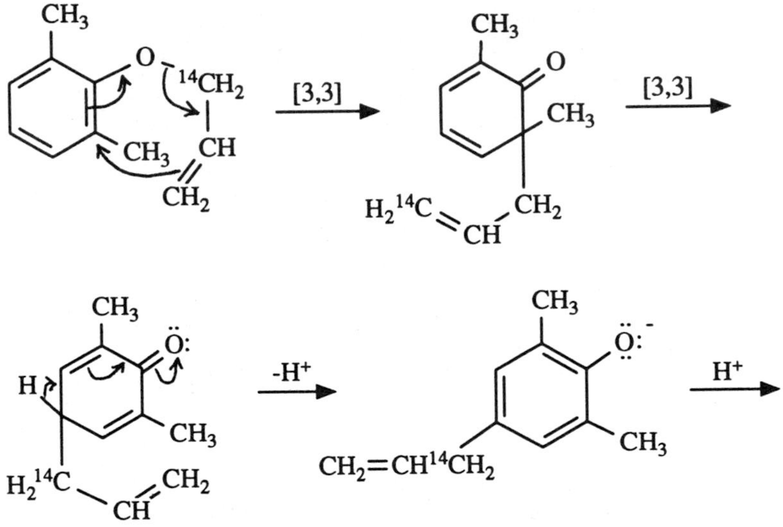

CH_3 O $^{14}CH_2$ CH_3 CH CH_2

[3,3]

CH_3 O CH_3 $H_2{}^{14}C$ CH CH_2

[3,3]

CH_3 Ö: H CH_3 $H_2{}^{14}C$ CH CH_2

-H⁺

CH_3 Ö:⁻ CH_3 $CH_2{=}CH^{14}CH_2$

H⁺

(c)

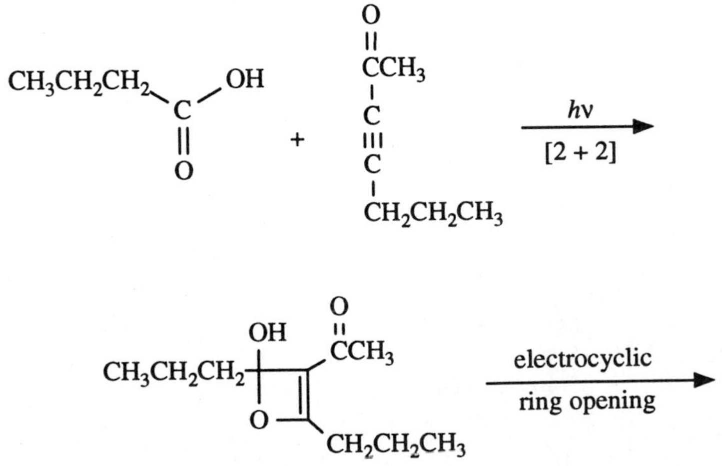

$CH_3CH_2CH_2$ C OH O + O CCH₃ C C $CH_2CH_2CH_3$

*h*ν

[2 + 2]

OH O $CH_3CH_2CH_2$ CCH₃ O $CH_2CH_2CH_3$

electrocyclic

ring opening

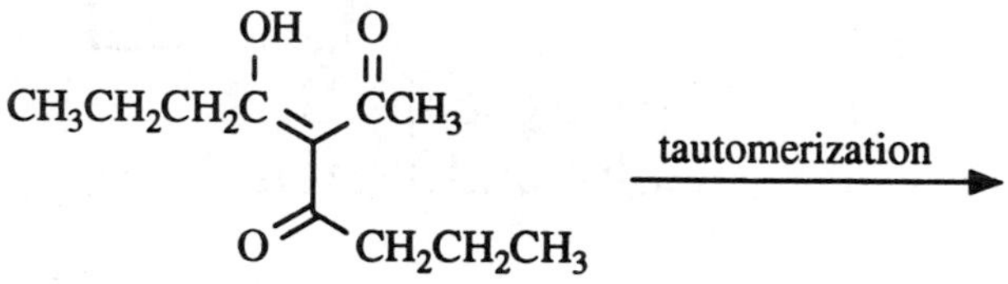

OH O

$CH_3CH_2CH_2C$ CCH_3

O $CH_2CH_2CH_3$

tautomerization

21.8 (a)

HO

an enol

H

O

(b) 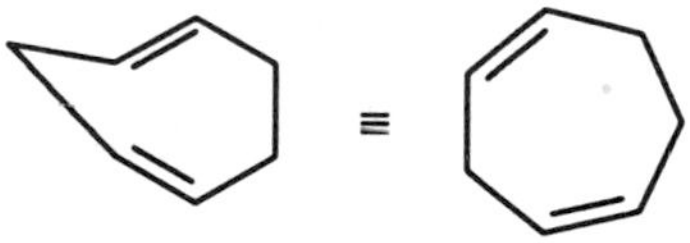 ≡ (c) 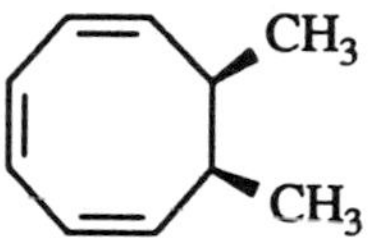 CH_3 CH_3

21.9

R′ HC R 3 3 2 C NH 1 1 2 N H

H^+

R′ C R NH$_2$ + N H

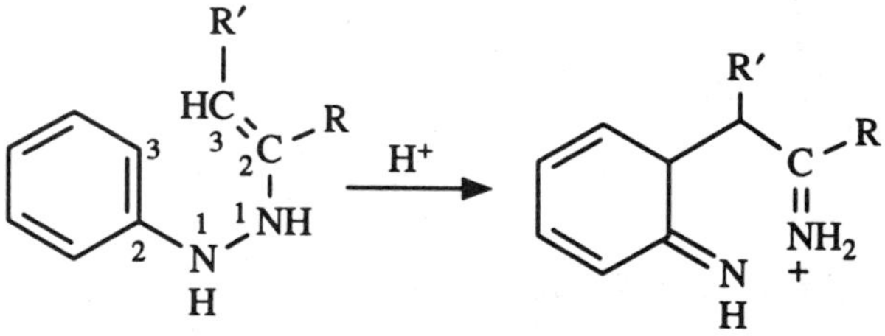

21.10 (a) The reaction sequence involves an initial dehydrohalogenation followed by a [4 + 2] cycloaddition.

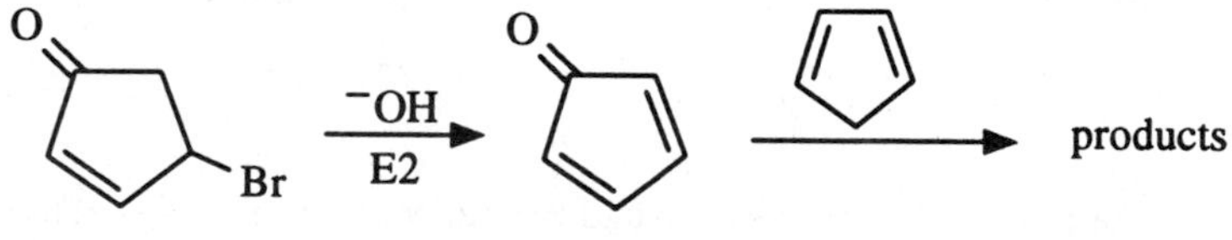

O Br ^-OH E2 O products

(b) The reaction is a 4*n* electrocyclic ring-closure reaction with disrotatory motion.

21.11 O $CH_2CH{=}CH_2$

CH_3C — CH_2

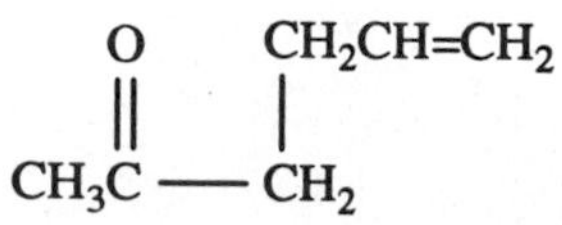

22

Spectroscopy II: Ultraviolet Spectra, Color and Vision, Mass Spectra

Answers to Additional Problems

22.1 (a) more (b) more

22.2 The equation that is used to solve these problems is

$$\Delta E = h\nu = \frac{hc}{\lambda}$$

Where h = Plank's constant = 6.6×10^{-27} erg-sec

c = speed of light = 3.0×10^{10} cm/sec

λ = wavelength of light in cm

(a) 1 nm = 10^{-7} cm; 230 nm = 230×10^{-7} cm

$$\Delta E = \frac{(6.6 \times 10^{-27})(3.0 \times 10^{10})}{(230 \times 10^{-7})}$$

$$= 8.6 \times 10^{-12} \text{ ergs}$$

(b) 1 μm = 10^{-4} cm; 6.2 μm = 6.2×10^{-4} cm

$\Delta E = 3.2 \times 10^{-13}$ ergs

22.3 (a) acetone (b) about 280 nm

(c) about 15 (d) $n \rightarrow \pi^*$ (280 nm); $\pi \rightarrow \pi^*$ (187 nm)

22.4 The base peak is at m/e = 96 and the molecular ion peak is at m/e = 150.

22.5 (a)

$$\left[CH_3CH_2CH_2\overset{\overset{:O\cdot}{\|}}{C}-CH_2CH_2CH_3\right]^{+\cdot} \xrightarrow{-\dot{C}H_2CH_2CH_3} \left[CH_3CH_2CH_2C{\equiv}\overset{+}{O}: \longleftrightarrow CH_3CH_2CH_2\overset{+}{C}{=}\ddot{O}:\right]^{+}$$

$$\left[\begin{matrix} H_2C & -H & \dot{O}: \\ | & & \| \\ H_2C & -\underset{H_2}{C}- & CCH_2CH_2CH_3 \end{matrix}\right]^{+\cdot} \xrightarrow{-CH_2{=}CH_2} \left[\begin{matrix} H\dot{O}: \\ | \\ CH_2{=}CCH_2CH_2CH_3 \end{matrix}\right]^{+\cdot}$$

(b)

$$\left[CH_3\overset{\overset{H}{|}}{C}H-\underset{\underset{CH_3}{|}}{C}H-\dot{\ddot{O}}H\right]^{+\cdot} \xrightarrow{-H_2\ddot{O}:} CH_3\overset{+}{C}H\dot{C}HCH_3$$

(c)

$$\left[CH_3CH_2-\underset{\underset{H}{\cdot}}{\overset{\overset{CH_3}{|}}{C}}-CH_2CH_2CH_3\right]^{+\cdot} \xrightarrow{-H\cdot} CH_3CH_2-\underset{+}{\overset{\overset{CH_3}{|}}{C}}-CH_2CH_2CH_3$$

$$\left[CH_3CH_2-\underset{\underset{H}{\cdot}}{\overset{\overset{CH_3}{|}}{C}}-CH_2CH_2CH_3\right]^{+\cdot} \xrightarrow{-CH_3CH_2\dot{C}H_2} CH_3CH_2-\underset{\underset{H}{|}}{\overset{\overset{CH_3}{|}}{C^+}}$$

$$\left[CH_3CH_2-\underset{\underset{H}{\cdot}}{\overset{\overset{CH_3}{|}}{C}}-CH_2CH_2CH_3\right]^{+\cdot} \xrightarrow{-\dot{C}H_3} CH_3CH_2-\underset{\underset{H}{|}}{\overset{+}{C}}-CH_2CH_2CH_3$$

$$\left[CH_3CH_2-\underset{\underset{H}{\cdot}}{\overset{\overset{CH_3}{|}}{C}}-CH_2CH_2CH_3\right]^{+\cdot} \xrightarrow{-CH_3\dot{C}H_2} {}^{+}\underset{\underset{H}{|}}{\overset{\overset{CH_3}{|}}{C}}-CH_2CH_2CH_3$$

(d) $\left[CH_3-CH(CH_3)-\dot{N}HCH_2CH_2CH_3\right]^{+\cdot} \xrightarrow{-\dot{C}H_3} \left[CH_3CH{=}\overset{+}{N}HCH_2CH_2CH_3\right]$

$\left[(CH_3)_2CH\dot{N}H-CH_2-CH_2CH_3\right]^{+\cdot} \xrightarrow{-\dot{C}H_2CH_3} \left[(CH_3)_2CH\overset{+}{N}H{=}CH_2\right]$

22.6 $\left[H_2N{=}CH_2\right]^+$, m/e = 30

$\left[HO{=}CH_2\right]^+$, m/e = 31

Both fragments arise from alpha fission.

22.7 (a) $\left[CH_3CH_2\overset{O}{\overset{||}{C}}OCH_3\right]^{+\cdot} \xrightarrow{-CH_3\dot{O}} \left[CH_3CH_2\overset{O}{\overset{||}{C}}\right]^+$ $m/e = 57$

$\xrightarrow{-CH_3\dot{C}H_2} \left[\overset{O}{\overset{||}{C}}OCH_3\right]^+$ $m/e = 59$

(b) $\xrightarrow{\text{alpha fission}}$ $\overset{+}{O}{=}CH_2$, $\dot{C}H_2$, O $\xrightarrow[-H_2C=O]{\text{alpha fission}}$ $\dot{C}H_2-CH_2-\overset{+}{O}{=}CH_2$ $m/e = 58$

22.8 (a) UV. One of the compounds is a conjugated diene while the other compound is a nonconjugated diene.

(b) ^1H NMR. 1-Naphthol would show aromatic absorption while the saturated alcohol would not.

(c) ^1H NMR. Toluene would show two singlets while ethylbenzene would show an upfield triplet and a downfield quartet, as well as one or more peaks from aryl protons.

(d) Mass Spectrometry. 3-Methylundccane would show a M-15 peak (methyl cleavage) and a M-29 peak (ethyl cleavage) while 2-methylundecane would show only the M-15 peak.

(e) Infrared. The ester would show C-O absorption while the ketone would not.

22.9 I, (c); II, (g); III; (f); IV, (h)

22.10 (a) 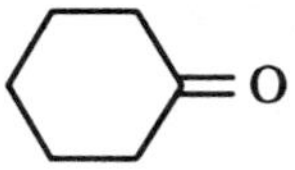 =O (b) $CH_3CHClCH_2Cl$

(c) 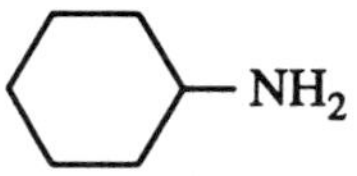 $-NH_2$ (d) $(CH_3)_2CHCH_2\overset{O}{\overset{\|}{C}}CH_3$

(e) 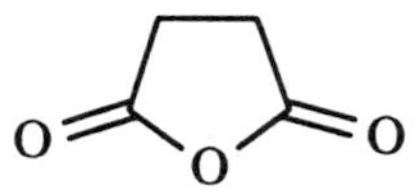 O= O =O

22.11 (a) $C_6H_5OCH_3$ (b) 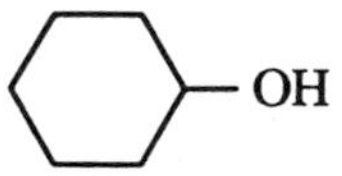 $-OH$

23

Carbohydrates

Answers to Additional Problems

23.1 (a) D-ribose (b) D-glucose

(c) D-fructose (d) D-galactose

23.2 (a) pyranose (b) pyranose (c) furanose

23.3 (a)

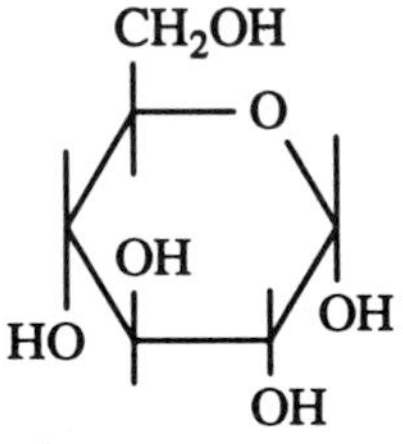

anomer 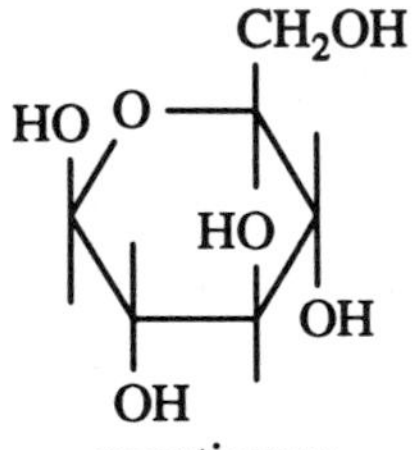 enantiomer

(b)

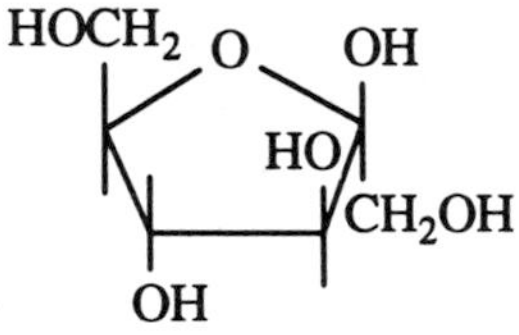

anomer 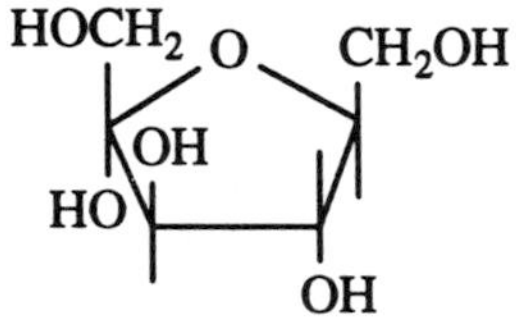 enantiomer

23.4 (a) propyl α-D-glucopyranoside

(b) 4-*O*-(β-D-galactopyranosyl)-α-D-glucopyranose (or α-lactose)

23.5 (a) The sugar is in the hemiacetal form; therefore, it is in equilibrium with the reactive aldehyde form and is a reducing sugar.

(b) is a glycoside: nonreducing

(c) The right-hand sugar unit of the disaccharide is in the hemiacetal form; consequently, that portion of the disaccharide is in equilibrium with its aldehyde form. The disaccharide is a reducing sugar.

(d) Both units are glycosidic; therefore, this sugar (like sucrose) is nonreducing.

23.6 (a)

$$\begin{array}{c} CHO \\ H-\!\!\!-\!\!OH \\ HO-\!\!\!-\!\!H \\ H-\!\!\!-\!\!OH \\ CH_2OH \end{array}$$

(b)

$$\begin{array}{c} CHO \\ H-\!\!\!-\!\!OH \\ HO-\!\!\!-\!\!H \\ H-\!\!\!-\!\!OH \\ H-\!\!\!-\!\!OH \\ CH_2OH \end{array}$$

(c)

$$\begin{array}{c} CHO \\ H-\!\!\!-\!\!OH \\ HO-\!\!\!-\!\!H \\ HO-\!\!\!-\!\!H \\ H-\!\!\!-\!\!OH \\ CH_2OPO_3^{2-} \end{array}$$

23.7 (a) $2\ \overset{O}{\overset{\|}{HCH}}$

(b) $CH_3\overset{O}{\overset{\|}{C}}H + H\overset{O}{\overset{\|}{C}}H$

(c) $CH_3\overset{O}{\overset{\|}{C}}H + HO\overset{O}{\overset{\|}{C}}H$

(d) $H\overset{O}{\overset{\|}{C}}H + 4\ H\overset{O}{\overset{\|}{C}}OH$

(e) (furanose ring: $HOCH_2$, O, OH, OH, OH) $\rightleftharpoons$

$$\begin{array}{c} CHO \\ H-\!\!\!-\!\!OH \\ H-\!\!\!-\!\!OH \\ H-\!\!\!-\!\!OH \\ CH_2OH \end{array} \xrightarrow{HIO_4} H\overset{O}{\overset{\|}{C}}H + 4\ H\overset{O}{\overset{\|}{C}}OH$$

23.8

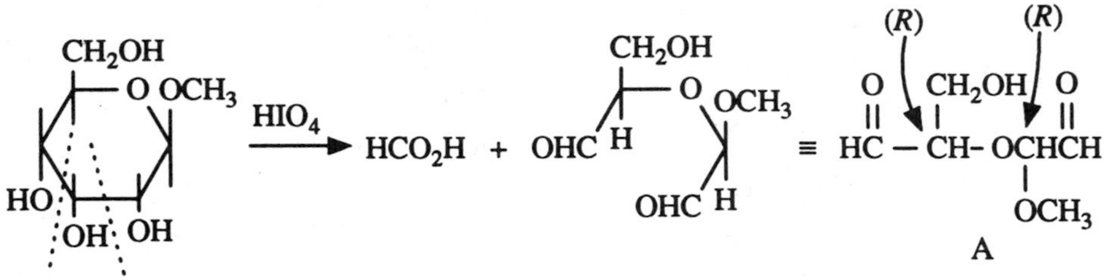

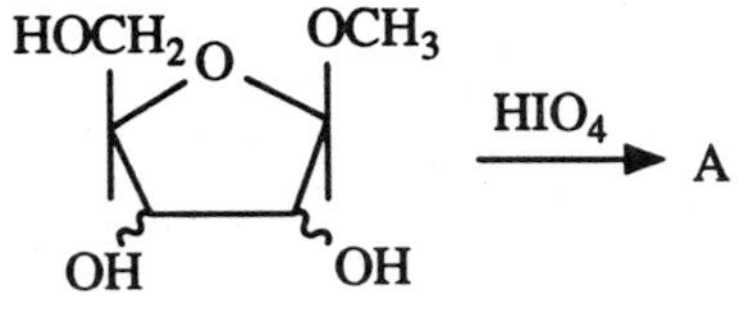

$\xrightarrow{HIO_4}$ HCO_2H + OHC ... ≡

$$HC(=O)-CH(CH_2OH)-OCH(OCH_3)CH=O$$

(R) (R)

A

B $\xrightarrow{HIO_4}$ A

In B, we cannot deduce the configurations at carbons 2 and 3 because the product aldehyde groups are achiral.

23.9 (a) (CHO; HO–H; H–OH; H–OH; CH_2OH) + excess CH_3COCCH_3 (two C=O) $\xrightarrow[-CH_3CO_2H]{CH_3CO_2^-\ Na^+}$ ($CH_3CO_2CH_2$, O, CH_3CO_2, O_2CCH_3, O_2CCH_3 furanose)

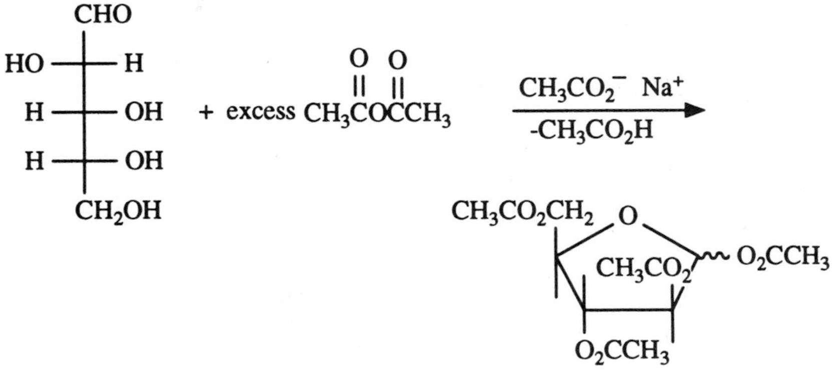

(b) (CHO; HO–H; H–OH; HO–H; H–OH; CH_2OH) $\xrightarrow[heat]{HNO_3}$ (CO_2H; HO–H; H–OH; HO–H; H–OH; CO_2H)

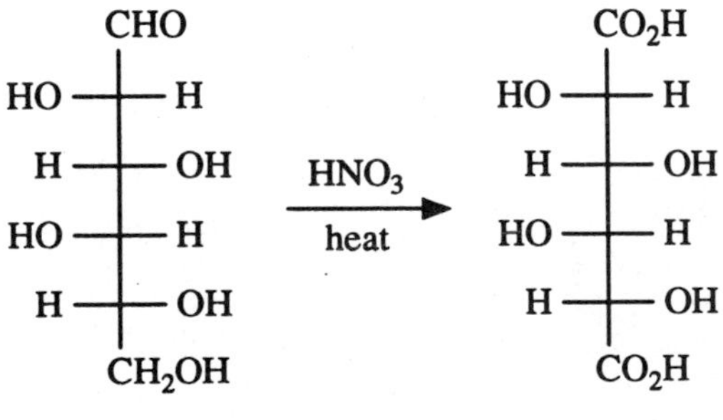

(c) 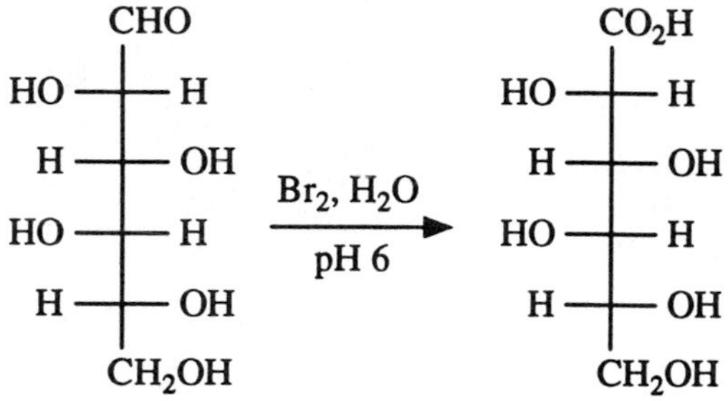 Fischer projection: CHO; HO—H; H—OH; HO—H; H—OH; CH_2OH $\xrightarrow[\text{pH 6}]{Br_2,\ H_2O}$ CO_2H; HO—H; H—OH; HO—H; H—OH; CH_2OH

(d) 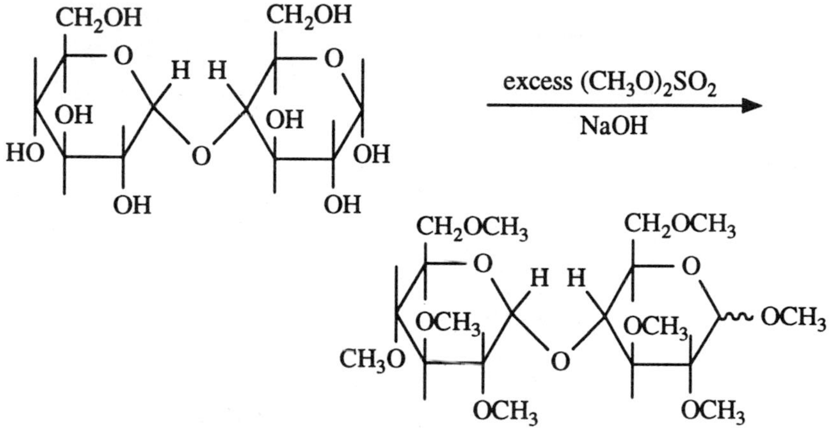 [disaccharide: two rings, each with CH_2OH, O, OH, HO, OH; H H; O bridge] $\xrightarrow[\text{NaOH}]{\text{excess } (CH_3O)_2SO_2}$ [product: CH_2OCH_3, CH_2OCH_3, O, O, H H, OCH_3, OCH_3, CH_3O, O, OCH_3, OCH_3, $\sim OCH_3$]

23.10 (a) 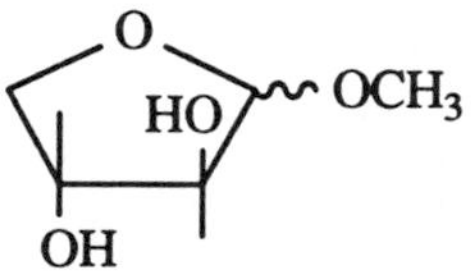 [five-membered ring: O, HO, $\sim OCH_3$, OH]

(b) 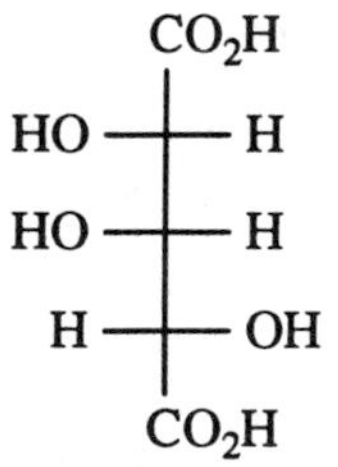 CO_2H; HO—H; HO—H; H—OH; CO_2H

(c) 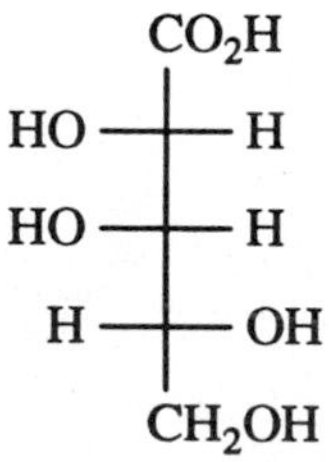 CO_2H; HO—H; HO—H; H—OH; CH_2OH

(d) 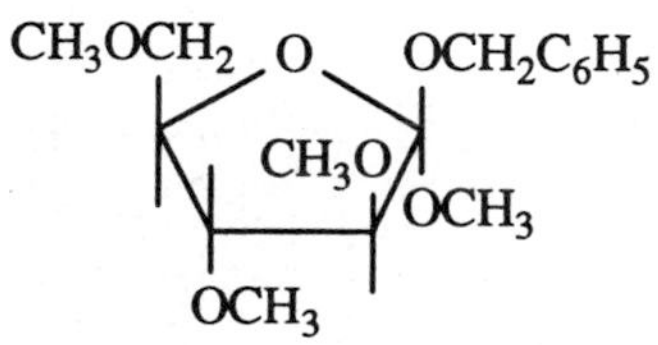 [five-membered ring: CH_3OCH_2, O, $OCH_2C_6H_5$, CH_3O, OCH_3, OCH_3]

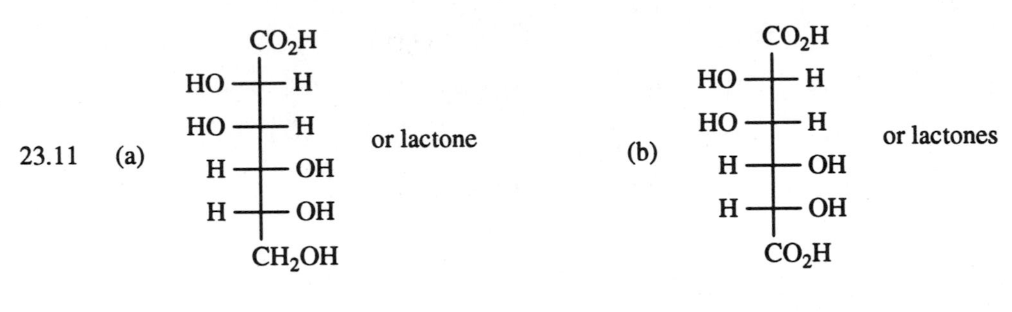

(c)

CH_2OH
O
OH OH
OC_2H_5
HO

(d)

CH_2OCH_3
O
OCH_3 OCH_3
OC_2H_5
CH_3O

(e) same as (d)

(f)

CH_2OAc
O
OAc OAc
OAc
AcO

$+ CH_3CO_2H$

Ac = CH_3CO—

(g) same as (f)

(h)

CH_2OH
HO — H
HO — H
H — OH
H — OH
CH_2OH

(i)

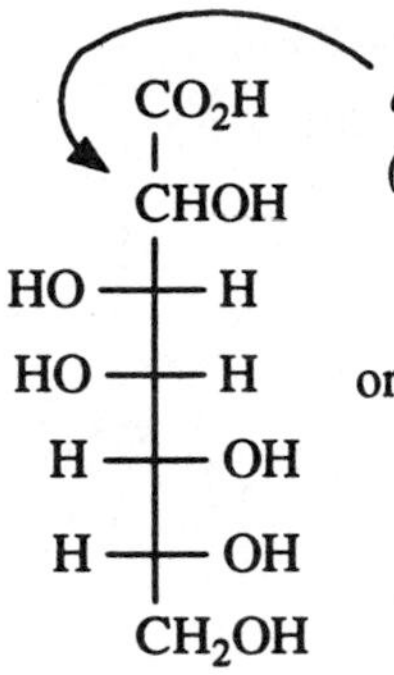

CO_2H
CHOH
HO—H
HO—H
H—OH
H—OH
CH_2OH

a mixture of (R) and (S)

or lactone

(j) same as (h)

(k) no reaction unless heat and pressure are used, in which case the product would be the same as in (h)

(l)

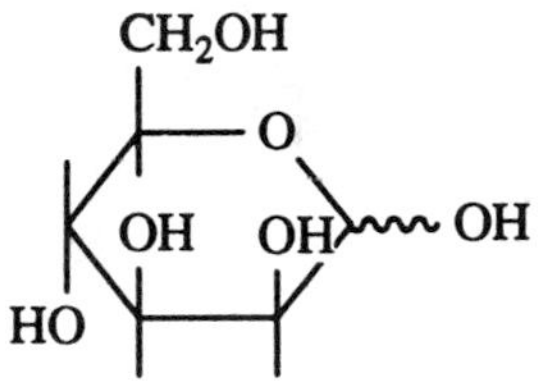

CH_2OH
O
OH OH OH
HO

23.12 (a)

CHO
H—OH
H—OH
CH_2OH

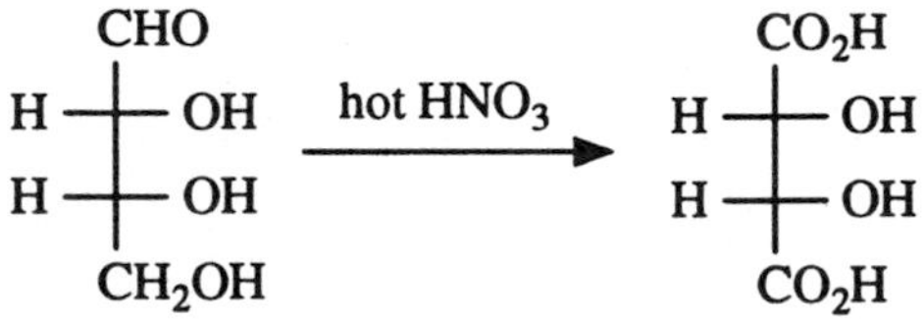

D-erythrose
(the D-tetrose)

hot HNO_3 →

CO_2H
H—OH
H—OH
CO_2H

meso-tartaric acid

(b)

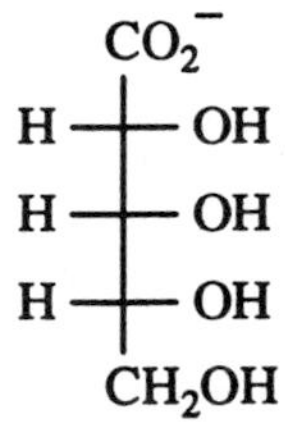

CO_2^-
H—OH
H—OH
H—OH
CH_2OH

D-ribonic acid

(c)

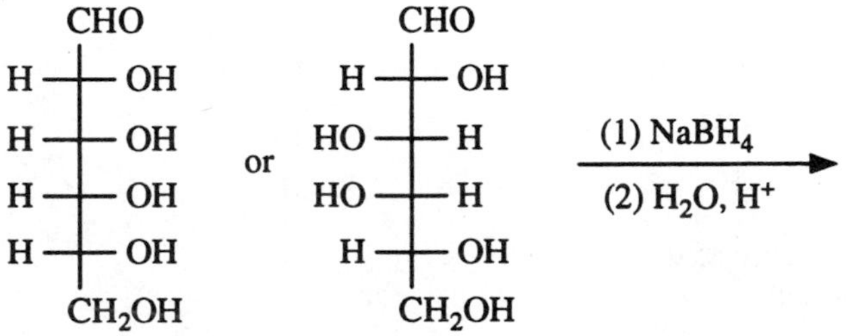

CHO
H—OH
H—OH
H—OH
H—OH
CH_2OH

D-allose

or

CHO
H—OH
HO—H
HO—H
H—OH
CH_2OH

D-galactose

(1) $NaBH_4$
(2) H_2O, H^+ →

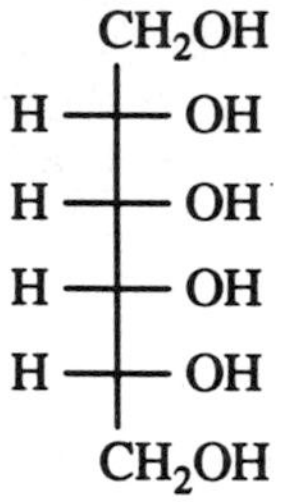

CH_2OH
H — OH
H — OH
H — OH
H — OH
CH_2OH

or

CH_2OH
H — OH
HO — H
HO — H
H — OH
CH_2OH

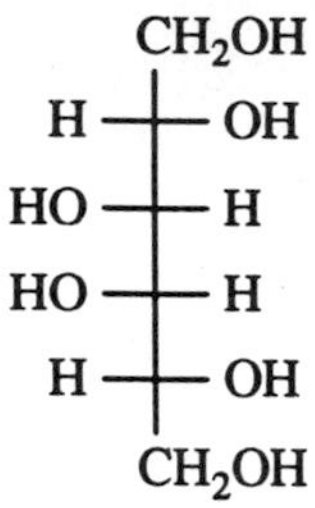

D-allitol D-galactitol

Both are *meso* alditols.

23.13 (a)

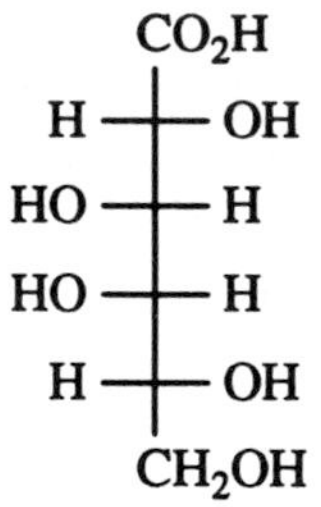

CO_2H
H — OH
HO — H
HO — H
H — OH
CH_2OH

D-galactonic acid

or

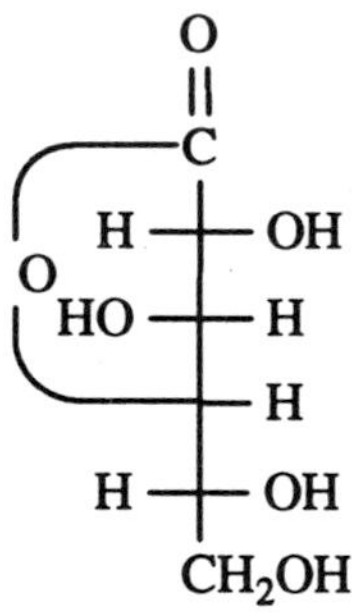

O
||
C
H — OH
O HO — H
— H
H — OH
CH_2OH

or

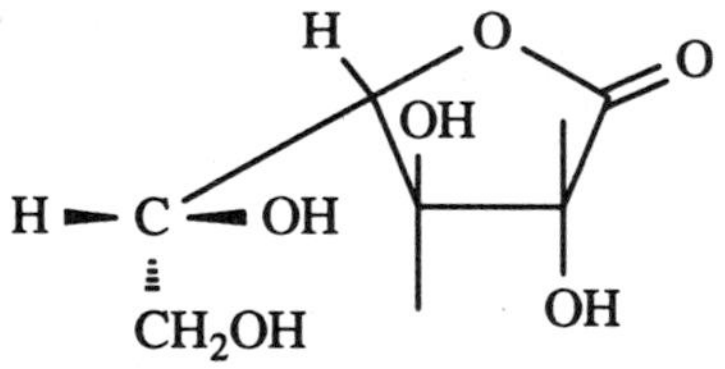

H O O
OH
H — C — OH
CH_2OH
OH

a lactone (Use models.)

(b)

CO_2H
H — OH
HO — H
HO — H
H — OH
CO_2H

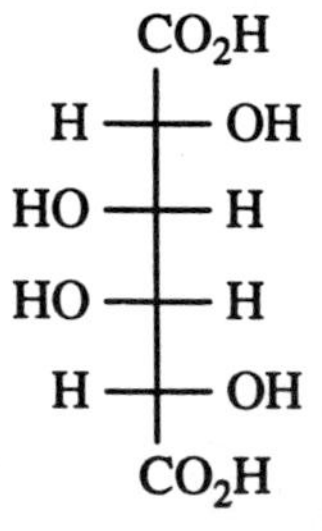

D-galactaric acid or lactones

(c)

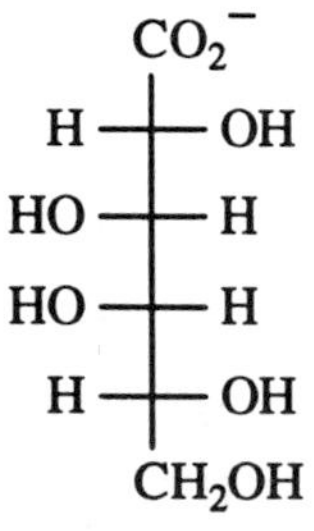

CO_2^-
H — OH
HO — H
HO — H
H — OH
CH_2OH

D-galactonate ion

(d)

CH_2OH
H — OH
HO — H
HO — H
H — OH
CH_2OH

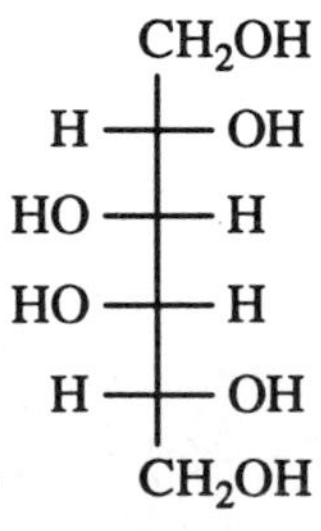

galactitol

23.14 (a)

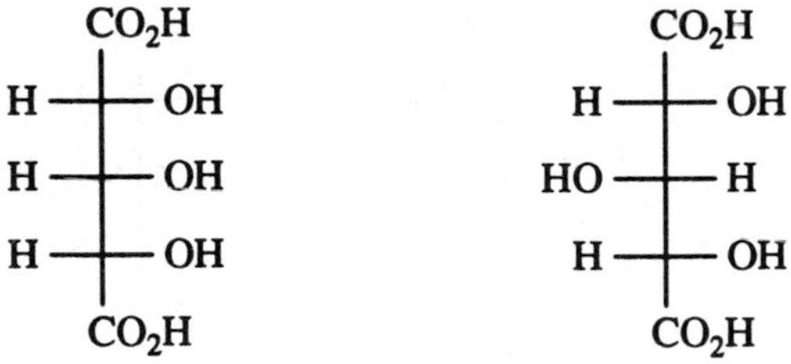

$$\begin{array}{c} CO_2H \\ H-\!\!\!-\!\!\!-OH \\ H-\!\!\!-\!\!\!-OH \\ H-\!\!\!-\!\!\!-OH \\ CO_2H \end{array} \qquad \begin{array}{c} CO_2H \\ H-\!\!\!-\!\!\!-OH \\ HO-\!\!\!-\!\!\!-H \\ H-\!\!\!-\!\!\!-OH \\ CO_2H \end{array}$$

meso *meso*

(b)

$$\begin{array}{c} CHO \\ H-\!\!\!-\!\!\!-OH \\ HO-\!\!\!-\!\!\!-H \\ H-\!\!\!-\!\!\!-OH \\ CH_2OH \end{array} \longrightarrow \begin{array}{c} CO_2H \\ HO-\!\!\!-\!\!\!-H \\ H-\!\!\!-\!\!\!-OH \\ CO_2H \end{array}$$

A D-(-)-tartaric acid

A $\xrightarrow{Br_2,\ H_2O}$ B:

$$\begin{array}{c} CO_2H \\ H-\!\!\!-\!\!\!-OH \\ HO-\!\!\!-\!\!\!-H \\ H-\!\!\!-\!\!\!-OH \\ CH_2OH \end{array}$$

B

A $\xrightarrow{HNO_3}$ C:

$$\begin{array}{c} CO_2H \\ H-\!\!\!-\!\!\!-OH \\ HO-\!\!\!-\!\!\!-H \\ H-\!\!\!-\!\!\!-OH \\ CO_2H \end{array}$$

C

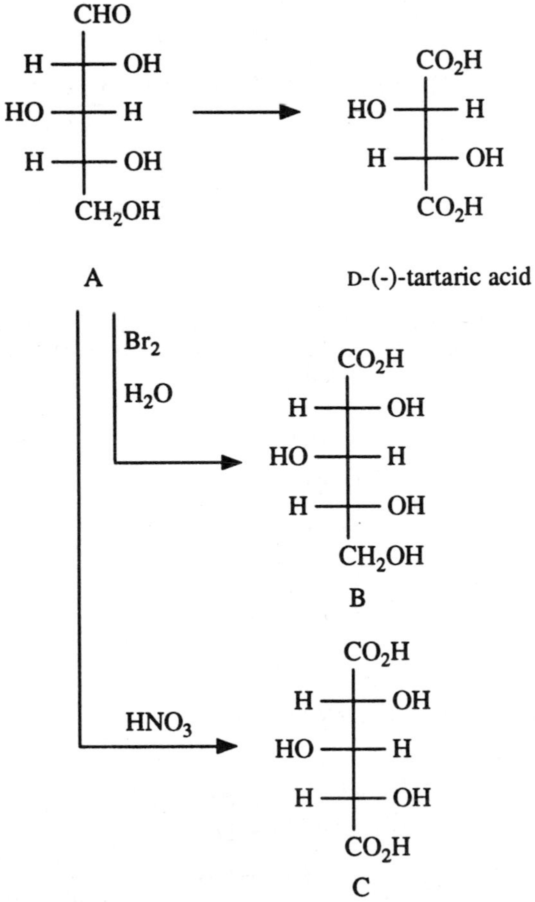

23.15 (a)

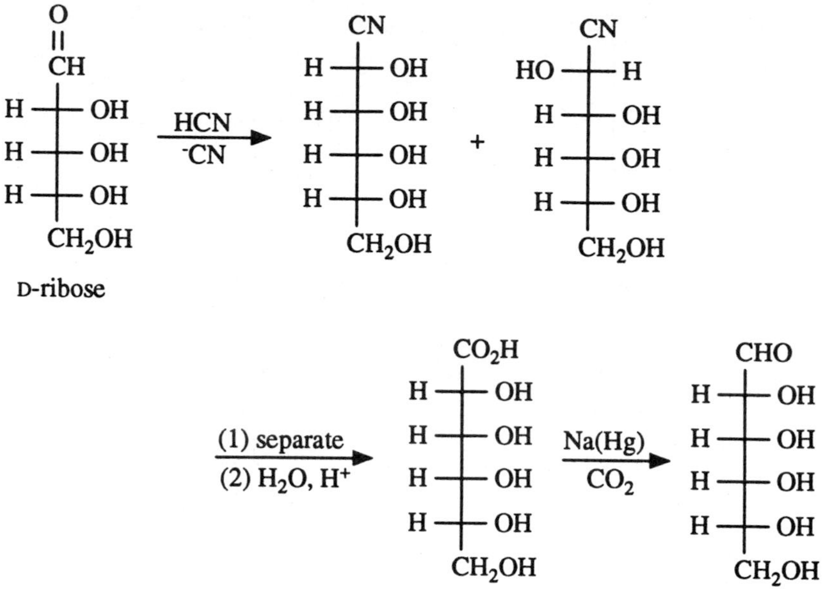

O
||
CH
H — OH
H — OH
H — OH
CH_2OH
D-ribose

$\xrightarrow[{}^{-}CN]{HCN}$

CN
H — OH
H — OH
H — OH
H — OH
CH_2OH

+

CN
HO — H
H — OH
H — OH
H — OH
CH_2OH

$\xrightarrow[\text{(2) } H_2O,\ H^+]{\text{(1) separate}}$

CO_2H
H — OH
H — OH
H — OH
H — OH
CH_2OH

$\xrightarrow[CO_2]{Na(Hg)}$

CHO
H — OH
H — OH
H — OH
H — OH
CH_2OH
D-allose

(b) Oxidation of D-allose with hot nitric acid would yield a *meso* aldaric acid, while oxidation of D-altrose would yield an optically active acid. (Reduction to an alditol could similarly be used.)

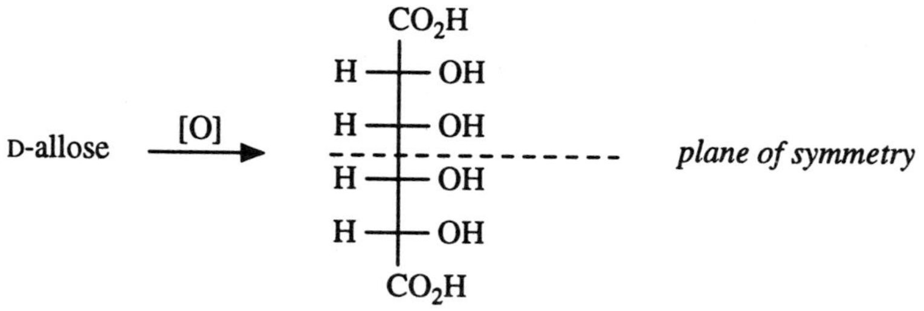

D-allose $\xrightarrow{[O]}$

CO_2H
H — OH
H — OH
- - - - - - - - - - *plane of symmetry*
H — OH
H — OH
CO_2H
meso

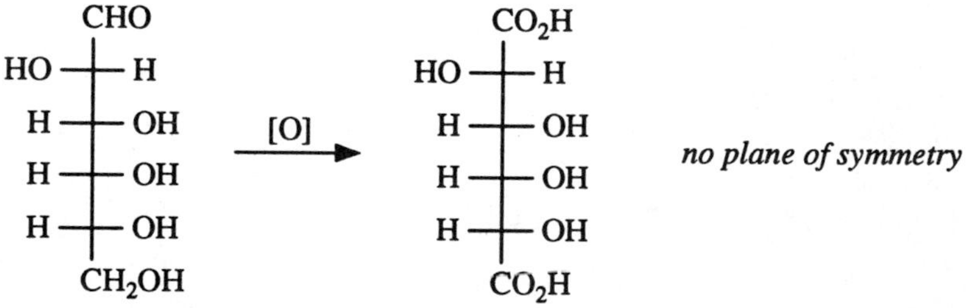

CHO
HO — H
H — OH
H — OH
H — OH
CH_2OH
D-altrose

$\xrightarrow{[O]}$

CO_2H
HO — H
H — OH
H — OH
H — OH
CO_2H

no plane of symmetry

23.16

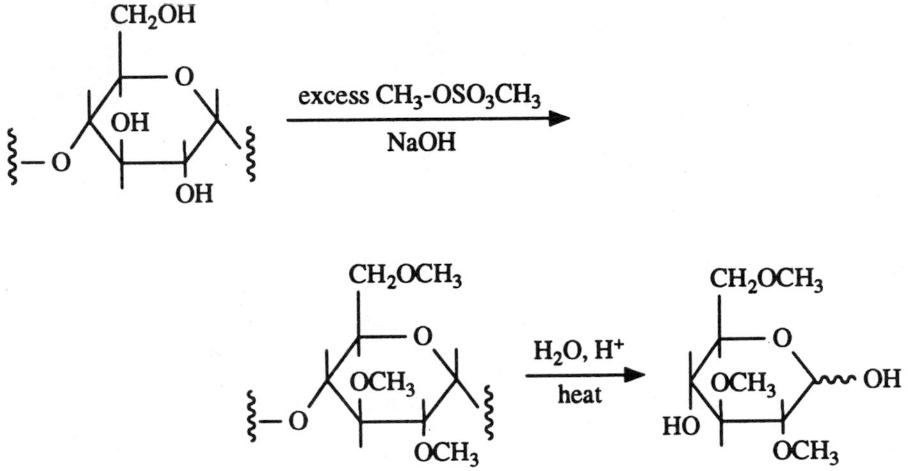

CH_2OH … O … OH … OH (structure) $\xrightarrow[\text{NaOH}]{\text{excess } CH_3\text{-}OSO_3CH_3}$

CH_2OCH_3 … O … OCH_3 … OCH_3 (structure) $\xrightarrow[\text{heat}]{H_2O,\ H^+}$ CH_2OCH_3 … O … OCH_3 … HO … OCH_3 … OH

23.17 eleven: α,α-1,1′; α,β-1,1′; β,β-1,1′; α-1,2′; β-1,2′; α-1,3′; β-1,3′; α-1,4′; β-1,4′; α-1,6′; β-1,6′. Two examples follow:

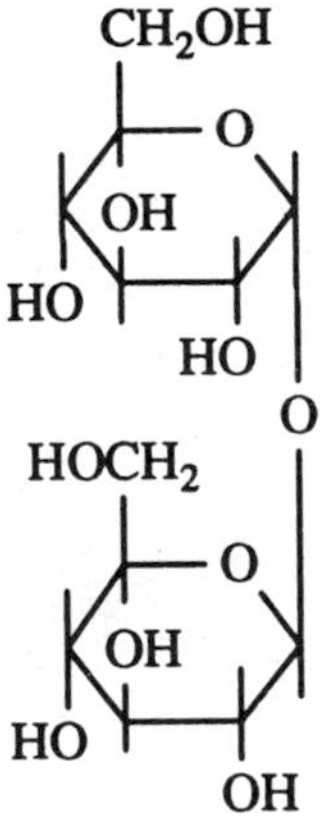

CH_2OH, O, OH, HO, HO, O, $HOCH_2$, O, OH, HO, OH

α,β-1,1′

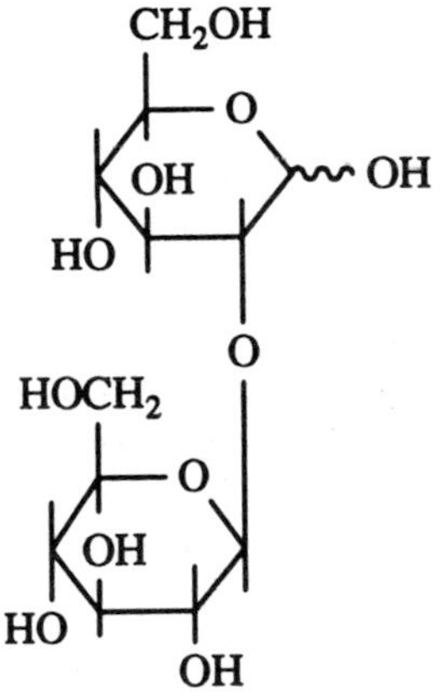

CH_2OH, O, OH, OH, HO, O, $HOCH_2$, O, OH, HO, OH

β-1,2′

24

Lipids

Answers to Additional Problems

24.1 (a) and (b) would be partly water soluble

(c) and (d) would be principally insoluble in water

24.2 Only compound (b) has an odd number of carbons and would not be expected to be found in a natural fat or oil.

24.3

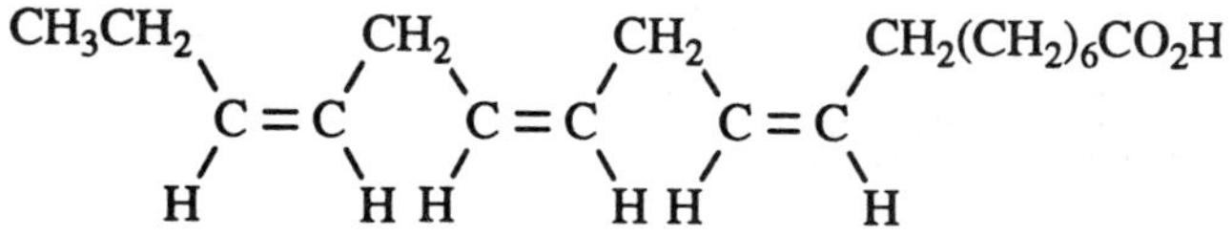

$$\begin{array}{c} CH_3CH_2 \quad\quad CH_2 \quad\quad CH_2 \quad\quad CH_2(CH_2)_6CO_2H \\ \backslash \quad / \quad \backslash \quad / \quad \backslash \quad / \\ C{=}C \quad C{=}C \quad C{=}C \\ / \quad \backslash \quad / \quad \backslash \quad / \quad \backslash \\ H \quad\quad H\;H \quad\quad H\;H \quad\quad H \end{array}$$

24.4

$$\begin{array}{l} CH_2OC(=O)(CH_2)_{16}CH_3 \\ H - C - OC(=O)(CH_2)_{16}CH_3 \\ CH_2OC(=O)(CH_2)_7CH{=}CH(CH_2)_7CH_3 \end{array} \qquad \begin{array}{r} CH_3(CH_2)_{16}C(=O)OCH_2 \\ CH_3(CH_2)_{16}C(=O)O - C - H \\ CH_3(CH_2)_7CH{=}CH(CH_2)_7C(=O)OCH_2 \end{array}$$

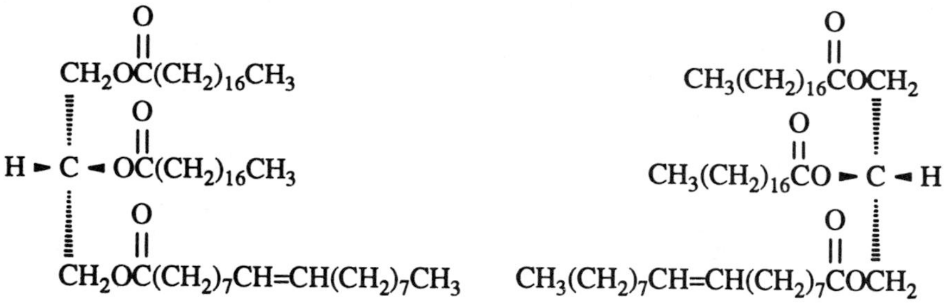

24.5 (a) is a sphingomyelin and (b) is a lecithin. A cerebroside and a cephalin are not shown.

24.6 A sphingomyelin is characterized by having a phosphate ester of an amide of sphingosine.

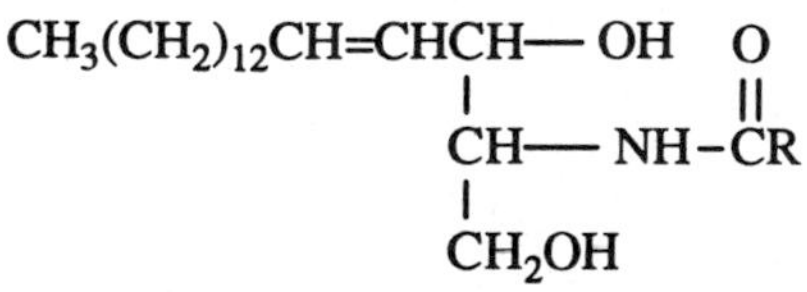

$$\begin{array}{l} CH_3(CH_2)_{12}CH{=}CHCH{-}OH \quad O \\ \quad\quad\quad\quad\quad\quad\quad\quad\quad | \quad\quad\quad\quad \| \\ \quad\quad\quad\quad\quad\quad\quad\quad\quad CH{-}NH{-}CR \\ \quad\quad\quad\quad\quad\quad\quad\quad\quad | \\ \quad\quad\quad\quad\quad\quad\quad\quad\quad CH_2OH \end{array}$$

an amide of sphingosine

A steroid has the characteristic steroid ring system.

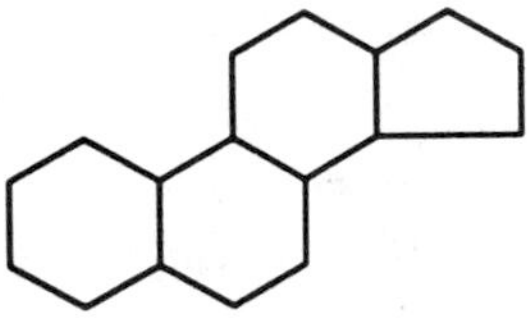

A cerebroside has a sugar in place of the phosphate ester of sphingomyelin.
A glyceride has three fatty esters of glycerol.

24.7 (a) androgenic (b) estrogenic

24.8 1 $\xrightarrow{NaOCH_3}$ A

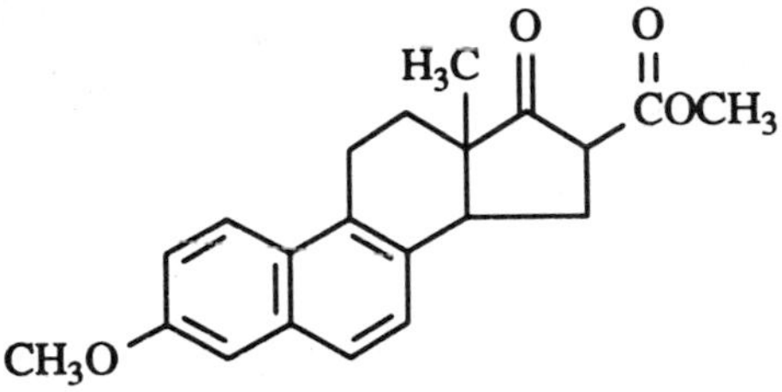

$\xrightarrow[H^+,\ -CO_2,\ -CH_3OH]{H_2O}$ equilenin

24.9 (a) $\xrightarrow{H_2CrO_4}$

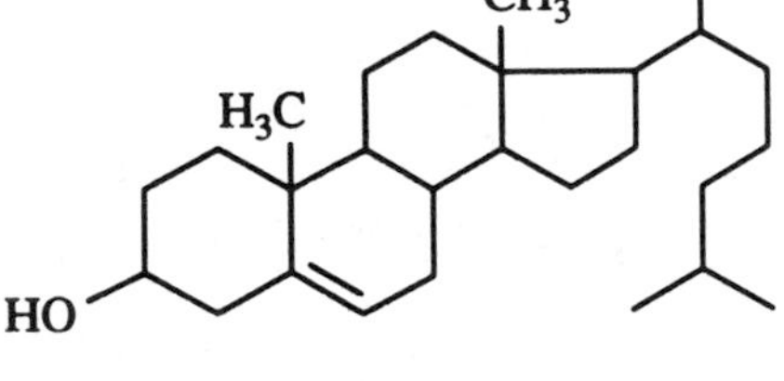

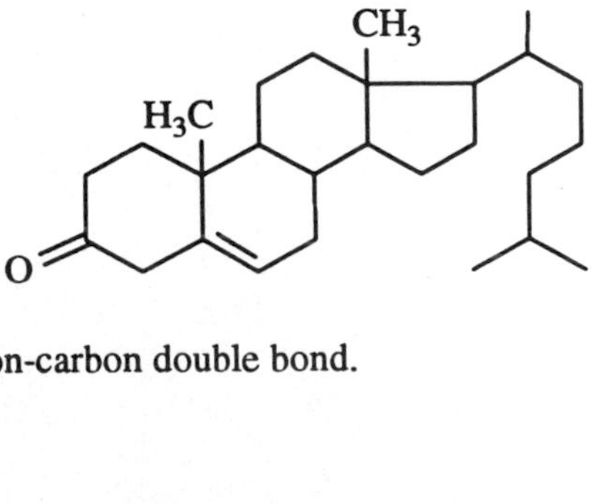

In (a), you might have also oxidized the carbon-carbon double bond.

(b) cholesterol $\xrightarrow[CCl_4]{Br_2}$

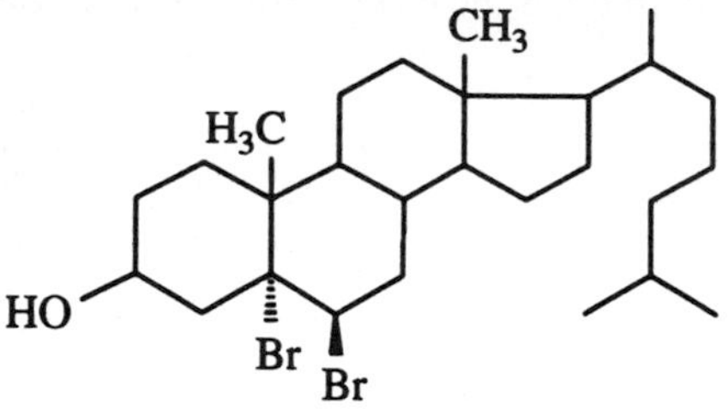

(c) cholesterol $\xrightarrow{NBS}$

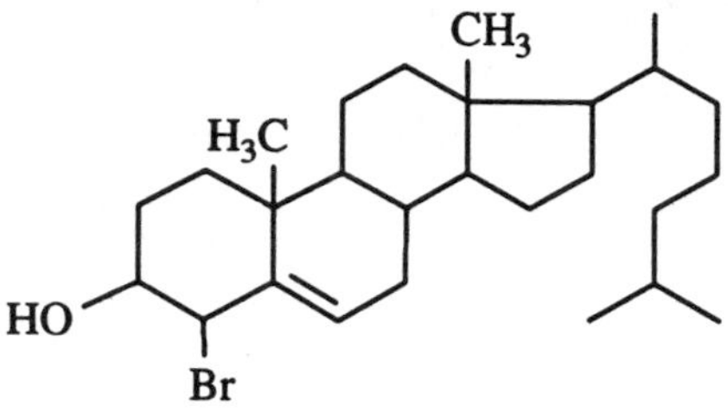

+ 7-bromocholesterol

(d) cholesterol $\xrightarrow{NaH}$

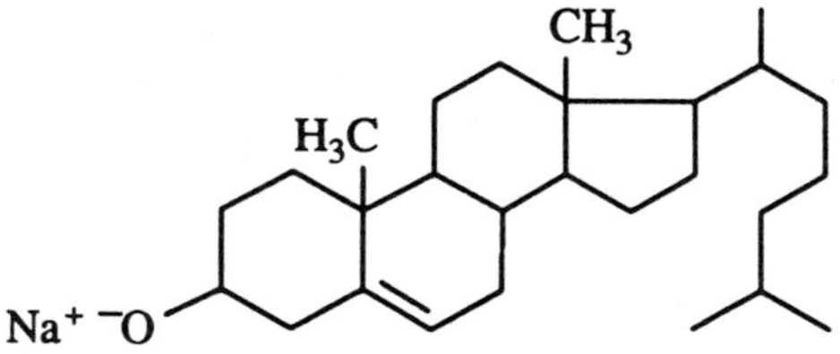

24.10 (a) The keto group at carbon 11 must be reduced to a hydroxyl group.

(b) The benzene ring must be reduced with one mole of hydrogen.

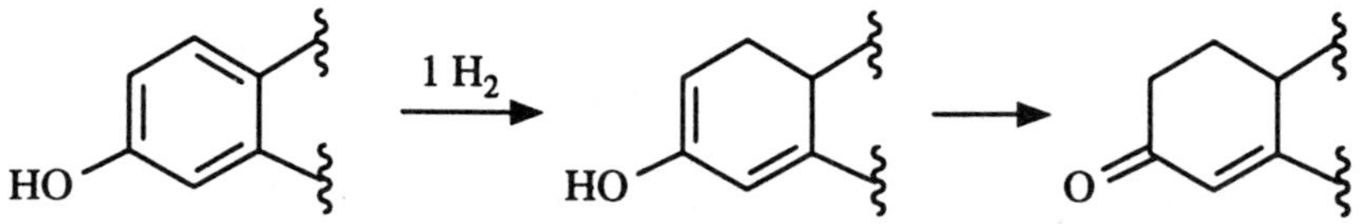

an enol

25

Amino Acids and Proteins

Answers to Additional Problems

25.1 (a) lys (b) cyS or cys (c) asn

(d) tyr (e) ser (f) phe

(g) pro (h) glu

25.2 (a) basic. Lysine has two amino groups and only one carboxyl group. Therefore, only one amino group will be neutralized by the carboxyl group; the other amino group will react with water to release a hydroxide ion.

$$H_2N(CH_2)_4\underset{\overset{|}{^+NH_3}}{CH}\overset{\overset{O}{||}}{C}O^- + H_2O \rightleftharpoons OH^- + H_3\overset{+}{N}(CH_2)_4\underset{\overset{|}{^+NH_3}}{CH}\overset{\overset{O}{||}}{C}O^-$$

(b) neutral. The α-amino group will be neutralized by the carboxyl group. The nitrogen of the amide group is not basic and will not undergo an acid-base reaction with water.

(c) acidic. Aspartic acid has two carboxyl groups and only one amino group.

(d) neutral. Leucine has one amino group and one carboxyl group.

25.3 (a)

$$(CH_3)_2CHCH_2CH_2CO_2H \xrightarrow[PBr_3\ \text{catalyst}]{Br_2} (CH_3)_2CHCH_2\overset{\overset{Br}{|}}{C}HCO_2H$$

$$\xrightarrow[(2)\ \text{neutralize}]{(1)\ \text{excess}\ NH_3} (R)\ (S)\text{-}\ (CH_3)_2CHCH_2\overset{\overset{NH_2}{|}}{C}HCO_2H$$

(b) Phthalimido-$NCH(CO_2C_2H_5)_2$ $\xrightarrow[\text{(2) } CH_3CH_2CHBrCH_3]{\text{(1) } NaOC_2H_5}$

Phthalimido-$N-C(CO_2C_2H_5)_2$ with $CHCH_2CH_3$ on C, and CH_3 on that CH; $\xrightarrow[\text{(2) neutralize}]{\text{(1) } H_2O,\ H^+,\ \text{heat}}$

$(R)\,(S)$-$CH_3CH_2CH(CH_3)CH(NH_2)CO_2H$

(c) $(CH_3)_2CHCH{=}O$ $\xrightarrow[\text{(2) HCN}]{\text{(1) } NH_3}$ $(CH_3)_2CHCH(NH_2)CN$ $\xrightarrow[\text{(2) neutralize}]{\text{(1) } H_2O,\ H^+,\ \text{heat}}$

$(R)\,(S)$-$(CH_3)_2CHCH(NH_2)CO_2H$

25.4 (a) $CH_3CH(NH_2)CO_2^-\ \overset{+}{N}H_4$ (b) $Cl^-\ \overset{+}{N}H_3CH_2CO_2H$ + $C_6H_4(CO_2H)_2$ (phthalic acid)

(c) $C_6H_5CH_2CH(\overset{+}{N}H_3\ Cl^-)CO_2H$ + $C_6H_4(CO_2H)_2$ (phthalic acid)

(d) $(CH_3)_2CHCH_2CH(NH_2)CN$ (e) $C_6H_5CH_2CH(\overset{+}{N}H_3\ Cl^-)CO_2CH_3$

25.5 (a) $H_3\overset{+}{N}CH(CH(CH_3)_2)CO_2^- + Cl^-$ (b) $H_2NCH((CH_2)_4NH_3^+)CO_2^- + Cl^-$

(c) $H_3\overset{+}{N}CH(CH(CH_3)_2)CO_2^- + Na^+ + H_2O$ (d) $H_3\overset{+}{N}CH((CH_2)_4NH_3^+)CO_2^- + Na^+ + H_2O$

In each case, look for the more acidic or more basic group in the reactant molecule.

25.6 (a) [Ruhemann's purple anion structure: two indane-1,3-dione units joined by =N–, one bearing O^-] $+ H_2N\overset{O}{\overset{\|}{C}}CH_2CH_2\overset{O}{\overset{\|}{C}}H + CO_2 + H_2O$

(b) $H_2N\overset{O}{\overset{\|}{C}}CH_2CH_2CH(NH\overset{\|}{\underset{O}{C}}CH_3)CO_2H + CH_3CO_2H$

(c) $CH_3\overset{O}{\overset{\|}{C}}O-C_6H_4-CH_2CH(NH\overset{\|}{\underset{O}{C}}CH_3)CO_2H + 2\ CH_3CO_2H$

(d) $CH_3\overset{O}{\overset{\|}{C}}OCH_2CH(NH\overset{\|}{\underset{O}{C}}CH_3)CO_2H + 2\ CH_3CO_2H$

(e) O O⁻ =N– O O $+ CH_3\overset{O}{\overset{||}{C}}H + CO_2 + H_2O$

25.7 (a) leucyllysine (b) lysylleucine

(c) tyrosylphenylalanylserine (d) serylphenylalanyltyrosine

25.8 (a) $CH_3\underset{\overset{+}{N}H_3}{CH}\overset{O}{\overset{||}{C}}-NH\underset{CH_2CH_2\overset{||}{\underset{O}{C}}NH_2}{CH}\overset{O}{\overset{||}{C}}O^-$

(b) $\overset{+}{N}H_2$ ring $-\overset{O}{\overset{||}{C}}\text{-}NH\underset{CH_3CHOH}{CH}\overset{O}{\overset{||}{C}}O^-$

(c) $H_3\overset{+}{N}CH_2\overset{O}{\overset{||}{C}} - NH\underset{CH_2CH_2\overset{||}{\underset{O}{C}}NH_2}{CH}\overset{O}{\overset{||}{C}}O^-$

(d) $H_2N(CH_2)_4\underset{NH_2}{CH}\overset{O}{\overset{||}{C}}-NH\underset{(CH_2)_4\overset{+}{N}H_3}{CH}\overset{O}{\overset{||}{C}}-NH\underset{CH_2}{CH}\overset{O}{\overset{||}{C}}O^-$ (indole ring, N H)

You might have protonated one of these amino groups.

25.9 (a) tyrosylserylphenylalanine (b) tryptophenylglycyltryptophan

25.10 (a) $(CH_3)_3CO\overset{O}{\overset{\|}{C}}NH\underset{CH_3}{\underset{|}{C}}H\overset{O}{\overset{\|}{C}}NHCH_2CO_2H \;+\; C_6H_{11}-NH\overset{O}{\overset{\|}{C}}NH-C_6H_{11}$

(b) $C_6H_5CH_2O\overset{O}{\overset{\|}{C}}NH\underset{CH_2C_6H_5}{\underset{|}{C}}H\overset{O}{\overset{\|}{C}}NHCH_2CO_2CH_2CH_3$

(c) $C_6H_5CH_3 \;+\; CO_2 \;+\; NH_2\underset{CH_2C_6H_5}{\underset{|}{C}}H\overset{O}{\overset{\|}{C}}NHCH_2CO_2CH_2CH_3$

25.11 (a) $H_2N\underset{CH_3}{\underset{|}{C}}HCO_2H \;+\; (CH_3)_3CO\overset{O}{\overset{\|}{C}}Cl \longrightarrow (CH_3)_3CO\overset{O}{\overset{\|}{C}}-NH\underset{CH_3}{\underset{|}{C}}H\overset{O}{\overset{\|}{C}}OH \;+\; HCl$

(b) $CH_3CO_2H \;+\; C_6H_{11}-N=C=N-C_6H_{11} \longrightarrow CH_3\overset{O}{\overset{\|}{C}}OC(-NH-C_6H_{11})=N-C_6H_{11}$

(c) $CH_3CO_2H \;+\; ClCO_2CH_2CH_3 \longrightarrow CH_3\overset{O}{\overset{\|}{C}}O\overset{O}{\overset{\|}{C}}OCH_2CH_3 \;+\; HCl$

(d) $H_2N\underset{CH_3}{\underset{|}{C}}HCO_2H \;+\; C_6H_5CH_2O\overset{O}{\overset{\|}{C}}Cl \longrightarrow C_6H_5CH_2O\overset{O}{\overset{\|}{C}}-NH\underset{CH_3}{\underset{|}{C}}HCO_2H \;+\; HCl$

25.12 (a) (2) (b) (2), (4) (c) (4), (6) (d) none (e) (6)

Bonds (1), (3), and (5) are not amide bonds and do not undergo hydrolysis reactions.

25.13 Trypsin catalyzes hydrolysis of the amide bond at the carboxy group of lysine or arginine.

(a) lys ⁄ asp-gly-ala-ala-glu-ser-gly $\xrightarrow{\text{trypsin}}$

lys + asp-gly-ala-ala-glu-ser-gly

(b) tyr-cys-lys ⁄ ala-arg ⁄ arg ⁄ gly $\xrightarrow{\text{trypsin}}$

try-cys-lys + ala-arg + arg + gly

(c) ala-ala-his-arg ⁄ glu-lys ⁄ phe-ile-gly-glu-gly-glu $\xrightarrow{\text{trypsin}}$

ala-ala-his-arg + glu-lys + phe-ile-gly-glu-gly-glu

26

Nucleic Acids

Answers to Additional Problems

26.1 (a) [structure: 2'-deoxyadenosine — adenine (NH_2) attached to a deoxyribose ring bearing $HOCH_2$, H, H, OH, H]

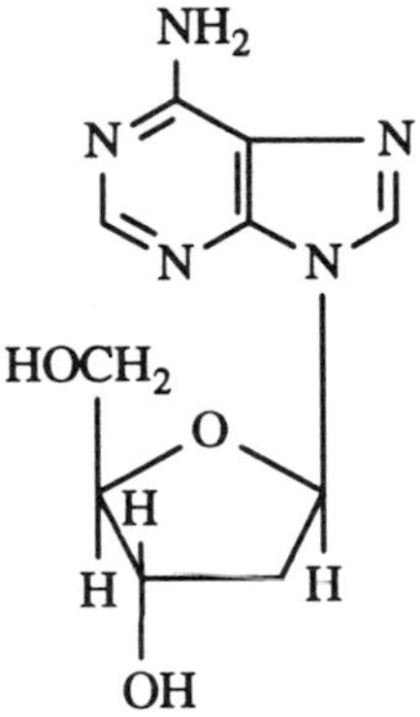

(b) [structure: adenosine — adenine (NH_2) attached to a ribose ring bearing $HOCH_2$, H, H, H, H, OH, OH]

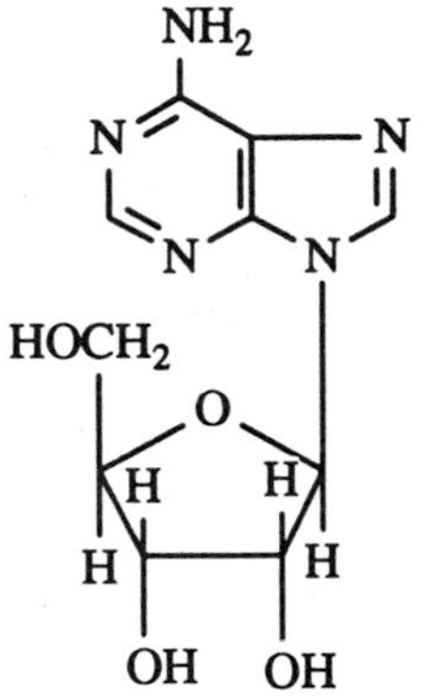

(c) [structure: guanine — O, HN, H_2N, N, N, N–H]

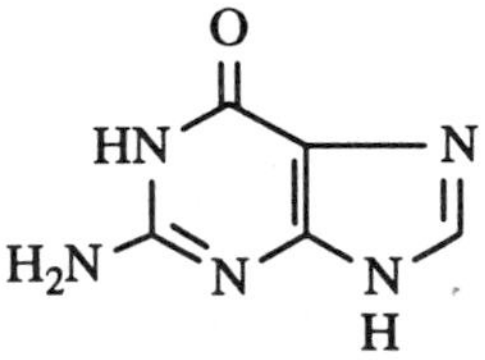

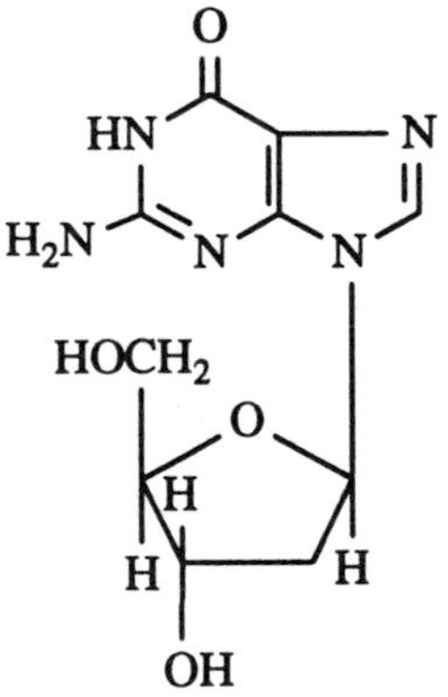

(d) [structure: 2'-deoxyguanosine — guanine (O, HN, H_2N) attached to a deoxyribose ring bearing $HOCH_2$, H, H, OH, H]

26.2 [structure: guanine (O, HN, H_2N) attached at N to CH_2–O–H_2C–$HOCH_2$]

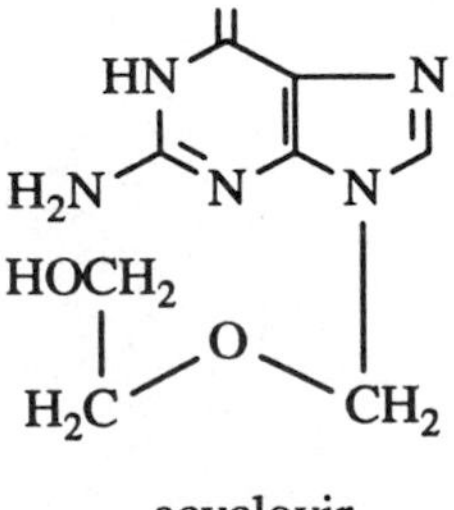

acyclovir

similar to deoxyguanosine, see Answer 26.1 (d)

26.3 Thymine hydrogen bonds with adenine. Therefore, there should be 19% adenine. Consequently, the combined percentage of guanine and cytosine should be [100% -2(19%)] or 62%. Since guanine and cytosine hydrogen bond, the percentage of each should be 31%.

26.4

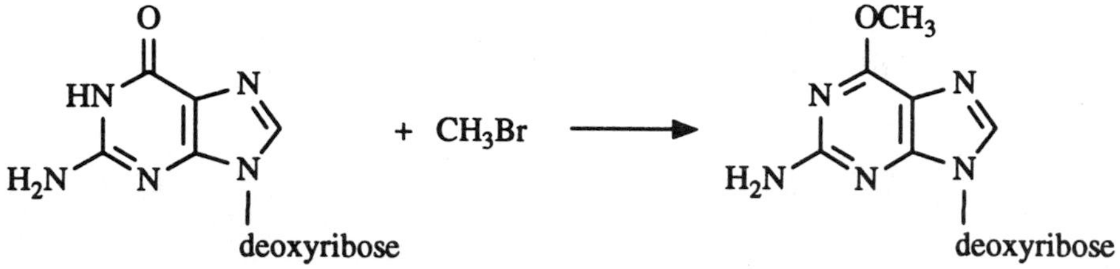

O
HN N
H$_2$N N N
deoxyribose
+ CH_3Br →
OCH$_3$
N N
H$_2$N N N
deoxyribose

The reaction prevents the guanosine from hydrogen bonding.

26.5

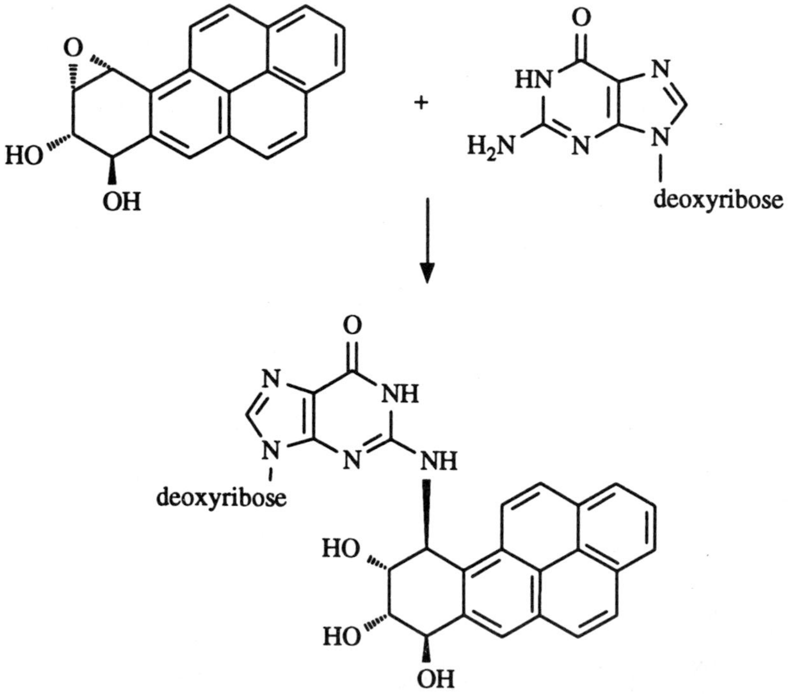

O
HO
OH
+
O
HN N
H$_2$N N N
deoxyribose

O
N NH
N N NH
deoxyribose
HO
HO
OH

26.6

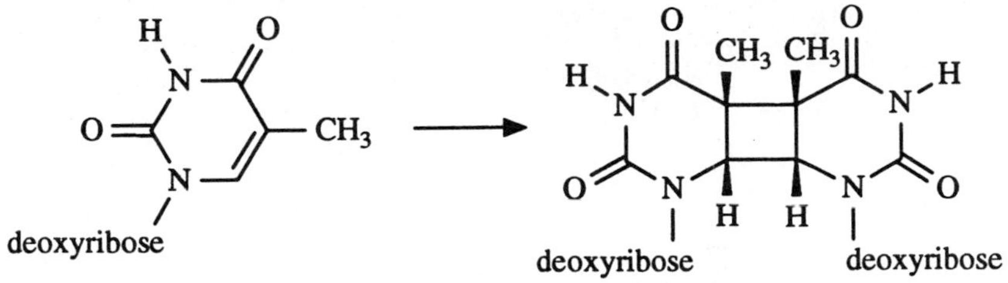

H O
N
O N CH_3
deoxyribose

O O
CH_3 CH_3
H N N H
O N N O
H H
deoxyribose deoxyribose

26.7 DNA contains the monosaccharide deoxyribose, while RNA contains the monosaccharide ribose. DNA contains thymine along with adenine, guanine, and cystosine. RNA contains uracil in place of thymine. DNA forms long double-helical chains. RNA molecules are shorter and do not form double helices.